# Food Structure Engineering and Design for Improved Nutrition, Health and Well-Being

# Food Structure Engineering and Design for Improved Nutrition, Health and Well-Being

Edited by

**Miguel Ângelo Parente Ribeiro Cerqueira**
International Iberian Nanotechnology Laboratory, Braga, Portugal

**Lorenzo Miguel Pastrana Castro**
International Iberian Nanotechnology Laboratory, Braga, Portugal

Academic Press is an imprint of Elsevier
125 London Wall, London EC2Y 5AS, United Kingdom
525 B Street, Suite 1650, San Diego, CA 92101, United States
50 Hampshire Street, 5th Floor, Cambridge, MA 02139, United States
The Boulevard, Langford Lane, Kidlington, Oxford OX5 1GB, United Kingdom

Copyright © 2023 Elsevier Inc. All rights reserved.

No part of this publication may be reproduced or transmitted in any form or by any means, electronic or mechanical, including photocopying, recording, or any information storage and retrieval system, without permission in writing from the publisher. Details on how to seek permission, further information about the Publisher's permissions policies and our arrangements with organizations such as the Copyright Clearance Center and the Copyright Licensing Agency, can be found at our website: www.elsevier.com/permissions.

This book and the individual contributions contained in it are protected under copyright by the Publisher (other than as may be noted herein).

**Notices**

Knowledge and best practice in this field are constantly changing. As new research and experience broaden our understanding, changes in research methods, professional practices, or medical treatment may become necessary.

Practitioners and researchers must always rely on their own experience and knowledge in evaluating and using any information, methods, compounds, or experiments described herein. In using such information or methods they should be mindful of their own safety and the safety of others, including parties for whom they have a professional responsibility.

To the fullest extent of the law, neither the Publisher nor the authors, contributors, or editors, assume any liability for any injury and/or damage to persons or property as a matter of products liability, negligence or otherwise, or from any use or operation of any methods, products, instructions, or ideas contained in the material herein.

ISBN 978-0-323-85513-6

For information on all Academic Press publications
visit our website at https://www.elsevier.com/books-and-journals

*Publisher:* Nikki P. Levy
*Acquisitions Editor:* Nina Bandeira
*Editorial Project Manager:* Kathrine Esten
*Production Project Manager:* Rashmi Manoharan
*Cover Designer:* Miles Hitchen

Typeset by STRAIVE, India

*To our families for their love and support*

# Contents

Contributors .................................................................................................................... xv
About the editors ............................................................................................................ xxi
Foreword ....................................................................................................................... xxiii
Preface ........................................................................................................................... xxv
Acknowledgments .......................................................................................................... xxvii

## PART I Introduction

### CHAPTER 1 Nutrition, health and well-being in the world: The role of food structure design ................................................................. 3
Miguel Ângelo Parente Ribeiro Cerqueira, David J. McClements, and Lorenzo Miguel Pastrana Castro

- 1.1 Food challenges and United Nations sustainable development goals ........ 3
- 1.2 Trends in human food consumption: The diet shift ................................... 8
- 1.3 Food structure design for nutrition and health benefits ............................. 11
- 1.4 Conclusions and future perspectives ......................................................... 13
  - Acknowledgments ..................................................................................... 13
  - References ................................................................................................. 13

### CHAPTER 2 New food structures and their influence on nutrition, health and well-being .................................................................. 17
D. Subhasri, J.A. Moses, and C. Anandharamakrishnan

- 2.1 Introduction ................................................................................................ 17
- 2.2 New food structuring techniques ............................................................... 19
- 2.3 New food structures ................................................................................... 20
  - 2.3.1 Energy density food ........................................................................ 20
  - 2.3.2 Customized food shape and structure ............................................. 22
  - 2.3.3 Modulating digestion through specially designed food microstructures ..... 23
  - 2.3.4 Enhanced bioavailability of encapsulated foods ............................. 26
  - 2.3.5 Tailored foods to meet nutritional needs ......................................... 27
- 2.4 Factors influencing the development of new food structures .................... 30
  - 2.4.1 Product research and development ................................................. 30
  - 2.4.2 Manufacturing process .................................................................... 31
  - 2.4.3 Post-production ............................................................................... 32
  - 2.4.4 Market analysis ............................................................................... 32
- 2.5 Demand and research gap .......................................................................... 32
- 2.6 Conclusion .................................................................................................. 33
  - References ................................................................................................. 33

# PART II  Strategies to modify structure/functionality/quality of foods

## CHAPTER 3  Electrotechnologies for the development of food-based structured systems ..... 43
Ricardo Nuno Pereira, Rui M. Rodrigues, and Antonio A. Vicente

- 3.1 Introduction ..... 43
- 3.2 Moderate electric fields technology ..... 45
  - 3.2.1 Effects of MEF at the molecular level ..... 46
  - 3.2.2 Effects of EF in nano and microstructures ..... 47
  - 3.2.3 Effects of EF in macrostructures ..... 49
- 3.3 Novel perspectives in biomolecular structures and functionality ..... 51
  - 3.3.1 Exploring EF technology to tailor biomaterials ..... 51
  - 3.3.2 Health implications ..... 52
- 3.4 Future perspectives ..... 54
- 3.5 Conclusions ..... 55
- Acknowledgments ..... 56
- References ..... 56

## CHAPTER 4  Encapsulation and colloidal systems as a way to deliver functionality in foods ..... 63
Cristian Dima, Elham Assadpour, and Seid Mahdi Jafari

- 4.1 Introduction ..... 63
- 4.2 Nutraceutical encapsulation as food quality improvement strategy ..... 65
- 4.3 Designing of colloidal delivery systems for food functionalization ..... 68
  - 4.3.1 Bioactives requirements ..... 69
  - 4.3.2 Physico-chemical characteristics of food-grade colloidal delivery systems ..... 73
- 4.4 Colloidal delivery systems in food functionalization ..... 86
  - 4.4.1 Lipid based-colloidal delivery systems ..... 87
  - 4.4.2 Polymeric colloidal delivery systems ..... 94
- 4.5 Conclusion and future perspective ..... 98
- Acknowledgment ..... 99
- References ..... 99
- Further reading ..... 111

## CHAPTER 5  How food structure influences the physical, sensorial, and nutritional quality of food products ..... 113
Meliza Lindsay Rojas, Mirian T.K. Kubo, Maria Elisa Caetano-Silva, Gisandro Reis Carvalho, and Pedro E.D. Augusto

- 5.1 Introduction ..... 113
- 5.2 Effect of food processing on food structure: Conventional and emerging technologies in food processing ..... 115

|  |  | 5.2.1 | Structural modifications in solid foods | 117 |
| --- | --- | --- | --- | --- |

5.2.1 Structural modifications in solid foods ............................................................... 117
5.2.2 Structural modifications in particulate foods ...................................................... 117
5.2.3 Structural modifications in food macromolecules ................................................ 118
**5.3** Structure modification and impact on physical properties, sensorial aspects and nutritional quality ........................................................................................... 120
5.3.1 Structure modification and impact on physical properties ................................... 120
5.3.2 Structure modification and impact on sensorial aspects ...................................... 122
5.3.3 Structure modification and impact on nutritional quality and health aspects ................................................................................................. 126
**5.4** Conclusion and future perspectives ............................................................................ 130
References .................................................................................................................... 130

## CHAPTER 6 Structure design for gastronomy applications ........................ 139
Alessandra Massa, Juan-Carlos Arboleya, Fabiola Castillo, and Eneko Axpe

**6.1** Introduction ................................................................................................................. 139
**6.2** Interaction between science and gastronomy: A good match for food product design ............................................................................................................. 140
6.2.1 Mammia (curd) ..................................................................................................... 141
**6.3** Food colloids in gastronomy ...................................................................................... 142
6.3.1 Types of food colloids .......................................................................................... 142
6.3.2 Developing and emerging applications of food colloids in gastronomy ....................................................................................................... 145
**6.4** Designing new food microstructures by biotechnology processes in the kitchen ............................................................................................................. 146
6.4.1 The role of fermentation: a revolutionary technology that always has been with us ................................................................................................ 146
6.4.2 A key player in fermentation: enzymes ............................................................... 146
6.4.3 Fermentation in the kitchen: relationship with sciences, and new food design ........................................................................................... 148
6.4.4 New fermented food for diet and health ............................................................. 149
**6.5** Structuring food for health and wellness .................................................................... 150
6.5.1 Introduction .......................................................................................................... 150
6.5.2 Reduction or replacement of fat through emulsions, hydrogels, oleogels and oleofoams ....................................................................................... 150
6.5.3 Reduction of fat in mayonnaise through different kinds of fat mimetics ................................................................................................... 151
6.5.4 How aerated food can help on the expected satiety ........................................... 151
**6.6** Conclusions ................................................................................................................. 152
References .................................................................................................................... 152

# PART III  Development of healthy products

### CHAPTER 7  Design of functional foods with targeted health functionality and nutrition by using microencapsulation technologies ............ 159
Guilherme de Figueiredo Furtado,
Juliana Domingues dos Santos Carvalho, Gabriela Feltre, and
Miriam Dupas Hubinger

- 7.1 Introduction ............................................................................................ 159
- 7.2 Strategies of microencapsulation ......................................................... 160
  - 7.2.1 Spray drying ............................................................................. 161
  - 7.2.2 Spray chilling ........................................................................... 162
  - 7.2.3 Ionic gelation ........................................................................... 164
- 7.3 Wall materials ........................................................................................ 166
- 7.4 Core materials ....................................................................................... 168
  - 7.4.1 Oil matrices ............................................................................. 168
  - 7.4.2 Bioactive compounds .............................................................. 170
  - 7.4.3 Probiotics ................................................................................. 171
- 7.5 Food applications .................................................................................. 172
  - 7.5.1 Spray drying ............................................................................. 172
  - 7.5.2 Spray chilling ........................................................................... 172
  - 7.5.3 Ionic gelation ........................................................................... 173
- 7.6 Final remarks ......................................................................................... 174
- References ..................................................................................................... 174

### CHAPTER 8  Strategies for the reduction of salt in food products .................. 187
Mirian dos Santos, Andrea Paola Rodriguez Triviño,
Julliane Carvalho Barros, Adriano G. da Cruz, and
Marise Aparecida Rodrigues Pollonio

- 8.1 Introduction ............................................................................................ 187
- 8.2 Salt, sodium, and health ........................................................................ 188
- 8.3 The role of sodium in food products .................................................... 189
  - 8.3.1 Effects on protein functional properties ............................... 190
  - 8.3.2 Salt as flavor enhancer: Impact on sensory properties ........ 193
  - 8.3.3 Microbial stability ................................................................... 194
- 8.4 Sodium reduction strategies for processed foods ............................... 195
  - 8.4.1 Use of flavor enhancers .......................................................... 199
  - 8.4.2 Use of other salts to substitute NaCl .................................... 200
  - 8.4.3 Crystal size modification ........................................................ 201
  - 8.4.4 Spray-dried salt particles ........................................................ 203
  - 8.4.5 Nonthermal processes for low/reduced sodium food products ............... 204
  - 8.4.6 Heterogeneous distribution of salt ........................................ 205

| | | |
|---|---|---|
| **8.5** | Challenges to reduce sodium in food products | 206 |
| **8.6** | Final considerations | 208 |
| | References | 209 |

**CHAPTER 9  Strategies for the reduction of sugar in food products ............... 219**

Ana Gomes, Ana I. Bourbon, Ana Rita Peixoto, Ana Sanches Silva, Ana Tasso, Carina Almeida, Clarisse Nobre, Cláudia Nunes, Claudia Sánchez, Daniela A. Gonçalves, Diogo Castelo-Branco, Diogo Figueira, Elisabete Coelho, Joana Gonçalves, José A. Teixeira, Lorenzo Miguel Pastrana Castro, Manuel A. Coimbra, Manuela Pintado, Miguel Ângelo Parente Ribeiro Cerqueira, Pablo Fuciños, Paula Teixeira, Pedro A.R. Fernandes, and Vitor D. Alves

| | | |
|---|---|---|
| **9.1** | Introduction | 219 |
| **9.2** | Functional and technological role of sugar in food products | 221 |
| **9.3** | Food reformulation to reduce free sugar intake | 222 |
| **9.4** | Sugar structure modification and encapsulation for enhanced sweet perception | 226 |
| **9.5** | Food grade alternatives to sugar | 227 |
| | 9.5.1 Nutritive sweeteners | 227 |
| | 9.5.2 High intensity sweeteners | 231 |
| **9.6** | Enzymatic and innovative methods to improve sweetening | 231 |
| | 9.6.1 Sugar reduction in milk and dairy products | 232 |
| | 9.6.2 Sugar reduction in juices and beverages | 233 |
| **9.7** | Products and market | 234 |
| **9.8** | Conclusions and future outlook | 236 |
| | Acknowledgments | 236 |
| | References | 236 |

**CHAPTER 10  New technological strategies for improving the lipid content in food products ......................................................... 243**

S. Cofrades and M.D. Alvarez

| | | |
|---|---|---|
| **10.1** | Introduction | 243 |
| **10.2** | Modification of the lipid fraction in food products | 244 |
| | 10.2.1 Decreasing fat and cholesterol contents by using ingredients that can serve as fat replacers | 246 |
| | 10.2.2 Improving the lipid profile in food products by using lipids with a healthy fatty acid profile | 250 |
| **10.3** | Decreasing fat digestibility in food products | 261 |
| **10.4** | Future perspectives | 263 |
| | Acknowledgments | 264 |
| | References | 264 |

## PART IV Health in vitro and in vivo studies

**CHAPTER 11  Understanding food structure modifications during digestion and their implications in nutrient release ................ 277**
Alejandra Acevedo-Fani, Debashree Roy, Duc Toan Do, and Harjinder Singh

- 11.1 Introduction ................................................................................ 277
- 11.2 Overview of the digestion process .......................................... 280
- 11.3 Digestion of macronutrients ..................................................... 282
  - 11.3.1 Proteins ........................................................................... 282
  - 11.3.2 Lipids ............................................................................... 282
  - 11.3.3 Starch .............................................................................. 283
- 11.4 Modification of plant-based food structures in the GIT ........ 284
  - 11.4.1 Starchy legumes ............................................................ 284
  - 11.4.2 Cereals ............................................................................ 287
  - 11.4.3 Vegetables and fruits .................................................... 288
  - 11.4.4 Tree nuts ......................................................................... 290
- 11.5 Modification of animal-based food structures in the GIT .... 293
  - 11.5.1 Meat ................................................................................ 293
  - 11.5.2 Milk .................................................................................. 298
- 11.6 Conclusions ................................................................................ 304
- Acknowledgments ............................................................................. 304
- References ......................................................................................... 304

**CHAPTER 12  Assessing nutritional behavior of foods through *in vitro* and *in vivo* studies ................................................ 315**
Didier Dupont and Olivia Ménard

- 12.1 Introduction ................................................................................ 315
- 12.2 *In vitro* oro-gastro-intestinal digestion models ..................... 316
  - 12.2.1 Static *in vitro* digestion models .................................. 316
  - 12.2.2 Dynamic *in vitro* digestion models ............................ 319
- 12.3 Absorption models .................................................................... 323
  - 12.3.1 Cellular models ............................................................. 323
  - 12.3.2 Ussing chambers .......................................................... 324
  - 12.3.3 Organoids ...................................................................... 325
- 12.4 *In vivo* models ........................................................................... 325
  - 12.4.1 Animal models .............................................................. 325
  - 12.4.2 Human subjects ............................................................ 326
- 12.5 Conclusion .................................................................................. 327
- References ......................................................................................... 328

**CHAPTER 13 Application of artificial neural networks (ANN) for predicting the effect of processing on the digestibility of foods** ...... 333
L.A. Espinosa Sandoval, A.M. Polanía Rivera, L. Castañeda Florez, and A. García Figueroa

- 13.1 Introduction ...... 333
- 13.2 Artificial intelligence in food processing ...... 334
  - 13.2.1 Artificial neural network application in food processing operations ...... 335
- 13.3 ANN in digestion modelling ...... 344
  - 13.3.1 *In vitro* models ...... 344
  - 13.3.2 ANN for the prediction of human digestion ...... 352
- 13.4 Conclusions ...... 354
- References ...... 355

# PART V Consumer's perception and acceptability

**CHAPTER 14 How to assess consumer perception and food attributes of novel food structures using analytical methodologies** ...... 365
Takahiro Funami and Makoto Nakauma

- 14.1 Introduction ...... 365
- 14.2 Texture perception and oral rheology and tribology ...... 366
  - 14.2.1 Extension rheology for texture evaluation with regard to swallowing physiology in humans ...... 366
  - 14.2.2 Tribology for texture evaluation with regard to moisture and fat related properties of foods ...... 367
  - 14.2.3 Soft machine mechanics for texture evaluation with regard to palatal size reduction ...... 369
  - 14.2.4 Bolus rheology ...... 371
- 14.3 Texture evaluation through human physiological responses ...... 374
  - 14.3.1 Tongue pressure measurement ...... 374
  - 14.3.2 Electromyography ...... 378
  - 14.3.3 Acoustic analysis of swallowing sound ...... 380
  - 14.3.4 Laryngeal movement measurement ...... 382
- 14.4 Texture and flavor interaction during food consumption ...... 382
  - 14.4.1 Flavor release control through texture modification ...... 382
  - 14.4.2 Enhanced aroma perception through inhomogeneous spatial distribution ...... 384
  - 14.4.3 Modification of human eating behavior by aroma perception ...... 385
- 14.5 Structure and formulation design of food products using hydrocolloids ...... 386
  - 14.5.1 Use of polysaccharides as a texture modifier in elderly foods ...... 386

|  | 14.5.2 Usefulness of xanthan gum as dysphagia thickener | 387 |
|---|---|---|
| **14.6** | Conclusion | 390 |
|  | References | 391 |

## CHAPTER 15 Designing and development of food structure with high acceptance based on the consumer perception .................. 399

Ricardo Isaías, Ana Frias, Célia Rocha, Ana Pinto Moura, and Luís Miguel Cunha

| **15.1** | Introduction | 399 |
|---|---|---|
| **15.2** | Determinants of acceptance of innovative food products from food structure design | 400 |
|  | 15.2.1 Sensory properties | 400 |
|  | 15.2.2 Health concerns | 402 |
|  | 15.2.3 Nutrition concerns | 404 |
|  | 15.2.4 Risk perception | 405 |
|  | 15.2.5 Convenience | 406 |
|  | 15.2.6 Price | 407 |
| **15.3** | Conclusions | 408 |
|  | Acknowledgments | 408 |
|  | References | 409 |

Index ........................................................................................ 415

# Contributors

**Alejandra Acevedo-Fani**
Riddet Institute, Massey University, Palmerston North, New Zealand

**Carina Almeida**
National Institute for Agricultural and Veterinary Research (INIAV), I.P., Vila do Conde, Portugal

**M.D. Alvarez**
Department of Characterisation, Quality and Safety, Institute of Food Science, Technology and Nutrition (ICTAN-CSIC), Madrid, Spain

**Vitor D. Alves**
Frulact SA, Maia, Portugal

**C. Anandharamakrishnan**
Computational Modeling and Nanoscale Processing Unit, National Institute of Food Technology Entrepreneurship and Management-Thanjavur, Ministry of Food Processing Industries, Government of India, Thanjavur, Tamil Nadu, India

**Juan-Carlos Arboleya**
Basque Culinary Center, Mondragon Unibertsitatea; BCC Innovation, Technological Center in Gastronomy, Donostia-San Sebastián, Gipuzkoa, Spain

**Elham Assadpour**
Department of Food Materials and Process Design Engineering, Gorgan University of Agricultural Sciences and Natural Resources, Gorgan, Iran; Nutrition and Bromatology Group, Analytical and Food Chemistry Department, Faculty of Food Science and Technology, University of Vigo, Ourense, Spain

**Pedro E.D. Augusto**
Department of Agri-food Industry, Food and Nutrition (LAN), Luiz de Queiroz College of Agriculture (ESALQ); Food and Nutrition Research Center (NAPAN), University of São Paulo (USP), São Paulo, Brazil; Université Paris-Saclay, CentraleSupélec, Laboratoire de Génie des Procédés et Matériaux, SFR Condorcet FR CNRS 3417, Centre Européen de Biotechnologie et de Bioéconomie (CEBB), Pomacle, France

**Eneko Axpe**
Basque Culinary Center, Mondragon Unibertsitatea, Donostia-San Sebastián, Gipuzkoa; Azti, Parque Tecnológico de Bizkaia, Derio; Ikerbasque, Basque Foundation for Science, Bilbao, Bizkaia, Spain

**Julliane Carvalho Barros**
State University of Campinas (UNICAMP), School of Food Engineering, Cidade Universitária Zeferino Vaz, São Paulo, Brazil

**Ana I. Bourbon**
International Iberian Nanotechnology Laboratory, Braga, Portugal

**Maria Elisa Caetano-Silva**
College of Agricultural, Consumer & Environmental Sciences, University of Illinois at Urbana-Champaign, Champaign, IL, United States

**Gisandro Reis Carvalho**
Department of Agri-food Industry, Food and Nutrition (LAN), Luiz de Queiroz College of Agriculture (ESALQ), University of São Paulo (USP), São Paulo, Brazil

**L. Castañeda Florez**
School of Food Engineering, Faculty of Engineering, Universidad del Valle, Tuluá, Valle del Cauca, Colombia

**Diogo Castelo-Branco**
Mendes Gonçalves SA, Golegã, Portugal

**Fabiola Castillo**
Basque Culinary Center, Mondragon Unibertsitatea, Donostia-San Sebastián, Gipuzkoa, Spain

**Elisabete Coelho**
LAQV-REQUIMTE, Department of Chemistry, University of Aveiro, Campus Universitário de Santiago, Aveiro, Portugal

**S. Cofrades**
Department of Products, Institute of Food Science, Technology and Nutrition (ICTAN-CSIC), Madrid, Spain

**Manuel A. Coimbra**
LAQV-REQUIMTE, Department of Chemistry, University of Aveiro, Campus Universitário de Santiago, Aveiro, Portugal

**Luís Miguel Cunha**
GreenUPorto/INOV4AGRO & DGAOT, Faculty of Sciences, University of Porto, Vila do Conde, Portugal

**Adriano G. da Cruz**
Food Department, Federal Institute of Education, Science and Technology of Rio de Janeiro (IFRJ), Rio de Janeiro, Brazil

**Guilherme de Figueiredo Furtado**
Department of Food Engineering, Faculty of Food Engineering, University of Campinas, Campinas, SP, Brazil

**Cristian Dima**
Faculty of Food Science and Engineering, "Dunarea de Jos" University of Galati, Galati, Romania

**Duc Toan Do**
Riddet Institute, Massey University, Palmerston North, New Zealand

**Mirian dos Santos**
State University of Campinas (UNICAMP), School of Food Engineering, Cidade Universitária Zeferino Vaz, São Paulo, Brazil

**Juliana Domingues dos Santos Carvalho**
Department of Food Engineering, Faculty of Food Engineering, University of Campinas, Campinas, SP, Brazil

**Didier Dupont**
STLO, INRAE—Institut Agro, Rennes, France

**L.A. Espinosa Sandoval**
GIPAB Group, School of Food Engineering, Faculty of Engineering, Universidad del Valle, Cali, Valle del Cauca; GIFTEX Group, Theorical and Experimental Physical-Chemistry Interdisciplinary Group, Universidad Industrial de Santander, Piedecuesta, Santander, Colombia

**Gabriela Feltre**
Department of Food Engineering, Faculty of Food Engineering, University of Campinas, Campinas, SP, Brazil

**Pedro A.R. Fernandes**
LAQV-REQUIMTE, Department of Chemistry, University of Aveiro, Campus Universitário de Santiago; CICECO, Department of Chemistry, University of Aveiro, Aveiro, Portugal

**Diogo Figueira**
Mendes Gonçalves SA, Golegã, Portugal

**Ana Frias**
GreenUPorto/INOV4AGRO & DGAOT, Faculty of Sciences, University of Porto, Vila do Conde, Portugal

**Pablo Fuciños**
International Iberian Nanotechnology Laboratory, Braga, Portugal

**Takahiro Funami**
San-Ei Gen F.F.I. Inc., Texture Design Research Laboratory, Toyonaka, Osaka, Japan

**A. García Figueroa**
School of Food Engineering, Faculty of Engineering, Universidad del Valle, Tuluá, Valle del Cauca, Colombia

**Ana Gomes**
CBQF-Centro de Biotecnologia e Química Fina—Laboratório Associado, Escola Superior de Biotecnologia, Universidade Católica Portuguesa, Porto, Portugal

**Daniela A. Gonçalves**
CEB—Centre of Biological Engineering, University of Minho, Braga, Portugal

**Joana Gonçalves**
CBQF-Centro de Biotecnologia e Química Fina—Laboratório Associado, Escola Superior de Biotecnologia, Universidade Católica Portuguesa, Porto, Portugal

**Miriam Dupas Hubinger**
Department of Food Engineering, Faculty of Food Engineering, University of Campinas, Campinas, SP, Brazil

**Ricardo Isaías**
GreenUPorto/INOV4AGRO & DGAOT, Faculty of Sciences, University of Porto, Vila do Conde, Portugal

**Seid Mahdi Jafari**
Department of Food Materials and Process Design Engineering, Gorgan University of Agricultural Sciences and Natural Resources, Gorgan, Iran; Nutrition and Bromatology Group, Analytical and Food Chemistry Department, Faculty of Food Science and Technology, University of Vigo, Ourense, Spain; College of Food Science and Technology, Hebei Agricultural University, Baoding, China

**Mirian T.K. Kubo**
Université de Technologie de Compiègne, UMR CNRS 7025, Enzyme and Cell Engineering Laboratory, Compiègne, France; Institute of Biosciences, Humanities and Exact Sciences (IBILCE), Department of Food Engineering and Technology (DETA), São Paulo State University (UNESP), São Paulo, Brazil

**Alessandra Massa**
Basque Culinary Center, Mondragon Unibertsitatea, Donostia-San Sebastián, Gipuzkoa, Spain

**David J. McClements**
Department of Food Science, University of Massachusetts Amherst, Amherst, MA, United States

**Olivia Ménard**
STLO, INRAE—Institut Agro, Rennes, France

**J.A. Moses**
Computational Modeling and Nanoscale Processing Unit, National Institute of Food Technology Entrepreneurship and Management-Thanjavur, Ministry of Food Processing Industries, Government of India, Thanjavur, Tamil Nadu, India

**Ana Pinto Moura**
GreenUPorto/INOV4AGRO & DCeT, Universidade Aberta, Rua do Amial, Porto, Portugal

**Makoto Nakauma**
San-Ei Gen F.F.I. Inc., Texture Design Research Laboratory, Toyonaka, Osaka, Japan

**Clarisse Nobre**
CEB—Centre of Biological Engineering, University of Minho, Braga, Portugal

**Cláudia Nunes**
LAQV-REQUIMTE, Department of Chemistry, University of Aveiro, Campus Universitário de Santiago; CICECO, Department of Chemistry, University of Aveiro, Aveiro, Portugal

**Lorenzo Miguel Pastrana Castro**
International Iberian Nanotechnology Laboratory, Braga, Portugal

**Ana Rita Peixoto**
Vieira de Castro Produtos Alimentares, S.A, V. N. Famalicão, Portugal

**Ricardo Nuno Pereira**
CEB—Centre of Biological Engineering; LABBELS—Associate Laboratory, University of Minho, Braga, Portugal

**Manuela Pintado**
CBQF-Centro de Biotecnologia e Química Fina—Laboratório Associado, Escola Superior de Biotecnologia, Universidade Católica Portuguesa, Porto, Portugal

**A.M. Polanía Rivera**
GIPAB Group, School of Food Engineering, Faculty of Engineering, Universidad del Valle, Cali; School of Food Engineering, Faculty of Engineering, Universidad del Valle, Tuluá, Valle del Cauca, Colombia

**Marise Aparecida Rodrigues Pollonio**
State University of Campinas (UNICAMP), School of Food Engineering, Cidade Universitária Zeferino Vaz, São Paulo, Brazil

**Miguel Ângelo Parente Ribeiro Cerqueira**
International Iberian Nanotechnology Laboratory, Braga, Portugal

**Célia Rocha**
GreenUPorto/INOV4AGRO & DGAOT, Faculty of Sciences, University of Porto, Vila do Conde; Sense Test, Lda., Vila Nova de Gaia, Portugal

**Rui M. Rodrigues**
CEB—Centre of Biological Engineering; LABBELS—Associate Laboratory, University of Minho, Braga, Portugal

**Meliza Lindsay Rojas**
Dirección de Investigación y Desarrollo, Universidad Privada del Norte (UPN), Trujillo, Peru

**Debashree Roy**
Riddet Institute, Massey University, Palmerston North, New Zealand

**Claudia Sánchez**
National Institute for Agricultural and Veterinary Research (INIAV), I.P., Alcobaça, Portugal

**Ana Sanches Silva**
National Institute for Agricultural and Veterinary Research (INIAV), I.P., Vila do Conde; Center for Study in Animal Science (CECA), University of Oporto, Oporto, Portugal

**Harjinder Singh**
Riddet Institute, Massey University, Palmerston North, New Zealand

**D. Subhasri**
Computational Modeling and Nanoscale Processing Unit, National Institute of Food Technology Entrepreneurship and Management-Thanjavur, Ministry of Food Processing Industries, Government of India, Thanjavur, Tamil Nadu, India

**Ana Tasso**
Mendes Gonçalves SA, Golegã, Portugal

**José A. Teixeira**
CEB—Centre of Biological Engineering, University of Minho, Braga, Portugal

**Paula Teixeira**
CBQF-Centro de Biotecnologia e Química Fina—Laboratório Associado, Escola Superior de Biotecnologia, Universidade Católica Portuguesa, Porto, Portugal

**Andrea Paola Rodriguez Triviño**
State University of Campinas (UNICAMP), School of Food Engineering, Cidade Universitária Zeferino Vaz, São Paulo, Brazil

**Antonio A. Vicente**
CEB—Centre of Biological Engineering; LABBELS—Associate Laboratory, University of Minho, Braga, Portugal

# About the editors

**Miguel Ângelo Parente Ribeiro Cerqueira** is a staff researcher in the Food Processing and Nutrition Group at the International Iberian Nanotechnology Laboratory, and his research is focused on the development of micro- and nanosized bio-based structures for food applications. He works on edible and biodegradable materials for packaging, encapsulation of functional compounds using emergent encapsulation technologies, and structuring gels, such as oleogels.

Miguel earned both a graduation and a PhD degree in biological engineering from the University of Minho. He received three scholar merit awards during graduation, and his PhD thesis was awarded for the best PhD thesis by the School of Engineering of the University of Minho. During his PhD, he performed scientific missions at the Federal University of Ceará, University of Aveiro and University College Cork. In 2011, he started as a postdoctoral researcher at UM and performed scientific missions at the University of Vigo, University of Campinas, and Institute of Agrochemistry and Food Technology—CSIC. He has authored more than 130 scientific articles, published 20 book chapters and 2 patents, and is editor of 3 books. Miguel has supervised more than 15 students (PhD and MSc). He is the cofounder of two start-ups (Improveat, Lda in 2013 and 2BNanoFood, Lda in 2020). In 2014, he won the Young Scientist Award, and in 2016 he was selected as Inaugural Member of the International Academy of Food Science and Technology, Early Career Scientist Section, both organized by the International Union of Food Science and Technology. In 2018, he joined the list of Highly Cited Researchers from Clarivate Analytics.

**Lorenzo Miguel Pastrana Castro** is currently Chair of the Research Office and Group Leader of the Food Processing and Nutrition Group, at the International Iberian Nanotechnology Laboratory (INL). He is also Full Professor of Food Science at the University of Vigo and Visiting Professor at the Universidade Federal Rural de Pernanbuco (Brasil) and Università Catollica del Sacro Cuore (Italy).

In 2015, he joined the INL as the Head of the Department of Life Sciences that included three Research Units, namely Food, Environment, and Health. At INL his research was marked oriented with a multidisciplinary approach integrating methods and concepts of biotechnology, nanotechnology, and mathematical modeling. Currently, he is working in three main research lines: food structure with an emphasis on 3D printing materials, encapsulation technologies for improving functional foods and food personalization, and active and intelligent food packaging. He is the author of more than 200 scientific contributions and holds 4 licensed patents relating to the development of new food products and processes. He was the principal investigator of more than 30 national and European research projects and contracts and the promoter of 2 food start-ups.

He also was the director of the Knowledge Transfer Office (2009–10) at the University of Vigo, and the founder of the Galician Agri-Food Technology Platform (2006). Currently, he is a member of the scientific board of the Portugal Foods innovation cluster.

# Foreword

Historically, the study of foods progressed from the early vision of cooks and some scientists to the analysis of chemical components in foods in the 19th century (also called bromatology). Later, it was realized that some food components (notably, vitamins) played an essential role in specific nutritional deficiencies and the science of nutrition was born. By the mid-20th century, the study of foods was centered on the processing of products of agricultural production (cereals, fruits and vegetables, meat and poultry, dairy products, etc.), and thematic academic departments flourished at several universities around the world. However, it was not until the latter part of the last century that the effects of the structural arrangement of chemical components and nutrients in foods—the food microstructure—on sensory properties and nutrition became evident. This was the foundation of what is now food materials science. Major food components, e.g., protein, polysaccharides, and lipids, did not differ from the already known physicochemical principles of molecular and polymer chemistry. The acceptance of this reality gave food technology a respectable site in the scientific world and signaled a departure from a long tradition of empirical conjectures and some myths.

Contrary to most consumer goods, our foods were not "designed" in an engineering sense. Fresh products and raw materials are "designed" by nature and the basic products that we eat (breads, cheeses, meat products, etc.) evolved mostly through serendipity and trial and error in culinary practice. In the mid-1900s, an expanding food industry catering to a booming urban population scaled up artisanal procedures into manufacturing processes, moving from small-batch quantities to high-volume, continuous operations. Food product development became a major preoccupation of this industry in order to expand product lines, create new foods, and utilize novel raw materials. Today, artificial intelligence and machine learning may assist human teams in R&D laboratories by analyzing complex data patterns acquired from multiple sources and providing guiding outputs to satisfy consumers' demands.

Why do we need designed foods? First, to make tasty and high-quality foods, low in salt, sugar, and fat, which are acceptable to people with dietary restrictions imposed by physiological and/or medical conditions (e.g., diabetes, hypertension, allergies and intolerances, and overweight); and second, to provide soft, nutritious, and savory foods to an expanding elderly population with special needs dictated by aging, and, in addition, to convey adequate nutrition supplementation to the many consumers that adopt restricted diets based on personal beliefs. From the environmental perspective, sustainable foods are needed to replace animal-based products with plant-based products and other sustainable raw materials (seaweed, insects, microbial sources, etc.). Last but not least, designing foods represents an opportunity to expand the variety of healthy culinary and gastronomic creations for the mass marketing of foods.

The volume in your hands, *Food Structure Engineering and Design for Improved Nutrition, Health and Well-Being* delves into these subjects in chapters written by experts in the relevant fields. The book is a comprehensive text covering the fundamental knowledge behind the design of food microstructures and food matrices, the processing of foods in our bodies (digestion), emerging technologies targeting nutrition improvement, the generation of reliable data to support health claims, and applications in gastronomy. This ample scope of subjects suggests that a transdisciplinary strategy for a rational design of foods for health and well-being should include the interactions of many disciplinary fields and the involvement of major stakeholders.

**José Miguel Aguilera**
*Department of Chemical and Bioprocess Engineering, Pontificia Universidad Católica de Chile, Vicuña Mackenna, 4860, Macul, Santiago, Chile*

# Preface

The study and understanding of Food Chemistry and Biochemistry and the development of Food Preservation and Processing Technologies were the main drivers of the scientific advancement in the 20th century. These developments allow to produce nutritious, safe, tasty, and convenient food products to meet consumer demands. However, food scientists realized this knowledge is not enough to explain some of the fascinating characteristics of food and its ingredients. For that reason, more recently, an increased attention has been paid to revealing how the unique food properties emerge from the food structure at micro- and nanoscale. Therefore, during the past years, understanding food structure has become the ultimate frontier between food science and soft matter. Nowadays, we know that food structures can be engineered and designed with the aim of creating food products with enhanced properties, opening the possibilities to develop a more precise food personalization. However, a book in which these new endeavors are presented and discussed is still missing.

This book gives new insights on the development of new healthy foods and the understanding of food structure effect on nutrition, health, and well-being, with the latest and updated vision of leading scientists in the area. The book is organized into 5 parts and 15 chapters focused on 5 main topics: (a) the new foods trends worldwide regarding sustainability goals and food structure; (b) innovative strategies for food structure development; (c) strategies to address some of the challenges for developing healthier food products, such as reduction of sugar, salt, and fats; (d) the assessment of health effect of foods by in vitro and in vivo tests; and (e) how the new food structure engineering and design determine the consumer choice and perception of the physiological process of digestion.

It is expected that the information presented in this book will increase the scientific understanding of the design and engineering of the food structure with the aim of a better nutrition. The authors are scientists from different areas that cover the principal aspects of food structure; the combination of these experts in areas such as nutritional aspects, food engineering, soft matter characterization and evaluation, and consumer perception with an industrial perspective gives a broad overview of the field to the readers, boosting their capability to integrate different aspects of product development.

# Acknowledgments

We thank all the authors for their contribution and for giving an excellent perspective on the best that has been done in the area. A big thanks to Kathrine Esten, editorial project manager, for her help during the editorial processes and for always giving us the best advice. We also thank Professor José Miguel Aguilera for his contribution to the foreword of this book.

This work was made possible with the support of the following projects:

cLabel+: Innovative natural, nutritious, and consumer-oriented clean label foods (POCI-01-0247-FEDER-046080), cofinanced by Compete 2020, Lisbon 2020, Portugal 2020, and the European Union, through the European Regional Development Fund (ERDF).

IceCare: Cardio-Healthy Functional Ice creams (NORTE-01-0247-FEDER-039927), cofinanced by Norte 2020 and the European Union, through the European Regional Development Fund (ERDF).

BetterFat4Meat: Development of structured fat for use in meat products by replacement of animal fat (POCI-01-0247- FEDER-039718), cofinanced by Compete 2020, Portugal 2020, and the European Union, through the European Regional Development Fund (ERDF).

FODIAC: Food for Diabetes and Cognition (reference number 778388) financed by the European Commission through the H2020-MSCA-RISE project.

# Introduction

PART I

# CHAPTER 1

# Nutrition, health and well-being in the world: The role of food structure design

Miguel Ângelo Parente Ribeiro Cerqueira[a], David J. McClements[b], and Lorenzo Miguel Pastrana Castro[a]

[a]International Iberian Nanotechnology Laboratory, Braga, Portugal, [b]Department of Food Science, University of Massachusetts Amherst, Amherst, MA, United States

## 1.1 Food challenges and United Nations sustainable development goals

A major challenge of the modern food and agricultural system is to feed a growing global population a healthy diet without damaging the environment. At present, however, this system is known to be having adverse effects on pollution, greenhouse gas emissions, land use, water use, and biodiversity loss, as well as on human health and well-being (Willett et al., 2019). An appreciable proportion of the global population suffers from undernutrition because they do not have sufficient food or are consuming low-quality diets, while another appreciable proportion suffers from overnutrition because they eat too much or the wrong kinds of foods. Undernutrition leads to hunger and diseases linked to micronutrient deficiencies, while overnutrition leads to obesity, diabetes, hypertension, coronary heart disease, stroke, and cancer. Consequently, there is a need to improve the global diet, which is only possible through a global change along the food chain that involves all relevant stakeholders (World Health Organization, 2021). These problems are a concern to both developed and developing countries (OECD, 2019). It has been estimated that the health care costs associated with diseases of overnutrition are similar to those of smoking or armed conflict and are predicted to double by 2030 (Giner & Brooks, 2019).

In 2015, the United Nations adopted 17 Sustainable Development Goals (SDG) for the Global Challenges (Fig. 1.1). The food industry greatly influences many of these goals but especially Goals: 2, 3, 12 and 13. SDG 2 aims to end hunger, achieve food security, improve nutrition, and promote sustainable agriculture. SDG 6 aims to ensure the availability and sustainable management of water and sanitation for all. This water is required for drinking and hygiene, but also to ensure food and agricultural productivity. SDG 12 aims to ensure sustainable consumption and production patterns. SDG 13 aims to ensure that the global climate is maintained in a state that will be suitable for the health and well-being of future generations.

According to Lillford and Hermansson, modern food science and technology should focus on creating a more diverse and sustainable food production and distribution system (Lillford & Hermansson, 2021). They highlighted seven missions that could help to redefine a new food system:

**FIG. 1.1**

Sustainable development goals of United Nations.

*The content of this publication has not been approved by the United Nations and does not reflect the views of the United Nations or its officials or Member States, https://www.un.org/sustainabledevelopment/.*

**Mission 1: Introduce more diverse and sustainable primary produce**. This mission is aimed to obtain optimal functional properties during the production of new raw materials, to understand and control the behavior of raw materials and ingredients during production, and to improve the knowledge for better control of the final product quality, including structure, material properties, sensory perception, and nutritional quality. Food structure design is of great importance to achieve this mission. In particular, understanding the properties and behavior of food ingredients is essential for creating a healthier and more sustainable food system.

**Mission 2: Develop new processes and systems to ensure sustainable manufacture**. This mission is aimed at developing precision engineering approaches to improve the efficiency of food manufacturing processes: reducing water use; recycling water; reducing waste; developing low-temperature processes, optimizing drying/rehydration processes; reducing fossil fuel and energy use; developing local manufacturing facilities; and promoting standardized methods for sustainability analysis and reporting. Food structure design can also have an impact on this mission because it can be used to optimize the formulation of more sustainable and healthy foods and ingredients, as well as novel processing technologies. An example of this approach is given in Chapter 3, where electrotechnologies are described that can be used to create new foods with unique characteristics. These technologies can be used to replace existing thermal processing methods, which may reduce energy use.

**Mission 3: Eliminate food and material waste in production, distribution, and consumption**. This mission is focused on reducing and recycling food wastes and byproducts from primary production to consumption. According to the Food and Agriculture Organization of the United Nations (FAO) approximately one-third of all food produced in the world in 2009, measured by weight, was lost or wasted (FAO, 2011). The aim of this mission is to improve the stability of primary produce to overcome problems during transportation and preservation. This can be achieved in various ways: developing low energy drying and freezing systems; developing sensors for monitoring foods from production to consumption; restructuring the ingredient and food production industries to add value to all sides. One of the examples is the use of food by-products such as whey protein for the development of food structures by additive manufacturing (i.e., 3D printing), which can result in new and healthy food products (Sager et al., 2021). This approach is discussed in Chapter 2. Also, in this mission, is mentioned the need of finding ways to reduce levels of petrochemical materials in packaging materials (e.g., by increasing recyclability and using biobased materials). This mission can be very different depending on the region of the world, while for low-income countries, the waste is higher in primary production due to inadequate post-harvest and distribution strategies in high-income countries, the losses are mostly related to the final product distribution and home consumption (World Resources Institute, 2019).

**Mission 4: Establish complete product safety and traceability**. This mission is focused on ensuring food safety. This objective can be addressed by developing rapid methods for identification and quantification of toxins, allergens, pathogenic organisms, and spoilage organisms across the food chain; understanding the epidemiology of microorganisms throughout the food environment; preventing the transfer of antimicrobial-resistant organisms to the food chain; providing traceability of products by introducing robust documentation of food products with different levels of information (e.g., primary source, processing methods, product composition and safety); and, using different methodologies and strategies to reduce microbiological contamination. Food structure design can have an important role in this mission by developing new antimicrobial delivery systems, creating innovative antimicrobial packaging materials, or controlling water activity in foods.

**Mission 5: Provide affordable and balanced nutrition to the malnourished**. The reformulation of food composition and processing can be used to create foods that are more nutritionally balanced and have a high bioavailability. In particular, the nutrition profile of foods can be targeted to the needs of specific populations, while maintaining food affordability, access, and quality. This mission is related to SDG2 and involves all regions from low-income countries where foods are less available to high-income countries where malnutrition can happen in some populations, such as the poor, elderly, or infants. Food structure design can be used to create tasty foods with desirable nutritional profiles and high nutrient bioavailability.

Fig. 1.2 shows the number of undernourished people around the world in 2019 and projections for 2030. The distribution of hunger is predicted to change over the next years, making Africa the region with the highest number of undernourished by 2030.

**Mission 6: Improve health through diet**. This mission impacts health and well-being with the aim of preventing non-communicable diseases. This will require reformulating foods so they have an appropriate nutritional profile, as well as controlling their behavior inside the gastrointestinal tract. Research has shown that food structural design can be used to control the digestibility and absorption of nutrients and nutraceuticals in foods, as well as to control their impact on the gut microbiome. However, further work is required to understand how specific bioactive components impact human health and well-being.

In particular, future research should focus on measuring the release of nutrients from whole foodstuffs throughout digestion. For example, the use of dietary biomarkers to assess food intake

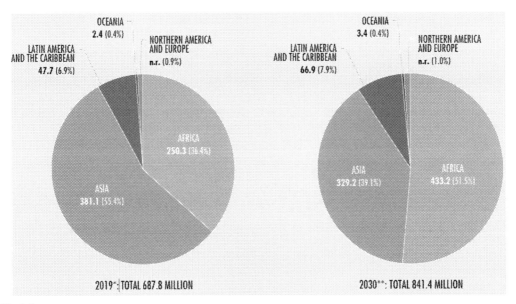

**FIG. 1.2**

Number of undernourished people in millions. Number of undernourished people in millions. * Projected values. ** Projections to 2030 do not consider the potential impact of the COVID-19 pandemic. n.r. = not reported, as the prevalence is less than 2.5%.

*From FAO, IFAD, UNICEF, WFP, & WHO. (2020). The State of Food Security and Nutrition in the World 2020. Transforming food systems for affordable healthy diets. doi:https://doi.org/10.4060/ca9692en. Reproduced with permission.*

among consumers and make use of metabolomics to detect responses to different foods and diets and, at the same time to define the nutrient needs of individuals within established nutritional groups for precise advice on diets. It will be essential to continue validating the impact of nutraceuticals on health (e.g., by using cohort studies and market data). Another important aspect will be to identify how macro and micronutrients can be combined and used on long term health via diet.

This mission will require the work of food science and technology, where food structure design can help increase bioavailability and the way that foods are digested, but also other areas of nutrition, medicine, neuroscience, and physiology need to be involved. Recently, FAO et al. (2020) presented how healthy diets [flexitarian (FLX), pescatarian (PSC), vegetarian (VEG) and vegan (VGN)] could impact the reduction of mortality in 2030 in relation to four non-communicable diseases: coronary heart disease, stroke, cancer and type-2 diabetes mellitus (Fig. 1.3).

**Mission 7: Integrate big data, information technology, and artificial intelligence throughout the food chain.** The main needs of this mission are to use multivariate data and machine learning to build reliable models for material/process interactions in food manufacture and identify statistical relationships between diet and health. It will also be important to guarantee valid data and develop secure methods to link information flows through the food chain (improving traceability,

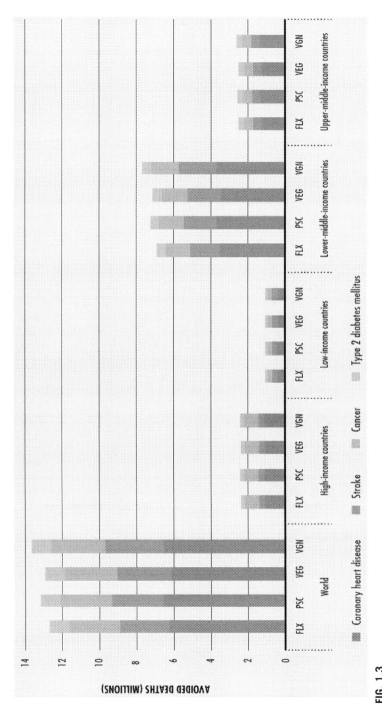

**FIG. 1.3**

Number of deaths avoided in 2030, related to four non-communicable diseases by moving from the benchmark diet of national average food consumption to the four healthy and sustainable dietary patterns. Number of deaths avoided in 2030, related to four non-communicable diseases (coronary heart disease, stroke, cancer and type-2 diabetes mellitus) by moving from the benchmark diet of national average food consumption to the four healthy and sustainable dietary patterns. The four alternative healthy diet patterns for the analysis include the flexitarian (FLX), the pescatarian (PSC), the vegetarian (VEG) and the vegan (VGN) diet.

*From FAO, IFAD, UNICEF, WFP, & WHO. (2020). The State of Food Security and Nutrition in the World 2020. Transforming food systems for affordable healthy diets. doi: https://doi.org/10.4060/ca9692en. Reproduced with permission.*

standardizing safety, and reducing costs and waste). The proper use of this data could contribute to all the previous missions, thereby helping to design a more efficient food system.

Globally, the food system has already made tremendous advances in helping to preserve and convert different types of natural resources into foods for human consumption, as well as ensuring they are delivered intact to the final consumer. The improvement in the sustainability and efficiency of the food chain will require a multidisciplinary approach and collaboration among different stakeholders. Food structure design can have an important role in several aspects. One of the highest impacts of this approach will be in the development of nutritionally-balanced foods, where the use of new or side stream materials, innovative processing technologies (Chapters 2–10) and mathematical and analytical approaches can highly impact these developments (Chapters 11–14).

## 1.2 Trends in human food consumption: The diet shift

People's diet is influenced by several factors, including culture, religion, climate, and traditions, which varies across countries, regions, and households. In 2017, it was shown that diets are changing rapidly, particularly with respect to fat, artificial sweeteners, and animal-sourced foods (Oberlander et al., 2017). In particular, it appears to be a global movement towards the so-called "Western diet" (Khoury et al., 2014). This diet is characterized by a high intake of refined carbohydrates, sugars, fats, processed foods, animal-sourced foods and an inadequate intake of fruits and vegetables (Popkin et al., 2012). In a study of the dietary changes of 118 countries from 1960 to 2010, it was shown that the total calorie intake increased for all countries and the nutritional quality decreased (Le et al., 2020) (Fig. 1.4).

The results show that all of the five dietary types are becoming less healthy but they also showed that this has happened because each diet has replaced carbohydrates with fats, which the authors claim reflect the transition from plant-based to animal-based foods (Fig. 1.5).

Recent studies suggest that global food consumption should shift to more plant-based foods, not only to increase the healthiness of the diets but also due to the environmental impact of animal-based foods. For instance, the EAT-Lancet commission presented global targets, based on available data and evidence for healthy diets and sustainable food production that could guarantee the UN SDGs and Paris Agreement are achieved (Willett et al., 2019). The commission presented a universal diet that could be used as a healthy reference diet and provided insights into how this diet could help improve the environment and human health. The healthy reference diet consists of vegetables, fruits, whole grains, legumes, nuts, and unsaturated oils, and includes a low to moderate amount of seafood and poultry, as well as a low quantity or red meat, processed meat, added sugar, refined grains, and starchy vegetables. The commission integrated the universal healthy diets and global scientific targets for sustainable food systems to provide scientific planetary boundaries to reduce environmental degradation caused by food production.

Recently, Smith et al. (2021) presented a computational model called DELTA that was able to determine the nutrient adequacy of current and proposed global food systems. With this model, it was possible to study several scenarios, such as the increase of population in 2030, and the change of the diet. For a scenario where the population increased to 8.6 billion people and there was a change in population structure (i.e., a higher ratio of adults to children and of women to men), the deficiency in calcium and vitamin E (already observed in 2018) will increase, and iron, potassium, riboflavin,

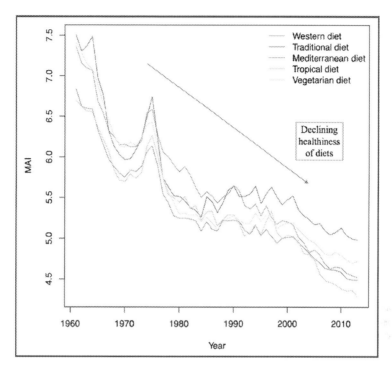

**FIG. 1.4**

Dietary quality from 1960 to 2010 based on the Mediterranean Adequacy Index for five dietary types.

*From Le, Disegna, Lloyd, (2020). Dietary quality from 1960 to 2010 based on the Mediterranean Adequacy Index (Fidanza et al. 2004) for five dietary types.*

vitamin A and vitamin B-12 will appear in the list of deficient nutrients. There may therefore be a need to fortify foods with bioavailable forms of these nutrients in the future. It is important to highlight that this deficiency is not related to a decrease in food consumption but mostly with an unbalanced diet. Another example is the no meat scenario in 2030, where all meat and seafood production was set to zero, and the remaining food groups were increased by 20% to have a similar total biomass production. In this scenario, the food available increased (due to reduced animal feed), however, the nutrient needs for iron, zinc and vitamin B-12 increased, which would require the consumption of more macronutrients to achieve the desired nutrients required and thus cause an excess intake of energy. In 2017, EFSA presented a technical report with the dietary reference values of nutrients for the population, covering water, fats, carbohydrates and dietary fiber, protein, energy, as well as 14 vitamins and 13 minerals (EFSA, 2017). Several reference values were reported that can be used to design foods and plan diets. These new approaches could help design foods that are nutritionally balanced without increasing the number of calories consumed.

In the last 10 years has been a large increase in consumers demanding organic and more "natural" products combined with an interest in more environmentally friendly products. These new trends have led to new challenges for the food industry, such as the reformulation of existing products using new

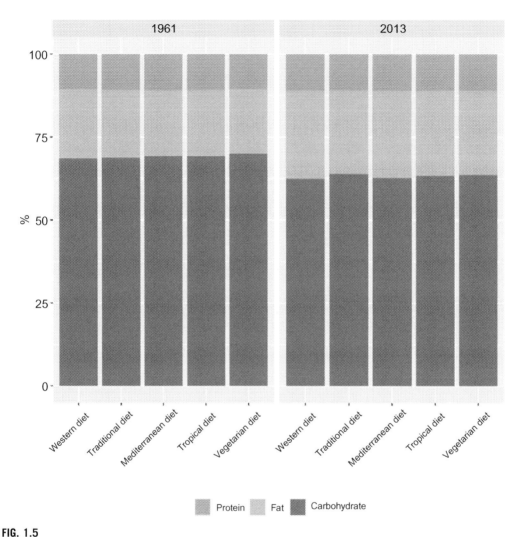

**FIG. 1.5**

Changes in macronutrient composition between 1961 and 2013.

*From Le, Disegna, Lloyd, (2020). Changes in macronutrient composition between 1961 and 2013.*

technologies and ingredients. This trend has been linked to the term "clean label". Clean label products demand arose, requesting that the use of ingredients and additives used for years by the food industry were replaced by more "natural" alternatives due to a shift in consumers' choices, where the natural had become one of the most important factors for food's selection (Asioli et al., 2017).

The trends in food consumption are being driven by various factors, including nutrient fortification (micro and macronutrients) to improve human health; the demand for healthy diets, and sustainability factors. The sustainability factors are perhaps one of the most challenging in the food industry and will only be possible through collaboration between all stakeholders.

## 1.3 Food structure design for nutrition and health benefits

The creation of nutritionally balanced and healthy processed foods requires careful ingredient selection and processing. Many processing operations can cause the degradation of important nutrients thereby reducing their efficacy, including dehydration, extrusion, thermal processing, and pasteurization. Moreover, operations such as homogenization and thermal processing can breakdown the structure of natural plant materials, which increases the digestibility of macronutrients (such as starch and fat), which can adversely affect the hormonal and metabolic systems. For these reasons, food processing has often been associated with unhealthy and unsustainable diets. These foods are often classified as processed foods (PF) or ultra-processed foods (UPF). In the past decade, several researchers and companies have shown that it is possible to create processed foods that have good nutritional balances (Ludwig et al., 2019; Monteiro et al., 2019).

Processed foods can also be classified according to their nutritional profiles (Sadler et al., 2021). However, the classification criteria used are often ambiguous, inconsistent, and opposed by many researchers, mostly because they do not fully consider the scientific evidence on the nutritional aspects of foods. One example is the NOVA classification system that proposed four food categories: unprocessed or minimally processed foods; processed culinary ingredients; processed foods; and ultra-processed foods (Monteiro et al., 2010, 2018). Several stakeholders around the world have used this classification to categorize their foods. However, in recent years, this classification system has been criticized by many researchers because it is too broad and does not consider the nutritional aspects of the foods (Gibney et al., 2017). For example, Vergeer et al. (2019) compared the nutritional quality of more-versus less-processed packaged foods and beverages in Canada. They used a large, branded food database and two processing classification systems: NOVA (Monteiro et al., 2018) and one proposed by Poti et al. (2015). They showed that most processed products under both systems are lower in protein and higher in total and free sugars, when compared with less-processed foods while the association of other nutrients/components and level of processing were less consistent. They concluded that calorie- and nutrient-dense foods exist across different levels of processing. They also suggested that food choices and dietary recommendations should not focus on processing classification but on energy or nutrient density. On the other side, Hall et al. (2019) conducted a randomized controlled trial examining the effects of ultra-processed versus unprocessed diets on ad libitum energy intake. The ultra-processed and unprocessed diets matched for calories, sugar, fat, fiber, and macronutrients, and the participants were instructed to consume as much or as little as desired. The results showed that the energy intake was higher in the case of the diet with ultra-processed foods (more 508 kcal/day), when compared to the unprocessed diet. They also observed that participants gained 0.9 kg on the ultra-processed diet and lost 0.9 kg on the unprocessed diet.

Several factors affect "healthy diets" but in the case of PF and UPF the design of foods that improve healthiness, and at the same time maintain consumer satisfaction, requires a detailed understanding of food microstructure and the interaction of food structures with physiological and behavioral processes occurring upon ingestion. These aspects are discussed in Chapters 11, 12, and 14. The oral processing and gastrointestinal behavior of foods are the main interfaces between food structure and their physiological effect on the consumer. In fact, the idea that food reformulation can improve the nutritional quality of food products and effectively promote healthier diets is not new and have been proposed by several stakeholders (Giner & Brooks, 2019). Food reformulation has focused on reducing or replacing trans fats and saturated fats, reducing sugar and salt content, and several programs were promoted by the governments and companies in this regard (Public Health England, 2015, 2017). However, this reformulation takes time, and it is not easy to remove some ingredients and additives that have been used in foods for decades to contribute to preservation, texture, and taste. Reducing salt, sugar, and fats are perhaps the most challenging ones. Poole and co-workers have presented strategies for replacing sugar and fats in foods (Poole et al., 2020), while Sun and co-workers have presented strategies to reduce salt in foods (Sun et al., 2021). The strategies to reduce salt, sugar and fats in foods are discussed in Chapters 8, 9 and 10, respectively. It should be noted that there is still much debate among nutritionists about the adverse effects of salts, sugars, and saturated fats on human health.

Food classifications can help stakeholders decide what foods and diets would be adequate in different parts of the world and in different contexts. However, the complexity of foods makes it difficult to use simple food processing classifications to define diets. In addition, the food processing level cannot be linked to the nutritional aspects or food's health benefits since there are several processed and ultra-processed foods that can accomplish all the nutritional requirements and bring health benefits to consumers. More sophisticated approaches should define the nutritional composition of foods, as well as the rate and extent of release of the nutrients during digestion. Processing may either increase or decrease nutrient bioavailability depending on the nutrient type, food type, and processing method.

The association between the level of food processing (as defined by the NOVA classification scheme) on the energy intake rate for 327 foods from 5 different studies has been examined (Forde et al., 2020). These authors showed that by increasing the level of food processing (i.e., from unprocessed to UPFs), the average energy intake rate goes from 35.5 kcal/min to $69.4 \pm 3.1$ kcal/min, but they observed that for each processing category, there is wide variability in the energy intake rate. They concluded that the relations between the level of food processing and obesity should also consider the differences in energy intake rates and that well-controlled human feeding trials are needed to define the causal mechanisms of ultra-processed foods that result in higher energy intakes.

Another factor that needs to be considered is the eating rate. Slowing down the eating rate to moderate food intake has been confirmed by several studies, which have shown that eating rate and chews per bite influence food intake (Ford et al., 2010; Galhardo et al., 2012). However, decreasing the eating rate of consumers under everyday conditions is hard, and can probably only be reliably achieved under controlled conditions. Food texture can be controlled to induce consumers to reduce their eating rate and, therefore their energy intake. Several studies have shown that harder foods are eaten more slow, which helps to reduce the energy intake of foods by modifying eating behavior (Forde et al., 2017; Wee et al., 2018). The effects of food texture on oral processing behavior and energy intake have been reviewed recently (Bolhuis & Forde, 2020). The authors confirmed that the bite size and chewing behavior (two oral processing characteristics) influenced both eating rate and food intake. The hardness and elasticity of solid foods increased chews per bite and decreased bite sizes, which resulted in a reduction of eating rate and food intake. Conversely, when the ability of foods to lubricate the mouth increases, it can stimulate faster eating rates by reducing the chews per bite required to agglomerate a swallowable bolus. The authors also concluded that the shape and size of foods could influence the eating rate and food intake since they influence bite sizes and surface area, which can change the moisture uptake, and influence bolus formation. In the case of semi-solid foods, the viscosity and particle size also affected the eating rate and food intake.

These results suggest that food design can help reduce energy intake by controlling the textural characteristics of foods and their behavior in the mouth. Chapter 14 discusses how food texture can be controlled and studied by analytical methods and how it can be used during food formulation.

## 1.4 Conclusions and future perspectives

Researchers and industry need to work together to find new strategies to tackle some of the challenges with the modern food system. The development of healthier and more sustainable foods will require people with different expertise to work together to address this problem, including agriculturalists, farmers, chemists, engineers, biologists, entrepreneurs, industrialists, nutritionists, and physiologists. The design and engineering of foods by changing their composition and structure can be part of the solution by developing foods with improved nutrition that can contribute to improve health and well-being and be sustainable with a low carbon footprint. Understanding and directing consumer behavior will also be crucial to overcome some of these challenges. Any newly designed foods that are healthier and more sustainable must also be affordable, delicious, and convenient, otherwise consumers will not adopt them.

Reverse engineering can be used to identify the ingredients and processes that should be used to create the desired attributes in foods, such as appearance, texture, shelf life, mouthfeel, nutritional profile, and consumer acceptance. This strategy relies on a good understanding of the molecular and physicochemical basis of food properties and their interactions with humans. Foods are extremely complex multicomponent materials, and the human mouth and digestive tract are also extremely complicated, which makes this difficult. However, big data and artificial intelligence algorithms may be useful for linking food properties to their composition and structure. This strategy could help design personalized foods based on consumers' nutritional and sensorial needs, as well as improve their environmental impact.

## Acknowledgments

This work was supported by the project cLabel+ (POCI-01-0247-FEDER-046080) co-financed by Compete 2020, Lisbon 2020, Portugal 2020 and the European Union, through the European Regional Development Fund (ERDF).

## References

Asioli, D., Aschemann-Witzel, J., Caputo, V., Vecchio, R., Annunziata, A., Næs, T., & Varela, P. (2017). Making sense of the "clean label" trends: A review of consumer food choice behavior and discussion of industry implications. *Food Research International, 99*, 58–71. https://doi.org/10.1016/j.foodres.2017.07.022.

Bolhuis, D. P., & Forde, C. G. (2020). Application of food texture to moderate oral processing behaviors and energy intake. *Trends in Food Science & Technology, 106*, 445–456. https://doi.org/10.1016/j.tifs.2020.10.021.

EFSA. (2017). Dietary reference values for nutrients. Summary Report. EFSA supporting publication, e15121. 98 pp. https://doi.org/10.2903/sp.efsa.2017.e15121.

FAO. (2011). *Global food losses and food waste—Extent, causes and prevention*. Rome: FAO.

FAO, IFAD, UNICEF, WFP, & WHO. (2020). *The State of Food Security and Nutrition in the World 2020. Transforming food systems for affordable healthy diets*. Rome: FAO, https://doi.org/10.4060/ca9692en.

Ford, A. L., Bergh, C., Södersten, P., Sabin, M. A., Hollinghurst, S., Hunt, L. P., & Shield, J. P. H. (2010). Treatment of childhood obesity by retraining eating behaviour: Randomised controlled trial. *BMJ (Online), 340* (7740), 250. https://doi.org/10.1136/bmj.b5388.

Forde, C. G., Leong, C., Chia-Ming, E., & McCrickerd, K. (2017). Fast or slow-foods? Describing natural variations in oral processing characteristics across a wide range of Asian foods. *Food and Function*, *8*(2), 595–606. https://doi.org/10.1039/c6fo01286h.

Forde, C. G., Mars, M., & de Graaf, K. (2020). Ultra-processing or oral processing? A role for energy density and eating rate in moderating energy intake from processed foods. *Current developments in Nutrition*. https://doi.org/10.1093/cdn/nzaa019.

Galhardo, J., Hunt, L. P., Lightman, S. L., Sabin, M. A., Bergh, C., Sodersten, P., & Shield, J. H. (2012). Normalizing eating behavior reduces body weight and improves gastrointestinal hormonal secretion in obese adolescents. *Journal of Clinical Endocrinology and Metabolism*, *97*(2), E193–E201. https://doi.org/10.1210/jc.2011-1999.

Gibney, M. J., Forde, C. G., Mullally, D., & Gibney, E. R. (2017). Ultra-processed foods in human health: A critical appraisal. *The American Journal of Clinical Nutrition*, *106*(3), ajcn160440. https://doi.org/10.3945/ajcn.117.160440.

Giner, C., & Brooks, J. (2019). Policies for encouraging healthier food choices. *Vol. 137. OECD food, agriculture and fisheries papers*. Paris: OECD Publishing. https://doi.org/10.1787/11a42b51-en.

Hall, K. D., Ayuketah, A., Brychta, R., Cai, H., Cassimatis, T., Chen, K. Y., Chung, S. T., Costa, E., Courville, A., Darcey, V., Fletcher, L. A., Forde, C. G., Gharib, A. M., Guo, J., Howard, R., Joseph, P. V., McGehee, S., Ouwerkerk, R., Raisinger, K., & Zhou, M. (2019). Ultra-processed diets cause excess calorie intake and weight gain: An inpatient randomized controlled trial of ad libitum food intake. *Cell Metabolism*, *30*(1), 67–77. e3. https://doi.org/10.1016/j.cmet.2019.05.008.

Khoury, C. K., Bjorkman, A. D., Dempewolf, H., Ramirez-Villegas, J., Guarino, L., Jarvis, A., Rieseberg, L. H., & Struik, P. C. (2014). Increasing homogeneity in global food supplies and the implications for food security. *Proceedings of the National Academy of Sciences*, *111*(11), 4001–4006. https://doi.org/10.1073/pnas.1313490111.

Le, T. H., Disegna, M., & Lloyd, T. (2020). National food consumption patterns: Converging trends and the implications for health. *EuroChoices*. https://doi.org/10.1111/1746-692x.12272.

Lillford, P., & Hermansson, A.-M. (2021). Global missions and the critical needs of food science and technology. *Trends in Food Science & Technology*, *111*, 800–811. https://doi.org/10.1016/j.tifs.2020.04.009.

Ludwig, D. S., Astrup, A., Bazzano, L. A., Ebbeling, C. B., Heymsfield, S. B., King, J. C., & Willett, W. C. (2019). Ultra-processed food and obesity: The pitfalls of extrapolation from short studies. *Cell Metabolism*, *30*(1), 3–4. https://doi.org/10.1016/j.cmet.2019.06.004.

Monteiro, C. A., Cannon, G., Moubarac, J.-C., Levy, R. B., Louzada, M. L. C., & Jaime, P. C. (2018). The UN decade of nutrition, the NOVA food classification and the trouble with ultra-processing. *Public Health Nutrition*, *21*(1), 5–17. https://doi.org/10.1017/s1368980017000234.

Monteiro, C. A., Cannon, G., Moubarac, J.-C., Levy, R. B., Louzada, M. L. C., & Jaime, P. C. (2019). Freshly prepared meals and not ultra-processed foods. *Cell Metabolism*, *30*(1), 5–6. https://doi.org/10.1016/j.cmet.2019.06.006.

Monteiro, C. A., Levy, R. B., Claro, R. M., de Castro, I. R. R., & Cannon, G. (2010). A new classification of foods based on the extent and purpose of their processing. *Cadernos de Saúde Pública*, *26*(11), 2039–2049. https://doi.org/10.1590/s0102-311x2010001100005.

Oberlander, L., Disdier, A.-C., & Etilé, F. (2017). Globalisation and national trends in nutrition and health: A grouped fixed-effects approach to intercountry heterogeneity. *Health Economics*, *26*(9), 1146–1161. https://doi.org/10.1002/hec.3521.

OECD. (2019). The heavy burden of obesity: The economics of prevention. *OECD health policy studies*. Paris: OECD Publishing. https://doi.org/10.1787/67450d67-en.

Poole, J., Bentley, J., Barraud, L., Samish, I., Dalkas, G., Matheson, A., Clegg, P., Euston, S. R., Kauffman Johnson, J., Haacke, C., Westphal, L., Molina Beato, L., Adams, M., & Spiro, A. (2020). Rising to the challenges:

Solution-based case studies highlighting innovation and evolution in reformulation. *Nutrition Bulletin*, *45*(3), 332–340. https://doi.org/10.1111/nbu.12456.

Popkin, B. M., Adair, L. S., & Ng, S. W. (2012). Global nutrition transition and the pandemic of obesity in developing countries. *Nutrition Reviews*, *70*(1), 3–21. https://doi.org/10.1111/j.1753-4887.2011.00456.x.

Poti, J. M., Mendez, M. A., Ng, S. W., & Popkin, B. M. (2015). Is the degree of food processing and convenience linked with the nutritional quality of foods purchased by US households? *The American Journal of Clinical Nutrition*, *101*(6), 1251–1262. https://doi.org/10.3945/ajcn.114.100925.

Public Health England. (2015). *Sugar reduction the evidence for action*. Public Health England. https://assets.publishing.service.gov.uk/government/uploads/system/uploads/attachment_data/file/470179/Sugar_reduction_The_evidence_for_action.pdf. (Accessed 20 September 2021).

Public Health England. (2017). *Sugar reduction: Achieving the 20%*. Public Health England. https://assets.publishing.service.gov.uk/government/uploads/system/uploads/attachment_data/file/604336/Sugar_reduction_achieving_the_20_.pdf. (Accessed 20 September 2021).

Sadler, C. R., Grassby, T., Hart, K., Raats, M., Sokolović, M., & Timotijevic, L. (2021). Processed food classification: Conceptualisation and challenges. *Trends in Food Science & Technology*, *112*, 149–162. https://doi.org/10.1016/j.tifs.2021.02.059.

Sager, V. F., Munk, M. B., Hansen, M. S., Bredie, W. L. P., & Ahrné, L. (2021). Formulation of heat-induced whey protein gels for extrusion-based 3D printing. *Foods*, *10*(1), 8. https://doi.org/10.3390/foods10010008.

Smith, N. W., Fletcher, A. J., Dave, L. A., Hill, J. P., & McNabb, W. C. (2021). Use of the DELTA model to understand the food system and global nutrition. *The Journal of Nutrition*, *151*(10), 3253–3261. https://doi.org/10.1093/jn/nxab199.

Sun, C., Zhou, X., Hu, Z., Lu, W., Zhao, Y., & Fang, Y. (2021). Food and salt structure design for salt reducing. *Innovative Food Science & Emerging Technologies*, *67*(102), 570. https://doi.org/10.1016/j.ifset.2020.102570.

Vergeer, L., Veira, P., Bernstein, J. T., Weippert, M., & L'Abbé, M. R. (2019). The calorie and nutrient density of more- versus less-processed packaged food and beverage products in the Canadian food supply. *Nutrients*, *11*, 2782. https://doi.org/10.3390/nu11112782.

Wee, M. S. M., Goh, A. T., Stieger, M., & Forde, C. G. (2018). Correlation of instrumental texture properties from textural profile analysis (TPA) with eating behaviours and macronutrient composition for a wide range of solid foods. *Food & Function*, *9*(10), 5301–5312. https://doi.org/10.1039/c8fo00791h.

Willett, W., Rockström, J., Loken, B., Springmann, M., Lang, T., Vermeulen, S., Garnett, T., Tilman, D., & Declerck, F. (2019). The lancet commissions food in the anthropocene : The EAT—Lancet commission on healthy diets from sustainable food systems. *Lancet*, *393*(18). https://doi.org/10.1016/S0140-6736. 31788-4.

World Resources Institute. (2019). *Creating a sustainable food future*. World Resources Report. https://research.wri.org/sites/default/files/2019-07/WRR_Food_Full_Report_0.pdf.

World Health Organization. (2021). Obesity and Overweight. https://www.who.int/news-room/fact-sheets/detail/obesity-and-overweight (Accessed 10 October 2021).

# CHAPTER 2

# New food structures and their influence on nutrition, health and well-being

**D. Subhasri, J.A. Moses, and C. Anandharamakrishnan**

*Computational Modeling and Nanoscale Processing Unit, National Institute of Food Technology Entrepreneurship and Management-Thanjavur, Ministry of Food Processing Industries, Government of India, Thanjavur, Tamil Nadu, India*

## 2.1 Introduction

Food is begun to be dealt with in terms of soft matter. Food scientists have started to realize the potential of soft matter and its importance in the science of dealing with food structure. In food processing, precise execution of unit processes is essential to ensure stable structure formation, newer gastronomic experience and the development of new food products for better nutrition. There is a strong transition towards healthier food structures based on natural and processed food ingredients. With health and wellness being of prior importance among consumers, there is a huge demand placed to the food industries for the production of new products which are nutritious, sustainable, and also economical.

Food that we consume is made up of numerous structures and phases. We can define food with the help of its structure. Individual foods vary depending on the components in food including the macronutrients and the way they are embedded in the matrix system. Chefs around the world are using several scientific techniques to create new structures for a better heuristic experience and nutrition. Food technologists are dealing with food structure in deeper perspectives. Research and development are being carried out on a wider scale to understand the food structure, its interaction with the food matrix, the role of structural changes undergone during the breakdown process, along with significant knowledge on nutrient bioavailability and bioaccessibility.

The history of food structuring dates back from the time humans started to develop products like bread and cheese. The three basic states of matter, as physicists and chemists describe, include gaseous, liquid, and solid-state. But in food, most dishes are composed of dispersed systems referred to as colloids. They consist of one phase dispersed in a different continuous phase. The study on food structuring was initially carried out by scientists like Herve This and Fito. Food can be described in terms of dispersions by the soft matter physics approach and can also be represented in the formula. For example, in the case of Chantilly cream, it can be represented as oil in water (O/W) emulsion + gaseous phase (G) →(G+O)/W; i.e., oil (milk) dispersed in water when purged with gas (whipping) will produce the whipped cream (This, 2005). Colloidal dispersions include emulsions, foams, gels like mayonnaise; meringue; jam; cooked egg; ice cream. Van der Sman introduced the discussion on four basic transformations which occurs in the processing of food including the destruction or destabilization of food structure, creation of new phases (by biochemical or phase transition), deformation and sizing of phases, and final stabilization of the new (de)formed state, i.e. the structure should be preserved till consumption (der Sman & der Goot, 2009).

The creation of food structure is so magnificent that a single food ingredient can be structured into various forms from gels, emulsions, powder, to solids. For example, milk can be structured into various forms like solid gels, emulsions, powder, and suspensions. To date, complex food structures are evolved utilizing the basic states of matter. From the encapsulated matrix, 3D printing of food-based ink composite, even specific foods capable of restructuring in the stomach for specifically targeted release. Initial studies on understanding the changes involved in the food system include individual component studies on various functionalities. From gelatinization and retrogradation in starch, protein aggregation, coagulation or unfolding and restructuring based on physical and mechanical conditions, free fatty acid release, and textural attributes of fat. An emphasis on a detailed understanding of the influence of mass transfer and heat transfer along with processing conditions for the fabrication of new food structures is needed. The interactions between the various structural elements can be studied based on thermodynamic, kinetic, statistical, and computational modeling to understand and arrive at a relation between food structure and physiochemical properties. Also, various analytical instrumental up-gradation and development of innovative assays will help in realizing the potential of food structure and attain comprehensive data of the food system. Specialized instruments include rheometer, confocal laser scanning microscopy, texture profile analyzer, X-ray diffraction, differential scanning calorimetry, X-ray tomography, NMR, ED-XRF. Also with innovations and the use of advanced technology in the field of food processing, there is wide scope for the development of new functional food structures.

In the development of new food products, the understanding of food component structure and functionalities is necessary. Starch, a simple carbohydrate can weaver the different structures through its amylopectin and amylose content. The branched molecules undergo intermolecular entanglements of the H bond during gelatinization and retrogradation. For the development of starch-based structures, studies on its gelatinization and swelling mechanisms are necessary. Understanding on melting and gelatinization in starch with the Flory Huggin model (Kugimiya & Donovan, 1981), the types of starches resistant, slowly digestible starch plays an important role in the design of customized structures (Hamaker, 2021). Breakfast and snack products are mostly comprised of starches, thus to help in weight management researchers and companies are putting up with the idea of a low carbohydrate-based diet, with low glycemic response, and glycemic load. Recently the utilization of starch to be digested at the ileum is also being researched to activate the gut-brain axis (Lim et al., 2021). This can give rise to the development of better nutritious starch-based food products.

Proteins are known as building blocks of energy. The development of protein-rich foods involves the basic conception of different types of protein and the wide functionalities they tend to offer. Protein reacts with other components to form protein interactions which can result in a wide variety of structures. Proteins are capable of aggregation in different dimensions. One dimensional aggregation of food proteins results in the formation of fibrillar aggregates, in two-dimensional aggregations, arrives at structures like foams and emulsions. The three-dimensional aggregation could result in the formation of products like gels (Mezzenga & Fischer, 2013). Protein gels are formed by solid transformation from liquid through the action of unfolding and re-association. When new products are derived from egg white they are fat-free. It was found that when tightly packed their proteins denature, like in the case of egg white they cause the thickening and form a gel (Garcés-Rimón et al., 2016); whereas when they denature in meat, they are found to shrink, solubilize, or gel, modifying the texture of the meat (Baldwin, 2012). Also, texturized protein products based on the unfolding and denaturation of proteins have led to the development of plant-based meat analogs. With the rising vegetarian population and trend towards sustainable agriculture, there is a transition towards plant-based diets. A protein-rich nutritious bar is one of the trending products which are commercialized by the companies. The protein-polymer interactions are also deeply studied for the fabrication of new products.

Fats play different roles in foods and are mainly responsible for the organoleptic properties, texture, and mouthfeel. Specific food structures slow down gastric emptying, thus the feeling of satiety or fullness can be achieved for appetite control. The nature of the ingredients and the processing conditions determine the characteristics of fat droplets. Another strong recommendation is to analyze the type of fat that we consume rather than the amount of fat. Also, it is found that emulsified lipids which are stable to aggregation and creaming were found to provide the effect of satiety with the reduced gastric emptying rate (McClements, 2015). Replacing saturated fat with unsaturated fat is found to reduce the risk of cardiovascular diseases. Increased intake of healthy fat from sources like fish and plants is being encouraged. Low-fat foods or fat substitutes are researched worldwide for the development of healthy fat-based products.

Food consumption trend changes with innovations, lifestyle and so nutrition. There is a continuous awareness of better health and nutrition for well-being. Consumption pattern (reduced sugar, salt, fat consumption, likewise increased plant-based diet), individual's health, work environment, and healthier lifestyle are among the overall recommendations from many health organizations throughout the world. Nutritionists and dietetician have come up with programs and meal plans for weight management. To know and be aware of what we consume is the major concern. On the other hand food technologists around the globe are researching on various food structures which can be developed or even modifications made in the existing structures during processing for fabrication of new food products with enhanced health and nutrition.

The design of food structures for nutrition and well-being requires a multidisciplinary approach. Food structuring is the possibility to attain a specific structure, for delivering specific functions. Chronic diseases have become a global concern. Governments have taken action on global health to ensure the safety of the citizens. Policies and actions are taken to ensure proper health and nutrition. People with different ailments require different diets. Hence the usage of novel food structures in diet plans and nutrition can aid in treatment and also help in the prevention of diseases. This chapter deals with various food structures which have been developed for better health and nutrition.

## 2.2 New food structuring techniques

This section will include a brief overview of the various structuring techniques available to render new food structures. The various food structuring techniques include gelation, spherification, texturization, encapsulation, multidimensional food printing, micronization, and other digitalized techniques. Gelation involves the formation of a 3D network system from the interaction between polymers. Spherification which is based on gelation involves the formation of spheres from alginate or calcium chloride which can also be used to encapsulate bioactives. Texturization is the process of using mechanical and structuring agents to modify the structure of the food components for better functionalities and also results in the formation of new textured products from the interaction between food components.

Encapsulation revolves around micro and nanoscale for the fabrication of active materials enclosed in a wall material. Encapsulation is the basis for the formation of various nutraceutical products with enhanced bioavailability and bioaccessibility. Multidimensional printing is a type of additive manufacturing that involves the deposition of food inks into customized shapes and textures. Micronization or size reduction is the usage of one among several existing techniques to reduce the size of the food matrix component. These techniques along with digitalization, modern techniques to fabricate equipment and demand have led to the formation of new food structures. These techniques along with digitalization, incorporation of computer aided tools and advanced processing methods for fabrication of equipments can lead to the commercialization of new food structures.

## 2.3 New food structures

In this section, the various new food structures developed to date are categorized and discussed based on their functionalities. The types of structured food discussed are represented in Fig. 2.1.

### 2.3.1 Energy density food

With changing diet habits and working patterns, the consumer mass is now demanding to consume low energy density foods for better nutrition and well-being. Understanding the energy intake of food is important to manage weight, and also to maintain good health. The risk of cardiovascular diseases

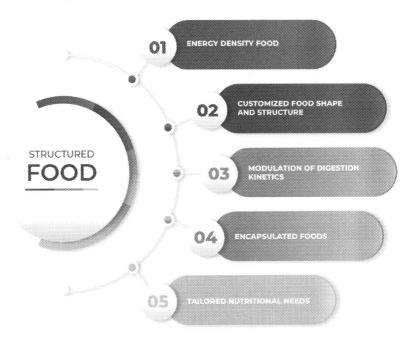

**FIG. 2.1**

Types of structured foods.

## 2.3 New food structures

in adults due to increased prolonged intake of high energy density foods has been reported in recent studies (Teo et al., 2020). Pediatric obesity is profoundly found to be associated with reduced healthy eating and increased consumption of high-energy foods (Epstein et al., 2008). Current research on calculating energy contents available through standardized metabolic condition and models have been carried out. This paves the way in the food industry sector to manipulate the interactions among food structures and reduce the energy available per unit of dry matter food.

An interesting approach to reduce the sugar, salt, and fat content in foods is devised in recent industrial food trends. Altering the food matrix design for reduced sugar, salt, and fat intake is researched in a wide spectrum. Lipids are known for their high energy density, hence when we are towards appetite control and weight management, we must consider reducing the uptake of trans or saturated fat. It is reported that the energy density of food plays a crucial role in consuming food beyond satiety resulting in obesity (Hoch, 2014). New food structures developed in this regard for low sugar, low-fat alternatives are also discussed.

Emerging food processing technologies include: high-pressure jet processing, microfluidization, ultrasound treatment, and hydrodynamic cavitation were explored to alter the functionality and reduce the use of stabilizers in the development of low-fat ice creams. High-pressure jet processing is found to alter the microstructure of low-fat ice cream. It leads to a reduction of air bubbles and ice crystal sizes while boosting the functionality of the destabilized fat droplets (Voronin et al., 2021). The particle size of hydrodynamic cavitation formed ice cream mix was larger than ice cream from the homogenized process, thus affecting the flow behavior to viscoelastic state. Hydrodynamic cavitation has shown the potential to develop new structured frozen desserts (Sim et al., 2021). With the increase in demand for healthy low-fat ice cream, there is a need to reduce extensive usage of stabilizers and fat replacers. In this regard additions of hydrocolloids and microparticulate proteins have been found to reduce the effect of ice crystal and structure with high nutritional profile. Microparticulated soy protein hydrolysate/xanthan gum complex when used as fat replacer altered the microstructure resulting in low fat ice cream. The use of hydrocolloids provided better emulsifying and foaming stability thus stabilized the microstructure of ice cream during freezing (Yan et al., 2021).

Sugar plays a multifaceted role in foods. It imparts flavor, controls shape and texture, mainly acts in the preservation and also as a humectant and anticoagulant. To replace sugar or lower the level of sugars is a major challenge without disrupting the hedonic scale. High molecular weight bulk sweeteners are used with a trial to mimic the structure of the food. The replacement of sugar in various food products leads to changes in the food matrix but enhances the nutritional profile with added replacers. For example, in a cake batter which is a complicated fat-in-water emulsion, the replacement of sugar with polydextrose caused decreased bubble size but allowed for 25% fat replacement and 22% sugar replacement (Kocer et al., 2007). Sugar plays a lot of functions in pastries, such as competing for water and thus lowering starch gelatinization and gluten development, functions as fluid during cooking, and recrystallizes on cooling which affects properties like aeration, texture, and mouthfeel. Spatial modulation of the food components distributed in a food matrix can serve as a potential way to reduce the use of salt, fat, and sugar contents in food (Stieger & Van de Velde, 2013).

The salt reduction can be done through various structural processes. The interaction of salt and food matrix before the perception of saltiness must be studied. The salt reduction can be achieved through the inhomogeneous distribution of salt in the food matrix and encapsulation (Busch et al., 2013). Milling and spray drying has been used for reducing the salt content in foods. Spray drying and electro-hydrodynamic atomized drying was experimented for sodium chloride size reduction and was found to decrease the sodium levels by 58.67 and 65.34%, respectively (Vinitha et al., 2021). In a study, it was found that when sodium chloride substitutes were combined with reduced particle size, it served as an effective way to reduce sodium consumption (Rodrigues et al., 2016).

### 2.3.2 Customized food shape and structure

Additive manufacturing or 3D printing is layer-by-layer deposition of the food ink into various desired shapes and textures. With the basics of usage of additive manufacturing in food industries, 3D printing has seen its emergence in this industry and is most explored for customization. Food structure is an important factor that influences the gastronomic experience of a person. Thus, beyond simply designing food products, 3D printing allows the incorporation of new dimensions and also bioactive compounds into the design of structures, hence enhancing the overall nutrient quality. 3D printing can meet the tailored needs of individuals and various food structures developed have been developed accordingly. Various new formulations are designed for specific target groups like dysphagic patients, military personnel, children, astronauts, sportsperson, and the elderly.

4D printing is being researched widely. It can give a real-time experience to the consumers. It has the advantages of shape memory, response to color, pH, and external stimuli. It can be deformation induced by water absorption, deformation induced by dehydration, nutrient change, and flavor change (Teng et al., 2021). Some of the 4D printing examples include the development of double-layer pumpkin paper where controlled deformation was achieved which can be used in modern cuisine (Chen, Zhang, Liu, & Bhandari, 2021). A mixture of red cabbage juice, vanillin, potato starch was used to study over time in response to internal and external pH stimuli, it was found that they are capable of new product development with food structures capable of pleasing persons with poor appetite (Ghazal et al., 2021). Purple sweet potato powder and mashed potato powder were used to study the controlled deformation which was brought about by the composition like edible salt exhibited during microwave dehydration (He et al., 2021). Ethyl cellulose and gelatin-based edible food composites were evaluated for 4D printing with osmotically derived structural changes, which paves the way for the design of new complex food structures which can be customized (Pulatsu et al., 2022). Turmeric (curcumin) incorporated sago flour-based 3D constructs can be developed in response to stimuli by pH-based triggers (Shanthamma et al., 2021).

The texture of 3D printed products can be varied involving spontaneous dehydration-triggered shape-change as in the study reported by Liu and his team (Liu et al., 2021). Here it was found that the starch-based edible structures were capable of changing shape upon dehydration. Thus, 4D printing can also be explored to create nutritious food structures. When 3D printed ergosterol incorporated sweet purple potato paste was exposed to UV-C irradiation, it was found to be converted into Vitamin D2 (Chen, Zhang, & Devahastin, 2021). Microcapsules were embedded with oil before printing, the inclusion of spherical structure changed to fusiform with 3D printing and it was found that upon heating the capsules released oil (Guo et al., 2021). Thus this study proved that combining 3D printing with microwave heating can have potential for the creation of 4D printed products.

Studies are being carried out to use 3D printing for producing new products to achieve an unprecedented sensory experience and to resolve the mastication and swallowing problems of dysphagic patients. Freeze-dried vegetable powders like garden pea and carrot have been used to create 3D food structures with the help of hydrocolloids to develop real food structures for easy-to-swallow food (Pant et al., 2021). Even fresh vegetables can be used in hospitals and nursing homes to develop food structures for elderly patients. Due to its large effect on the viscosity of pastes, which causes significant changes in printability, water content is also the most effective ingredient to be considered for 3D printing in the fruit and vegetable sector. Cereal-based snacks can also be customized via 3D printing by changing the porosity fraction and a detailed study on internal voids was carried out (Derossi et al., 2020).

Meat and meat alternatives can also be printed using layer-by-layer printing. 3D-printed chicken nuggets were experimented with the addition of wheat flour to increase the printability of the fibrillar protein network of chicken (Wilson et al., 2020). This can be further researched with varying fat content for better health and nutrition. 3D printed cooked pork paste with the incorporation of hydrocolloids showed that they increased the water retention and enhanced viscous-like behavior. Thus this has the potential to act as transitional foods for dysphagic patients (Dick et al., 2020). The 3D printed egg yolk was found to be more structurally stable than egg white which can be explained by the better starch-protein interaction in egg yolk material supply, thus it was able to withstand the shape (Anukiruthika et al., 2020). Customized 3D-printed mushroom-based snacks can be used as sustainable alternative sources of nutrition (Keerthana et al., 2020).

A fruit-based formulation with required iron, calcium, and vitamin D for 3–10 years old was experimented with using 3D printing to form customized structures that can be of nutritious appetizer to children (Derossi et al., 2018). Data-driven delights or snacks can be produced automatically with a multidimensional printing approach with a different spectrum of customization.

### 2.3.3 Modulating digestion through specially designed food microstructures

Consumers are becoming increasingly aware of the nutrients which are contained in food, and also whether they can be synthesized or can be absorbed by the body. The availability of nutrients, in general, varies with the changes undergone by the food structures post-consumption. The food matrix undergoes several breakdowns and reacts with various gastrointestinal components. The fate of food in the gastrointestinal tract also depends on the structure of the food matrix. For modulation of hydrolysis of macronutrients and understanding kinetics of nutrient bioavailability, study on food and its structure is essential to develop tailored products with better nutrition (Somaratne et al., 2020). First, there is a mastication process in the mouth with the help of salivary enzymes resulting in bolus formation. This step is followed up by bolus acted upon by gastric enzymes and digestive juices. At last, it reaches the intestine; where nutrient absorption takes place in the villi region; even fermentation takes place in the colon region.

Food structure can be altered for better digestion kinetics and nutrition. Each component of food, proteins, carbohydrate, and fat was modified for varied physiological responses. Starch is the foremost source of glucose in the human diet. Glycaemic index and/or glycaemic load are frequently used to describe glucose responses to starchy foods. Depending on the rate and extent of digestion in the small intestine, it can have a range of effects on postprandial glycemic response, which is connected to blood glucose irregularities, insulin levels, and even insatiable hunger and food intake over time (Priyadarshini et al., 2021).

The nutritional significance of different starches is highlighted by categorizing starch bioavailability into rapidly digestible starch, slowly digestible starch, and resistant starch. Food bioavailability can be modified by selecting food matrices with varying degrees of amylolysis susceptibility and food processing to maintain or produce new matrices, as starch is the principal structure-building macro constituent of foods. The low glycaemic impact of slowly digested starches is expected to be beneficial in the dietary prevention and control of type 2 diabetes (Augustin et al., 2015). In addition to increasing satiety and decreasing food intake, starch that resists digestion in the upper gastrointestinal system has shown to improve satiety and decrease food intake.

The starch structure can be modified by using techniques based on moisture, temperature, and shear which can influence gelatinization and retrogradation. Detailed studies on the kinetics of starch digestion have been carried out to understand the basics of starch involved in new product development, to design nutritional starch based products according to their functionalities for targeted groups The schematic representation of the effect of processing on native starch structures is represented in Fig. 2.2 (Pellegrini et al., 2020). Multifaceted factors like crystal characteristics, degree of branching, and amylose content determine the overall glycemic index (GI) level. The ability to predict the glycemic response can act as a critical factor for the creation of carbohydrate-based products with a more controlled energy distribution, lower glucose blood concentrations, and having health benefits of reduction in the risk of hyperglycemia in diabetes patients or athletes (Bellmann et al., 2018). Glycemic and insulinemic responses were studied for durum and cereal-based meals, it was found that the large differences in the glycemic response may be explained due to the physical structure of the starchy foods, whether durum wheat flour or the use of parboiled rice (Järvi et al., 1995). Processing also plays an important role in determining the GI index, it is found that in the case of oats substituted products,

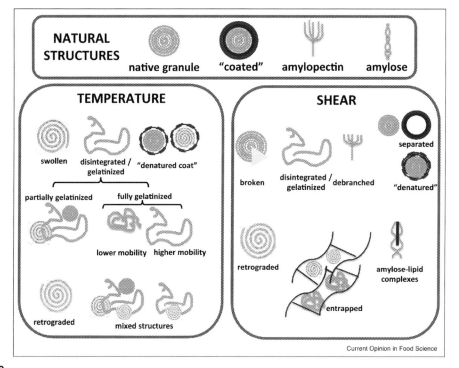

**FIG. 2.2**

Schematic representation of the effect of processing on native starch structures (Pellegrini et al., 2020).

cooling is employed for highly gelatinized products, as it can enhance retro gradation of starch to generate resistant starch and thus reduced GI level (Zhang et al., 2021).

Food synergism which is the interaction of food components during eating and digestion can also affect digestibility. A study involving fruit proteases like actinidin from kiwifruit could help in protein breakdown (Montoya et al., 2014). The digestion of more complex foods, in which a multitude of nutrients is released simultaneously, possibly with synergistic or competing interactions, is one of the next concerns. For example, the lipolysis kinetics of oil-in-water emulsions entrapped in a protein network is highly influenced by proteolysis kinetics, although both reactions appear to be mostly independent when the same emulsion is dispersed in a liquid protein solution (Mat et al., 2016).

Likewise, lipid type and the environment encountered during lipolysis in the gastrointestinal tract affect the rate of lipolysis and free fatty acid release. Legume based powder was used as a wheat flour substitute in a study to decrease the starch digestibility of biscuits. This behavior was explained by the reduced wheat starch gelatinization caused by the leguminous substitute which affected hindrance in water uptake by the starch granules during the biscuit making process (Delamare et al., 2020). These types of food products can help in the development of foods for dietary management or prevention of type 2 diabetes.

Gluten-free food structures have been developing around these days for people with celiac disease. Findings have reported that with higher amylose content the glycemic response is low. In general gluten-free cereal-based products with higher amylose starches were found to have higher resistant starch content and lower *in vitro* GI. Also, heat—moisture treatment, and modification of starches can act as potential approaches to increase the resistant starch content in the development of gluten-free food products (Giuberti & Gallo, 2018). The need for understanding the mass transfer properties was realized to describe the bioavailability and bioaccessibility of the macronutrients of the digested food. It is reported that the processing conditions of the final product can alter the accessibility of the starch hydrolyzing enzymes (Santhi Rajkumar et al., 2020). The shape and size of the pasta act as a factor in digestion with varied accessibility to starch by amylase due to the different surface-to-weight ratios.

Understanding the key dynamics of structural changes in starch helps to design new starch-based products. Controlling the integrated functions of ready-to-eat starchy health-promoting foods involves determining the microstructural changes of starch and their linkages to enzymatic digestion and sensory qualities of starchy foods during household cooking (e.g., boiling, steaming, and stir-frying) (Shen et al., 2021).

Sucrose is added in the preparation of confectionary products. High sucrose content products have a higher GI. Sucrose alternatives like fructooligosaccharides, inulin, coconut sugar, sugar alcohols were tested in the preparation of new confectionary products with low GI. Granola bars were prepared with sucrose alternatives and it was found that fructooligosaccharides can be used to develop medium glycemic load and low GI-based snacks (Sethupathy et al., 2020).

Soft matter physics, mass transfer and structural changes involved during oral processing and digestion have been studied to arrive at an understanding of nutrient absorption and bioavailability. When heat moisture treatment and high hydrostatic pressure were investigated in starch, it was found that the dual modification showed its potential to be used in the preparation of low GI foods. With increased moisture content in moisture treatment, it was found that modified starch exhibited a lower glucose release rate (Colussi et al., 2020). The legume mung bean was researched under hydrothermal processing to develop composites with lower starch digestibility. The structure digestion relationship was obtained when mung bean starch was studied under controlled gelatinization conditions. It was found that the mung composite noodles can be cooked up to 70°C for a longer time to reduce the GI (Awais et al., 2020).

In a study with two dairy-based products it was found that the food matrix had a strong effect on the gastric behavior, proteolysis, and also lipolysis till the amount of nutrient absorbed (Mulet-Cabero et al., 2017). The difference in milk fat globule structure of emulsion in infant milk formula can affect the rate of casein proteolysis and lipolysis, hence the surface area was found to be the determining factor in controlling the rate of gastric lipolysis. Thus the focus on the development of new minimal processes is required for better infant food formulation (Bourlieu et al., 2015).

### 2.3.4 Enhanced bioavailability of encapsulated foods

At the molecular, nano, micro, and macro scales, nutrients are "grouped," or molecules are integrated into food structures. Encapsulation is one of the developed techniques to alter and enhance the bioavailability of the bioactive components which are integrated. Various new structures developed through encapsulation include: spray-dried powder formulations, beverages, chocolates with encapsulated probiotics, and fortified dairy products.

Encapsulation involves the entrapment of bioactive inside wall material. These structures are comprised of varied core materials depending on the usage. The active agent is the core material or the payload phase in encapsulated structure, while the outer wall material is known as the coating membrane or carrier material or the external phase (Ezhilarasi et al., 2013). Lipids, food-grade polymers including polysaccharides and proteins, as well as their mixtures, are some of the common wall components (Vimala Bharathi et al., 2018).

Encapsulation protects its core material from thermal heat, pH, and moisture until the release, thus it can be used in the creation of products where a delayed release is required. Like encapsulated flavor released during chewing in chewing gum, encapsulated probiotics are released in the intestine surviving gastric phase, as leavening agents where it is released and acts during baking (Augustin & Hemar, 2009).

Fruit juices are commonly consumed foods and are nutritious because of the bioactive compounds and their health-benefiting effects. They have certain limitations in solubility, stability, thus when encapsulated structures are used it is proven to enhance their solubility and even provide a controlled release (Ephrem et al., 2018).

Nutraceuticals are known to have multi-target and multi-therapeutic efficacy against several chronic clinical conditions. It is said that the synergistic combinations of nutraceuticals have higher activity when compared to the individual compounds (Leena et al., 2020). Encapsulation can be combined with other delivery systems. The synergistic nutraceuticals curcumin and resveratrol were encapsulated using core-shell nanoparticles and delivered through a 3D printed gelatin hydrogel. It was found that the zein-PEG-based core-shell uniformly distributed nanoparticles were able to protect the encapsulated bioactives and thus these structures can be used as a potential customized oral delivery system (Leena et al., 2022). Encapsulations of vegetable and essential oils have been explored at the industry level, as it has advantages in rendering oxidative stability and shelf life. For reducing storage loss, essential oils are microencapsulated or nano encapsulated before their application in food beverages (Bakry et al., 2016). Also in the promotion of healthy foods, the encapsulation of flavor and aromas can act as an added advantage.

Novel oil structuring methods are researched for unsaturated fat alternatives like oleogels. It can be used specifically to tailor foods for cardiovascular disease in bakery products, confectionery, and even meat-based products (Pehlivanoğlu et al., 2018). It is reported that carotenoid bioavailability can be increased when consumed with lipids, thus when carotene encapsulated oleogel structures have been fabricated the bioavailability of carotene was enhanced. It was found that when oil was structured with monoglyceride crystals, the oleogel matrix was capable of entrapping curcumin and a controlled release was obtained (Palla et al., 2022). Emulsion templated medium chain triglyceride based oleogel was proved to act as a carrier for synergistic delivery of curcumin and resveratrol (Kavimughil et al., 2022). Oleogels have also been tested to replace cream in ice creams without any significant changes in viscosity and textural properties (Silva-Avellaneda et al., 2021). The application of oleogel in chocolate to lower trans fatty acid was also researched. Oleogel from oryzanol/stearic acid/lecithin was used to prepare functional chocolate with β-sitosterol and corn oil (Sun et al., 2021). The microstructure of oleogel can also play a factor in determining the hardness of the cookies prepared from emulsifier-based oleogels. Chitosan-based edible oleogel was produced with emulsion templated method with vanillin crosslinking modification, making it suitable for resulting in trans-fat free potential food structures (Brito et al., 2022). The structure of encapsulating like cross-linked core-shell biopolymer nanoparticles were found as suitable delivery systems for curcumin and piperine. The structure developed can be applied in functional food development (Chen et al., 2020).

Probiotics are the live microorganism included structures that are used in the wider spectrum of food applications. The efficiency of the probiotics is dependent on the food matrix in which it is entrapped, as the probiotics are prone to digestive conditions. Encapsulated probiotics help to protect the probiotics in beverages or other food products. Probiotics are used by gastroenterologists to treat GI disease, and nutritionists utilize them to prevent disease or maintain good health (Roobab et al., 2020). Encapsulation of probiotics through various approaches like spray drying and spray freeze drying have found to offer advantages, for example, retention of quality, stability and also can be oriented for specific target delivery (Yoha et al., 2020).

### 2.3.5 Tailored foods to meet nutritional needs

Another food structure that has been developed for specific diets has been pointed out. Diet can be used to provide disease-specific nutrition, like iron-rich supplements for anemic patients, sugar alternatives for diabetic individuals; low-fat foods for heart patients, calcium incorporated products for osteoporosis, and enhanced vitamins in food for all age groups. Protein is known for its ability to increase satiety and reduce appetite. Obesity among school children and adults is a serious issue that countries are dealing with for better nutrition and well-being. Loss of control over eating and changes in the dietary pattern is found to cause several diseases. In a study, it was found that when there is a protein preload before consumption of ultra-processed food, there can be a reduced intake in consuming ultra-processed snacks with proximity to further eating (Nasser et al., 2022). Hence the amount of food intake, the nature of food consumed, the effect of the matrix in which it is included, co-ingested food plays an important role in the development of tailored foods. The various food structures to be developed for tailored needs are represented in Fig. 2.3.

Increased intake of carbohydrates with a higher glycemic load is reported to contribute to cardiovascular diseases. There is an increasing trend to design and fabricate low GI foods tailored for patients with diabetes and other cardio vascular disease (CVD). Pectin is found to have the potential in replacing

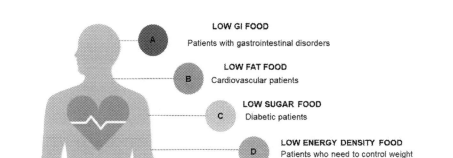

**FIG. 2.3**

Food structuring for tailored needs.

sucrose and reducing the postprandial glycemic response, thus diets based on low methoxyl pectins are to be encouraged and more awareness should be created on the health benefits (Muñoz-Almagro et al., 2021).

Pasta with high nutritional value and lower GI has increased demand at the industry level. It is among the common food in several cuisines. They can be developed to tailor the needs of patients with celiac diseases or type 2 diabetics patients (Di Pede et al., 2021). The gluten-free consumption trend is increasing nowadays as a result of increased cases of coeliac and also the need to be precautious of the allergenic protein in the diet. There are several strategies to attain gluten-free food products, including modification or using alternative flours, incorporating new functional ingredients, or altering the structure through processing (Gao et al., 2017). Gluten is mainly responsible for the structure formation in the making of noodles and bread by the structural organization. So, care should be taken to obtain similar textural attributes in the development of gluten-free products. Starch gelatinization and protein denaturation build the basic firm texture for gluten-free pasta. Extrusion serves as a varied approach to obtain gluten-free structured food products; When replacing gums with proteins, care should be taken about the texture, it is said that other structure forming agents like transglutaminase should be used (Gao et al., 2018).

Fats have a unique role in food processing, hence to mimic the properties of fat replacer in food formulations should be given consideration. In the food industry, there is a continuous search for replacements of Trans and saturated fatty acids for people with high cholesterol levels who need better nutritional lipid formulations. One among them is the three-dimensional network food structure of thermo-reversible gels in which an organic continuous phase entrapped is termed organogels. These types of structured emulsions can be used as substitutes in meat products, margarine, and spreads. A study involving margarine formulated using organogels from combined soybean oil, candelilla wax, fully hydrogenated palm oil, and monoacylglycerols. It was found to have similar technical properties with reduced saturated fatty acids (Silva et al., 2021). Salad dressing is an oil-in-water emulsion with high-fat content; the by-products from the cold press industry (cumin, coconut, pumpkin, and flaxseed) were evaluated for their potential use in the fabrication of low-fat dressing (Tekin-Cakmak et al., 2021). Pregelatinized amaranth flour was used to replace oil in amaranth cakes, it was found that the *in vitro* digestibility of starch was improved (Carmona-Garcia et al., 2022). There is an increased need for healthier meat products, with low fat or reduced fat despite the high protein nature of the food product. This is suggested for people with a high risk of CVD (Kumar, 2021).

When low-methoxyl pectin was used in the preparation of low fate set yogurt it improved its firmness and rheology due to increased cross-linkage with casein (Khubber et al., 2021). A new technique like lipase immobilization was used to process low-fat cheese, which involves the addition of lipase nanotubes-based milk oil in water Pickering emulsions in place of milk fat during the manufacture of cheese. Thus the flavor defects of low-fat cheese can also be dealt with by promoting the free fatty acid release (Guan et al., 2021). A list of some of the recently developed food structures is listed in Table 2.1.

**Table 2.1 New food structures developed for customization.**

| Food structure | Tailored need | Findings | References |
| --- | --- | --- | --- |
| Yogurt type plant-based gels | Prevention of chronic degenerative diseases by plant protein incorporation | Fermentative gelation of pea protein in yogurt-based gel could increase plant protein intake and help in healthy aging | Klost and Drusch (2019) |
| Fortified curcumin nanoemulsion incorporated dairy gels | Therapeutic foods for specific populations | The isocaloric structures helped in the release of bioactive components from functional foods which can be altered with the complex transition in the gastrointestinal tract | Qazi et al. (2021) |
| HIPE (high internal phase emulsion) based gel systems | To design low fat-based food products | HIPE based gels can be fabricated using food-based hydrocolloids | Patel et al. (2014) |
| Low glycemic gluten-free noodles | Gluten-free products for celiac patients | The addition of inulin and defatted rice bran was found to lower the glycemic index | Raungrusmee et al. (2020) |
| Inulin-based emulsion filled gel | Low fat-based snacks for CHD and CVD patients | Replacement of fat in shortbread cookies inulin based emulsion filled gels exhibited similar physicochemical properties | Paciulli et al. (2020) |
| 3D printed vegetable paste | Dysphagia patients | Facilitated nutritious intake of vegetables in par with International Dysphagia Diet Standardization Initiative (IDDSI) framework for easy swallow ability | Pant et al. (2021) |
| 3D printed gelatin hydrogel | Therapeutic customization | Enhanced bioaccessibility and intestinal permeability of the synergistic bioactives can be delivered through customized shapes | Leena et al. (2022) |

*Continued*

Table 2.1 New food structures developed for customization—cont'd

| Food structure | Tailored need | Findings | References |
| --- | --- | --- | --- |
| 3D printed encapsulated probiotics | Gastrointestinal clinical condition patients | Freeze-dried 3D printed encapsulated probiotics were found to have a better survival rate under simulated *in vitro* digestion and could act as a potential probiotic delivery system | Yoha et al. (2021) |
| Morphed flat pasta | Sustainable packed pasta for consumers who want greener solutions | Traditional ingredients fabricated into customized shapes based on diffusion mechanism | Tao et al. (2021) |
| Probiotics in the cereal food matrix | Gastrointestinal and digestive system disorders | The beta-glucans containing food matrices help in transporting the probiotic strains in the GI tract | Sharma et al. (2021) |
| Probiotic incorporated in HIPE | Better microbial gut health | It can ensure enhanced storage and viability of the probiotic strain because of the increased oil fraction | Qin et al. (2021) |

## 2.4 Factors influencing the development of new food structures

Various factors influence the development of new food structures. Novel techniques and ingredients are incorporated in the development of new food structures. Wide studies on the need and the demand of the consumers should be studied before the design of the food product. Consumers should be aware of the techniques used for the production of the products or at least the raw ingredients used.

The developed food products are designed for whole groups or specific targeted groups, hence the overall safety and quality of the structure should be evaluated. In this section, factors influencing the development of new food structures have been categorized into four types. Factors influencing product research development, the manufacturing process, post-production, and finally market analysis.

From conceptualization to commercialization, the development of new food structures is a lengthy process. This requires a systematic approach that includes the knowledge of the target groups, the specific health requirement, the food structure responsible, and the properties of the targeted bioactive or product should be analyzed, along with the study on the bioavailability and bioaccessibility of the required ingredients.

### 2.4.1 Product research and development

The development of food products has witnessed several technological signs of progress. However, food technological transformation is debated and addressed to ensure consumers' trust and reliance. The main factor behind developing personalized food structures is to understand the need of the different products. Several studies on the target group should be done before the development of the new product structure. This is the most vital stage to be considered.

Consumers have a strong likeliness towards the naturalness of the product. People have different interests and values, which could explain some of the observed disparities in agri-food technology acceptability. Prior research has largely examined the effect of food technology phobia, disgust sensitivity, and cultural factors on risk perception and acceptability in individuals (Siegrist & Hartmann, 2020). Industry players recognize the importance that consumers give towards the naturalness of the product and utilize the same during the market campaigns. Consumer demand for clean labels continues to grow (Maruyama et al., 2021).

There is changing lifestyle trend seen in today's aging population and people on a preferred diet that offers long-term prevention to chronic diseases. The knowledge of functional foods became increasingly aware among consumers nowadays. Functional foods to be more successful must still meet the consumer's demand. We have several new food structures developed, and several ongoing types of research for better nutritious products. It is emphasized that there is a need for a common regulated database accessible to the consumers on the information of the alternatives used apart from the regulating bodies since they are the end-users. They should be more aware of the technology and processing involved in the product consumed. With increased demand to alter the food structure for better nutrition, it is said to consider the correlation between texture, structure, and also oral processing. Hence there is a need for more studies on the textural attributes brought about during the oral breakdown of food consumed (Pascua et al., 2013).

### 2.4.2 Manufacturing process

The main consideration during the production of new food structures is the feasibility and cost involved. With consumers' preferences towards sustainable products, care should be taken during the production of new food structures, on the knowledge of ingredients used and processing techniques involved. Lifecycle assessment of newly developed food structures should be brought under consideration. The manufacturer should consider the following criteria before designing the functional food structure. It includes target population, health target, target bioactive, target dose, target delivery site, and target claim (Sun-Waterhouse, 2011).

Certain food structures require more research to commercialize the product on large scale. Like 3D printing is suitable for mass customization but not yet capable of mass manufacturing due to limitations in production speed and other technological impediments. Care should be taken on the printing speed and mass-scale manufacturing conditions. Whereas this technology has the potential to allow restaurant, cafe, and bakery owners to create a unique pattern of edible customized food structures to their customers' tastes and preferences. Aside from printer advances (printing speed, precision, and sample productivity), a rational supply chain structure and optimization of material flow and logistic costs can aid in 3D printing becoming more sustainable. Products like gluten-free pasta, modified confectionary items can be produced in the existing production line, the requirement of the substitutes should be taken care of.

There is a need to understand the interaction of oleogel with the food matrix, as it is used in several food applications (López-Pedrouso et al., 2021). In the case of meat products formulated with fat substitutes, care should be given to adopting new packaging systems as oxidative stability must be considered by the industry stakeholders (Kumar, 2021). Hence the overall approach of selection of raw materials, ease of manufacturing, the nature of packaging and storage must be incorporated to design and fabricate the final customized product.

### 2.4.3 Post-production

Several customized products have been developed and are available in the market, from nutrition bars, functional beverages, nutraceuticals encapsulated powders, and texture modified foods for dysphagic patients. The main concern revolves around the safety and quality of the produce. Regulatory agencies are emphasized to design a database for tailored food structures, containing data on various nutritional capacities, shelf life, toxicity studies, and also its bioaccessibility and bioavailability of incorporated nutrients. There are existing *in vitro* and *in vivo* model systems, which are elaborately used to study and analyze the structural breakdown under simulated gastrointestinal conditions.

The regulatory agencies ensure the safety of the product released in the market. For example, several sweeteners have come into play in the production of sugar-free beverages and other products. With safety concerns, it is noted that only few low-calorie sweeteners have been approved by the FDA (Kroger et al., 2006). The viability of the food products containing probiotics must be considered, due to the physiochemical changes involved during the long-term storage and also the fate of probiotics during the gastrointestinal passage must be considered. The sorption properties of the packaging material with newly designed food must be analyzed to ensure the safety of the final product. Also, toxicology studies along with elemental analysis should be evaluated for commercializing the product.

### 2.4.4 Market analysis

Market analysis is the most influencing and deciding factor in bringing a product to the consumer for end-use. The availability and price of the product should be analyzed before distribution and sales. Some of the food structures are capable to be delivered in a regular restaurant or café, while some of the food structures require processing and storage facilities; hence the necessary production lines should be created and monitored to ensure delivery of proper nutrition and safety. The global wellness market is trending and it is reported that in past couple of years there has been significant awareness about health and well-being (Singh, 2022).

There is a need for more research on the *in vitro* and *in vivo* correlations in commonly arrived testing model and system worldwide to understand the effect of novel gluten-free food structures (Giuberti & Gallo, 2018). Some of the products containing nutraceuticals benefit significantly for a specific population and are often expensive. The cost of production and sales is being analyzed and cheaper methods with ensuring safety are being researched. Even though the products are healthy, consumers give importance to the organoleptic properties of the product, hence policymakers, stakeholders, industrialists, nutritionists, and government bodies should ensure creating awareness on the benefits of health and well-being.

## 2.5 Demand and research gap

With increasing awareness of healthy lifestyle and better nutrition, the demand for functional food products increased worldwide. It was forecasted that from the period 2021 to 2027, the functional food market is expected to grow at a CAGR of 6.7%, from $177,770.0 million in 2019 to $267,924.4 million in 2027 (Kamble & Deshmukh, 2022). There is a massive need for the development of personalized food structures that are designed specifically to cater to the needs of the elderly, children, athletes, military personnel, and clinical condition patients. The personalized nutrition market is expected to grow at a CAGR of 16% from 2021 to 2030 (Aritzon, 2021). These forecasts have shown that the demand for personalized foods is increasing continuously and there is a demand for new technological innovative food structures.

The research gap which is of major concern is the need for universal databases on the information about customization of nutrients to be delivered for clinical patients depending on their health. Also, the main situation to be addressed includes safety, quality, and other attributes. The changes undergone during digestion must be taken into account with the help of simulated digestive models. Information concerning the toxicity and bioavailability of nutrients should be studied comprehensively. An integrated approach to studying food in terms of soft matter, nutrient composition, the effect of digestive processing, the release of nutrients, and textural perception is emphasized in the development of new food structures for better health and nutrition.

## 2.6 Conclusion

The macronutrients exist in various structural states in nature; the interaction among them along with processing conditions leads to the formation of food structures. Several existing food structures can be produced using structuring techniques and processing methods. With the increase in demand for better nutrition, health, and well-being several new food structures are being researched and fabricated. The influence of these structures in delivering the nutrients along with the gastronomic experience has been discussed in this chapter. The factors which influence the development of new food structures have been listed to ensure the understanding of the need for products to be commercialized. Specific target groups including the elderly, children, sportspersons and clinical condition patients can be benefited from the developed food structures. The UN SDG goals of 3 and 12 are aimed to lie on par with the development of food structures for better health and well-being.

## References

Anukiruthika, T., Moses, J. A., & Anandharamakrishnan, C. (2020). 3D printing of egg yolk and white with rice flour blends. *Journal of Food Engineering*, *265*, 109691. https://doi.org/10.1016/j.jfoodeng.2019.109691 (April 2019).

Aritzon. (2021). *Personalized nutrition market*. GlobeNewswire, Inc. https://www.globenewswire.com/en/news-release/2021/07/02/2257018/0/en/Personalized-Nutrition-Market-is-Expected-to-Grow-at-a-CAGR-of-16-0-from-2021-to-2030.html.

Augustin, M. A., & Hemar, Y. (2009). Nano- and micro-structured assemblies for encapsulation of food ingredients. *Chemical Society Reviews*, *38*(4), 902–912. https://doi.org/10.1039/b801739p.

Augustin, L. S. A., Kendall, C. W. C., Jenkins, D. J. A., Willett, W. C., Astrup, A., Barclay, A. W., Björck, I., Brand-Miller, J. C., Brighenti, F., Buyken, A. E., Ceriello, A., La Vecchia, C., Livesey, G., Liu, S., Riccardi, G., Rizkalla, S. W., Sievenpiper, J. L., Trichopoulou, A., Wolever, T. M. S., & Poli, A. (2015). Glycemic index, glycemic load and glycemic response: An international scientific consensus summit from the international carbohydrate quality consortium (ICQC). *Nutrition, Metabolism and Cardiovascular Diseases*, *25*(9), 795–815. https://doi.org/10.1016/j.numecd.2015.05.005.

Awais, M., Ashraf, J., Wang, L., Liu, L., Yang, X., Tong, L. T., Zhou, X., & Zhou, S. (2020). Effect of controlled hydrothermal treatments on mung bean starch structure and its relationship with digestibility. *Foods*, *9*(5). https://doi.org/10.3390/foods9050664.

Bakry, A. M., Abbas, S., Ali, B., Majeed, H., Abouelwafa, M. Y., Mousa, A., & Liang, L. (2016). Microencapsulation of oils: A comprehensive review of benefits, techniques, and applications. *Comprehensive Reviews in Food Science and Food Safety*, *15*(1), 143–182. https://doi.org/10.1111/1541-4337.12179.

Baldwin, D. E. (2012). Sous vide cooking: A review. *International Journal of Gastronomy and Food Science*, *1*(1), 15–30. https://doi.org/10.1016/j.ijgfs.2011.11.002.

Bellmann, S., Minekus, M., Sanders, P., Bosgra, S., & Havenaar, R. (2018). Human glycemic response curves after intake of carbohydrate foods are accurately predicted by combining in vitro gastrointestinal digestion with in silico kinetic modeling. *Clinical Nutrition Experimental*, *17*, 8–22. https://doi.org/10.1016/j.yclnex.2017.10.003.

Bourlieu, C., Ménard, O., De La Chevasnerie, A., Sams, L., Rousseau, F., Madec, M. N., Robert, B., Deglaire, A., Pezennec, S., Bouhallab, S., Carrière, F., & Dupont, D. (2015). The structure of infant formulas impacts their lipolysis, proteolysis and disintegration during in vitro gastric digestion. *Food Chemistry*, *182*, 224–235. https://doi.org/10.1016/j.foodchem.2015.03.001.

Brito, G. B., Peixoto, V. O. D. S., Martins, M. T., Rosário, D. K. A., Ract, J. N., Conte-Júnior, C. A., Torres, A. G., & Castelo-Branco, V. N. (2022). Development of chitosan-based oleogels via crosslinking with vanillin using an emulsion templated approach: Structural characterization and their application as fat-replacer. *Food Structure*, *32*, 100264. https://doi.org/10.1016/J.FOOSTR.2022.100264.

Busch, J. L. H. C., Yong, F. Y. S., & Goh, S. M. (2013). Sodium reduction: Optimizing product composition and structure towards increasing saltiness perception. *Trends in Food Science and Technology*, *29*(1), 21–34. https://doi.org/10.1016/j.tifs.2012.08.005.

Carmona-Garcia, R., Agama-Acevedo, E., Pacheco-Vargas, G., Bello-Perez, L. A., Tovar, J., & Alvarez-Ramirez, J. (2022). Pregelatinised amaranth flour as an ingredient for low-fat gluten-free cakes. *International Journal of Food Science & Technology.*. https://doi.org/10.1111/IJFS.15589.

Chen, S., Li, Q., McClements, D. J., Han, Y., Dai, L., Mao, L., & Gao, Y. (2020). Co-delivery of curcumin and piperine in zein-carrageenan core-shell nanoparticles: Formation, structure, stability and in vitro gastrointestinal digestion. *Food Hydrocolloids*, *99*, 105334. https://doi.org/10.1016/j.foodhyd.2019.105334 (May 2019).

Chen, J., Zhang, M., & Devahastin, S. (2021). UV-C irradiation-triggered nutritional change of 4D printed ergosterol-incorporated purple sweet potato pastes: Conversion of ergosterol into vitamin D2. *LWT*, *150*, 111944. https://doi.org/10.1016/j.lwt.2021.111944.

Chen, F., Zhang, M., Liu, Z., & Bhandari, B. (2021). 4D deformation based on double-layer structure of the pumpkin/paper. *Food Structure*, *27*, 1–8. https://doi.org/10.1016/j.foostr.2020.100168 (August 2020).

Colussi, R., Kringel, D., Kaur, L., da Rosa Zavareze, E., Dias, A. R. G., & Singh, J. (2020). Dual modification of potato starch: Effects of heat-moisture and high pressure treatments on starch structure and functionalities. *Food Chemistry*, *318*, 126475. https://doi.org/10.1016/j.foodchem.2020.126475.

Delamare, G. Y. F., Butterworth, P. J., Ellis, P. R., Hill, S., & Warren, F. J. (2020). Incorporation of a novel leguminous ingredient into savoury biscuits reduces their starch digestibility : Implications for lowering the glycaemic index of cereal products. *Food Chemistry: X*, *5*, 100078. https://doi.org/10.1016/j.fochx.2020.100078.

der Sman, R. G. M., & der Goot, A. J. (2009). The science of food structuring. *Soft Matter*, *5*(3), 501–510. https://doi.org/10.1039/B718952B.

Derossi, A., Caporizzi, R., Azzollini, D., & Severini, C. (2018). Application of 3D printing for customized food. A case on the development of a fruit-based snack for children. *Journal of Food Engineering*, *220*, 65–75. https://doi.org/10.1016/j.jfoodeng.2017.05.015.

Derossi, A., Caporizzi, R., Paolillo, M., & Severini, C. (2020). Programmable texture properties of cereal-based snack mediated by 3D printing technology. *Journal of Food Engineering*, *289*. https://doi.org/10.1016/j.jfoodeng.2020.110160.

Di Pede, G., Dodi, R., Scarpa, C., Brighenti, F., Dall'asta, M., & Scazzina, F. (2021). Glycemic index values of pasta products: An overview. *Foods, 10*(11). https://doi.org/10.3390/foods10112541.

Dick, A., Bhandari, B., Dong, X., & Prakash, S. (2020). Feasibility study of hydrocolloid incorporated 3D printed pork as dysphagia food. *Food Hydrocolloids, 107*, 105940. https://doi.org/10.1016/j.foodhyd.2020.105940.

Ephrem, E., Najjar, A., Charcosset, C., & Greige-Gerges, H. (2018). Encapsulation of natural active compounds, enzymes, and probiotics for fruit juice fortification, preservation, and processing: An overview. *Journal of Functional Foods, 48*, 65–84. Elsevier https://doi.org/10.1016/j.jff.2018.06.021.

Epstein, L. H., Paluch, R. A., Beecher, M. D., & Roemmich, J. N. (2008). Increasing healthy eating vs. reducing high energy-dense foods to treat pediatric obesity. *Obesity, 16*(2), 318–326. https://doi.org/10.1038/oby.2007.61.

Ezhilarasi, P. N., Karthik, P., Chhanwal, N., & Anandharamakrishnan, C. (2013). Nanoencapsulation techniques for food bioactive components: A review. *Food and Bioprocess Technology, 6*(3), 628–647. https://doi.org/10.1007/s11947-012-0944-0.

Gao, Y., Janes, M. E., Chaiya, B., Brennan, M. A., Brennan, C. S., & Prinyawiwatkul, W. (2017). *Invited Review Gluten-free bakery and pasta products: Prevalence and quality improvement.*. https://doi.org/10.1111/ijfs.13505.

Gao, Y., Janes, M. E., Chaiya, B., Brennan, M. A., Brennan, C. S., & Prinyawiwatkul, W. (2018). Gluten-free bakery and pasta products: Prevalence and quality improvement. *International Journal of Food Science and Technology, 53*(1), 19–32. https://doi.org/10.1111/ijfs.13505.

Garcés-Rimón, M., Sandoval, M., Molina, E., López-Fandiño, R., & Miguel, M. (2016). Egg protein hydrolysates: New culinary textures. *International Journal of Gastronomy and Food Science, 3*, 17–22. https://doi.org/10.1016/j.ijgfs.2015.04.001.

Ghazal, A. F., Zhang, M., Bhandari, B., & Chen, H. (2021). Investigation on spontaneous 4D changes in color and flavor of healthy 3D printed food materials over time in response to external or internal pH stimulus. *Food Research International, 142*. https://doi.org/10.1016/j.foodres.2021.110215.

Giuberti, G., & Gallo, A. (2018). Reducing the glycaemic index and increasing the slowly digestible starch content in gluten-free cereal-based foods: A review. *International Journal of Food Science and Technology, 53*(1), 50–60. https://doi.org/10.1111/ijfs.13552.

Guan, T., Liu, B., Wang, R., Huang, Y., Luo, J., & Li, Y. (2021). The enhanced fatty acids flavor release for low-fat cheeses by carrier immobilized lipases on O/W pickering emulsions. *Food Hydrocolloids, 116*, 106651. https://doi.org/10.1016/j.foodhyd.2021.106651 (September 2020).

Guo, C., Zhang, M., & Devahastin, S. (2021). Color/aroma changes of 3D-printed buckwheat dough with yellow flesh peach as triggered by microwave heating of gelatin-gum Arabic complex coacervates. *Food Hydrocolloids, 112*, 106358. https://doi.org/10.1016/j.foodhyd.2020.106358 (August 2020).

Hamaker, B. R. (2021). Current and future challenges in starch research. *Current Opinion in Food Science, 40*, 46–50. https://doi.org/10.1016/j.cofs.2021.01.003.

He, C., Zhang, M., & Devahastin, S. (2021). Microwave-induced deformation behaviors of 4D printed starch-based food products as affected by edible salt and butter content. *Innovative Food Science & Emerging Technologies, 70*, 102699. https://doi.org/10.1016/j.ifset.2021.102699.

Hoch, T. (2014). *Fat_carbohydrate ratio but not energy density determines snack food intake and activates brain reward areas _ enhanced reader.pdf*.

Järvi, A. E., Karlström, B. E., Granfeldt, Y. E., Björck, I. M. E., Vessby, B. O. H., & Asp, N. G. L. (1995). The influence of food structure on postprandial metabolism in patients with non-insulin-dependent diabetes mellitus. *American Journal of Clinical Nutrition, 61*(4), 837–842. https://doi.org/10.1093/ajcn/61.4.837.

Kamble, A., & Deshmukh, R. (2022). *Functional food market size and share with industry overview by 2027*. Allied Market Research https://www.alliedmarketresearch.com/functional-food-market.

Kavimughil, M., MariaLeena, M., Moses, J. A., & Anandharamakrishnan, C. (2022). 3D printed MCT oleogel as a co-delivery carrier for curcumin and resveratrol. *Biomaterials*, (121616), 121616. https://doi.org/10.1016/j.biomaterials.2022.121616.

Keerthana, K., Anukiruthika, T., Moses, J. A., & Anandharamakrishnan, C. (2020). Development of fiber-enriched 3D printed snacks from alternative foods: A study on button mushroom. *Journal of Food Engineering*, *287*, 110116. https://doi.org/10.1016/j.jfoodeng.2020.110116.

Khubber, S., Chaturvedi, K., Thakur, N., Sharma, N., & Yadav, S. K. (2021). Low-methoxyl pectin stabilizes low-fat set yoghurt and improves their physicochemical properties, rheology, microstructure and sensory liking. *Food Hydrocolloids*, *111*, 106240. https://doi.org/10.1016/j.foodhyd.2020.106240.

Klost, M., & Drusch, S. (2019). Structure formation and rheological properties of pea protein-based gels. *Food Hydrocolloids*, *94*, 622–630. https://doi.org/10.1016/j.foodhyd.2019.03.030.

Kocer, D., Hicsasmaz, Z., Bayindirli, A., & Katnas, S. (2007). Bubble and pore formation of the high-ratio cake formulation with polydextrose as a sugar- and fat-replacer. *Journal of Food Engineering*, *78*(3), 953–964. https://doi.org/10.1016/J.JFOODENG.2005.11.034.

Kroger, M., Meister, K., & Kava, R. (2006). Low-calorie sweeteners and other sugar substitutes: A review of the safety issues. *Comprehensive Reviews in Food Science and Food Safety*, *5*(2), 35–47. https://doi.org/10.1111/j.1541-4337.2006.tb00081.x.

Kugimiya, M., & Donovan, J. W. (1981). Calorimetric determination of the amylose content of starches based on formation and melting of the amylose-lysolecithin complex. *Journal of Food Science*, *46*(3), 765–770. https://doi.org/10.1111/j.1365-2621.1981.tb15344.x.

Kumar, Y. (2021). Development of low-fat/reduced-fat processed meat products using fat replacers and analogues. *Food Reviews International*, *37*(3), 296–312. https://doi.org/10.1080/87559129.2019.1704001.

Leena, M. M., Anukiruthika, T., Moses, J. A., & Anandharamakrishnan, C. (2022). Co-delivery of curcumin and resveratrol through electrosprayed core-shell nanoparticles in 3D printed hydrogel. *Food Hydrocolloids*, *124*(PA), 107200. https://doi.org/10.1016/j.foodhyd.2021.107200.

Leena, M. M., Silvia, G., Kannadasan, V., Jeyan, M., & Anandharamakrishnan, C. (2020). Synergistic potential of nutraceuticals: Mechanisms and prospects for futuristic medicine. *Food & Function*.. https://doi.org/10.1039/d0fo02041a.

Lim, J., Ferruzzi, M. G., & Hamaker, B. R. (2021). Dietary starch is weight reducing when distally digested in the small intestine. *Carbohydrate Polymers*, *273*, 118599. https://doi.org/10.1016/J.CARBPOL.2021.118599.

Liu, Z., He, C., Guo, C., Chen, F., Bhandari, B., & Zhang, M. (2021). Dehydration-triggered shape transformation of 4D printed edible gel structure affected by material property and heating mechanism. *Food Hydrocolloids*, *115*, 106608. https://doi.org/10.1016/j.foodhyd.2021.106608.

López-Pedrouso, M., Lorenzo, J. M., Gullón, B., Campagnol, P. C. B., & Franco, D. (2021). Novel strategy for developing healthy meat products replacing saturated fat with oleogels. *Current Opinion in Food Science*, *40*, 40–45. https://doi.org/10.1016/j.cofs.2020.06.003.

Maruyama, S., Streletskaya, N. A., & Lim, J. (2021). Clean label: Why this ingredient but not that one? *Food Quality and Preference*, *87*, 104062. https://doi.org/10.1016/j.foodqual.2020.104062.

Mat, D. J. L., Le Feunteun, S., Michon, C., & Souchon, I. (2016). In vitro digestion of foods using pH-stat and the INFOGEST protocol: Impact of matrix structure on digestion kinetics of macronutrients, proteins and lipids. *Food Research International*, *88*, 226–233. https://doi.org/10.1016/j.foodres.2015.12.002.

McClements, D. J. (2015). *Reduced-fat foods: The complex science of developing diet-based Stratergies for tackling overweight and obesity*.

Mezzenga, R., & Fischer, P. (2013). The self-assembly, aggregation and phase transitions of food protein systems in one, two and three dimensions. *Reports on Progress in Physics*, *76*(4). https://doi.org/10.1088/0034-4885/76/4/046601.

Montoya, C. A., Rutherfurd, S. M., Olson, T. D., Purba, A. S., Drummond, L. N., Boland, M. J., & Moughan, P. J. (2014). Actinidin from kiwifruit (Actinidia deliciosa cv. Hayward) increases the digestion and rate of gastric emptying of meat proteins in the growing pig. *British Journal of Nutrition*, *111*(6), 957–967.

Mulet-Cabero, A. I., Rigby, N. M., Brodkorb, A., & Mackie, A. R. (2017). Dairy food structures influence the rates of nutrient digestion through different in vitro gastric behaviour. *Food Hydrocolloids*, *67*, 63–73. https://doi.org/10.1016/J.FOODHYD.2016.12.039.

Muñoz-Almagro, N., Montilla, A., & Villamiel, M. (2021). Role of pectin in the current trends towards low-glycaemic food consumption. *Food Research International*, *140*, 109851. https://doi.org/10.1016/j.foodres.2020.109851.

Nasser, J. A., Albajri, E., Lanza, L., Gilman, A., Altayyar, M., Thomopoulos, D., & Bruneau, M. (2022). Interaction of protein preloads and physical activity on intake of an ultra-processed, high sugar/high fat food/low protein food. In. *Nutrients*, *14*(4), 884. https://doi.org/10.3390/nu14040884.

Paciulli, M., Littardi, P., Carini, E., Paradiso, V. M., Castellino, M., & Chiavaro, E. (2020). Inulin-based emulsion filled gel as fat replacer in shortbread cookies: Effects during storage. *LWT*, *133*, 109888. https://doi.org/10.1016/j.lwt.2020.109888.

Palla, C. A., Aguilera-Garrido, A., Carrín, M. E., Galisteo-González, F., & Gálvez-Ruiz, M. J. (2022). Preparation of highly stable oleogel-based nanoemulsions for encapsulation and controlled release of curcumin. *Food Chemistry*, *378*, 132132. https://doi.org/10.1016/J.FOODCHEM.2022.132132.

Pant, A., Lee, A. Y., Karyappa, R., Lee, C. P., An, J., Hashimoto, M., Tan, U. X., Wong, G., Chua, C. K., & Zhang, Y. (2021). 3D food printing of fresh vegetables using food hydrocolloids for dysphagic patients. *Food Hydrocolloids*, *114*, 106546. https://doi.org/10.1016/j.foodhyd.2020.106546.

Pascua, Y., Koç, H., & Foegeding, E. A. (2013). Food structure: Roles of mechanical properties and oral processing in determining sensory texture of soft materials. *Current Opinion in Colloid and Interface Science*, *18*(4), 324–333. https://doi.org/10.1016/j.cocis.2013.03.009.

Patel, A. R., Rodriguez, Y., Lesaffer, A., & Dewettinck, K. (2014). High internal phase emulsion gels (HIPE-gels) prepared using food-grade components. *RSC Advances*, *4*(35), 18136–18140. https://doi.org/10.1039/c4ra02119c.

Pehlivanoğlu, H., Demirci, M., Toker, O. S., Konar, N., Karasu, S., & Sagdic, O. (2018). Oleogels, a promising structured oil for decreasing saturated fatty acid concentrations: Production and food-based applications. *Critical Reviews in Food Science and Nutrition*, *58*(8), 1330–1341. https://doi.org/10.1080/10408398.2016.1256866.

Pellegrini, N., Vittadini, E., & Fogliano, V. (2020). Designing food structure to slow down digestion in starch-rich products. *Current Opinion in Food Science*, *32*, 50–57. https://doi.org/10.1016/j.cofs.2020.01.010.

Priyadarshini, S. R., Moses, J. A., & Anandharamakrishnan, C. (2021). Determining the glycaemic responses of foods: Conventional and emerging approaches. *Nutrition Research Reviews*, 1–27.

Pulatsu, E., Su, J.-W., Lin, J., & Lin, M. (2022). Utilization of ethyl cellulose in the osmotically-driven and Anisotropically-actuated 4D printing concept of edible food composites. *Carbohydrate Polymer Technologies and Applications*, *3*, 100183. https://doi.org/10.1016/j.carpta.2022.100183.

Qazi, H. J., Ye, A., Acevedo-Fani, A., & Singh, H. (2021). In vitro digestion of curcumin-nanoemulsion-enriched dairy protein matrices: Impact of the type of gel structure on the bioaccessibility of curcumin. In. *Food Hydrocolloids*, *117*. https://doi.org/10.1016/j.foodhyd.2021.106692.

Qin, X. S., Gao, Q. Y., & Luo, Z. G. (2021). Enhancing the storage and gastrointestinal passage viability of probiotic powder (lactobacillus plantarum) through encapsulation with pickering high internal phase emulsions stabilized with WPI-EGCG covalent conjugate nanoparticles. *Food Hydrocolloids*, *116*, 106658. https://doi.org/10.1016/J.FOODHYD.2021.106658.

Raungrusmee, S., Shrestha, S., Sadiq, M. B., & Anal, A. K. (2020). Influence of resistant starch, xanthan gum, inulin and defatted rice bran on the physicochemical, functional and sensory properties of low glycemic gluten-free noodles. *LWT*, *126*, 109279. https://doi.org/10.1016/j.lwt.2020.109279.

Rodrigues, D. M., de Souza, V. R., Mendes, J. F., Nunes, C. A., & Pinheiro, A. C. M. (2016). Microparticulated salts mix: An alternative to reducing sodium in shoestring potatoes. *LWT - Food Science and Technology*, *69*, 390–399. https://doi.org/10.1016/j.lwt.2016.01.056.

Roobab, U., Batool, Z., Manzoor, M. F., Shabbir, M. A., Khan, M. R., & Aadil, R. M. (2020). Sources, formulations, advanced delivery and health benefits of probiotics. *Current Opinion in Food Science*, *32*, 17–28. https://doi.org/10.1016/j.cofs.2020.01.003.

Santhi Rajkumar, P., Suriyamoorthy, P., Moses, J. A., & Anandharamakrishnan, C. (2020). Mass transfer approach to in-vitro glycemic index of different biscuit compositions. *Journal of Food Process Engineering, 43*(12), e13559. https://doi.org/10.1111/jfpe.13559.

Sethupathy, P., Suriyamoorthy, P., Moses, J. A., & Chinnaswamy, A. (2020). Physical, sensory, in-vitro starch digestibility and glycaemic index of granola bars prepared using sucrose alternatives. *International Journal of Food Science and Technology, 55*(1), 348–356. https://doi.org/10.1111/ijfs.14312.

Shanthamma, S., Preethi, R., Moses, J. A., & Anandharamakrishnan, C. (2021). 4D printing of sago starch with turmeric blends: A study on pH-triggered spontaneous color transformation. *ACS Food Science and Technology, 1*(4), 669–679. https://doi.org/10.1021/acsfoodscitech.0c00151.

Sharma, R., Mokhtari, S., Jafari, S. M., & Sharma, S. (2021). Barley-based probiotic food mixture: Health effects and future prospects. *Critical Reviews in Food Science and Nutrition*. https://doi.org/10.1080/10408398.2021.1921692. Taylor & Francis.

Shen, S., Chi, C., Zhang, Y., Li, L., Chen, L., & Li, X. (2021). New insights into how starch structure synergistically affects the starch digestibility, texture, and flavor quality of rice noodles. *International Journal of Biological Macromolecules, 184*, 731–738. https://doi.org/10.1016/j.ijbiomac.2021.06.151.

Siegrist, M., & Hartmann, C. (2020). Consumer acceptance of novel food technologies. *Nature Food, 1*(6), 343–350. https://doi.org/10.1038/s43016-020-0094-x.

Silva, T. J., Fernandes, G. D., Bernardinelli, O. D., Silva, E. C.d. R., Barrera-Arellano, D., & Ribeiro, A. P. B. (2021). Organogels in low-fat and high-fat margarine: A study of physical properties and shelf life. *Food Research International, 140*. https://doi.org/10.1016/j.foodres.2020.110036 (October 2020).

Silva-Avellaneda, E., Bauer-Estrada, K., Prieto-Correa, R. E., & Quintanilla-Carvajal, M. X. (2021). The effect of composition, microfluidization and process parameters on formation of oleogels for ice cream applications. *Scientific Reports, 11*(1), 1–10. https://doi.org/10.1038/s41598-021-86233-y.

Sim, J. Y., Enteshari, M., Rathnakumar, K., & Martínez-Monteagudo, S. I. (2021). Hydrodynamic cavitation: Process opportunities for ice-cream formulations. *Innovative Food Science and Emerging Technologies, 70*, 102675. https://doi.org/10.1016/j.ifset.2021.102675.

Singh, P. (2022). *Future of health and wellness industry*. Retrieved February 27, 2022, from https://appinventiv.com/blog/wellness-market-statistics-for-future-growth/.

Somaratne, G., Ferrua, M. J., Ye, A., Nau, F., Floury, J., Dupont, D., & Singh, J. (2020). Food material properties as determining factors in nutrient release during human gastric digestion: A review. *Critical Reviews in Food Science and Nutrition, 60*(22), 3753–3769. https://doi.org/10.1080/10408398.2019.1707770.

Stieger, M., & Van de Velde, F. (2013). Microstructure, texture and oral processing: New ways to reduce sugar and salt in foods. *Current Opinion in Colloid and Interface Science, 18*(4), 334–348. https://doi.org/10.1016/j.cocis.2013.04.007.

Sun, P., Xia, B., Ni, Z. J., Wang, Y., Elam, E., Thakur, K., Ma, Y. L., & Wei, Z. J. (2021). Characterization of functional chocolate formulated using oleogels derived from β-sitosterol with γ-oryzanol/lecithin/stearic acid. *Food Chemistry, 360*, 130017. https://doi.org/10.1016/J.FOODCHEM.2021.130017.

Sun-Waterhouse, D. (2011). The development of fruit-based functional foods targeting the health and wellness market: A review. *International Journal of Food Science and Technology, 46*(5), 899–920. https://doi.org/10.1111/j.1365-2621.2010.02499.x.

Tao, Y., Lee, Y.-C., Liu, H., Zhang, X., Cui, J., Mondoa, C., Babaei, M., Santillan, J., Wang, G., Luo, D., et al. (2021). Morphing pasta and beyond. *Science Advances, 7*(19), eabf4098. https://doi.org/10.1126/sciadv.abf4098.

Tekin-Cakmak, Z. H., Karasu, S., Kayacan-Cakmakoglu, S., & Akman, P. K. (2021). Investigation of potential use of by-products from cold-press industry as natural fat replacers and functional ingredients in a low-fat salad dressing. *Journal of Food Processing and Preservation, 45*(8), 1–13. https://doi.org/10.1111/jfpp.15388.

Teng, X., Zhang, M., & Mujumdar, A. S. (2021). 4D printing: Recent advances and proposals in the food sector. *Trends in Food Science & Technology.*. https://doi.org/10.1016/j.tifs.2021.01.076.

Teo, P. S., van Dam, R. M., Whitton, C., Tan, L. W. L., & Forde, C. G. (2020). Consumption of foods with higher energy intake rates is associated with greater energy intake, adiposity, and cardiovascular risk factors in adults. *The Journal of Nutrition Nutritional Epidemiology*, *388*, 539–547. chrome-extension://dagcmkpagjlhakfdhnbomgmjdpkdklff/enhanced-reader.html?openApp&pdf=https%3A%2F%2Fwatermark.silverchair.com%2Fnxaa344.pdf%3Ftoken%3DAQECAHi208BE49Ooan9kkhW_Ercy7Dm3ZL_9Cf3qfKAc485ysgAAAsAwggK8BgkqhkiG9w0BBwagggKtMIICqQIBADCCAqIGCSqGSIb3DQ.

This, H. (2005). Modelling dishes and exploring culinary 'precisions': The two issues of molecular gastronomy. *British Journal of Nutrition*, *93*(S1), S139–S146. https://doi.org/10.1079/bjn20041352.

Vimala Bharathi, S. K., et al. (2018). Nano and Microencapsulation using food grade polymers. *Polymers for Food Applications.*. https://doi.org/10.1007/978-3-319-94625-2.

Vinitha, K., Maria Leena, M., Moses, J. A., Anandharamakrishnan, C., et al. (2021). Size-dependent enhancement in salt perception: Spraying approaches to reduce sodium content in foods. *Powder Technology Journal*, *378*, 237–245. https://doi.org/10.1016/j.powtec.2020.09.079. In this issue.

Voronin, G. L., Ning, G., Coupland, J. N., Roberts, R., & Harte, F. M. (2021). Freezing kinetics and microstructure of ice cream from high-pressure-jet processing of ice cream mix. *Journal of Dairy Science*, *104*(3), 2843–2854. https://doi.org/10.3168/jds.2020-19011.

Wilson, A., Anukiruthika, T., Moses, J. A., & Anandharamakrishnan, C. (2020). Customized shapes for chicken meat–based products: Feasibility study on 3D-printed nuggets. *Food and Bioprocess Technology*, *13*(11), 1968–1983. https://doi.org/10.1007/s11947-020-02537-3.

Yan, L., Yu, D., Liu, R., Jia, Y., Zhang, M., Wu, T., & Sui, W. (2021). Microstructure and meltdown properties of low-fat ice cream: Effects of microparticulated soy protein hydrolysate/xanthan gum (MSPH/XG) ratio and freezing time. *Journal of Food Engineering*, *291*, 110291. https://doi.org/10.1016/j.jfoodeng.2020.110291.

Yoha, K. S., Anukiruthika, T., Anila, W., Moses, J. A., & Anandharamakrishnan, C. (2021). 3D printing of encapsulated probiotics: Effect of different post-processing methods on the stability of Lactiplantibacillus plantarum (NCIM 2083) under static in vitro digestion conditions and during storage. *LWT*, *146*, 111461. https://doi.org/10.1016/j.lwt.2021.111461.

Yoha, K. S., Moses, J. A., & Anandharamakrishnan, C. (2020). Effect of encapsulation methods on the physicochemical properties and the stability of Lactobacillus plantarum (NCIM 2083) in synbiotic powders and in-vitro digestion conditions. *Journal of Food Engineering*, *283*, 110033. https://doi.org/10.1016/j.jfoodeng.2020.110033.

Zhang, K., Dong, R., Hu, X., Ren, C., & Li, Y. (2021). *Oat-based foods_ Chemical constituents, glycemic index, and the effect of processing*.

# PART II

# Strategies to modify structure/functionality/quality of foods

# CHAPTER 3

# Electrotechnologies for the development of food-based structured systems

Ricardo Nuno Pereira[a,b], Rui M. Rodrigues[a,b], and Antonio A. Vicente[a,b]

[a]CEB—Centre of Biological Engineering, University of Minho, Braga, Portugal, [b]LABBELS—Associate Laboratory, University of Minho, Braga, Portugal

## 3.1 Introduction

The food industry has come a long way from a commodity industry organized for the productivity and focused on producing large amounts of cheap food, to a high added-value product industry driven by the consumers' demands and committed with societal changes. Consequently, the increasing demand for high quality foods, new and improved functional properties, along with the need for more sustainable ingredients for efficient and integrated processes has driven technological advances into the production of new food structured systems, often supported by the incorporation of innovative technologies.

It is generally recognized that chemical composition alone does not reflect the nutritional quality and physiological response to foods. The organization of the different structural elements (e.g., proteins, polysaccharides and fat/oils) and their interactions (among them or with other minor constituents) critically define a given food structure, from molecular scale to its macroscopic properties. Therefore, the structure of the food matrix plays a major role in the definition of its sensorial attributes and stability, as well as influences the digestibility and bioavailability of nutrients and bioactive agents (McClements et al., 2015). It is clear then that for a proper understanding of the food behavior and to attain a rational design and control of its functionality, a fundamental knowledge about the composition, arrangement and interaction of the different food components is necessary. During the last few decades, food scientists have been focusing their efforts not only on understanding structure-function relationship, but also on the development of new functional and innovative structured systems to promote sensorial attributes to food, promote better nutrition and ultimately health and well-being. To do so, a myriad of new processing technologies have been implemented in food technology and research arena, with the goal of effectively control food structure at multiple levels. These processing methods often involve at some extent of physical or chemical stress, or even the combination of both effects—i.e. mechanical, pressure, thermal, electrical, reduction/oxidation, complexation—to achieve an intended and controlled structural or functional changes in foods or food components.

Initially aiming at replacing the high energy demanding and highly aggressive thermal processes, emerging technologies such as high hydrostatic pressure processing, sonication, or electric fields (EF)

processing have been arising as alternatives to potentially improve foodstuff quality and functionality (de Vries, Axelos, et al., 2018; Ma et al., 2019). A particular branch of these innovative technologies, known as electrotechnologies, relies on the use of electric current as a processing strategy. By the application of an external EF on bio-based materials, several phenomena can be triggered out, such as: (i) electroporation of cells and tissues (Kotnik et al., 2015); (ii) heat dissipation through ohmic heating (OH) (Sastry, 2008); (iii) changes in macromolecules (Rodrigues, Avelar, et al., 2020); (iv) charge-related phenomena such electrophoretic movements of molecule/particle alignments (Suscillon et al., 2013; Takashima & Schwan, 1985); (v) and in extreme cases, arching and plasma channels formation resulting in light emission, shockwaves and cavitation (Li et al., 2019). These EF-related phenomena, namely thermal, nonthermal or in conjugation, can be used for technological purposes and exponentiated according to different technology variations. The most relevant electrotechnologies in food science are the Pulsed Electric Fields (PEF) and Moderate Electric Field (MEF) technologies latter one also associated with OH (Rocha et al., 2018). PEF uses high intensity EF pulses with short duration (i.e., 1–50 kV/cm pulses of nano or microseconds) to explore the electric field driven phenomena and minimize heat dissipation. PEF technology has been extensively used to promote the permeabilization of cells and tissues, resulting in microbial inactivation, extraction of intercellular compounds or textural changes in biological tissue (Bhat, Morton, Mason, & Bekhit, 2019; Vorobiev & Lebovka, 2016). This technology is recognized in the food industry as an effective way to achieve tissue permeabilization and nonthermal pasteurization (Alexandre et al., 2019). The presence of high EF also demonstrated to affect macromolecules, such as proteins and polysaccharides, resulting in inactivation of enzymes, changes in protein functional properties or starch modifications and hydrolysis (Zhao et al., 2014; Zhu, 2018).

MEF technology relies on the passage of alternating EF, usually comprehended between 1 and 1000 V/cm, with an unrestricted time application. The application of MEF may result in a mild electroporation effect, or disturbances in macromolecules folding, and depending on its intensity, time of the application, as well as electrical properties of the solutions, promote heat dissipation through the Joule's effect (EF) (Rodrigues, Genisheva, et al., 2019). Heat dissipation can be restricted by controlling electrical variables and electrical conductivity of the target product; however, the industrial application of this technology is still focused on thermal processing by OH. OH is recognized by its fast and homogeneous way of heating, as well as by its high energetic efficiency, bringing significant advantages when compared with other conventional thermal processing technologies (Ramaswamy et al., 2016). Furthermore, the conjugation of thermal and nonthermal effects revealed to bring interesting advantages in the microbial and enzymatic inactivation, extraction processes and macromolecule modifications (Fasolin et al., 2021; Jaeger et al., 2016; Jakób et al., 2010). Fig. 3.1 illustrates a diagram of an electrical circuit of MEF technology and its main features.

It is important to highlight that OH effect is a natural consequence of applying an electric field to a semi-conductor material—food material can allow the passage of electric current without occurring OH. In this sense, for a broader definition it could be more accurate to designate this technology by MEF rather than OH.

MEF technology has recently contributed to important developments on the control and manipulation of biomaterials structure and functionality. Due to its growing relevance and increasing evidence of its use on tailoring food structure and function, this chapter will be devoted to the review of the most recent findings and implications of this technology.

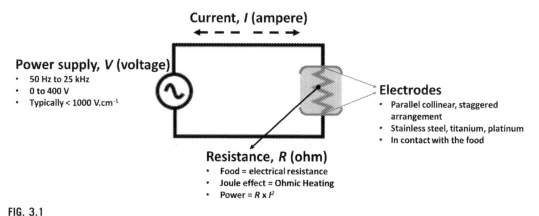

**FIG. 3.1**

Electric circuit diagram and main features of MEF technology.

## 3.2 Moderate electric fields technology

The use of MEF technology in the food industry has been long experienced. The first industrial application of the process occurred at the beginning of the 20th century for milk pasteurization (Anderson & Finkelstein, 1919). Despite the early recognized potential, technical and operational limitations related with lack of process control, the cost of electricity and contamination from the electrode's material, prevented its implementation (Rodrigues, Genisheva, et al., 2019). It was not until the 1980s that technological developments in power supply systems and stable electrode materials contacting the foods brought the opportunity to make electricity-based processing technologies viable. Soon, OH processing found its successful way into the food industry, bringing particular advantages on the processing of high viscous foods, particulate solid-liquid mixtures and sensitive products such as fruit based or high protein foods, such as egg and dairy products (Kaur & Singh, 2016; Knirsch et al., 2010). With the growing implementation and a renewed research interest, OH processing was applied to a variety of processes such as pasteurization, blanching, thawing, dehydration and fermentation and biomass bioprocessing and extraction. Additionally to the operational advantages (in terms of energetic efficiency, operational flexibility and improved product's quality), the occurrence of nonthermal effects due to the presence of an external EF started to draw food researchers attention (Pereira et al., 2011; Sastry, 2008). Table 3.1 summarizes some milestones of OH in food processing.

Increased inactivation kinetics of microorganisms and enzymes or permeabilization of cellular tissue have been reported to occur during OH. These effects proved to be advantageous in food processing, either increasing inactivation efficiencies (or allowing the thermal processing to occur at lower temperatures achieving similar inactivation rates as conventional processing) and by leveraging extraction processes in different biotechnological applications. In some cases, the extent of MEF nonthermal effects even allowed to achieve microbial inactivation (Machado et al., 2010) or protein unfolding (Bekard & Dunstan, 2014) at room temperature. All of this has led to an increasing interest in the EF effects at the molecular level and a better understanding about its interactions with the food constituents and other biomaterials. This also allowed to start establishing a basis for a better understanding about how to influence and control structural and functional properties of different materials using an electric field.

**Table 3.1 Overview of some milestones and its approximate dates regarding fundamental and applied aspects of MEF technology.**

| Milestone | Year |
|---|---|
| Patents and scientific reports regarding electric pasteurization of milk | 1897–1968 |
| Blanching using electroconductive heating principle | 1975 |
| Technology licensing from APV (Baker Perkins) | 1980 |
| OH pilot processing facility (Land-O'Lakes Company, USA) | 1990 |
| OH for fermentation applications | 1996 |
| Advances in power supply—development of Integrated Gate Bipolar Transistor (IGBT) | 1997 |
| Effects of MEF on enzyme and microbiological inactivation | 2004 |
| Exploring MEF effects on the development of structured protein-based systems (e.g., micro nano structures, gels, and films) | 2010 |
| Influence of electric frequency on enzyme activation/inactivation | 2016 |
| Impact of OH on structural and immunological aspects of whey proteins | 2020 |
| Use of OH on development of biomaterials | 2021 |

### 3.2.1 Effects of MEF at the molecular level

Proteins were naturally the first macromolecules to be studied under EF effects. They not only present important functional and structural properties in biological systems as well as they are highly responsive to external stimuli. Also, the application of OH in food processing leads to the study of proteins as quality indicators by determining protein denaturation and endogenous enzymes' activity. Proteins thermally processed under OH showed different physicochemical and functional properties than those processed by conventional methods. Whey proteins showed different denaturation kinetics under OH, traduced by lower denaturation reaction order and rate constant (Pereira et al., 2011) and soy proteins evidenced additional modifications by MEF during OH when compared to conventional thermal treatment (Cha, 2011). Several authors also reported a general tendency for an enhanced inactivation kinetics of enzymes subjected to OH, suggesting the occurrence of mechanisms involving the MEF action and the movement of charged species, such metal ions (Castro et al., 2006; Içier et al., 2008; Makroo et al., 2017). Recently, complex mechanisms were suggested to be associated with MEF, involving not only the MEF intensity and charge movement, but also its alternation (electric frequency) and temperature dependence. MEF action resulted in enhancement of enzymatic activity close to the activation temperatures and on an increased inactivation observed at higher temperatures (Brochier & Domeneghini, 2016; Samaranayake & Sastry, 2016a). These effects were also demonstrated to be frequency dependent with higher effectiveness at lower frequency values (i.e., <100 Hz), reaching up to 41% increase of α-amylase activity at optimum temperature (Samaranayake & Sastry, 2016b, 2018). The authors suggested that the translational electrophoretic displacement of the enzyme and its relationship with the frequency applied is the factor affecting the enzymatic activation or inactivation. Under low frequencies, protein molecules align with the field direction and rotate as the field is reversed, while under higher frequencies, the faster field reversal results in molecular oscillation instead of full rotations. These molecular motions imply additional energy dissipation on the protein molecule, possibly acting as a local thermal elevation. These effects caused by the oscillating MEF, also proved to destabilize protein structure and conformation independently of thermal action. Bekard and Dunstan, (2014) demonstrated that the energy dissipated in the proteins through frictional drag under an oscillating EF is sufficient to break H-bonds and result in protein unfolding. These MEF effects at the molecular level were also studied on the denaturation and interaction of β-Lactoglobulin (β-Lg). Evaluating the thermal unfolding process under MEF at different intensities and frequencies, Rodrigues and co-workers demonstrated a destabilizing effect of the MEF on the protein fold. MEF action resulted on the decrease of the protein's melting temperature up to 20 °C under 10 V/cm and 50 Hz but little effects were observed at high frequencies (i.e., $\geq 2$ kHz) (Rodrigues, Avelar, et al., 2020). The unfolding process was dependent on the EF intensity and frequency applied, being favored under low frequencies and by the increase of the EF strength. The conformation of the unfolded forms of the protein was affected by the MEF action regardless of the electric frequency applied, resulting in differentiated secondary and tertiary structural features and different interactions with ligands, namely the hydrophobic probe ANS and the ligand retinol. Despite the occurrence of MEF effects in protein structure and conformation, they were mostly observer in synergy with thermal action and were dependent on the environmental conditions such as temperature, ionic strength, protein concentration and pH (Rodrigues, Vicente, et al., 2019).

MEF effects have also been studied in other biomolecules such as starch. MEF treatment at subgelatinization temperatures of potato starch resulted in a more stable and homogeneous internal structure of the biopolymer (Cha, 2012, 2015). MEF treatments of rice starch resulted in a decrease of the onset gelatinization temperature, thus requiring less energy to cook, and resulting in a gel with higher stability (An & King, 2007). Moreover, the effect of MEF varies depending on the type of starch and its composition, such as fats, proteins, and amylose. On the contrary, conventional heating simply showed to promote higher thermal resistance and narrower gelatinization temperature ranges regardless of starch source. Through these results, it was also suggested that the MEF effects during OH can be used to alter thermal characteristics of rice starch and flours, differently to what is observed to conventional heating, depending on applied voltage and frequency.

The demonstration of the MEF effects at molecular scale and unraveling the fundamental factors of its action/operating mechanism opens an interesting perspective to use MEF as a tool to control biomolecule's physiochemical and functional properties at a molecular scale. This may enable new strategies on the development of novel or improved systems with applicability in areas such as food, biotechnology, biomaterials, pharmaceutical and biomedical. However, further research is needed to fully elucidate the parameters involved in MEF action, namely the EF intensity, current density, wave shape and frequency dependence, time scale of application, occurrence of electrochemical reactions, and synergic effects with thermal action. Furthermore, the MEF treatment specificities and interaction with the biomolecules is dependent on the composition of the media, which determines the electric conductivity as well as the chemical environment, thus critically defining the molecule's reactivity, interactions and stability.

### 3.2.2 Effects of EF in nano and microstructures

Food proteins of several sources (e.g., animal, vegetable, microbial) have recognized potential to be used as building blocks in de development of functional supramolecular structures such as aggregates, gels or emulsions (Foegeding, 2015). Often, to induce and control protein functionality and potentiate their capacity to interact, among each other or with other molecules, a driving force is required in order to change protein conformation and expose their reactive groups—i.e., free thiols, hydrophobic regions or charged species—usually hindered within the protein fold (Mirmoghtadaie et al., 2016). Interaction and aggregation of globular proteins is a well-established and widely explored field of food science and technology, however the MEF action can contribute with new insights on the development of protein systems aiming at improving and controlling functionality. As already discussed in the previous section, MEF has the capacity to change protein structure by changing the natural balance of the interactions, which in turn can greatly influence both the protein's functionality and ability to form structured systems (Abaee et al., 2017).

MEF has demonstrated to significantly increase the surface hydrophobicity of β-Lg and change the free sulfhydryl (SH) profile by disrupting disulfide bonds or exposing free cysteine residues berried within the protein fold. These changes were also influenced by the environmental conditions, such as pH, ionic strength and temperature (Rodrigues, Vicente, et al., 2019). Therefore, MEF along with the control of the physicochemical conditions, can influence supramolecular formation, since hydrophobic interactions and disulphide crosslinking are known to be decisive in protein-protein interactions and aggregation. In fact, MEF effects during OH of whey proteins, compared with conventional heating, result in distinctive aggregation behaviors. MEF treatments at 12 V/cm resulted in the reduction of free thiol groups in about 30%, in size reduction of the formed aggregates—from 300 nm of the conventional process to 120 nm under MEF—and on morphological changes favoring the formation of elongated structures (Pereira et al., 2016). The increase of the surface hydrophobicity, reduction in free SH content, and consequent reduction in aggregates size and size distribution were also reported by Rodrigues and coworkers (Rodrigues, Avelar, et al., 2020; Rodrigues, Fasolin, et al., 2020). In these studies, it was also demonstrated that the electrical variables of MEF, namely EF strength and electric frequency, can influence the interaction and aggregation process, where the EF increase and use of low frequencies magnified the reported effects. Subaşı and coworkers reported MEF effect on the processing of caseinate and sunflower protein (Subaşı et al., 2021). MEF affected the secondary, tertiary structures and thermal properties of both protein systems. Furthermore, particle size was reduced as well as homogeneity, interfacial tension at the air/water interface were also changed by MEF action.

MEF ability to influence whey protein aggregation and functionality during OH was also evidenced using a cold gelation strategy. In this two-step process, the protein denaturation is initially induced by conventional and OH, followed by the addition of $Fe^{2+}$ at room temperature, resulting in further aggregation (Pereira et al., 2017). The presence of the MEF (10 V/cm) resulted in the reduction of the particle size distribution and changes in physical stability, rheological behavior and microstructure of the obtained microgels. Furthermore, it was possible to increase the amount of $Fe^{2+}$ incorporation and modulate the rheological properties of the gels. Similarly, MEF processing of lactoferrin results in differentiated physicochemical properties, i.e., reduction of aggregates size, polydispersity, free SH content and surface hydrophobicity (Furtado et al., 2018). The production of emulsions from heat-treated lactoferrin, by MEF and conventional heating, resulted in a gel-like substance with different properties. The MEF treated samples showed to have a finer and less rigid microstructure with a more fluid behavior.

Despite several reports, the mechanisms of MEF action in protein aggregation are not yet fully elucidated. However, the reports are consistent on the increase of the surface hydrophobicity, reduction of free SH groups, aggregates size reduction and narrower size distribution. A possible explanation for MEF action (see Fig. 3.2) may be related to the differentiated structural changes induced by the EF

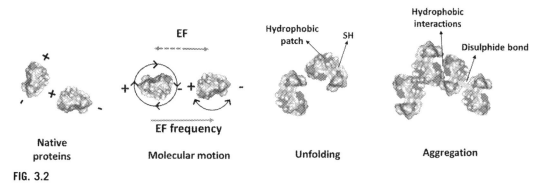

**FIG. 3.2**

Proposed mechanism of MEF action in protein molecules.

effects (i.e., reorganization of intramolecular bonds and reactive groups of amino acids, such as exposure of hydrophobic patches and SH groups), favoring protein interactions. Once the hydrophobic interactions take place, the disulphide crosslinking is facilitated by the molecular rearrangements, presence of paring redox groups (within the protein or free radicals produced by EF action) and consequently resulting in a faster stabilization of the aggregation process, which in turn give rise to smaller aggregates, with a lower amount of free SH groups.

MEF effects in supramolecular structure formation involving other macromolecules (e.g., polysaccharides) were not described yet. Nonetheless, interaction between proteins and polysaccharides are well recognized and acknowledged as a potential strategy to develop food's nutritional and organoleptic properties. Maillard reactions, for example, may play a decisive role in functional and biological properties such as antioxidative, antimicrobial, antihypertensive, mutagenic or carcinogenic of foods (Jaeger et al., 2010; Perusko et al., 2015). The formation of protein-polysaccharide complexes, aggregates and coacervates also have a great potential as carriers for health-promoting bioactive compounds, allowing an improved physical stability and adequate digestion within the gastrointestinal (GI) tract (Semenova, 2017; Wei & Huang, 2019). Although there is still a lack of reports on the EF action in protein-polysaccharide interactions and structure formation, it is clear the reported effects in protein structural features and interactions potential may have a significant impact on the formation of such structures. Therefore it is expected that in the near future, MEF can be used to induce and control such processes.

### 3.2.3 Effects of EF in macrostructures

With the fundamental effects of MEF on the structure of proteins and polysaccharides, as well as in protein aggregation previously discussed, it is expected that these can impact the macro properties and functionality of food systems. MEF effects were described in thermal and cold gelation of whey proteins (Rodrigues et al., 2015; Rodrigues, Fasolin, et al., 2020). The obtained gels presented differentiated structural features characterized by a finer stranded network. These gels also presented differences on viscoelastic properties and functionality with MEF action being linked to softer gels that were more elastic (reduction of Young's modulus of 25% and increase of breaking strain by 15%), and increasing water retention capacity from 88% to 93% and swelling capacity from 6% to 13%. These modifications of gels microstructure and functionality may result from the lower content in free SH groups of the aggregates, leading to lower disulphide crosslinking and a higher proportion of hydrogen bonds and hydrophobic interactions on the protein network stabilization. Processing caseinate solutions by conventional heating or MEF, followed by acidification until the isoelectric point, resulted in the formation of self-sustained gels. The processing method did not change the structure or mechanical properties of the gels, however the modification in protein structure imposed by MEF resulted in higher protein solubility and lower water retention on the obtained gels (Moreira et al., 2019).

Other protein-based structures were produced under MEF effects such as protein films. Similarly to the observed for the protein gels, MEF influenced the unfolding and aggregation mechanisms of whey proteins, and consequently the formation of protein-based film (Pereira et al., 2010). The films obtained under MEF action presented similar mechanical properties to the control, however they were thinner and with lower permeability to water vapor. On another work, Tinoco et al. used OH to produce keratin films (Tinoco et al., 2020). The effects of MEF during the processing of keratin solutions were once again linked to structural modifications that resulted on the reduction of the free thiol content, solubility increase, and films with lower swelling capacity, morphology changes and lower opacity. Protein-lipid film formation from soy-milk was also affected by OH processing (Lei et al., 2007). The protein incorporation into the films was higher under MEF action, increasing the film formation rate and process yield. Additionally, the obtained films presented higher water absorption capacity.

Polysaccharide-based films were also produced under MEF action. Chitosan film forming solution processed under MEF resulted on the formation of films with higher crystallinity, a more homogeneous structure, improved mechanical properties and increased gas barrier properties (Souza et al., 2009, 2010). Also, MEF action during OH processing of film forming solution of starch and chitosan, reinforced with microcrystalline cellulose, resulted on the modulation of the film properties (Coelho et al., 2017). MEF action allowed the production of chitosan films with higher hydrophobicity and lower water vapor permeability, but no apparent changes in structural and mechanical properties were observed. In starch films, MEF resulted on the reduction of the film's hydrophobicity, lower tensile strength and young's modulus.

It is now obvious that MEF can interfere with the structural features of macromolecules and play a decisive role on the network formation and techno-functional properties of food systems based on proteins and polysaccharides. However, foods are often complex matrices where a large variety of biomolecules interact to form complex structures. Nonetheless, some examples of OH processing in complex foods address the effects of MEF in structure formation. An example of these effects was observed on the acid gelation of milk, where OH successfully improved the acid milk gels (Caruggi et al., 2019). The application of MEF between 25 and 55 V/cm resulted on an increase of up to 21% on the gel strength and allowed to reduce the gelation time up to 40% when compared with conventional heating under similar conditions. The increase of gel firmness was also accompanied by a reduction in syneresis and a more compact structure with smaller pore size. OH as a processing method for tofu production has revealed advantages on products' texture (Shimoyamada et al., 2015; Wang et al., 2018) and it is even commercially available at industrial scale by the manufacturer Yanagiya (http://ube-yanagiya.com/). Despite the reported effects of MEF in soy proteins and soy milk properties (Liu & Chang, 2007; Shimoyamada et al., 2015) it is yet not clear if the reported improvements in tofu's quality are the result of nonthermal effects or from the fast and homogeneous heating provided by the technology. The production of surimi production, a fish-based gel, is another well-established example of the use of OH with operational advantages on the product's quality (Fowler & Park, 2015; Shahidi et al., 1998). The MEF variables used—i.e., variation of the electric field intensity and heating rate—resulted on the differentiation of the microstructure and mechanic properties of the product. Corroborating some fundamental studies described previously, MEF action influenced the free SH content and resulted on the modulation of the gel's structure, as well as on the reduction of the activity of endogenous enzymes (Tadpitchayangkoon et al., 2012).

The reported studies of MEF effects in protein aggregation, gel and film formation as well as on the development of complex food structures such as milk gels, tofu and surimi, validate the potential to influence and control protein-based structures. There are also some evidences on the MEF's potential to modulate polysaccharide molecular properties and structure formation. Despite the early development stage of the technology in modulation and control of food structured systems, the perspectives are favorable for a rapid development of functionalization protocols and methodologies based in OH and MEF action. The MEF processing specifications—inexistence of diffusional limitations, rapid and homogeneous heating, no moving parts—as well as the existence of nonthermal effects in molecular structure, interactions and structure formation place this processing technology on an advantaged position to structure sensitive and added value products with tailored functionality.

## 3.3 Novel perspectives in biomolecular structures and functionality

It is now established that MEF, either during OH or applied with the focus of exploiting its nonthermal effects, can effectively modify the structure and functionality of macromolecules. This potentiality opens new perspectives on the development of tailored food-grade biomaterials based in biological macromolecules whose manipulation requires solely physical methods and does not allow the use of any harmful chemical agent or chemical modification. Furthermore, MEF processing is a versatile process, where thermal and electrical parameters can be easily adjusted and act as a control method for the biophysical interactions occurring between the biomolecules and the solvent. The input of MEF processing capacity comes at a decisive time, where the global nutritional supply is changing towards new and sustainable sources. These shifts bring new challenges to the development of functional food systems, with added value associated with nutritional and functional properties, contributing to an adequate nutritional intake and health benefits.

### 3.3.1 Exploring EF technology to tailor biomaterials

The ability of promoting and control macromolecular interactions and suprastructure formation is decisive on the development of functional foods but also other applications such as biomedicine or material science. In this sense, proteins from several different sources can be used as the basis for the development of biomaterials (Choi et al., 2018). The particular relation between protein structure and its function is well established and gains particular importance not only on the mentioned protein-protein interaction and structure formation, but also on the molecular affinities and interactions established within a specific environment (Foegeding, 2015). A particular consequence of these structural properties is the interaction between food proteins and bioactive compounds or protein-based carriers of drugs in biomedicine (Betz et al., 2012; Ramos et al., 2017; Semenova, 2017; Tarhini et al., 2017). The MEF effects in protein interactions, were previously discussed, such as the increase of hydrophobic interaction, conservation of retinol binding in unfolded forms of β-Lg, or higher incorporation of iron in fluid gel systems (Pereira et al., 2017; Rodrigues, Avelar, et al., 2020). The ability of a protein network to entrap or establishing molecular interaction with bioactive compounds or other macromolecules have notorious advantages in functional food systems, but equal potential can be achieved in other areas. The modulation of a solution's viscosity or the viscoelastic properties of hydrogels can equally bring opportunities to develop innovative food structuring agents or biomedical devices, where the viscoelastic properties are controlled by the process and not by the composition. In this context, Rodrigues and coworkers developed a solid-based protein structures by promoting the gelation of BSA and caseinate, followed by solvent removal (Rodrigues et al., 2021). Processing the protein solutions with MEF at 20 and 40 V/cm modified the protein denaturation, aggregation and gelation profiles. The solid structures obtained also presented differentiated structural features, such as higher porosity and different pore geometry, which resulted in altered mechanical properties and interactions with aqueous media such as higher contact angle and swelling capacity. These structures were tested as scaffolds to support the growth on human fibroblast cells, demonstrating an increased ability to support cellular adhesion and increase proliferation.

Another example of current research in innovative protein systems with multiple applications is the development of protein fibrillar systems. Protein-based amyloid fibrils are highly order structures resultant from the self-assembly of proteins and have been the focus of increasing interest due to the parallelism with medical conditions but also recognized as potential in nanotechnology, material and environmental sciences (Cao & Mezzenga, 2019; Knowles & Mezzenga, 2016; Loveday et al., 2017). The highly ordered structures of protein fibrils are achieved by a highly specific aggregation and for most globular proteins a partial unfolding or hydrolysis is required to expose the protein segments susceptible of fibrillization (van der Linden & Venema, 2007). Although the MEF effects in protein fibril formation were not reported yet, the established conformational changes and interaction modulation by MEF action opens interesting perspectives in this field. The reported ability of MEF to interfere with protein unfolding and increase the surface hydrophobicity of proteins (Rodrigues, Avelar, et al., 2020; Subaşı et al., 2021), as well as to influence protein-protein interactions and aggregation may help to tune and potentiate the fibrillation process. A preliminary indication of this potential was given by the work of Pereira and co-workers, where the study of WPI aggregation at pH 3 demonstrated that MEF action resulted in the formation of elongated aggregates resembling amyloid fibrils (Pereira et al., 2016).

The whole potential of MEF in material science is yet to be explored, however the existing reports and recognized potential allow foreseeing promising developments in this field. The perspective to induce and control protein and polysaccharide functionalization solely by physical methods is fundamental to the development of innocuous systems with increased biocompatibility and cleaner production strategies.

### 3.3.2 Health implications

Food structure, from the molecular to the macroscopic level, has naturally implications in consumers' health. The manipulation of food structure at any level can potentially change the eating process, the digestibility, bioavailability and bioaccessibility and ultimately the biochemical interaction of the food components and the human body.

Macrostructural properties have long been on the spotlight of food research due to their close relationship with the product's organoleptic characteristics. However, a renewed interest on the control of food structure has appeared associated with health concerns. An example of this is the growing efforts to develop foods targeting patients with dysphagia (Alsanei & Chen, 2019). Emerging processing technologies have been identified as potential alternatives to modulate food texture through macromolecule modification—e.g., protein structural modifications—without the need to change the formulation (Pure et al., 2021; Sungsinchai et al., 2019). As previously discussed in Section 3.2, MEF processing has the capacity to change the texture profile of protein-based fluids and gels. A general tendency to reduce viscosity and the strength of gel was reported in independent studies; additionally the control of EF intensity and frequency electively contributed to the texture profile of the processed protein systems. All of this places MEF processing as a potential tool to modulate food structure in accordance with the application needs.

Another big trend is the caloric reduction in foods and increased satiety, usually achieved by restricting fat and sugar content and replacing them with proteins and polysaccharides (Campbell et al., 2017; Palzer, 2009). These changes in the composition of foods will naturally result on structural and sensorial changes. Some of the better established processes in fat replacements are in the dairy products, where the incorporation of protein-based formulations such as micro-aggregates and gels have successfully been used to emulate the sensorial attribute of fat (Kew et al., 2020). Furthermore, the physicochemical attributes of the replacement systems effectively influence the structure and texture profile of dairy products (e.g., cheese and yogurt), with structural aspects (such as protein conformation, aggregates size, and hydrophobicity) playing decisive roles in the achievement of the desired properties (Gamlath et al., 2020; Torres et al., 2018). The reported ability to control protein's aggregates size, shape or hydrophobicity as well as starch gelatinization through MEF processing also show their potential application on the design of fat replacements and low caloric density foods.

Ultimately, food will undergo digestion of the GI tract, which defines the satiety, nutritional intake and interactions with the human body such as bioactivity or immune response. Digestion is a complex series or physical and biochemical processes and food structure plays a decisive role on the outcomes of the process. The action of emerging food processing technologies—e.g., PEF and high hydrostatic pressure (HHP)—have demonstrated to affect the digestibility of protein-based foods, such as muscle tissue or vegetable proteins (Bhat, Morton, Mason, Bekhit, & Mungure, 2019; Chian et al., 2019; Gharibzahedi & Smith, 2021). The relatively new approach to food structure modification with resort to MEF has not yet been verified on the digestibility profile of foods. Nonetheless, susceptibility to the digestion of food components is related with the accessibility of GI enzymes to the structure of the food constituents molecules, such as fats, polysaccharides and proteins (Golding, 2019). The structural changes imposed by MEF, as well as the changes in starch gelatinization or protein aggregation and gel crosslinking, can certainly influence their digestibility. The place where the hydrolysis takes place (e.g., stomach, ileum jejunum duodenum) and its extent can also define the bioavailability of amino acids and release of bioactive peptides (Nasri, 2017; Rutherfurd-Markwick, 2012; Wang et al., 2018). Additionally, some food proteins are known for partially resisting GI digestion, such as β-Lg from milk and β-conglycinin from soy, thus contributing to allergenicity issues of these proteins in humans (De Angelis et al., 2017; Macierzanka et al., 2012; Nguyen et al., 2015). MEF action was demonstrated to increase soy protein digestibility, possibly by inducing changes in protein structure and allowing enzymatic action, but also by increasing of trypsin inhibitors inactivation (Shivaji et al., 2021). In fact, a recent work has demonstrated that MEF processing of soy bean protein can reduce its immunoreactivity up to 36% (Pereira et al., 2021). In this study, it was also demonstrated that the immunoreactivity changes were associated with protein structural changes induced by MEF and with electrochemical effects and protein-metal interaction, occurring under low frequencies (at 50 Hz). On a related study, authors have explored MEF action in β-Lg immunoreactivity under different pasteurization protocols (Pereira et al., 2020). Protein structural changes and aggregation profiles were highly dependent on the time-temperature binomials applied, but also on the MEF variables used. These results highlight the potential of MEF processing in controlling protein functionality and the possible implication in the quality of the food products and ultimately in the health and well-being of the population.

## 3.4 Future perspectives

Electrotechnologies are rapidly gaining notoriety in food science and technology research in addition to its industrial applications in food processing. Not only the technological advantages of their utilization—i.e., potential to increase safety, quality, and functionality of food—are driving this change, but also the rise of a new panorama on the public opinion, legislation and industry approach towards the production process. Priority is being given to more environmentally friendly solutions that can contribute to process intensification and reduction of the environmental footprint. Alongside, it is a priority to maximize the product's value by increasing the quality, safety, and health benefits. In this context, electrotechnologies present a unique combination of applicability on the current food production arena, operational advantages in terms of energetic efficiency, maintenance, and resources consumption (De Vries, Mikolajczak, et al., 2018; Pereira & Vicente, 2010) (i.e., fossil fuels, water) and the potential to enhance and differentiate the functionalization and structuring processes.

Particularly, MEF processing holds promising perspectives on the improvement and development of new functionalities through the modification of macromolecules and structuring foods and biomaterials. The increasing literature addressing MEF effect in macromolecule modification settles the basis for the development of strategies to control their functionality with impacts in multiple areas of biotechnology. On the previous sections, it was reviewed and discussed its potential to influence and control structural modifications, interactions and complex formation, aggregation and network formation (see Fig. 3.3).

This establishes MEF processing capacity to intervene in multiple processes and multiple scales and the impact of the MEF action on the particular relationship between food structure, health and well-being. Based on these findings and perspectives, it is only expected that MEF technology will increase its presence in research and industry. Certainly, there are still limiting aspects on the technological implementation and a larger body of research regarding validation is needed to fully access the dynamic and complex behavior of different functionalization protocols, products, and applications. Additionally, operational and technical benefits as well as drawbacks, must be evaluated through a comprehensive approach. To do so and to fully reveal MEF's potentialities and limitations in structuring food systems, along with other potential applications in biotechnology in general, the knowledge gap must be approached on (i) improved understanding of the fundamental EF action on different biomacromolecules and their interaction in complex foods; (ii) an enhanced understanding and distinction between thermal and nonthermal effects associated with MEF as well as their synergy; (iii) synergic or cumulative effects with other food processing methods (e.g., enzymatic processes, fermentation, cooking); (iv) improved knowledge on the previously mentioned effects on the digestion process, bioaccessibility and biochemical interaction with the human body; and (v) analysis of its feasibility, scalability and impact through the life cycle and techno-economical assessment tools. All of this will allow establishing MEF as a bioprocessing tool to control food structure and functionality.

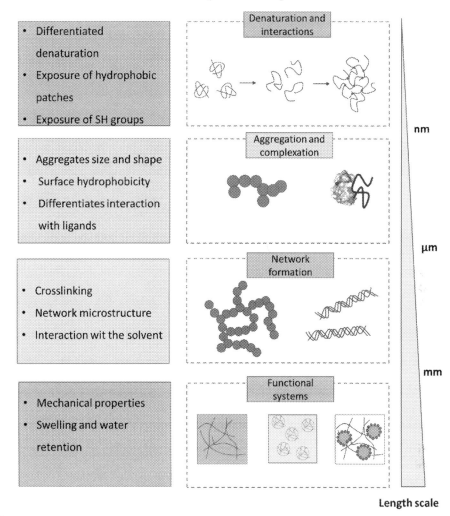

**FIG. 3.3**

MEF effects in structural and functional properties of protein-based systems at different length scales.

## 3.5 Conclusions

Electrotechnologies are now a reality in food and bioprocessing industries. Their implementation brings interesting advantages in processing efficiency, quality and safety of foods, and reduced environmental impact. Additionally, the action of nonthermal effects has the potential to interact with macromolecules such as polysaccharides and proteins impacting structural aspects from the molecular to macroscopic scale. EF effects, particularly during MEF processing and functionalization protocols have been reviewed and related with the potential to control functional properties and structuring food-based systems and other bio-based structures. However, the technical knowledge to achieve a rational design of structured systems for incorporation or development of foods conjugating functionality with health/wellness is still far from being achieved. Furthermore, the development of MEF as a tool to tune food structure is still at an early stage of development and more fundamental knowledge is needed to fully elucidate the technological potentialities and limitations of this approach.

The path to come must be steeled on a systematic and interdisciplinary approach on the several aspects affected by MEF, covering from the fundamental aspects at molecular scale until the interaction and biochemical interaction with the human body. This must be integrated on the farm to fork strategy, allowing the production, processing and ultimately consumption of food with low environmental impact, nutritionally rich and ideally with added benefits on the consumer's health. This is particularly critical at the current time, where the consumption patterns and production strategies are pushing the boundaries of the planetary resources and the society is going through important changes motivated by environmental awareness as well as the fragility and large-scale implications on public health.

## Acknowledgments

This work was supported by the Portuguese Foundation for Science and Technology (FCT) under the scope of the strategic funding of UIDB/04469/2020 unit and by the I&D&I AgriFood XXI project (NORTE-01-0145-FEDER-000041) co-funded by the European Regional Development Fund under the scope of Norte 2020 (Programa Operacional Regional do Norte 2014/2020).

## References

Abaee, A., Mohammadian, M., & Jafari, S. M. (2017). Whey and soy protein-based hydrogels and nano-hydrogels as bioactive delivery systems. *Trends in Food Science & Technology*, *70*, 69–81. https://doi.org/10.1016/j.tifs.2017.10.011.

Alexandre, E. M. C., Pinto, C. A., Moreira, S. A., Pintado, M., Saraiva, J. A., & Galanakis, C. M. (2019). 5. Non-thermal food processing/preservation technologies. In *Saving food* (pp. 141–169). https://doi.org/10.1016/B978-0-12-815357-4.00005-5.

Alsanei, W. A., & Chen, J. (2019). Chapter 19. Food structure development for specific population groups. In *Handbook of food structure development* (pp. 459–479). https://doi.org/10.1039/9781788016155-00459.

An, H. J., & King, J. M. (2007). Thermal characteristics of ohmically heated rice starch and rice flours. *Journal of Food Science*, *72*(1), C084–C088. https://doi.org/10.1111/j.1750-3841.2006.00239.x.

Anderson, A. K., & Finkelstein, R. (1919). A study of the electro-pure process of treating milk. *Journal of Dairy Science*, *2*(5), 374–406. https://doi.org/10.3168/jds.S0022-0302(19)94338-4.

Bekard, I., & Dunstan, D. E. (2014). Electric field induced changes in protein conformation. *Soft Matter*, *10*(3), 431–437. https://doi.org/10.1039/c3sm52653d.

Betz, M., García-González, C. A., Subrahmanyam, R. P., Smirnova, I., & Kulozik, U. (2012). Preparation of novel whey protein-based aerogels as drug carriers for life science applications. *The Journal of Supercritical Fluids*, *72*, 111–119. https://doi.org/10.1016/j.supflu.2012.08.019.

Bhat, Z. F., Morton, J. D., Mason, S. L., & Bekhit, A. E. D. A. (2019). Pulsed electric field improved protein digestion of beef during in-vitro gastrointestinal simulation. *LWT*, *102*(December 2018), 45–51. https://doi.org/10.1016/j.lwt.2018.12.013.

Bhat, Z. F., Morton, J. D., Mason, S. L., Bekhit, A. E.-D. A., & Mungure, T. E. (2019). Pulsed electric field: Effect on in-vitro simulated gastrointestinal protein digestion of deer Longissimus dorsi. *Food Research International*, *120*(July 2018), 793–799. https://doi.org/10.1016/j.foodres.2018.11.040.

Brochier, B., & Domeneghini, G. (2016). Influence of moderate electric fi eld on inactivation kinetics of peroxidase and polyphenol oxidase and on phenolic compounds of sugarcane juice treated by ohmic heating. *LWT - Food Science and Technology*, *74*, 396–403. https://doi.org/10.1016/j.lwt.2016.08.001.

Campbell, C. L., Wagoner, T. B., & Foegeding, E. A. (2017). Designing foods for satiety: The roles of food structure and oral processing in satiation and satiety. *Food Structure*, *13*, 1–12. https://doi.org/10.1016/j.foostr.2016.08.002.

Cao, Y., & Mezzenga, R. (2019). Food protein amyloid fibrils: Origin, structure, formation, characterization, applications and health implications. *Advances in Colloid and Interface Science*, *269*, 334–356. https://doi.org/10.1016/j.cis.2019.05.002.

Caruggi, N., Lucisano, M., Feyissa, A. H., Rahimi Yazdi, S., & Mohammadifar, M. A. (2019). Effect of ohmic heating on the formation and texture of acid milk gels. *Food Biophysics*, *14*(3), 249–259. https://doi.org/10.1007/s11483-019-09578-y.

Castro, I., Macedo, B., Teixeira, J. A., & Vicente, A. A. (2006). The effect of electric field on important food-processing enzymes: Comparison of inactivation kinetics under conventional and ohmic heating. *Journal of Food Science*, *69*(9), C696–C701. https://doi.org/10.1111/j.1365-2621.2004.tb09918.x.

Cha, Y.-H. (2011). Effect of ohmic heating on characteristics of heating denaturation of soybean protein. *The Korean Journal of Food and Nutrition*, *24*(4), 740–745. https://doi.org/10.9799/ksfan.2011.24.4.740.

Cha, Y.-H. (2012). Effect of ohmic heating at subgelatinization temperatures on thermal-property of potato starch. *The Korean Journal of Food and Nutrition*, *25*(4), 1068–1074. https://doi.org/10.9799/ksfan.2012.25.4.1068.

Cha, Y.-H. (2015). Effect of ohmic heating on external and internal structure of starches. *The Korean Journal of Food and Nutrition*, *28*(1), 126–133. https://doi.org/10.9799/ksfan.2015.28.1.126.

Chian, F. M., Kaur, L., Oey, I., Astruc, T., Hodgkinson, S., & Boland, M. (2019). Effect of pulsed electric fields (PEF) on the ultrastructure and in vitro protein digestibility of bovine longissimus thoracis. *LWT*, *103*(October 2018), 253–259. https://doi.org/10.1016/j.lwt.2019.01.005.

Choi, S. M., Chaudhry, P., Zo, S. M., Han, S. S., Chun, H. J., Park, C. H., Kwon, I. K., & Khang, G. (2018). Advances in protein-based materials: from origin to novel biomaterials. In *Cutting-edge enabling technologies for regenerative medicine* (pp. 161–210).

Coelho, C. C. S. S., Cerqueira, M. A., Pereira, R. N., Pastrana, L. M., Freitas-Silva, O., Vicente, A. A., Cabral, L. M. C. C., & Teixeira, J. A. (2017). Effect of moderate electric fields in the properties of starch and chitosan films reinforced with microcrystalline cellulose. *Carbohydrate Polymers*, *174*(July 2017), 1181–1191. https://doi.org/10.1016/j.carbpol.2017.07.007.

De Angelis, E., Pilolli, R., Bavaro, S. L., & Monaci, L. (2017). Insight into the gastro-duodenal digestion resistance of soybean proteins and potential implications for residual immunogenicity. *Food & Function*, *8*(4), 1599–1610. https://doi.org/10.1039/c6fo01788f.

de Vries, H., Axelos, V. M. A., Sarni-Manchado, P., & O'Donohue, M. (2018). Meeting new challenges in food science technology: The development of complex systems approach for food and biobased research. *Innovative Food Science and Emerging Technologies*, *46*(April), 1–6. https://doi.org/10.1016/j.ifset.2018.04.004.

De Vries, H., Mikolajczak, M., Salmon, J.-M., Abecassis, J., Chaunier, L., Guessasma, S., Lourdin, D., Belhabib, S., Leroy, E., & Trystram, G. (2018). Small-scale food process engineering—Challenges and perspectives. *Innovative Food Science & Emerging Technologies*, *46*, 122–130. https://doi.org/10.1016/j.ifset.2017.09.009.

Fasolin, L. H., Rodrigues, R. M., & Pereira, R. N. (2021). Effects of electric fields and electromagnetic wave on food structure and functionality. In *Food structure and functionality* (pp. 95–113). https://doi.org/10.1016/B978-0-12-821453-4.00009-0.

Foegeding, E. A. (2015). Food protein functionality—A new model. *Journal of Food Science*, *80*(12), C2670–C2677. https://doi.org/10.1111/1750-3841.13116.

Fowler, M. R., & Park, J. W. (2015). Effect of salmon plasma protein on Pacific whiting surimi gelation under various ohmic heating conditions. *LWT - Food Science and Technology*, *61*(2), 309–315. https://doi.org/10.1016/j.lwt.2014.12.049.

Furtado, G., Pereira, R. N. C., Vicente, A. A., & Cunha, R. L. (2018). Cold gel-like emulsions of lactoferrin subjected to ohmic heating. *Food Research International*, *103*, 371–379. https://doi.org/10.1016/j.foodres.2017.10.061.

Gamlath, C. J., Leong, T. S. H., Ashokkumar, M., & Martin, G. J. O. (2020). Incorporating whey protein aggregates produced with heat and ultrasound treatment into rennet gels and model non-fat cheese systems. *Food Hydrocolloids*, *109*, 106103. https://doi.org/10.1016/j.foodhyd.2020.106103.

Gharibzahedi, S. M. T., & Smith, B. (2021). Effects of high hydrostatic pressure on the quality and functionality of protein isolates, concentrates, and hydrolysates derived from pulse legumes: A review. *Trends in Food Science & Technology*, *107*, 466–479. https://doi.org/10.1016/j.tifs.2020.11.016.

Golding, M. (2019). Exploring and exploiting the role of food structure in digestion. In *Interdisciplinary approaches to food digestion* (pp. 81–128). https://doi.org/10.1007/978-3-030-03901-1_5.

Içier, F., Yildiz, H., & Baysal, T. (2008). Polyphenoloxidase deactivation kinetics during ohmic heating of grape juice. *Journal of Food Engineering*, *85*(3), 410–417. https://doi.org/10.1016/j.jfoodeng.2007.08.002.

Jaeger, H., Janositz, A., & Knorr, D. (2010). The Maillard reaction and its control during food processing. The potential of emerging technologies. *Vieillissement*, *58*(3), 207–213. https://doi.org/10.1016/j.patbio.2009.09.016.

Jaeger, H., Roth, A., Toepfl, S., Holzhauser, T., Engel, K. H., Knorr, D., Vogel, R. F., Bandick, N., Kulling, S., Heinz, V., & Steinberg, P. (2016). Opinion on the use of ohmic heating for the treatment of foods. *Trends in Food Science and Technology*, *55*, 84–97. https://doi.org/10.1016/j.tifs.2016.07.007.

Jakób, A., Bryjak, J., Wójtowicz, H., Illeová, V., Annus, J., Polakovič, M., & Polakovic, M. (2010). Inactivation kinetics of food enzymes during ohmic heating. *Food Chemistry*, *123*(2), 369–376. https://doi.org/10.1016/j.foodchem.2010.04.047.

Kaur, N., & Singh, A. K. (2016). Ohmic heating: Concept and applications—A review. *Critical Reviews in Food Science and Nutrition*, *56*(14), 2338–2351. https://doi.org/10.1080/10408398.2013.835303.

Kew, B., Holmes, M., Stieger, M., & Sarkar, A. (2020). Review on fat replacement using protein-based microparticulated powders or microgels: A textural perspective. *Trends in Food Science & Technology*, *106*, 457–468. https://doi.org/10.1016/j.tifs.2020.10.032.

Knirsch, M. C., Alves dos Santos, C., de Oliveira, M., Soares Vicente, A. A., & Vessoni Penna, T. C. (2010). Ohmic heating - a review. *Trends in Food Science and Technology*, *21*(9), 436–441. https://doi.org/10.1016/j.tifs.2010.06.003.

Knowles, T. P. J., & Mezzenga, R. (2016). Amyloid fibrils as building blocks for natural and artificial functional materials. *Advanced Materials*, *28*(31), 6546–6561. https://doi.org/10.1002/adma.201505961.

Kotnik, T., Frey, W., Sack, M., Haberl Meglič, S., Peterka, M., & Miklavčič, D. (2015). Electroporation-based applications in biotechnology. *Trends in Biotechnology*, *33*(8), 480–488. https://doi.org/10.1016/j.tibtech.2015.06.002.

Lei, L., Zhi, H., Xiujin, Z., Takasuke, I., & Zaigui, L. (2007). Effects of different heating methods on the production of protein-lipid film. *Journal of Food Engineering*, *82*(3), 292–297. https://doi.org/10.1016/j.jfoodeng.2007.02.030.

Li, Z., Fan, Y., & Xi, J. (2019). Recent advances in high voltage electric discharge extraction of bioactive ingredients from plant materials. *Food Chemistry*, *277*, 246–260. https://doi.org/10.1016/j.foodchem.2018.10.119.

Liu, Z.-S., & Chang, S. K. C. (2007). Soymilk viscosity as influenced by heating methods and soybean varieties. *Journal of Food Processing and Preservation*, *31*(3), 320–333. https://doi.org/10.1111/j.1745-4549.2007.00128.x.

Loveday, S. M., Anema, S. G., & Singh, H. (2017). β-Lactoglobulin nanofibrils: The long and the short of it. *International Dairy Journal*, *67*, 35–45. https://doi.org/10.1016/j.idairyj.2016.09.011.

Ma, H., Jia, J., Liu, D., Jia, J., Liu, D., & Ma, H. (2019). *Advances in food processing technology*. https://doi.org/10.1007/978-981-13-6451-8.

Machado, L. F., Pereira, R. N., Martins, R. C., Teixeira, J. A., & Vicente, A. (2010). Moderate electric fields can inactivate Escherichia coli at room temperature. *Journal of Food Engineering*, *96*(4), 520–527. https://doi.org/10.1016/j.jfoodeng.2009.08.035.

Macierzanka, A., Böttger, F., Lansonneur, L., Groizard, R., Jean, A. S., Rigby, N. M., Cross, K., Wellner, N., & MacKie, A. R. (2012). The effect of gel structure on the kinetics of simulated gastrointestinal digestion of bovine β-lactoglobulin. *Food Chemistry*, *134*(4), 2156–2163. https://doi.org/10.1016/j.foodchem.2012.04.018.

Makroo, H. A., Rastogi, N. K., & Srivastava, B. (2017). Enzyme inactivation of tomato juice by ohmic heating and its effects on physico-chemical characteristics of concentrated tomato paste. *Journal of Food Process Engineering*, *40*(3), e12464-n/a. https://doi.org/10.1111/jfpe.12464.

McClements, D. J., Zou, L., Zhang, R., Salvia-Trujillo, L., Kumosani, T., & Xiao, H. (2015). Enhancing nutraceutical performance using excipient foods: Designing food structures and compositions to increase bioavailability. *Comprehensive Reviews in Food Science and Food Safety*, *14*(6), 824–847. https://doi.org/10.1111/1541-4337.12170.

Mirmoghtadaie, L., Shojaee Aliabadi, S., & Hosseini, S. M. (2016). Recent approaches in physical modification of protein functionality. *Food Chemistry*, *199*, 619–627. https://doi.org/10.1016/j.foodchem.2015.12.067.

Moreira, T. C. P., Pereira, R. N., Vicente, A. A., & da Cunha, R. L. (2019). Effect of Ohmic heating on functionality of sodium caseinate—A relationship with protein gelation. *Food Research International*, *116*(April 2018), 628–636. https://doi.org/10.1016/j.foodres.2018.08.087.

Nasri, M. (2017). Protein hydrolysates and biopeptides: production, biological activities, and applications in foods and health benefits. A review. In *Vol. 81. Advances in food and nutrition research* (1st ed.). https://doi.org/10.1016/bs.afnr.2016.10.003.

Nguyen, T. T. P., Bhandari, B., Cichero, J., & Prakash, S. (2015). Gastrointestinal digestion of dairy and soy proteins in infant formulas: An in vitro study. *Food Research International*, *76*, 348–358. https://doi.org/10.1016/j.foodres.2015.07.030.

Palzer, S. (2009). Food structures for nutrition, health and wellness. *Trends in Food Science and Technology*, *20*(5), 194–200. https://doi.org/10.1016/j.tifs.2009.02.005.

Pereira, R. N., & Vicente, A. A. (2010). Environmental impact of novel thermal and non-thermal technologies in food processing. *Food Research International*, *43*(7), 1936–1943. https://doi.org/10.1016/j.foodres.2009.09.013.

Pereira, R. N., Costa, J., Rodrigues, R., Villa, C., Machado, L., Mafra, I., & Vicente, A. (2020). Effects of ohmic heating on the immunoreactivity of β-lactoglobulin—A relationship towards structural aspects. *Food & Function*, *11*(5), 4002–4013. https://doi.org/10.1039/C9FO02834J.

Pereira, R. N., Rodrigues, R. M., Altinok, E., Ramos, Ó. L., Xavier Malcata, F., Maresca, P., Ferrari, G., Teixeira, J. A., & Vicente, A. A. (2017). Development of iron-rich whey protein hydrogels following application of ohmic heating—Effects of moderate electric fields. *Food Research International*, *99*(January), 435–443. https://doi.org/10.1016/j.foodres.2017.05.023.

Pereira, R. N., Rodrigues, R. M., Machado, L., Ferreira, S., Costa, J., Villa, C., Barreiros, M. P., Mafra, I., Teixeira, J. A., & Vicente, A. A. (2021). Influence of ohmic heating on the structural and immunoreactive properties of soybean proteins. *LWT*, *148*(May), 111710. https://doi.org/10.1016/j.lwt.2021.111710.

Pereira, R. N., Rodrigues, R. M., Ramos, Ó. L., Xavier Malcata, F., Teixeira, J. A., & Vicente, A. A. A. (2016). Production of whey protein-based aggregates under ohmic heating. *Food and Bioprocess Technology*, *9*(4), 576–587. https://doi.org/10.1007/s11947-015-1651-4.

Pereira, R. N., Souza, B. W. S., Cerqueira, M. A., Teixeira, A., Teixeira, J. A., & Vicente, A. A. A. (2010). Effects of electric fields on protein unfolding and aggregation: Influence on edible films formation. *Biomacromolecules*, *11*(11), 2912–2918. https://doi.org/10.1021/bm100681a.

Pereira, R. N., Teixeira, J. A. J. A., & Vicente, A. A. A. (2011). Exploring the denaturation of whey proteins upon application of moderate electric fields: A kinetic and thermodynamic study. *Journal of Agricultural and Food Chemistry*, *59*(21), 11589–11597. https://doi.org/10.1021/jf201727s.

Perusko, M., Al-Hanish, A., Cirkovic Velickovic, T., & Stanic-Vucinic, D. (2015). Macromolecular crowding conditions enhance glycation and oxidation of whey proteins in ultrasound-induced Maillard reaction. *Food Chemistry*, *177*, 248–257. https://doi.org/10.1016/j.foodchem.2015.01.042.

Pure, A. E., Yarmand, M. S., Farhoodi, M., & Adedeji, A. (2021). Microwave treatment to modify textural properties of high protein gel applicable as dysphagia food. *Journal of Texture Studies*, May, 1–9. https://doi.org/10.1111/jtxs.12611.

Ramaswamy, H. S., Marcotte, M., & Sudhir Sastry, K. A. (2016). *Ohmic heating for food processing*. Retrieved from: http://www.crcnetbase.com/doi/abs/10.1201/b12112-22.

Ramos, O. L., Pereira, R. N., Martins, A., Rodrigues, R., Fuciños, C., Teixeira, J. A., Pastrana, L., Malcata, F. X., & Vicente, A. A. (2017). Design of whey protein nanostructures for incorporation and release of nutraceutical compounds in food. *Critical Reviews in Food Science and Nutrition*, *57*(7), 1377–1393. https://doi.org/10.1080/10408398.2014.993749.

Rocha, C. M. R. R., Genisheva, Z., Ferreira-Santos, P., Rodrigues, R., Vicente, A. A., Teixeira, J. A., & Pereira, R. N. (2018). Electric field-based technologies for valorization of bioresources. *Bioresource Technology*, *254* (November 2017), 325–339. https://doi.org/10.1016/j.biortech.2018.01.068.

Rodrigues, R. M., Martins, A. J., Ramos, O. L., Malcata, F. X., Teixeira, J. A., Vicente, A. A., & Pereira, R. N. (2015). Influence of moderate electric fields on gelation of whey protein isolate. *Food Hydrocolloids*, *43*, 329–339. https://doi.org/10.1016/j.foodhyd.2014.06.002.

Rodrigues, R. M., Avelar, Z., Vicente, A. A., Petersen, S. B., & Pereira, R. N. (2020). Influence of moderate electric fields in β-lactoglobulin thermal unfolding and interactions. *Food Chemistry*, *304*(March 2019), 125442. https://doi.org/10.1016/j.foodchem.2019.125442.

Rodrigues, R. M., Fasolin, L. H., Avelar, Z., Petersen, S. B., Vicente, A. A., & Pereira, R. N. (2020). Effects of moderate electric fields on cold-set gelation of whey proteins—From molecular interactions to functional properties. *Food Hydrocolloids*, *101*(November 2019), 105505. https://doi.org/10.1016/j.foodhyd.2019.105505.

Rodrigues, R. M., Genisheva, Z., Rocha, C. M. R., Teixeira, J. A., Vicente, A. A., & Pereira, R. N. (2019). Ohmic heating for preservation, transformation, and extraction. In *Green food processing techniques* (pp. 159–191). https://doi.org/10.1016/B978-0-12-815353-6.00006-9.

Rodrigues, R. M., Pereira, R. N., Vicente, A. A., Cavaco-Paulo, A., & Ribeiro, A. (2021). Ohmic heating as a new tool for protein scaffold engineering. *Materials Science and Engineering: C*, *120*, 111784. https://doi.org/10.1016/j.msec.2020.111784.

Rodrigues, R. M., Vicente, A. A., Petersen, S. B., & Pereira, R. N. (2019). Electric field effects on β-lactoglobulin thermal unfolding as a function of pH—Impact on protein functionality. *Innovative Food Science & Emerging Technologies*, *52*(October 2018), 1–7. https://doi.org/10.1016/j.ifset.2018.11.010.

Rutherfurd-Markwick, K. J. (2012). Food proteins as a source of bioactive peptides with diverse functions. *British Journal of Nutrition*, *108*(Suppl. 2), 149–157. https://doi.org/10.1017/S000711451200253X.

Samaranayake, C. P., & Sastry, S. K. (2016a). Effect of moderate electric fields on inactivation kinetics of pectin methylesterase in tomatoes: The roles of electric field strength and temperature. *Journal of Food Engineering*, *186*(Suppl. C), 17–26. https://doi.org/10.1016/j.jfoodeng.2016.04.006.

Samaranayake, C. P., & Sastry, S. K. (2016b). Effects of controlled-frequency moderate electric fields on pectin methylesterase and polygalacturonase activities in tomato homogenate. *Food Chemistry*, *199*(Suppl. C), 265–272. https://doi.org/10.1016/j.foodchem.2015.12.010.

Samaranayake, C. P., & Sastry, S. K. (2018). In-situ activity of α-amylase in the presence of controlled-frequency moderate electric fields. *LWT, 90*(October 2017), 448–454. https://doi.org/10.1016/j.lwt.2017.12.053.

Sastry, S. (2008). Ohmic heating and moderate electric field processing. *Food Science and Technology International, 14*(5), 419–422. https://doi.org/10.1177/1082013208098813.

Semenova, M. (2017). Protein-polysaccharide associative interactions in the design of tailor-made colloidal particles. *Current Opinion in Colloid and Interface Science, 28*, 15–21. https://doi.org/10.1016/j.cocis.2016.12.003.

Shahidi, F., Ho, C.-T., van Chuyen, N., Shahidi, F., Ho, C.-T., & van Chuyen, N. (1998). Process-induced chemical changes in food. In *Vol. 434. Advances in experimental medicine and biology*. https://doi.org/10.1007/978-1-4899-1925-0.

Shimoyamada, M., Itabashi, Y., Sugimoto, I., Kanauchi, M., Ishida, M., Tsuzuki, K., Egusa, S., & Honda, Y. (2015). Characterization of soymilk prepared by ohmic heating and the effects of voltage applied. *Food Science and Technology Research, 21*(3), 439–444. https://doi.org/10.3136/fstr.21.439.

Shivaji, J. B., Mohan, R. J., Rawson, A., & Pare, A. (2021). Effect of ohmic heating on protein characteristics and beany flavor of soymilk. *International Journal of Chemical Studies, 9*(1), 248–252. https://doi.org/10.22271/chemi.2021.v9.i1e.11658.

Souza, B. W. S., Cerqueira, M. A., Martins, J. T., Casariego, A., Teixeira, J. A., & Vicente, A. A. (2010). Influence of electric fields on the structure of chitosan edible coatings. *Food Hydrocolloids, 24*(4), 330–335. https://doi.org/10.1016/j.foodhyd.2009.10.011.

Souza, B. W. S. S., Cerqueira, M. A., Casariego, A., Lima, A. M. P. P., Teixeira, J. A., & Vicente, A. A. (2009). Effect of moderate electric fields in the permeation properties of chitosan coatings. *Food Hydrocolloids, 23*(8), 2110–2115. https://doi.org/10.1016/j.foodhyd.2009.03.021.

Subaşı, B. G., Jahromi, M., Casanova, F., Capanoglu, E., Ajalloueian, F., & Mohammadifar, M. A. (2021). Effect of moderate electric field on structural and thermo-physical properties of sunflower protein and sodium caseinate. *Innovative Food Science & Emerging Technologies, 67*, 102593. https://doi.org/10.1016/j.ifset.2020.102593.

Sungsinchai, S., Niamnuy, C., Wattanapan, P., Charoenchaitrakool, M., & Devahastin, S. (2019). Texture modification technologies and their opportunities for the production of dysphagia foods: A review. *Comprehensive Reviews in Food Science and Food Safety, 18*(6), 1898–1912. https://doi.org/10.1111/1541-4337.12495.

Suscillon, C., Velev, O. D., & Slaveykova, V. I. (2013). Alternating current-dielectrophoresis driven on-chip collection and chaining of green microalgae in freshwaters. *Biomicrofluidics, 7*(2), 024109. https://doi.org/10.1063/1.4801870.

Tadpitchayangkoon, P., Park, J. W., & Yongsawatdigul, J. (2012). Gelation characteristics of tropical surimi under water bath and ohmic heating. *LWT - Food Science and Technology, 46*(1), 97–103. https://doi.org/10.1016/j.lwt.2011.10.020.

Takashima, S., & Schwan, H. P. (1985). Alignment of microscopic particles in electric fields and its biological implications. *Biophysical Journal, 47*(4), 513–518. https://doi.org/10.1016/S0006-3495(85)83945-X.

Tarhini, M., Greige-Gerges, H., & Elaissari, A. (2017). Protein-based nanoparticles: From preparation to encapsulation of active molecules. *International Journal of Pharmaceutics, 522*(1–2), 172–197. https://doi.org/10.1016/j.ijpharm.2017.01.067.

Tinoco, A., Rodrigues, R. M., Machado, R., Pereira, R. N., Cavaco-Paulo, A., & Ribeiro, A. (2020). Ohmic heating as an innovative approach for the production of keratin films. *International Journal of Biological Macromolecules, 150*, 671–680. https://doi.org/10.1016/j.ijbiomac.2020.02.122.

Torres, I. C., Amigo, J. M., Knudsen, J. C., Tolkach, A., Mikkelsen, B.Ø., & Ipsen, R. (2018). Rheology and microstructure of low-fat yoghurt produced with whey protein microparticles as fat replacer. *International Dairy Journal, 81*, 62–71. https://doi.org/10.1016/j.idairyj.2018.01.004.

van der Linden, E., & Venema, P. (2007). Self-assembly and aggregation of proteins. *Current Opinion in Colloid and Interface Science, 12*(4–5), 158–165. https://doi.org/10.1016/j.cocis.2007.07.010.

Vorobiev, E., & Lebovka, N. (2016). Extraction from foods and biomaterials enhanced by pulsed electric energy. In *Innovative food processing technologies: Extraction, separation, component modification and process intensification*. https://doi.org/10.1016/B978-0-08-100294-0.00002-X.

Wang, X., Qiu, N., & Liu, Y. (2018). Effect of different heat treatments on in vitro digestion of egg white proteins and identification of bioactive peptides in digested products. *Journal of Food Science*, *83*(4), 1140–1148. https://doi.org/10.1111/1750-3841.14107.

Wei, Z., & Huang, Q. (2019). Assembly of protein-polysaccharide complexes for delivery of bioactive ingredients: A perspective paper. *Journal of Agricultural and Food Chemistry*, *67*(5), 1344–1352. https://doi.org/10.1021/acs.jafc.8b06063.

Zhao, W., Tang, Y., Lu, L., Chen, X., & Li, C. (2014). Review: Pulsed electric fields processing of protein-based foods. *Food and Bioprocess Technology*, *7*(1), 114–125. https://doi.org/10.1007/s11947-012-1040-1.

Zhu, F. (2018). Modifications of starch by electric field based techniques. *Trends in Food Science & Technology*, *75*, 158–169. https://doi.org/10.1016/j.tifs.2018.03.011.

# CHAPTER 4

# Encapsulation and colloidal systems as a way to deliver functionality in foods

Cristian Dima[a], Elham Assadpour[b,c], and Seid Mahdi Jafari[b,c,d]

[a]Faculty of Food Science and Engineering, "Dunarea de Jos" University of Galati, Galati, Romania, [b]Department of Food Materials and Process Design Engineering, Gorgan University of Agricultural Sciences and Natural Resources, Gorgan, Iran, [c]Nutrition and Bromatology Group, Analytical and Food Chemistry Department, Faculty of Food Science and Technology, University of Vigo, Ourense, Spain, [d]College of Food Science and Technology, Hebei Agricultural University, Baoding, China

## 4.1 Introduction

With the development of human civilizations, food has been required to have new attributes that meet the increasingly complex and diverse requirements of consumers. People have understood that "eating" is no longer just a matter of survival, but has become a matter of health, well-being and last but not least, a matter of education and culture. Unfortunately, food has become a national security issue and one of the most acute problems of globalist politics today. In this context, food is no longer considered a simple product resulting from man's friendly relationship with nature, but has become a complex system, obtained through laborious research by scientists in different fields (e.g., engineering, chemistry, physics, biology, medicine). An eloquent example is the development of nanotechnologies that have changed many paradigms in all fields of activity, including in the agri-food sector (Bajpai et al., 2018).

The food industry has benefited from the performance of nanotechnologies, developing the concept of nanofood. This concept refers to foods that, in their production, processing and packaging, nanotechnologies and nano-materials have been used in order to improve their quality, safety and nutritional value (Assadpour et al., 2020; Sekhon, 2010). Nanotechnologies have allowed the adoption of new axes for the development of the agri-food sectors, such as:

- design of bionanomaterials, such as bionanopesticides, bionanofertilizers, bionanostimulators, etc., which contribute to increasing plant and animal production;
- design and preparation of innovative colloidal nanobiosystems for the delivery of nutraceuticals, able to contribute to increasing the health benefits and quality and safety of food;
- modification of food micro- and nanostructures;

- manufacture of intelligent, active and edible packaging, based on bionanomaterials, which ensures a high protection of food;
- design of high-performance nanosensors capable of detecting traces of contaminants, mycotoxins and microorganisms in food (Handford et al., 2014).

Interdisciplinary research in the field of food science led to the definition of new concepts such as: functional foods, nutraceuticals, dietary supplements, medical foods and food excipients (Dima et al., 2020a; Dima et al., 2020b; He et al., 2019). Although the term functional food is considered a marketing term, several organizations have defined and regulated the concept of functional food (Arihara, 2014; Bigliardi & Galati, 2013; Crowe-White & Francis, 2013). Thus, the Academy of Nutrition and Dietetics has defined functional foods "as whole foods along with fortified, enriched, or enhanced foods that have a potentially beneficial effect on health when consumed as part of a varied diet on a regular basis at effective levels" while the European Commission distinguishes between functional foods and dietary supplements: a functional food is "a food that beneficially affects one or more target functions in the body, beyond adequate nutritional effects, in a way that is relevant to either an improved state of health and well-being and/or reduction of risk of disease. It is part of a normal food pattern. It is not a pill, a capsule or any form of dietary supplement". Also, functional foods are not identified with nutraceuticals because the term nutraceuticals refer to any bioactive components that manifest their health benefits delivered primarily as dietary supplements, while functional foods refer exclusively to foods, either in the form of fruits and vegetables rich in biocomponents, either as food matrices fortified with biocomponents (Crowe-White & Francis, 2013).

Functional foods are not medical foods either. According to the Food and Drug Administration (FDA), a medical food is defined as "a food which is formulated to be consumed or administered enterally under the supervision of a physician and which is intended for the specific dietary management of a disease or condition for which distinctive nutritional requirements, based on recognized scientific principles, are established by medical evaluation" (Lewis et al., 2019). These foods are prepared and delivered according to special regulations and are recommended for vulnerable people such as infants and the elderly or people with diseases that require special nutrition. Medical foods contain all the nutrients except the disease-causing component, such as phenyl alanine for people with phenylketonuria.

In order to increase the bioavailability of some biocomponents, food excipients are used. They are consumed either together with a functional food rich in biocomponents, such as fruits, vegetables, nuts, seeds, grains, or together with dietary supplements in the form of capsules, pills, or syrup. For example, lettuce, carrots or tomatoes are consumed in the presence of vegetable oil (Martínez-Huélamo et al., 2015; Nagao et al., 2013). Lipids contribute both to the extraction of hydrophobic bioactives from plant cells and to their solubilization in mixed micelles consisting of triglycerides, monoglycerides and free fatty acids (Salvia-Trujillo et al., 2017). Also, some dairy products, such as yogurt, can be eaten with fruit or nutraceuticals (McClements & Xiao, 2014).

The development of functional foods and other types of nutrient biosystems defined above, was achieved in correlation with the development of research in the field of food colloids. Science of colloids togheter soft matter has provided food scientists with laws, principles, and information useful for understanding the physicochemical properties of food, such as stability and texture. It was also possible to explain the interphase phenomena that play an important role in food processing (e.g., adsorption, capillarity, surface tension), as well as the mechanisms involved in food digestion and the bioavailability of biocomponents in food. Colloid science has played a key role in the use of classical colloidal systems (emulsions, microemulsions, gels) as nutraceutical delivery systems or in the development of innovative colloidal systems, such as liposomes, niosomes, hexosomes, solid lipid nanoparticles (SLNs), mixed nanoparticles (NPs), nanoclusters, coated NPs and embedded NPs that can ensure the encapsulation, protection and controlled release of biocomponents, helping to improve the nutritional, sensory and health benefits of food (McClements, 2017a).

Research on the encapsulation of bioactive compounds in various colloidal delivery systems (CDSs) with applications in the agri-food, pharmaceutical, cosmetic and other industrial fields had developed enormously in the last ten years, when over 70,000 scientific papers and patents were published. The results of this work highlighted those studies on colloidal micro/nanobiosystems loaded with biocomponents used in food fortification, have made an important contribution to:

- improvement and diversification of techniques for encapsulating bioactive compounds;
- development of new biomaterials and capitalization of biopolymers from by-products as encapsulation materials;
- improving the physicochemical properties of biocomponents (e.g., solubility, chemical stability);
- increasing shelf life and reducing transport and storage costs for food additives and nutraceuticals;
- improving the bioaccessibility and bioavailability of bioactive compounds;
- improving the nutritional, sensory and health benefits of functional foods;
- reducing the risk of toxicity and the impact of colloidal micro/nanobiosystems on the environment.

This chapter provides an overview of the principles and factors involved in the design of CDSs for food ingredients and nutraceuticals in order to increase the quality and health benefits of food. The physicochemical characteristics of micro/nanoparticles and their influence on the functionality of biocomponents incorporated into food matrices and their impact on digestion, absorption and metabolism are also discussed.

## 4.2 Nutraceutical encapsulation as food quality improvement strategy

Food functionalization is done by incorporating the active components in the food matrices. The incorporation of biocomponents into food is often a challenge due to its low solubility, low chemical stability, interaction with other substances present in the food, specific odor and taste that could alter the quality of the food. To reduce these shortcomings, the researchers investigated the possibility of using CDSs for active components. There is a wide variety of CDSs that ensure the encapsulation, protection, transport and controlled release of active components, described in detail in various works (Aditya et al., 2017; Cerqueira et al., 2014; McClements, 2015).

Encapsulation is an effective method of functionalizing food ingredients and nutraceuticals, helping to improve the nutritional, sensory and health attributes of food. Encapsulation is the entrapping of a solid, liquid or gaseous substance called an encapsulated substance (active, core, payload, fill) into another substance, called an encapsulating material (shell material, wall material, carrier material, coating material), in the form of colloidal systems with different structure and dimensions, in the range of micro- or nanometers. Some CDSs are in the liquid state and have a complex structure, in which bioactive compounds are incorporated into liquid colloidal particles, dispersed in the liquid external phase, such as emulsions, nanoemulsions, microemulsions, double emulsions, vesicles, lipids, micelles, among others (Robert et al., 2019). Other CDSs are in the solid state, such as

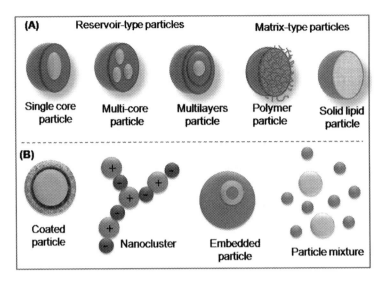

**FIG. 4.1**

Schematic representation of colloidal particles (A) and complex structured colloidal delivery systems (B).

micro/nanoparticles. Particles loaded with bioactive components from CDSs have different shapes, structures and sizes, which differentiate them and randomly classify them into different groups (Fig. 4.1). Thus, from a morphological point of view, the following types of particles are differentiated:

- reservoir-type particles, having in their structure one or more homogeneous nuclei loaded with bioactive substances, covered with one or more layers of encapsulating material (shel material);
- matrix-type particles, also called micro- or nanospheres, whose morphology is homogeneous or heterogeneous, consisting either of solidified lipids (SLN, NLC) or of a three-dimensional polymer network (polymer particles).

Size is an important feature that differentiates particles from CDSs into: microparticles (100 nm < $d$ < 1000 μm) and NPs (10 < $d$ < 100 nm). Some authors classified CDSs according to their complexity and grouped them into: simple CDSs and complex structured CDSs (Fig. 4.1) (McClements, 2017a; McClements, 2018a). The first group included classical colloidal systems, such as emulsions, nanoemulsions, microemulsions, liposomes, polymer and hydrogel particles, while the second group included complex CDSs obtained by coating, embedding, clustering and mixing of simple CDSs. These operations modulated the functional properties of active delivery systems, such as: surface charge, hydrophobicity, physical stability, release rate, rheological properties, flow properties, as well as digestibility and bioavailability of bioactive compounds, among others.

Thus, in order to modify the surface characteristics of micro/nanoparticles, most authors used the Layer-by-Layer (LBL) technique. By this technique the electrically charged colloidal particles are coated with polyelectrolytes with electric charge opposite to the surface charge of the particles, such as polycations (e.g., chitosan), polyanions (e.g., alginate, carrageenan, pectin, hyaluronic acid). By successively adding other layers of polyelectrolytes, the surface electric charge can be modified as needed.

For example, increasing the electrokinetic potential of micro/nanoparticles increases the stability of colloidal systems due to electrostatic repulsive forces, and changes in surface electrical charge can influence the cellular absorption of biocomponents encapsulated in various CDSs.

Some works have shown that negative NPs generally showed a lower cellular uptake than positive NPs due to electrostatic repulsions that occur between particles and the mucus layer of the intestine (Li et al., 2015). In this regard, da Silva Santos et al. (2019) published an interesting paper in which they studied the effect of chitosan layer on bioaccessibility and cellular uptake of curcumin nanoemulsions. The authors analyzed the effect of the chitosan layer on the behavior of curcumin nanoemulsions during in vitro digestion, and observed that under gastric conditions at pH 2, nanoemulsions were stable due to protonation of amino groups that raised the electrokinetic potential to high positive values (+36 mV). On the contrary, under intestinal conditions, at pH 7.4, the degree of protonation of the amino groups in the chitosan layer decreased, the electrokinetic potential was reduced to +22 mV, and curcumin nanoemulsions were destroyed by flocculation or coalescence. There is numerous research in the literature on the use of the LBL method for the functionalization of different types of lipid-based CDSs, such as: SLNs (Mohamed et al., 2021) and liposomes (Lai et al., 2020). Other authors prepared Pickering emulsions using solid NPs such as protein-based particles, silicon dioxide, calcium carbonate, and starch granules, which adsorbed on the surface of the droplets forming a protective layer (Dickinson, 2017; Shi et al., 2020).

Complex coacervation is another method of manipulating the surface properties of polymeric micro/nanoparticles by coating them with a layer of biopolymers with an opposite electric charge. Thus, Dima et al. (2014) encapsulated *Pimenta dioica* essential oil in chitosan microparticles that were coated with a layer of k-carrageenan in order to increase particle solubility and modulate the release rate under different pH conditions. The authors showed that at pH < 4.0 the two biopolymers have opposite charges and interact by electrostatic forces forming coacervates or complex polyelectrolytic systems covering the microparticles.

Another group of complex CDSs is obtained by including simple colloidal systems into large solid or liquid structures. For example, O/W emulsions loaded with hydrophobic active components can be incorporated into polymeric particles obtained by gelling. This ensures an increase in emulsion stability, good protection and a controlled release of bioactive components. Complex CDSs are also the double W/O/W emulsions obtained by including the W/O emulsion in an external aqueous phase. These colloidal systems allow the simultaneous encapsulation of hydrophilic and hydrophobic compounds (Dima & Dima, 2018).

Cyclodextrins are cyclic oligosaccharides, derived from starch, that allow the capture of hydrophobic molecules (guest molecules) with the formation of guest molecules-cyclodextrin inclusion complex. Encapsulation in cyclodextrins is one of the most effective techniques for functionalizing flavors and aromas used in the food industry (Kfoury et al., 2016). This technique increases the water solubility of hydrophobic compounds, reduces the rate of photodegradation of photosensitive compounds and ensure the gradual release of biocomponents. Some authors incorporated complex bioactive-cyclodextrins into polymer particles, edible films or composite materials for smart packaging, contributing to the improvement of their functional properties (Szente & Fenyvesi, 2018).

Recently, some researchers have studied increasing the bioavailability of active components in food by using two or more types of CDSs simultaneously. Thus, it has been shown that a mixture of nanoemulsions and polymeric NPs can improve the bioavailability of active biocomponents, such as polyphenols, because by digesting lipids in nanoemulsions, monoglycerides and free fatty acids are formed.

These, together with other surfactant species in the intestinal fluid (i.e., bile acids, bile salts, phospholipids), form mixed micelles that contribut to the increase of solubility and cellular uptake of biomolecules (Zhang, Zhang, Zou, & McClements, 2016).

## 4.3 Designing of colloidal delivery systems for food functionalization

For the production of functional foods whose attributes must meet the requirements of consumers, it is necessary to apply a quality management that includes analysis and control in all phases of end-product development, including research, production, marketing and consumer satisfaction. The design of CDSs for food fortification is an important stage in which specialists from different fields participate: food science, medicine, materials science, engineering, etc. To this end, food scientists adapted the principles of designing a end-product used in other fields of research. An eloquent example is the design of functional foods based on the principles of reverse engineering, which consists in establishing the characteristics required of a product by analyzing, disassembling, testing and selecting alternative solutions capable of generating a product with attributes as close as possible to the required product (Cerqueira et al., 2014).

The reverse engineering design of CDSs used in food functionalization is based on the investigation of end-product quality in correlation with consumer preferences or physician recommendations (Thomopoulos et al., 2019). The concept of reverse engineering is a modern concept of innovative food design (nanofoods, functional foods, medical foods, nutragenomics) and its application to the design of CDS-based functional foods involves the following aspects (Dima et al., 2020c):

- achieving the nutritional, sensory and functional profile of the food based on consumer requirements;
- achieving the genetic profile of the consumer, according to which the food is designed (nutrigenomics);
- selection of bioactive components (nutraceuticals), depending on the required health benefits;
- selection of encapsulating materials and types of colloidal micro/nanobiosystems as nutraceutical bionanocarriers;
- selection of the encapsulation technique, the food matrix and the food excipient;
- selection of the type of packaging and storage and transport conditions;
- analysis of the perception of food for consumption;
- study of the bioavailability of nutraceuticals incorporated in food;
- cost analysis;
- study of the risk of toxicity and the impact on the environment.

According to the principles of reverse engineering, the design of colloidal micro/nanobiosystems begins by analyzing the nutritional and health attributes of functional food and identifying the target group of consumers: children, the elderly, consumers with religious restrictions, consumers with special nutrition, etc. Depending on the attributes of the functional food, bioactive compounds are selected and the characteristics of colloidal micro/nanobiosystems are defined, which in turn depend on the nature of the encapsulating material and the encapsulation technique (Cerqueira et al., 2014).

To identify and select the most appropriate CDSs for food application, other authors have used the concept of Delivery by Design (DbD) (Kharat & McClements, 2019; McClements, 2018a).

This concept is an adaptation of the concept Quality by Design (QbD) which is part of the principles of Total Quality Management (TQM) and Total Product Quality (TPQ) applied to the production of quality pharmaceutical products (Q8 (R2) Pharmaceutical Development, 2009). The authors proposed a rapid design of CDSs with applications in food functionalization going through the following stages:

- defining the physicochemical and structural characteristics of bioactive compounds;
- analysis of physico-chemical characteristics and nutritional and sensory attributes of functional foods;
- specification of CDS attributes in correlation with the attributes of the end-product;
- selection of CDS manufacturing techniques and process optimization;
- establishment of CDS analysis and control protocols;
- cost analysis and optimization of CDS performance.

Both design models for CDSs, described above, are based on the attributes of the end- product that influence the selection of bioactives, encapsulating materials and encapsulation techniques and require a critical analysis of the physicochemical characteristics of micro/nanoparticles in CDSs used in food functionalization.

### 4.3.1 Bioactives requirements

#### 4.3.1.1 Molecular characteristics

Molecular characteristics of bioactives, such as molecular weight, chemical structure, lipophilicity, refractive index and surface tension, plays an important role in the preparation of CDSs and in the bioavailability of bioactives encapsulated. Thus, the size of the molecules and the molar volume influence the encapsulation efficiency, retention, release rate, solubility and bioavailability. In general, there is a correlation between the size of bioactive molecules and the type of CDSs used in encapsulation. For example, lipid-based CDSs (nanoemulsions, SLNs, liposomes) encapsulate low-molecular-weight lipophilic bioactives which can be incorporated into small oil droplets, and the encapsulation of higher-molecular-weight biomolecules is recommended to make in polymeric micro/nanoparticles, so that their loss during storage or release into the GI fluid to be limited by particles' pore size (Li et al., 2021; McClements, 2017b). A special behavior is manifested by peptides and proteins that, in aqueous solution, can adopt different conformational structures, which gives them different molecular dimensions. Thus, it was shown that in aqueous solution, at an appropriate pH, proteins with the same molecular weight have different molar volume, which varies in the order: globular proteins < random coil < rigid rod. This behavior requires specific conditions for protein encapsulation (McClements, 2018b; Perry & McClements, 2020).

Also, the large size of the bioactives limits the solubilization in the micelles and reduces cellular absorption. To avoid this shortcoming, it is recommended to use carrier oils with long-chain triglycerides, because by hydrolysis, they form long-chain monoglycerides and long-chain free fatty acids involved in the formation of mixed micelles. The hydrophobic chain of these amphiphilic compounds is long enough to form a high volume hydrophobic core in which can be included high molecular size bioactives such as $\beta$-carotene (Qian et al., 2012).

The chemical structure of bioactives plays an essential role in the bioaccessibility and bioavailability of encapsulated bioactives. The presence of hydroxyl, carboxyl, amino and other polar functional groups determines the acid and base properties, increases the solubility of the bioactives, and favors their interaction with other food components, such as the binding of polyphenols or metal ions to proteins. An important structural feature of bioactives is lipophilicity, which plays a major role in cellular uptake. The transport of molecules across the cell membrane is limited by the lipophilic domain of the bilayer formed by the hydrophobic chains of phospholipid molecules. Therefore, only lipophilic molecules can pass through the cell membrane by passive transport (diffusion).

The evaluation of lipophilicity is done using the $\log P$ parameter, where $P$ is the partition coefficient defined as the ratio between the concentration of the biocomponent distributed in the nonpolar phase (e.g., octanol) and in the polar phase (aqueous phase). Research results have shown that, for good bioavailability, bioactives must have an average value of $\log P$. Very low values of $\log P$ correspond to a high polarity and do not allow the penetration of molecules through the lipophilic layer of the cell membrane, such as vitamin C which has $\log P = -2.15$, while very high values of $\log P$ express a high lipophilicity which hinders the solubility of bioactives in GI fluids, such as β-carotene with $\log P = 17.62$ (Table 4.1). According to Lipinski's rules, the best cell absorption is for molecules with $M \leq 500$ g/mol and $\log P \leq 5$ (Dima et al., 2020d; Lipinski, 2004).

The partition coefficient, surfactant HLB and surface tension can provide information about the location of bioactives in CDS, in bulk phases or at interface. For example, Dima and Dima (2018) studied the antioxidant capacity of chlorogenic acid encapsulated in W/O/W double emulsions. By determining the partition coefficient of chlorogenic acid between linseed oil and aqueous phase ($K_{\text{linseed oil/water}} = 4.7 \times 10^{-1} \pm 0.13$), the authors demonstrated that some of the chlorogenic acid molecules are distributed at the W/O interface and inhibit the processes oxidative degradation of linseed oil.

**Table 4.1 Physicochemical properties of some bioactive compounds (PubChem, DrugBank).**

| Compound | Molar mass (g/mol) | Melting temperature $T_m$ (°C) | Partition coefficient octanol/water ($\log P$) | Water solubility $C_{SW}$ (g/L) |
|---|---|---|---|---|
| **Carotenoids** | | | | |
| β-carotene | 536.9 | 183 | 17.62 | $6 \times 10^{-4}$ |
| Lycopene | 536.9 | 175 | 15.6 | $7.3 \times 10^{-4}$ |
| Lutein | 568.9 | 196 | 7.9 | $7.3 \times 10^{-4}$ |
| Zeaxanthin | 568.9 | 215.5 | 10.9 | $8.0 \times 10^{-5}$ |
| Astaxanthin | 596.8 | 182.5 | 13.27 | $7.9 \times 10^{-13}$ |
| **Phytosterols** | | | | |
| β-sitosterol | 414.7 | 143.5 | 10.5 | 10 |
| Stigmasterol | 412.7 | 170 | 9.43 | $1.1 \times 10^{-8}$ |
| Campesterol | 400.7 | 157.5 | 9.97 | $2.0 \times 10^{-6}$ |
| **Vitamins** | | | | |
| Vitamine C | 176.12 | 191 | −2.15 | 330 |
| Vitamin D3 | 384.64 | 84.5 | 7.5 | $1.3 \times 10^{-8}$ |
| Chlorogenic acid | 354.31 | 205–209 | −0.3 | 40 |
| **Polyphenols** | | | | |
| Quercetin | 302.236 | 316.5 | 1.48 | $6 \times 10^{-2}$ |
| Curcumin | 368.12 | 183 | 3.29 | $3.12 \times 10^{-3}$ |
| Resveratrol | 228.24 | 254 | 3.1 | $3 \times 10^{-2}$ |

The polarity of bioactive molecules is an important criterion for selecting the encapsulating material and the type of CDS. Thus, hydrophobic bioactives are encapsulated mainly in lipid-based CDSs, or polymeric CDSs while hydrophilic bioactives are encapsulated in complex CDSs, such as double emulsions, liposomes, colloidosomes and SLNs (McClements, 2015; Sedaghat Doost et al., 2020).

### 4.3.1.2 Physical state and solubility

At room temperature, some bioactives are in the liquid state (e.g., polyunsaturated fatty acids, essential oils and α-tocopherol), and others are in the solid state, such as polyphenols (e.g., curcumin, catechin, quercetin, resveratrol), phytosterols, carotenoids, etc. (McClements, 2017a). Their encapsulation is done in different CDSs using specific methods. Most liquid biocomponents are hydrophobic (e.g., ω-3 polyunsaturated acids, essential oils, carotenoids) and consequently can be encapsulated in CDS liquids with hydrophobic structural domains, such as emulsions, microemulsions, liposomes. Sometimes these liquid CDSs are introduced into solid polymer particles resulting in more stable delivery systems, which ensure a controlled release of biocomponents. When encapsulating solid biocomponents, the physical state in which they are found is taken into account: amorphous or crystalline state. Because amorphous solid biocomponents are less thermodynamically stable, they have higher chemical reactivity, are more soluble, and consequently have a higher bioavailability than crystalline ones. Also, some crystalline solids can exist in different forms of crystallization (polymorphs) that differ in stability, solubility, refractive index and even bioactivity.

The solid biocomponents are encapsulated either in the solid state, by coating micro or solid NPs with encapsulating material, or in the liquid state, by solubilizing them in different solvents and then introducing them into different types of CDSs. For example, in order to improve the solubility and bioavailability of curcumin, some researchers prepared nanocurcumin by coating solid curcumin NPs with biopolymers (carboxymethylcellulose sodium salt, polyvinyl alcohol, polyvinyl pyrrolidone) (Rachmawati et al., 2013), and others solubilized curcumin in ethanol and encapsulated it in various polymeric materials (maltodextrin, starch, pectin, chitosan, etc.) using the spray drying technique (Guo et al., 2020; Lucas et al., 2020). Hydrophilic solid biocomponents, such as peptides, proteins, vitamins and salts, are first solubilized in water, under certain conditions of pH and ionic strength and then introduced into CDSs with hydrophilic structural domains, such as: W/O/W emulsions, W/O emulsions, liposomes, biopolymer microgels, among others (McClements, 2015). The encapsulation of hydrophobic crystalline solid biocomponents in liquid CDSs is achieved by their solubilization in a vegetable oil, or in other solvents (ethanol), at room temperature, followed by the preparation of various colloidal dispersed systems. Solubilization is done below the saturation concentration to avoid the formation of crystals that cause CDS destabilization and alter the solubility, bioaccessibility and bioavailability of bioactive components. Many authors reported the loss of crystallinity and increased bioavailability of some bioactives, such as quercetin, resveratrol, curcumin by encapsulation in polymer NPs using zein protein, soy protein isolate, whey protein isolate as wall materials (Liu et al., 2016; Patel et al., 2012; Pujara et al., 2017).

The increase of the hydrophobic biocomponent content in the lipid phase can be achieved by melting the bioactive compound and mixing it with the heated lipid at a temperature higher than the melting temperature of the biocomponent. The hot liquid mixture is used in the preparation of SLNs or nanostructured lipid carriers (NLCs). The heating should be done carefully to avoid possible polymorphic transformations that can affect the morphology of the particles and the bioactivity of the component. The solubility of biocomponents in CDS phases is different depending on the molecular characteristics (i.e., chemical structure, polarity, molecular mass, partition coefficient, number of hydrogen donor/acceptor) and physicochemical conditions (i.e., pH, ionic strength, temperature, phase polarity). Knowledge of the solubility of biocomponents is necessary because it has a major influence on the retention, encapsulation efficiency, loading degree, release rate and CDS stability (McClements, 2018a).

### 4.3.1.3 Chemical stability

Many biocomponents are sensitive to changes in physicochemical parameters during storage, processing and digestion. Degradation of biocomponents leads to decreased bioactivity and the appearance of compounds that cause alterations in food attributes or even increase the risk of toxicity. The main intrinsic and external factors that contribute to the chemical degradation of biocomponents are: pH, ionic strength, enzymatic composition, temperature, oxygen content, content of prooxidants (enzymes, metal ions), light, etc. These factors can cause for example: chemical degradation by hydrolysis, isomerization, autooxidation and oxidation. Thus, the pH variation causes the protonation or deprotonation of acid-base functional groups, such as hydroxyl, carboxyl, amino, sulfate and, phosphate, as a result of which the solubility of biocomponents changes or their transformation into more stable compounds (Honda et al., 2019).

A relevant example in this regard is curcumin. Curcumin is a polyphenol in the structure of which there is an alpha beta-unsaturated beta-diketone moiety and two aromatic ring systems containing $O$-methoxy phenolic groups. When the pH changes, the beta-diketone undergoes a keto-enol tautomerism. In acidic or neutral conditions the ketone form predominates, while in alkaline conditions the enolate form predominates. These tautomers have different water solubility and chemical stability (Zheng & McClements, 2020). The ketone form explains the poor water solubility of curcumin under acidic or neutral conditions, while the enol form, which occurs at alkaline pH, corresponding to the p$K_a$ values of the hydroxyl groups (8.38 for the enol hydroxyl group and 9.88/10.51 for the phenolic groups), has a high solubility in water, but unfortunately has low chemical stability. In an alkaline environment, curcumin degrades rapidly to more stable compounds, such as ferulic acid, feruloylmethane, and vanillin (Bhatia et al., 2016; Zheng et al., 2018). Also, the changes in the chemical structure of curcumin, depending on the pH value, leads to a change in the color of the curcumin solution. Thus, in the food range pH of 2–7, curcumin is golden yellow, while under alkaline conditions (pH 7 to 8.5), the curcumin solution is brownish-orange, or even reddish at higher pH, due to degradation products.

The chemical stability of bioactive proteins and peptides is an important feature in the efficacy of CDSs. The size and 3D structure of proteins can change irreversibly when happens the variation of pH, ionic strength and solvent polarity. Some studies shown that in GI fluids, the larger peptides including, insulin, calcitonin, secretin, glucagon, and somatostatin were metabolized rapidly, compared to the smaller peptides that showed good stability (Wang et al., 2015). Insulin is known to be a peptide involved in glucose regulation. Insulin deficiency in the human body causes one of the most common diseases, diabetes, which causes many deaths. Due to enzymatic degradation in gastric conditions, insulin is delivered to the human body parenterally, causing high discomfort to patients. Therefore, in recent decades, a large number of researchers studied the encapsulation of insulin in various CDSs to ensure a possible oral administration of insulin (Alai et al., 2015; Mansoor et al., 2019; Ramesan & Sharma, 2014).

Temperature affects the chemical stability of bioactives depending on their structure, physical condition and the presence of other compounds that may interact with the biocomponents under certain conditions of pH and ionic strength. Thus, the increase of temperature determines the increase of the degradation rate of biaoctives both by physical transformations (change of conformation, crystallization, melting) and by chemical reactions (decarboxylation reaction, Maillard reaction) (Peleg et al., 2012).

Therefore, it is required to analyze the thermal behavior of bioactives before encapsulation, so that, for temperature-sensitive bioactives, encapsulation techniques that do not involve high temperatures are used.

The presence of oxygen or pro-oxidants (metal ions, oxidizing enzymes, hydroperoxides) in CDSs or functional foods generates reactive species that cause the breaking of double bonds and the formation of radical species, causing oxidative degradation of polyunsaturated biocomponents, such as ω-3 polyunsaturated fatty acids. Prevention of oxidation reactions can be achieved in several ways, such as: reducing the oxygen content of CDSs or food matrix using vacuum packs, or packs under nitrogen pressure, complexation of pro-oxidant metal ions with chelating agents (EDTA) or co-encapsulation of oxidation-sensitive bioactives, together with antioxidant agents (Liu et al., 2020).

Electromagnetic radiation (light) is an important factor that can change the chemical structure, respectively the biological activity of light-sensitive bioactives, such as polyphenols, carotenoids, polyunsaturated fatty acids, vitamins. The effect of electromagnetic radiation on biocomponents is different. While for some bioactives, such as carotenoids, or curcumin, UV-VIS radiation has a positive effect, improving antioxidant or antimicrobial activity (Chen et al., 2020; Ioannou et al., 2020), for other bioactives, such as vitamins (C, B12, B6, B2, and folic acid), lipids (polyunsaturated fatty acids, phospholipids), UV-VIZ radiation causes photochemical degradation (Guneser & Karagul Yuceer, 2012; Pramanik et al., 2017).

### 4.3.2 Physico-chemical characteristics of food-grade colloidal delivery systems

After defining the properties of bioactive compounds, it is required to specify the physico-chemical characteristics of CDSs that correspond to the attributes of the functional food. In this regard, food-grade CDSs are designed so that they do not chemically interact with the components of the food matrix and do not alter the texture, stability, and sensory properties of the food. Physical state, rheological properties, optical properties, dispersibility and solubility, as well as chemical composition are important characteristics that influence the quality and efficiency of CDSs. For example, to fortify liquid foods (soft drinks), low-viscosity liquid CDSs or soluble powders are used that do not alter the viscosity of the food, while semi-solid foods (purees, yogurt, creams) are fortified using viscous delivery systems such as hydrogel particles (McClements, 2018a, 2018b). To fortify solid foods, both liquid CDSs that mix easily during processing and solid particles whose size does not alter the texture and sensory properties of the food can be used.

#### *4.3.2.1 Wall material requirements*

The chemical composition of food-grade CDSs influences the general characteristics of CDSs, the physicochemical properties of micro/nanoparticles in CDSs and the bioavailability of encapsulated bioactive components. Natural and synthetic, organic or inorganic materials that play different roles are used in the manufacture of food-grade CDSs. Some of them are encapsulating materials that ensure the incorporation of bioactives (hydrophobe or hydrophile bioactives), and others are adjuvant materials, used in the encapsulation process, such as surfactants, co-surfactants, chelating agents, cryoprotective agents, etc. The selection of encapsulating materials used in the food and pharmaceutical industries requires strict compliance with specialized organizations such as the FDA, the European Food Safety Authority (EFSA). Some requirements relate to food quality, safety and security, such as toxicity risk, chemical stability, influence on food texture and sensory properties, environmental impact, others relate to functional properties such as solubility, surface activity, viscosity, protection of bioactives, and other requirements relate to the accessibility of materials and the cost price.

In general, a material can be used to encapsulate nutraceuticals and food ingredients if it is considered a Generally Recognized as Safe (GRAS) material, is not toxic or presents a minimal risk of toxicity, does not change the texture, color, taste and smell of food, in water forms solutions or suspensions with low viscosity, has a good surface tension corresponding to emulsification, provides good protection of biocomponents against temperature, light, oxygen, pH variations, enzymes, has a low cost and has a friendly impact on the environment. Some encapsulating materials, such as chitosan, alginates, inulin etc., have different biological properties (antioxidant, antimicrobial, anti-inflammatory, anticancer, etc.) that contribute to increasing the health benefits of functional foods (Lizardi-Mendoza et al., 2016). The most used materials for encapsulating bioactives with applications in food functionalization are natural materials such as polysaccharides, proteins and lipids (Sobel et al., 2014), as summarized in Table 4.2.

**Table 4.2 Encapsulating materials for food applications.**

| Wall materials used for encapsulation of hydrophilic bioactives ||| Wall materials used for encapsulation of hydrophobic bioactives |||
| --- | --- | --- | --- | --- | --- |
| Carbohydrate polymers | Proteins | Other polymers | Lipids | Wax | Polymers |
| **Unmodified carbohydrate polymers**: sugar, starch, glucose syrup, maltodextrins **Modified carbohydrate polymers**: dextrins, cyclodextrins, OSA-starch, cellulose derivatives **Gums**: gum Arabic, alginate, pectin, carrageenan, galactomannans, etc. | **Animal proteins**: gelatin, casein, whey proteins **Vegetable proteins**: soy proteins, wheat protein, corn proteins | Polyethylene glycol (PEG), Polyvinyl acetate (PVA), Poylactic acid (PLA), Poly(DL-lactic-*co*-glycolic acid) (PLGA), Chitosan | Glycerides, Phospholipids, Fatty acids, Fatty alcohols, Plant sterols, Hard fat, Hydrogenated fat | Beeswax, Paraffin wax, Carnauba wax | Shellac Ethyl cellulose |

The main functional properties that influence the quality of encapsulating materials are: solubility, gelling and viscoelasticity, emulsifying properties, phase transition and polymorphism.

## Solubility

In general, the encapsulating material must form solutions at high concentrations and low viscosities in order to be easily mixed and transported through pipes. Some encapsulating materials are soluble in water, others form suspensions, and others, such as lipids, are insoluble in water. The solubility of the wall material depends on the chemical structure, temperature, pH and ionic strength of the medium. For example, nonionic polysaccharides (starch, cellulose, hemicelluloses) are insoluble in water, while ionic polysaccharides (gum arabic, pectins, alginates, etc.) are soluble in water. Chitosan is an amino-polysaccharide, with poor solubility in water and alkaline solutions, but is soluble in acidic solutions due to protonation of deacylated amino groups.

The solubility of proteins in aqueous solution varies between 0% and 35% (vol) and is strongly influenced by chemical structure, pH, ionic strength of the solvent and temperature (Walstra et al., 1999). Thus at pH values corresponding with pI the polypeptide chains are coiled in a random coil, with minimal distances between opposite charges and the strong electrostatic forces are manifested. The random coil makes it difficult for water molecules to penetrate, which means that the solubility of the protein is minimal and the phenomenon of isoelectric aggregation (precipitation) occurs. At values higher or lower than pI there is an excess charges (negative net charges at pH>pI or positive net charges at pH<pI). Repulsive forces which occur between identical electric charges, keep the chains in an extended state, and are able to interact with water molecules. At these pH values, the solubility of the proteins becomes significant. This behavior explains the variation of the optical, rheological properties and especially of the surfactant properties of proteins as well as the interaction of proteins with other polyelectrolytes such as anionic polysaccharides.

The presence of salts in protein solutions influences their solubility. This is because ions, in certain concentrations, can change the thickness of the diffuse layer that is created around the polypeptide chain, causing changes in its conformation. A low salt content (0.25–1 M) causes an increase in protein solubility by the phenomenon of *salting in*. As the salt concentration increases by more than 1 M, the solubility of the protein decreases through the so-called salting out phenomenon. This is due to the change in the thickness of the diffuse layer and the conformation of the polyion which causes a different exposure of the polypeptide chain to water molecules.

## Gelling and viscoelasticity

Gelling and viscoelasticity are important properties that are taken into account in the selection of encapsulating materials. This is because the gelling process is mainly found in the preparation of polymeric particles and influences the physico-chemical and sensory characteristics of functional foods. Polysaccharides and proteins form gels through various mechanisms, such as temperature variation, pH variation, the presence of salts or by complex coacervation (Zhang et al., 2015). Some gels are the result of physical interactions (hydrogen interactions, hydrophobic interactions, electrostatic interactions), while other gels are formed by covalent crosslinking. An important feature of hydrogels is the water retention capacity that influences the viscoselasticity of CDSs, the degree of swelling, the kinetics and release mechanisms of encapsulated biocomponents. Some polysaccharides form thermoreversible gels by cooling (agar-agar, k-carageenan, weakly methoxylated pectin) or by heating (methylcellulose, hydroxypropylmethyl cellulose), and others form thermoreversible gels by crosslinking with bivalent cations (e.g., alginates and strongly methoxylated pectin).

The solutions of biopolymers used as wall materials are viscoelastic and are characterized by the mechanical modules: storage modulus ($G'$) and loss modulus ($G''$). The storage module ($G'$) measures the elasticity of the material and represents its ability to retain the received energy, while the loss module ($G''$) characterizes the viscosity of the material and expresses the ability of the material to dissipate energy. Some polysaccharides, such as xanthan, form weak gels, due to the easy opening of the junction areas, even at very low shear rates. For these gels, both modules have a low frequency dependence and have $G' > G''$ over a wide frequency range. Other polysaccharides, such as amylose, agarose, and gellan form firm gels, due to the strong attractive interactions between macromolecular chains. In these systems, the junction areas are very stable, both modules are independent of frequency, and $G' \gg G''$ (Razavi & Irani, 2019). The texture of the gels varies depending on the type of gel: elastic, fragile, firm, etc. For example, pectin, alginate and carrageenan form fragile gels in the presence of $Ca^{2+}$, $Fe^{2+}$, $Mg^{2+}$ or $K^+$ ions (Mierczyńska et al., 2015).

## Emulsifying properties

Because emulsification is an important step in the encapsulation process, it is necessary for the encapsulation material to have emulsifying properties. Among polysaccharides, gum arabic has the best emulsifying properties, due to its mixed composition of arabinogalactan and protein (Ebrahimi et al., 2020). The hydrophobic part of the molecule, due to the protein chain, is adsorbed on the surface of the oil droplets, while the polar groups, located on the polysaccharide chain, are in contact with water. Thus, the amphiphilic molecules of gum arabic contribute both to emulsification and to the steric and electrostatic stabilization of the emulsion. The improvement of the emulsifying properties of polysaccharides is achieved by chemical changes to them. For example, native starch, lacking surface activity, becomes a good emulsifier if it is modified by reaction with octenyl succinic anhydride. Thus, the octenyl hydrocarbon radical gives the starch polysaccharide chain a hydrophobic character, which allows it to be adsorbed at the O/W interface just like an emulsifier.

Due to the amphiphilic structure, the vast majority of proteins have surfactant properties. The surface activity of proteins is influenced by: origin, structure, solubility, extraction method and storage conditions. They are adsorbed at the Liquid/Air or Liquid/Liquid interface via hydrophobic amino acid chains. As shown above, proteins can adopt different configurations in aqueous solutions and at the O/W interface, depending on the intensity of electrostatic, hydrophobic and steric interactions, or van der Waals forces and hydrogen bonds. These interactions are influenced by several factors, such as: pH, ionic strength, dielectric constant and temperature. Changes in experimental conditions cause conformational changes that involve changes in surface activity, emulsifying and stabilizing properties of emulsions (McClements & Gumus, 2016). The most common conformations of surfactant proteins correspond to the globular structure and the random coil structure. Globular proteins, such as whey proteins, egg proteins, soy proteins, have an almost spherical, compact geometry, in which most non-polar groups are inside, while polar groups are present on the surface. However, as the temperature increases, or at certain salt concentrations, the helical or β-pleated sheet structures of the polypeptide chains of the globular proteins unfold in such a way that the hydrophobic domains can be adsorbed at the O/W interface. Proteins whose polypeptide chains are organized random coil structure have a higher availability in the adsorption of hydrophobic groups, due to the flexibility of the chains.

Solubility is an important factor that influences the surface activity of proteins. For example, proteins with maximum values of hydrophilicity (gelatin) and hydrophobicity (zein) have low values of surface activity. Studies shown that among milk proteins there is the following order of variation in surface activity: β-casein > micellar casein > BSA > α-La > αs-casein = κ-casein > β-Lg (Singh & Chaudhary, 2010). In general, proteins have a surface tension between 22 and 42 mN/m and the interfacial tension between 8 and 22 mN/m, depending on the nature of the oil (McClements & Gumus, 2016). The adsorption of proteins at the O/W interface gives them very good emulsifying properties. Proteins contribute on the one hand to the formation of emulsions (by decreasing the O/W interfacial tension) and on the other hand to the stabilization of emulsions, by forming on the surface of oil droplets a protective layer of different thickness and electric charge, depending on molecular mass, pH and ionic strength of the aqueous phase (Lam & Nickerson, 2013).

## Phase transition and polymorphism

Phase transition and polymorphism are two important properties of encapsulating materials, such as polymeric materials and lipids. These properties are due to the different organization of molecules under certain conditions of temperature and concentration and have a major influence on the characteristics of CDSs. Some polymers, such as synthetic polymers, have the chains organized in a solid crystalline structure, while polymers, called amorphous polymers, have frozen" chains, and randomly distributed at low temperature (vitreous state). By heating the amorphous polymer, the segments of the "frozen" chains, corresponding to the vitreous state, become mobile, and the polymeric material is soft and flexible. This physical state of the amorphous polymer is called the rubbery state.

The temperature at which the transition of the amorphous polymer from the vitreous to the rubbery state takes place is called the glass transition temperature ($T_g$) (Zanotto & Mauro, 2017). Polysaccharides can have either an amorphous structure or a mixed, semi-crystalline structure, in which there are both crystalline and amorphous structures, such as starch, cellulose, alginates, etc. These polymers have better elastic properties because they combine the firmness of crystalline polymers with the flexibility of amorphous ones. The value of $T_g$ increases with increasing molecular weight of the polymer and decreases with increasing water content of the material (Dima et al., 2020e). The $T_g$ affects the physico-chemical characteristics of micro/nanoparticles, the chemical stability of bioactive compounds, respectively the sensory attributes of food. In general, foods are stable at temperatures below $T_g$ (Le Meste et al., 2002). The $T_g$, also controls the dehydration processes of materials, which occur in the processes of preparing solid CDSs in powder form, such as freeze-drying, spray-drying, extrusion, crystallization, etc. The glassy state of the dry particles prevents the loss of volatile biocomponents during storage in low humidity conditions. Decreasing $T_g$ reduces the retention of encapsulated biocomponents and causes agglomeration of particles into powders. The mechanism and kinetics of the release of biocomponents from micro/nanoparticles are also influenced by the $T_g$ of the wall material. Thus, in the presence of water, which acts as a plasticizer, there is a transition from the glassy to the rubbery state, because water molecules enter the outer layer of the particle and destroy the "frozen" structure of the polymer, allowing the release of encapsulated biomolecules (Maderuelo et al., 2011).

Phase transitions are also found in lipids that show the polymorphism phenomenon, due to the organization of molecules in geometric structures with different stability. Thus, triglyceride molecules can be organized into hexagonal, orthorhombic and triclinic structures, corresponding to polymorphs α, β′ and β whose melting points and thermodynamic stability decrease in the order of β > β′ > α. Some fats, such as cottonseed oil, palm oil, talow oils and milk fat crystallize into the more stable polymorphic form β, while soybean, sunflower, peanut, corn, and olive oils crystallize into the polymorphic form β' (Ribeiro et al., 2015). Lipid crystallization plays an important role in the preparation of NLCs and in the preparation of Pickering emulsions (Ghosh & Rousseau, 2011).

Phospholipids form mesophase structures in which molecules are organized into aggregates with different geometries. In water, the amphiphilic molecules of glycerophospholipids self-assemble spontaneously to form different mesophase structures depending on water content or temperature. (Dima et al., 2020e). The property of phospholipids and other lipid amphiphilic molecules to self-assemble into mesophase structures has been widely used in encapsulating bioactive compounds in different types of CDSs, such as: liposomes, niosomes, hexosomes, cubosomes, etc. (Meikle et al., 2017; Rostamabadi et al., 2019; Sarabandi et al., 2019; Tavakoli et al., 2018).

### 4.3.2.2 Particles characteristics

In the design of micro/nanocolloidal systems loaded with nutraceuticals, an important objective is the analysis of their physico-chemical characteristics. Knowledge of the characteristics of micro/nanoparticles is important because they influence the stability of CDSs, the quality and safety of functional foods. The main characteristics of micro/nanoparticles in CDSs are: morphology and structure, appearance and color, size and particle size distribution, surface electric charge, dispersion or aggregation capacity, hydrophobicity and solubility, chemical composition and chemical reactivity. These features play a key role in:

- compatibility of CDSs (taste, smell) in the food matrices in which they are incorporated, without modifying the physical (color, appearance, texture) and sensory attributes;
- protection of bioactive compounds against the aggression of physical factors (temperature, light, oxygen, humidity), or biochemical factors (pH, ionic strength, enzymes)
- controlled release of bioactives in certain sectors of the GI tract;
- bioaccessibility and bioavailability of nutraceuticals;
- controlling the risk of toxicity of NPs due their accumulation in tissues and organs (Assadpour & Jafari, 2019; Dima et al., 2020e).

#### Particle shape and microstructure

The shape of the particles in a CDS is generally difficult to control. Usually, when preparing lipid-based NPs, such as emulsions, nanoemulsions, microemulsions, spherical particles are formed due to the effect of surface tension. Both polymeric NPs and especially inorganic NPs can have different shapes (spheres, rods, wires) depending on the preparation methods (García-Rodríguez et al., 2018; McClements & Xiao, 2017).

The internal structure of micro/nanoparticles plays an important role both in the efficiency of encapsulation, the degree of retention of biocomponents, the release rate, mechanical strength, and in the food microstructure. Compared to the shape of the particles, their internal structure can be controlled by selecting encapsulating materials and preparation techniques. For example, the emulsion microstructure can be modified by using crystallizable lipids or by Pickering stabilization, with solid NPs (Alehosseini et al., 2021; Burgos-Díaz et al., 2020). Also, the microstructure of emulsions is influenced by the type of emulsifiers and the presence of other components (phospholipids, proteins, polysaccharides, salts) that can change the surface of particles and can cause the formation of aggregates (clusters, flocons) that affect the food texture. Micro/nanoparticles prepared by gelling or coacervation have different internal structure depending on the gelling mechanism, the chemical structure of the biopolymer, the gelling temperature, the nature and concentration of the ions involved in gelling, etc. (Jeong et al., 2020; Zeng et al., 2012).

### Particle size and particle size distribution

Knowing the particle size and particle size distribution in a CDS is important due to their impact on the stability and physico-chemical properties of CDS, such as optical properties, rheological properties and implicitly on the sensory attributes of functional foods. Most CDSs are prepared by methods that require high energy consumption. The colloidal particles obtained from a discontinuous phase with a very large interphase surface which gives the colloidal system a high energy state, and respectively a low thermodynamic stability. Therefore these CDSs tend to be destroyed either by the separation of particles due to external forces, such as gravitational force, centrifugal force, electric force, or by the formation of aggregates (clusters, flocons) due to the internal forces, such as *van der Waals* forces, hydrophobic interactions, electrostatic forces, depletion forces (McClements, 2016).

The velocity ($v$) at which particles move through the dispersion medium due to gravity measures the kinetic stability of colloidal systems and is calculated with the Eq. (4.1):

$$v = \pm \frac{g}{18} \frac{(\rho_2 - \rho_1)d^2}{\eta_1} \tag{4.1}$$

where: $d$ is the diameter of the colloidal particle, $\rho_1$ and $\rho_2$ are the densities of the continuous phase and colloidal particle, respectively, $\eta$ is the shear viscosity of continuous phase and $g$ is the gravitational constant. The sign (+) is applied if the particle density is higher than that of the continuous phase and the particles move downwards and separate by sedimentation, while the sign (−) is applied if the particle density is higher smaller than that of the continuous phase, and the particles move upwards and separate by creaming.

According to Eq. (4.1), the kinetic stability of CDSs is mainly influenced by the particle size and the viscosity of the continuous phase. Thus, the decrease of the particle size and the increase of the viscosity of the continuous phase ensure a good kinetic stability of CDSs. These factors can be easily controlled. For example, particle size can be manipulated by selecting preparation methods and optimizing technological parameters, while the viscosity of the continuous phase can be changed using thickening agents, such as hydrocolloids.

The particle size strongly influences the optical properties of CDSs, respectively, the appearance of functional foods. Some CDSs with a particle diameter of less than 50 nm, such as nanoemulsions, are optically transparent due to poor light scattering and do not change the appearance of beverages, while particles with a diameter of more than 100 nm cause strong light scattering, and CDS have a cloudy or opaque, affecting the appearance of the food. Incorporating CDS into food matrices may affect the texture of the functional food. The perception of particles in the mouth is different depending on their size. It has been shown that particles <50 μm are poorly felt in the mouth, in the form of smooth microcorpuscules, while larger particles are perceived as rough or gritty microcorpuscules.

Particle size also influences the retention and release of encapsulated biocomponents. Compared to microparticles, the nanoparticle CDSs have poor retention and rapid release of biocomponents. For example, the release of biocomponents from NPs is faster because the distance that bioactive molecules must traverse during diffusion is smaller than in the case of microparticles. Therefore, the loss of encapsulated biocomponents during storage is lower at microparticles than in NPs. It has been shown that the half-life of the biocomponents into a spherical homogeneous particle increases in proportion to the square of the particle diameter, according to the Eq. (4.2) (McClements, 2018b):

$$t_{1/2} = \frac{0.014 \cdot k_{12} \cdot d^2}{D} \tag{4.2}$$

where: $D$ is the diffusion coefficient, $k_{12}$ is the biocomponent partition coefficient between the particle and the liquid external phase.

When passing through the GI tract, micro/nanoparticles in food change their size due to digestive conditions (e.g., pH, ionic strength, the presence of enzymes and surfactants). Digestion rate is affected by several factors, of which particle size plays an important role. Thus, lipid-based CDSs with very small particles have a high digestion rate due to the large interphase surface that allows access to a larger amount of lipase. The decrease of the digestion rate is done by restricting the access of lipase to the lipid surface of the particle which can take place by aggregating the particles into clusters. This process can be achieved by controlling the repulsive and attractive forces between the particles. In this regard, CDSs with particles with opposite electric charges (e.g., polymeric particles, SLN and NLCs) can be used, or conditions can be created to increase attractive interactions such as changing pH, ionic strength and the addition of polyelectrolytes (McClements, 2018b).

It has also been shown that particle size influences the absorption of nutraceuticals through the intestinal wall, affecting their bioavailability (Salvia-Trujillo et al., 2017). Studies shown that the optimal size of micro/nanoparticles absorbed through tight junctions of the epithelial layer is 50–100 nm (Behzadi et al., 2017; Brandelli, 2020). The cellular uptake of larger particles occurs through different mechanisms. Thus, in vitro experiments showed that NPs with a size between 250 nm and 3 µm are involved in the phagocytosis process, while NPs with a size in the range of 120–200 nm are internalized by endocytosis (Foroozandeh & Aziz, 2018).

### Particles surface charge

Surface charge is an important feature of particles in a CDS because it influences the physical and chemical stability of CDSs, the interaction of particles with other components in the food matrix or GI fluids, and the cellular absorption of particles. The electrical properties of micro/nanoparticles are evaluated by measuring the electrokinetic zeta potential ($\zeta$). This represents the potential that occurs at the contact between the fixed layer of ions, counterions and solvent molecules adsorbed on the surface of the particles and the diffuse layer containing ions and counterions in the continuous phase of CDSs. The sign and the values of the electrokinetic potential of the particles can be manipulated by using ionic surfactants or biopolymers containing ionogenic functional groups.

In general, the carboxyl, sulfate, sulfonate groups provide the particles with a negative electrokinetic potential, while the amino groups lead to a positive electrokinetic potential. pH is an important factor influencing the sign and value of the electrokinetic potential of particles in CDSs. Most biopolymers are pH-sensitive, which can affect the degree of ionization of functional groups and the charge density on the particle surface, respectively. For example, pectin and alginate, due to carboxyl groups, have a negative electrokinetic potential in the pH range of 3–7. while chitosan, with degree of deacetylation (DDA) 75–90% has a positive electrokinetic potential at pH 3–7.5 due to protonation of free amino groups. Proteins, such as β-lactoglobulin, which has an isoelectric point (pI ≈ 5.5) have a positive electrokinetic potential at pH < pI and a negative electrokinetic potential at pH > pI.

Superficial electrical charges cause repulsive interactions between colloidal particles and ensure the physical stability of CDSs. The results showed that the optimal values of the electrokinetic potential that provide the physical stability of CDSs are in the range −30 to +30 mV (McClements, 2016). In order to increase the bioavailability of hydrophobic bioactive compounds, some authors prepared CDSs complex by mixing two CDSs with charged particles with opposite electric charges. For example, Chen et al. (2015) prepared mixed CDSs using tangeretin-loaded zein NPs (positive NPs) with lipid NPs (negative NPs), and Zou et al. (2016) mixed curcumin-loaded zein NPs with lipid NPs. Both studies reported an increase in the bioavailability of hydrophobic biocomponents to an increase in the lipid content of NPs in mixed CDSs. Authors explained the results obtained by highlighting the different role that the two types of colloidal particles have. Thus, zein NPs provide the protection of bioactive components, while lipid NPs, under the action of lipase in the intestinal fluid, form monoglycerides and free fatty acids that contribute to the solubilization of hydrophobic biocomponents.

The surface charges of the particles can reduce the oxidative degradation of the compounds by repelling the prooxidant ions. In this regard, Dima and Dima (2018) showed that chitosan present in the internal or external aqueous phases of W/O/W emulsions reduced the oxidation of flaxseed oil due to the rejection of $Fe^{2+}$ and $Fe^{3+}$ ions by the positive charges of chitosan adsorbed on W/O interface. An important role is played by the superficial charge in the cellular uptake of NPs and implicitly in the bioavailability of bioactive compounds. Most papers published in recent years have revealed that neutral or positively charged particles are better absorbed than negatively charged particles (Foroozandeh & Aziz, 2018; Fröhlich, 2012). On the one hand, this is due to the negative charge of the mucin in the mucus layer and on the other hand to the negative charges of the cell membrane that allow the internalization of positive NPs and oppose negatively charged NPs. Positive NPs have also been shown to fluidize the cell membrane, while negatively charged NPs cause membrane gelation (Arvizo et al., 2010; Li & Malmstadt, 2013).

### 4.3.2.3 Loading performance of colloidal particles

An important characteristic of CDSs that influences both the attributes of functional foods and the cost price of CDSs is the loading of colloidal particles with biocomponents. This characteristic is evaluated in terms of loading capacity, encapsulation efficiency, retention and release rate. Knowing the loading characteristics of colloidal particles is important in selecting CDSs for different applications and determining the amounts of CDSs used to make food functional so that the required effect is maximum.

Loading capacity (LC)
LC is the maximum amount of bioactive agent that is inside the colloidal particles. *LC* is calculated as the percentage of bioactive compound in the total mass of the colloidal particles (Eq. 4.3):

$$LC = \frac{m_B}{m_P} \cdot 100 \tag{4.3}$$

where $m_B$ is the mass of the bioactive and $m_P$ is the total mass of the colloidal particles (bioactive +particle).

LC varies depending on the type of CDS. Some CDSs, such as surfactant micelles, have low LC values ($\approx 5\%$), while O/W emulsions and nanoemulsions have $LC > 90\%$ (McClements, 2018a).

## Encapsulation efficiency (EE)

EE is calculated as the ratio of the mass of encapsulated bioactive ($m_{BE}$) to the total mass of bioactive in the CDS ($m_{BT}$ = mass of encapsulated bioactive + mass of non-encapsulated bioactive) (Eq. 4.4):

$$EE\% = \frac{m_{BE}}{m_{BT}} \cdot 100 \tag{4.4}$$

EE values are expressed as a percentage and vary between 0% and 100% depending on the nature of the bioactive, the type of particles, the wall material and the preparation technique.

## Retention and release

Retention is the ability of CDSs to retain as much bioactive agent within colloidal particles during storage, transport and incorporation into food matrices, while the release is the ability of CDSs to allow bioactive molecules to exit from colloidal particles after a certain time and under certain conditions. Retention and release can be expressed in terms of retention efficiency (RE) and release level (RL). RE is defined as the percentage of bioactive remaining within the colloidal particle after a certain storage or treatment time (Eq. 4.5), while the RL is the percentage of bioactive released from the colloidal particles after a certain period of time, and under certain conditions of storage or treatment (Eq. 4.6):

$$RE = \frac{m_{B,E(t)}}{m_{B,E(t=0)}} \cdot 100 \tag{4.5}$$

$$RL = \frac{m_{B,NE(t)}}{m_{B,E(t=0)}} \cdot 100 \tag{4.6}$$

here, $m_{B,E(t)}$ is the mass of bioactive that remain inside the particles (encapsulated) at time $t$; $m_{B,E(t=0)}$ is the encapsulated bioactive mass; $m_{B,NE(t)}$ is the mass of nonencapsulated (released) bioactive at time $t$. RL may range from zero (none released) to 100% (all released) depending on the CDS type and releasing conditions tested (e.g., temperature, pH and enzymes).

Volatile compounds such as flavors most easily leave colloidal particles during storage. To study the kinetics of aroma retention in different colloidal particles, the Avrami equation is used as a mathematical model (Eq. 4.7):

$$R = \exp(-k_R \cdot t^n) \tag{4.7}$$

where $R$ is the retention of bioactive in colloidal particles; $t$ is storage time; $k_R$ is release rate constant and $n$ is a parameter representing the release mechanism. For example, in spherical coloidal particles, the value $n=0.54$ corresponds to a diffusion-controlled release mechanism, and for $n=1$, the release is made according to a first-order mechanism (Soottitantawat et al., 2015).

The release of bioactive compounds is an important feature of colloidal particles that must be taken into account in the design of CDSs. Thus, depending on the specific applications, controlled release CDS can be designed, such as CDSs with prolonged release, with a fast release, or with on-target release. Prolonged release of bioactive compounds provide the long-term biological effect, helping to increase food quality, such as increasing the shelf life of food when using CDSs with gradual release of compounds with preservative properties (i.e., antioxidant and antibacterial properties). Some CDSs are designed so that the release is fast at first (burst release), followed by a slow release. These CDSs encapsulate the dyes, which must ensure the desired color since their incorporation into the food matrix, and then maintain it during treatment or storage. CDSs that transport and release biocomponents to a specific target are also of major importance both in food functionalization and in the manufacture of innovative medicines. For example, biocomponents that are absorbed in the colon, such as polyphenols, sensitive to the conditions of gastric and intestinal fluids, or drugs that are used to treat colon diseases, are encapsulated in pH-sensitive polymer NPs using natural and synthetic polymers such as pectins, inulin, alginates, chitosan, Eudragit L100 and Eudragit S100) or in lipid-based CDSs [e.g., liposomes, SLNs, self-microemulsifying drug delivery system (SMEDDS)] (Auriemma et al., 2013; Lee et al., 2020).

The main mechanisms by which bioactive compounds are released from CDSs are: diffusion, swelling, particle erosion, particle fragmentation. These mechanisms depend on the type of CDS (liquid-, semi-solid-, solid CDSs), the properties of bioactive compounds, wall materials and environmental triggers (e.g., pH, ionic strength, temperature, light, enzyme activity) (McClements, 2018a). The kinetic profile and release mechanisms of bioactive compounds encapsulated in different kinds of CDSs have been extensively studied using simple or sophisticated mathematical models and empirical, semiempirical, and analytical techniques (Flores & Kong, 2017; Katouzian et al., 2017; Malekjani & Jafari, 2021; Siepmann & Siepmann, 2012).

In the following are briefly presented the mathematical models corresponding to the diffusion-controlled release for two types of colloidal particles: core-shell colloidal particles and monolitic colloidal particles.

### Release from core-shell colloidal particles

The core-shell colloidal particles have a core as a reservoir in which the bioactive compound is completely or partially dissolved in a suitable solvent (polar solvent for hydrophilic bioactive compounds and non-polar solvent for hydrophobic bioactive compounds). The reservoir core is covered by a shell made of encapsulating material. The release of biocomponents from the core-shell colloidal particles consists in the passage of molecules through the shell of the particle due to the concentration gradient. Although the release is the result of several physical processes such as diffusion of the liquid into the particle, solubilization of biocomponents, diffusion of biocomponents in the release liquid. The slowest process, which limits the release rate, is the diffusion of bioactive molecules from the reservoir through wall material. There are two cases of core-shell colloidal particles, studied with two different kinetic equations. One case is the core-shell colloidal particles in which the biocomponent is completely dissolved (reservoir solution type). In this case, the concentration of the bicomponent in the reservoir is less than its solubility and decreases during release. The second case refers to the core-shell colloidal particles in which the biocomponent is partially dissolved (reservoir dispersion type), and the concentration of the biocomponent in the reservoir is higher than its solubility. During the release, the biocomponent concentration remains constant because the amount of biocomponent released is replaced by the same amount of dissolved biocomponent (Malekjani & Jafari, 2021).

The kinetic study of the release by diffusion of the molecules of biocomponent dissolved integrally in core phase (reservoir solution type) was performed using a first-order kinetic equation of the following form (McClements, 2015; Siepmann et al., 2012; Siepmann & Siepmann, 2012):

$$\frac{M_{(t)}}{M_\infty} = 1 - \exp\left[-\frac{3 \cdot r_O \cdot D_S \cdot K_{SC}}{r_I^2 \cdot r_O - r_I^3} \cdot t\right] = 1 - \exp\left[-\frac{3 \cdot (r_I + \delta) \cdot P_S}{r_I^2} \cdot t\right] \tag{4.8}$$

here, $r_O$ and $r_I$ are the inner and outer radii of the shell, $D_S$ is the diffusion coefficient of the bioactive component within the shell; $K_{SC}$ is the shell-core partition coefficient of the bioactive component;

$\delta = (r_O - r_I)$ is the thickness of the shell and $P_S = \frac{D_S \cdot K_{SC}}{\delta}$, is the permeability of the shell. $M_{(t)}$ and $M_\infty$ are the amount of bioactive compound released at time $t$, respectively the amount of bioactive compound released when the system reaches equilibrium (which can be considered equal to the initial amount of biocomponent within particle). Eq. (4.8) expresses the release in time of biocomponents totally solubilized in particle core (reservoir solution). In this case the concentration of biocomponents in the reservoir solution decreases during release. Eq. (4.8) is valid by assuming the following conditions (Malekjani & Jafari, 2021):

- the particle is considered spherical and the particle core is homogeneous (reservoir solution);
- the rate-determining step is the diffusion of biocomponent molecules through the shell;
- the coefficient of diffusion is constant;
- the shell thickness is very small compared to the particle size and it remains constant during release;
- perfect sink condition is maintained during release (the concentration of biocomponent in the acceptor fluid is negligible, because the concentration in the core particles is much higher than the concentration in the fluid surrounding the particle);
- the effect of the liquid layers on the surface of the particles is negligible

Eq. (4.8) shows that the rate of release of biocomponents from core-shell colloidal particles with homogeneous particles core (reservoir solution) increases as the diffusion coefficient and the shell-core partition coefficient increase and the thickness of the shell decreases (Zandi et al., 2017).

Sometimes bioactive compounds are not completely dissolved in the solvent in the reservoir so that in the particle core there is a dispersion in which the bioactive compound exists in both soluble and insoluble form. In this case the concentration of the bioactive component in the reservoir dispersion ($C$) is higher than the equilibrium solubility of the bioactive component (saturation concentration $C_S^*$). If the dissolution rate of the insoluble fraction is higher than the diffusion of molecules through the particle shell, then the concentration of the solubilized component remains constant during release because the released molecules are replaced by the dissolved ones. For core-shell spherical particles, of the "reservoir dispersion" type, the release rate is given by a zero-order equation, of the following form (Eq. 4.9):

$$M(t) = \frac{4\pi \cdot D_S \cdot K_{SC} \cdot C_S^* \cdot r_O \cdot r_I}{r_O - r_I} \cdot t = 4\pi \cdot P_S \cdot C_S^* \cdot (r_I + \delta) \cdot r_I \cdot t \qquad (4.9)$$

In accord with the Eq. (4.9), the increase of some parameters such as the diffusion coefficient ($D_S$), the shell-core partition coefficient ($K_{SC}$), and the saturation concentration of the bioactive in the particle core ($C_S^*$) determine increase of the release rate while the increase of thickness of the shell determines lower release rate of bioactive from the nanoparticles of "reservoir" type. The two equations highlight the main factors that can be manipulated in such a way that the release is prolonged or rapid. For example, to design a CDSs with prolonged release, the conditions for decreasing the parameters such as the diffusion coefficient, core-shell partition coefficient and the increase of the shell thickness of core-shell particles are studied. Reduction of the diffusion coefficient of the bioactive component through the shell material is studied according to the Stokes-Einstein equation applied to spherical particles moving through a homogeneous fluid:

$$D_S = \frac{k_B \cdot T}{6\pi \cdot \eta_S \cdot r} \qquad (4.10)$$

where $k_B$ is the Boltzmann's constant, $T$ is the absolute temperature, $\eta_S$ is the viscosity of the shell material, and $r$ is the radius of the bioactive molecule.

According to Eq. (4.10), the reduction of the diffusion coefficient can be achieved by increasing the viscosity of the shell material. It should also be borne in mind that large bioactive molecules have a lower diffusion coefficient than small molecules. For example, D-limonene has a molecular radius of 0.402 nm and an oil diffusion coefficient of $1.09 \times 10^{-11}$ m²/s, while vitamin C (ascorbic acid) has a molecular radius of 0.330 nm and an oil diffusion coefficient of $1.32 \times 10^{-11}$ m²/s (McClements, 2015).

Shell-core partition coefficient ($K_{SC}$) is defined as the ratio between the concentration of the biocomponent in the particles shell and the concentration of the biocomponent in the particles core. Therefore, its reduction is achieved by decreasing the concentration of biocomponent in the particle shell. In the case of hydrophilic biocomponents, the reduction of shell-core partition coefficient can be made by using hydrophobic coating material such as lipids with high melting points (e.g., triglycerides and waxes) or hydrophobic polymers below their $T_g$ (e.g. zein, gliadin, gum copal and gum damar) in which the solubility of the hydrophilic biocomponent is very low (Abu-Diak et al., 2012; McClements, 2015; Wadher et al., 2011). The thickness of the shell ($\delta$) can be controlled by core-shell particle preparation techniques. Thus, for liposomes, the layer can be a few nanometers thick, while for emulsions prepared by the LBL technique, $\delta$ can be several hundred nanometers depending on the number of layers added, and for particles obtained by spray-drying technique, $\delta$ can reach several hundred micrometers (McClements, 2015).

### Release from homogeneous colloidal particles

Some CDSs have homogeneous liquid particles such as emulsions or solids particles such as polymer microparticles, fat microparticles and SLNs. Their preparation is performed by solubilizing the bioactive component in a hydrophilic (for hydrophilic compounds) or lipophilic (for hydrophobic compounds) base, followed by uniform dispersion in a matrix (polymer or fat). Spray chilling or extrusion are techniques for preparing homogeneous colloidal particles loaded with nutraceuticals used in food fortification (Oxley, 2012). The following kinetic equation can be used to study the release of bioactive compounds from homogeneous colloidal particles (Siepmann & Siepmann, 2012):

$$\frac{M_{(t)}}{M_\infty} = 1 - \exp\left[-\frac{1.2\pi^2 \cdot D}{K_{DC} \cdot r^2} \cdot t\right] \tag{4.11}$$

where $D$ is the diffusion coefficient of the biocomponent inside the colloidal particle; $r$ is the particle radius and $K_{DC}$ is the dispersed phase-continuous phase partition coefficient and is defined as the ratio of the concentration of biocomponent inside the particle to the concentration of biocomponent in the liquid in which the biocomponent is released.

Eq. (4.11) is applied by assuming the following conditions:

- the colloidal particles are spherical and homogeneous;
- the biocomponent is completely dissolved in the particle;
- the initial concentration of the biocomponent in the continuous phase (specific environmental trigger) is zero;
- dispersion phase-continuous phase partition coefficient ($K_{DC}$) has high values.

According to Eq. (4.11), the diffusion coefficient and the particle radius influence differently the release rate of biocomponents from homogeneous colloidal particles. Thus, the decrease of the diffusion coefficient determines the decrease of the release rate. Therefore, the release of bioactive from liquid particles is faster than from solid particles, because the latter has a high viscosity which makes it difficult for molecules to diffuse through the particles. On the contrary, the decrease of the radius of the homogeneous colloidal particles determines the increase of the release rate. The release of biomolecules from large particles is slower than the release from small particles, due to the large distance that the molecules have to travel to get out of the particle. The kinetic curves show that the release rate of biocomponents from homogeneous particles increases rapidly in the first minutes (burst release), after which the release is slow for a long time, as the concentration gradient of the biocomponent in the particle and the outer liquid decreases.

### Release from porous colloidal particles

Many types of CDSs used in food functionalization contain polymeric particles in which the polymer chains form a three-dimensional network in which the solubilized biocomponent molecules are trapped. The release of molecules from porous colloidal particles, such as polymeric particles can be achieved by simple diffusion, swelling and diffusion or fragmentation (Zhang et al., 2015). For release from hydrogel particles, the molecules of the bioactive agent diffuse first through the solvent in which it was solubilized (usually water for hydrophilic compounds) located in the pores of the network, then through the polymeric network of the particle. The release rate of bioactives from hydrogel particles is studied with an equation similar to Eq. (4.11), in which the diffusion coefficient ($D_{gel}$) is defined according to the polymer structure and pore size, using Eq. (4.12) (Chan & Neufeld, 2009).

$$D_{gel} = D_{Water} \exp\left[-\pi\left(\frac{r_H + r_f}{\xi + 2r_f}\right)\right] \quad (4.12)$$

where $D_{Water}$ is the translational diffusion coefficient of the biocomponent through pure water, $\xi$ is pore size, $r_f$ is the cross-sectional radius of the polymer chain and $r_H$ is the hydrodynamic radius of polymer.

Knowledge of the mechanisms and kinetics of the release of biocomponents from colloidal particles in CDSs aids food scientists assess the bioaccessibility of biocomponents during food digestion, which is an important step in their bioavailability. Also, the study of the retention and release of biocomponents provides useful information for determining the optimal doses used to fortify foods, so that biocomponents have maximum biological effects and avoid their uncontrolled release that could cause an unpleasant odor or taste in food.

## 4.4 Colloidal delivery systems in food functionalization

To this end, along with traditional techniques (formulation and blending), recently innovative techniques have been developed for food functionalization, such as: micro- and nanoencapsulation of bioactives using different CDSs, obtaining edible films with bioactives, vacuum impregnation and production of the personalized foods (nutrigenomics) (Betoret et al., 2011).

There are a large number of types of CDSs used in food functionalization, starting with the simple types of CDSs and continuing with the complex CDSs mentioned at the beginning of this chapter. This section will briefly describe the main types of CDSs and their applications in fortifying different foods.

### 4.4.1 Lipid based-colloidal delivery systems
#### 4.4.1.1 Emulsions and microemulsions

Emulsions are colloidal dispersions obtained by mixing of two immiscible liquids, such as oil and water, in the presence of an emulsifier whose molecules are adsorbed at the O/W interface favoring the breaking of the film between the two liquids and the transformation of a liquid into droplets (Fig. 4.2). Depending on the O/W ratio and the nature of the emulsifier, one of the liquids is dispersed in the form of small droplets (discontinuous phase), inside the other liquid which represents the continuous phase. In general, hydrophilic emulsifiers (HLB > 10) favor the formation of O/W emulsions, while lipophilic emulsifiers (HLB < 8) favor the formation of W/O emulsions. The packing parameter of the surfactants also influences the type of emulsion formed. This allows the use of emulsions as systems for transporting and delivering both hydrophobic and hydrophilic biocomponents. O/W emulsions are the most used CDSs in the food industry and will be discussed in this section.

Depending on the size of the oil droplets, O/W emulsions are: conventional emulsions, which contain droplets with an average diameter >200 nm and nanoemulsions, which contain droplets with a diameter between 20 and 200 nm (Choi & McClements, 2020). Both conventional emulsions and nanoemulsions are thermodynamically unstable systems and are destroyed by gravitational separation, flocculation, coalescence and Ostwald ripening. The main differences are that nanoemulsions have better stability to gravitational separation and flocculation than conventional emulsions, and the appearance of nanoemulsions is transparent to translucent, while conventional emulsions are turbid to opaque.

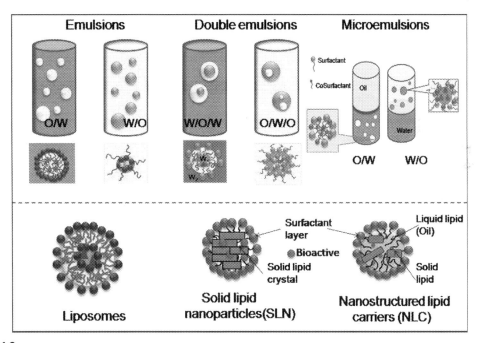

**FIG. 4.2**

Lipid-based colloidal delivery systems.

This allows the use of nanoemulsions to fortify transparent drinks (waters, soft drinks) (McClements, 2015; Piorkowski & McClements, 2014).

Conventional emulsions and nanoemulsions are prepared by high-energy method or low-energy method. The first group of methods uses devices that produce strong mechanical forces, such as high-pressure homogenizers, microfluidizers, or ultrasonicators, which are capable of breaking the two immiscible liquids or dividing the larger droplets into smaller droplets (Jafari et al., 2007). Obtaining emulsions or nanoemulsions by low-energy methods does not require mechanical devices, because they are based on spontaneous emulsification, phase inversion temperature, and phase inversion composition methods (Komaiko & McClements, 2015). Due to the high content of surfactants, which could increase the risk of toxicity, low-energy methods are not recommended for the production of nanoemulsions with applications in the food industry.

In recent decades, an impressive number of papers on emulsions and especially on nanoemulsions have been published in the literature. Some authors studied the influence of factors such as oil type, O/W ratio, emulsifier type, preparation technique on physico-chemical characteristics (Choi & McClements, 2020; McClements & Rao, 2011; Ozturk, 2017) while other researchers used nanoemulsions as food excipients in order to improve the bioavailability of nutraceuticals (McClements & Xiao, 2014). Luo et al. (2017) studied the effects of emulsifier concentration (quillaja saponins and whey protein isolates) and microfluidization pressure on the stability of β-carotene-loaded nanoemulsions, and Kumar et al. (2017) prepared resveratrol-loaded nanoemulsions with 99% LC using a mixture of lecithin and Tween 80 as emulsifiers, and a sonicator as a homogenizer. Other researchers evaluated the influence of oil phase composition on the bioavailability of nutraceuticals in nanoemulsions (Qian et al., 2012; Salvia-Trujillo et al., 2013).

Zhang, Zhang, Kumosani, et al. (2016) prepared β-carotene-loaded nanoemulsions, using long-chain triglycerides, medium-chain triglycerides and orange oil as the oily phase, and Ahmed et al. (2012), prepared curcumin-loaded nanoemulsions using long-, medium-, or short-chain triglycerides as the carrier oil and evaluated the bioavailability of biocomponents in simulated GI fluids. The results revealed that the bioavailability of nutraceuticals was much higher in nanoemulsions prepared with long-chain triglycerides than in those prepared with medium-chain triglycerides, because long-chain triglycerides formed mixed micelles with larger hydrophobic domains in which were solubilized a more amount hydrophobic biocomponent. A number of researchers investigated the possibility of fortifying foods such as beverages, cheese, milk, yogurt, meat and meat products, using nanoemulsions loaded with vitamins, polyphenols, carotenoids, polyunsaturated fatty acids (PUFAs), plant extracts, among others (Table 4.3).

An interesting type of emulsion is double emulsions. They are obtained by embedded a type of emulsion (W/O or O/W emulsions) in a continuous external phase which may be aqueous or oily phase. Depending on the structure, the double emulsions can be W/O/W and O/W/O double emulsions. The advantage of these CDSs is their ability to encapsulate both hydrophilic and hydrophobic bioactive compounds. W/O/W double emulsions can also be used in the preparation of reduced-fat foods (Eisinaite et al., 2017). The W/O/W emulsions are used more to fortify food due to their better stability (Muschiolik & Dickinson, 2017). The preparation of the W/O/W emulsions is done in two steps. A W/O emulsion is first prepared using a lipophilic emulsifier. The bioactive component is solubilized in the internal aqueous phase and the hydrophobic one is solubilized in oil. The W/O emulsion is added in portions, under gentle stirring in the external aqueous phase containing a hydrophilic emulsifier. Several researchers studied the W/O/W emulsions as delivery systems of different compounds, such

Table 4.3 Overview on representative colloidal delivery systems used in food functionalization.

| Colloidal delivery system (CDS) | Bioactive component | Encapsulation material | Encapsulation technique | Food matrix/functional food | References |
|---|---|---|---|---|---|
| Nanoemulsions | Nisin, Thymol | Soybean oil, Starch octenyl succinate | High-speed homogenization followed by high-pressure homogenization | Cantaloupe juice | Sarkar et al. (2017) |
| Conventional emulsion | Vitamins E, A, Q10 | Flaxseed oil | Mechanical homogenization followed by microfluidization | Cheese milk | Stratulat et al. (2014) |
| Nanoemulsions | Pear juice concentrate | Medium chain triglyceride Tween 20, Starch | Sonication method | Pear juice | Chaudhari et al. (2015) |
| Nanoemulsion | *Mentha piperita* L. Essential oil | *Mentha piperita* L. Essential oil, distilled water and ethanol | Emulsion phase inversion method | Guava and mango juices | de Carvalho et al. (2018) |
| Bilayer emulsions | ω-3 polyunsaturated fatty acids | Chia oil, sunflower lecithin, chitosan | Mecanical omogenization LBL method | Oil fortified | Julio et al. (2018) |
| Water/oil/water double emulsions | Olive leaves extract | Olive oil, linseed oil, fish oil | Two steps method | Fresh pork meat | Robert et al. (2019) |
| Nanostructured lipid carriers (NLC) | Vitamin D3 | Glyceryldistearate Glyceryldibehenate Caprylic/caprictriglycerides | Hot mecanical omogenization | Food Beverages | Mohammadi et al. (2017) |
| NLC | Docosahexaenoic acid (DHA), Eicosapentaenoic acid (EPA), Astaxanthin | Palm stearin Krill oil Lecithin | Mechanical homogenization and sonication | Simulate beverage | Zhu et al. (2015) |
| Liposomes | Vitamin D3 | Lecithin, Maltodextrin, Gum Arabic, Modified starch, Milk protein concentrate | Spray drying | Yogurt | Jafari et al. (2019) |
| Liposomes | Fish oil | Soy lecithin, Sunflower oil | Thin film hydration/Sonication | Yogurt | Ghorbanzade et al. (2017) |

*Continued*

Table 4.3 Overview on representative colloidal delivery systems used in food functionalization—cont'd

| Colloidal delivery system (CDS) | Bioactive component | Encapsulation material | Encapsulation technique | Food matrix/ functional food | References |
|---|---|---|---|---|---|
| Liposomes | Ascorbic acid | Dipalmitoylphosphatidylcholine (DPPC) Cholesterol | Dehydration-rehydration | Orange juice | Wechtersbach et al. (2012) |
| Liposomes | *Satureja* plant essential oil | Lecithin, cholesterol, chitosan | Thin-film hydration and sonication | Lamb meat | Pabast et al. (2018) |
| Liposomes | Vitamins C, E | Soybean phosphatidylcholine Soybean phosphatidylcholine-Stearylamine | Dehydration-rehydration | Orange juice | Marsanasco et al. (2011) |
| Polymer microparticles | Chia oil | Maltodextrin (10 DE) and modified starch (Hi-Cap100 | Spray drying | Meat product | Noello et al. (2016) |
| Polymer microparticles | Allicin | β-Cyclodextrin and porous starch | Spray drying | Tofu, bread cooked chicken and cooked pork | Wang et al. (2018) |
| Polymer microparticles | Thyme essential oil | Chitosan | Emulsion/ionic gelation | Beef burger | Ghaderi-Ghahfarokhi et al. (2016) |
| Polymer nanoparticles | Curcumin | Myofibrillar protein extracted from minced chicken breast | Electrostatic complexation | Marinated chicken meat | Wu et al., 2019 |
| Polymer nanoparticles | Curcumin | Skim milk | Spray drying | Milk fortification | Neves et al. (2019) |
| Polymer nanoparticles | Vitamin D | Potato protein | Precipitation | Food beverage | David and Livney (2016) |
| Polymer microparticles | Curcumin | Gelatin, Porous starch | Spray drying | Tofu, cooked pork and bread | Wang et al., 2012 |
| Milk protein complexes | Vitamin A | Sodium caseinate Succinylated sodium caseinate | Assembled and reassembled polymers | Milk fortification | Rana et al. (2019) |
| Casein micelles | Vitamin D3 | Casein | Reassembly of casein micelles | Fat-free yogurt | Levinson et al. (2016) |

as vitamins (Dima & Dima, 2018; Dima & Dima, 2020), resveratrol (Matos et al., 2014), curcumin and catechin (Aditya et al., 2015), or minerals (Herzi & Essafi, 2018; Zhu et al., 2017).

Microemulsions are thermodynamically stable and isotropic colloidal systems that contain the association micelles (5–100 nm) formed by the self-assembly of surfactants. Depending on the structure, the microemulsions can be O/W and W/O microemulsions which allows the solubilization and delivery of hydrophobic or hydrophilic biocomponents. Research in recent years shown that microemulsions can be used successfully as nutraceutical delivery systems, given the following advantages (Roohinejad et al., 2018):

- is formed spontaneously, without energy consumption;
- are transparent, and do not change the appearance of food;
- the size of the drops is very small, which allows absorption through the intestinal wall;
- they have low viscosity, being easy to administer;
- they are thermodynamically stable, which allows them a long storage time;
- have a high solubilization capacity, which allows the incorporation and transport of a large amount of bioactive compound.

In this regard, different types of microemulsions have been prepared in order to solubilize flavor compounds and dyes, extract bioactive compounds and increase the bioavailability of lipophilic nutraceuticals (Amiri-Rigi & Abbasi, 2019; Garti & Aserin, 2012; Uchiyama et al., 2019). The main disadvantages of microemulsions are the low stability to temperature variation and the high content of surfactants which increases the risk of toxicity and cost price.

### 4.4.1.2 Solid lipid nanoparticles and nanostructured lipid carriers
In last decades, two lipid-based CDSs, namely SLNs and NLCs were widely investigated. These are solid lipid particles with a size between 30 and 1000 nm and are prepared by the same high-energy methods as nanoemulsions, with the difference that the oily phase consists of solid lipids with a melting point higher than 40 °C. The preparation of SLNs, and NLCs is done in the following stages (Duong et al., 2020):

- melting of the solid lipid and incorporation of the hydrophobic biocomponent in the oily phase;
- mixing the hot oil phase with the hot aqueous phase in the presence of emulsifiers using an appropriate technique;
- the controlled cooling of the obtained nanoemulsion, when the total or partial crystallization of the lipid from the discontinuous phase droplets takes place;
- separation of SLNs or NLCs.

For the preparation of SLNs and NLCs, lipids with a high content of saturated fatty acids are used, such as: purified stearic acid, triglycerides, diglycerides, monoglycerides, glyceryl behenate, carnauba and candelilla waxes, hydrogenated sunflower and rapeseed oils and coconut oil. The emulsifiers used in the preparation of SLNs and NLCs are: mono- and diacylglycerols, sorbitan esters, sucrose esters and phospholipids (da Silva Santos et al., 2019). Along with emulsifiers, co-emulsifiers are sometimes used, such as: sodium taurodeoxycholate, sodium glycolate, sodium dodecyl sulfate, which contribute to the further decrease of interfacial tension, increase of particle stability, inhibition of lipid crystallization and increase system viscosity. The SLNs size and morphology are influenced by numerous factors, such as: lipid properties, surfactant type, surfactant/oil/water ratio, bioactive component properties, preparation technique (Gordillo-Galeano & Mora-Huertas, 2018). The characteristics of the encapsulated bioactive compound influence the internal structure of SLN. Thus, bioactive components, which are liquid at room temperature, with melting points below the melting point of the oily phase, form capsule-like particles, in which one or more droplets of the liquid bioactive component are surrounded by a solid lipid coating. Bioactive compounds with high melting points are found inside or on the surface of SLN in crystals form dispersed in a solid lipid matrix. SLN have a number of disadvantages, such as: the relatively low nutraceutical loading capacity and the dispersion of biomolecules on the particle surface and their expulsion due to the polymorphic transition of the lipid: $\alpha \rightarrow \beta' \rightarrow \beta$.

In order to eliminate these disadvantages, some researchers studied the possibility of changing the structure of solid lipid particles to obtain a new type of solid lipid particles, used as carriers of bioactive compounds, called NLCs (Müller et al., 2002 ). Therefore, NLCs are SLNs with a modified structure by adding an oil such as medium-chain triglycerides (MCT), oleic acid, sunflower oil, flaxseed oil and soybean oil, flaxseed oil and palm oil to the molten lipid. The oil in the solid lipid matrix modifies the internal structure of the NPs and improves their qualities (stability and release rate). Thus, the hydrophobic bioactive component is solubilized in the liquid lipid inside the lipid nanoparticle, contributing to the increase of their loading capacity. Also, in the mixture of oil and solid lipid, the crystallinity of fats decreases, avoiding the expulsion of biomolecules from the particles. NLCs are more efficient CDSs than SLNs because: they are more stable, have a higher loading capacity, provide better protection of bioactive compounds and their controlled release, are easily digestible and improve the bioavailability of bioactive compounds (Das et al., 2012).

Several researchers encapsulated bioactives in SLNs and NLCs with possible applications in food fortification (Katouzian et al., 2017). For example, Pandita et al. (2014) prepared resveratrol-loaded SLNs using stearic acid and poloxamer 188, by solvent diffusion-solvent evaporation method, and Couto et al. (2017) encapsulated vitamin $B_2$ in fully hydrogenated canola oil NPs using the supercritical co-injection process and production of particles from gas-saturated solutions.

Although only two decades have passed since the first studies on NLCs, they soon became one of the most studied CDSs, with great potential for application both in the food industry and in the cosmetics and pharmaceutical industries. Many researchers studied the encapsulation, retention and release, bioaccessibility and bioavailability of a wide range of hydrophobic biocomponents encapsulated in NLCs, using different solid and liquid lipids, surfactants and preparation techniques. Thus, Liu and Wu (2010) encapsulated lutein in NLCs using tripalmitin as solid lipid and corn oil as liquid lipid. Other authors used as liquid lipids even hydrophobic biocomponents, such as α-tocopherol which has been incorporated into lipid NPs as a concentrated plant extract or in the form of oils containing it in significant amounts, such as soybean, sunflower, and canola oil (Tamjidi et al., 2013).

Many researchers have also encapsulated polyunsaturated fatty acids in NLCs to prevent oxidative degradation and studied the effects of solid lipids, surfactants and preparation techniques on physical stability, encapsulation efficiency, retention and oxidation during production, packaging, storage, and treatment (Lacatusu et al., 2013; Salminen et al., 2013). Some researchers applied nutraceutical-loaded SLNs or NLCs to fortify different foods (Table 4.3). For example, Zhu et al. (2015) prepared NLCs containing high content of krill oil using palm stearin as a solid lipid and lecithin as a surfactant and polyunsaturated fatty acids as the bioactive. The authors optimized the parameters related to the composition of the mixture (krill oil and lecithin levels) and the parameters related to sonication technique (power and time sonication), using response surface methodology. The reported results showed that lipid NPs with a size of approximately 140 nm and the encapsulation efficiency of bioactives [docosahexaenoic acid (DHA), eicosapentaenoic acid (EPA) and astaxanthin] of 96–97% were obtained. NLCs dispersed in water or in a simulated beverage were physically and chemically stable during storage for 70 days at room temperature and 4 °C and showed increased protection of DHA, EPA and astaxanthin inside the particles against the photooxidation reaction.

Other researchers prepared vitamin D3-loaded NLCs dispersion using glyceryldistearate, glyceryldibehenate and caprylic/caprictriglycerides, as solid lipids, Miglyol and octyloctanoat as liquid lipids, and Tween 80, Tween 20 and Poloxamer 407 as surfactants (Mohammadi et al., 2017). The authors obtained vitamin $D_3$ loaded-NLCs with a size ranged between 77 and 2504 nm, depending on the type and amount of lipids and surfactants used. *In vivo* studies on the bioavailability of vitamin $D_3$ revealed that the incorporation of vitamin $D_3$ into NLCs increased the absorption of vitamin $D_3$ by oral administration.

### *4.4.1.3 Liposomes*

Liposomes are spherical lipid aggregates, with a size between 20 nm and a few μm obtained by self-closing phospholipid bilayers around an aqueous core. Since 1965, when Alec Banghan and his team published the first works on liposomes, and to this day, many researchers teams studied the mechanisms of formation, structure and especially their applications (Qu et al., 2017; Sala et al., 2017). Liposome formation is the result of intermolecular lipid-lipid and lipid-water interactions. The energy introduced into the system by different techniques (homogenization, sonication or heating) contributes to the self-assembly of amphiphilic lipid molecules in bilayers and the formation of bilayered vesicles. The curvature and closure of lipid bilayers depend on the one hand on the energy supplied and on the other hand on the structure of the lipids present in the membrane. For example, sterols play an important role in liposome formation and stabilization (Mozafari et al., 2008). In the preparation of liposomes, neutral phospholipids (1-monostearoyl-rac-glycerol, 1,2-dipalmitoyl-sn-glycerol, tripalmitin, cholesterol) and ionic phodpholipids (1-a-phosphatidylcholine, 3-sn-phosphatidylethanolamine, l-a-phosphatide-l-serine, l-a-phosphatidylinositol, lyso-phosphatidylcholine, sphingomyelin) are used.

Vesicles obtained using synthetic nonionic surfactants, such as: alkyl ethers, alkyl glyceryl ethers, fatty acid esters with sorbitan, fatty acid esters with polyoxyethylenes are called niosomes. They are more stable than liposomes, are obtained in milder conditions and have a higher encapsulation efficiency. Therefore, niosomes are often preferred as CDSs for foods fortification with vitamins, minerals, nutraceuticals (Dima et al., 2020c; Mozafari et al., 2008). The literature mentions several criteria for classifying liposomes. Depending on the structure, liposomes (and niosomes) are grouped into unilamellar vesicles (ULVs, >20 nm), consisting of a single phospholipid bilayer surrounding the aqueous phase and multilamellar vesicles (MLVs, ≥400 nm), consisting of several bilayers concentric separated by aqueous domains. The unilamellar vesicles can be subclassified by size into giant unilamellar vesicles (GUVs, >1 μm), large unilamellar vesicles (LUVs, 80 nm to 1 μm), and small unilamellar vesicles (SUVs, 20–80 nm) (Liu et al., 2015). Liposomes and niosomes are prepared by multiple methods, such as: lipid layer hydration, reverse phase evaporation, transmembrane pH gradient method, microfluidization, extrusion and ultrasound (Rostamabadi et al., 2019).

The selection of the liposome preparation method is made taking into account certain factors, such as: type of liposomes, properties of phospholipids, the relationship of phospholipids and environment in which they are dispersed, properties of encapsulated bioactive molecules, etc. The presence of cholesterol in the structure of liposomes determines an increase in liposome stability, but also a decrease in encapsulation efficiency (da Silva Malheiros et al., 2010). The improvement of the functional properties of liposomes has been achieved by grafting on their surface some polymeric compounds, such as polyethylene glycol, chitosan and its derivatives or bio-recognition agents such as DNA probe, gangliosides and antibodies, which ensure the steric stabilization of liposomes, recognition of enzymes, and good adhesion on the surface of biological membranes, contributing to increasing the bioavailability of nutraceuticals (Caddeo et al., 2018).

Researchers today talk about four generations of liposomes, with multiple applications in medicine starting with the classic first generation of liposomes, and continuing with liposomes with more complex structure and functionality, such as: stealth liposomes and stimulus-responsive liposomes, ligand-targeted liposomes and theranostic liposomes (Islam Shishir et al., 2019). The use of liposomes in the food industry is limited by their low stability, high cost price of pure phospholipids, low encapsulation efficiency, presence of residues of organic solvents, and difficulty in their production on an industrial scale. However, liposomes also have some advantages, such as the presence in their composition only of natural substances, the possibility of encapsulating both hydrophilic and hydrophobic biocomponents and excellent compatibility with many food matrices. Some companies have distributed various liposome-based commercial products on the market, such as: Liposomal turmeric with fulvic acid, Liposomal curcumin syrup, Liposomal vitamin C tablets, Liposomal glutathione capsules, etc., which bring important health benefits to consumers (Shukla et al., 2017). Although numerous papers have been published in the literature on the encapsulation of biocomponents in liposomes, the stability and behavior of liposomes in simulated GI fluids, the kinetics of release or cellular absorption of liposomes, there are few studies on the food functionalization using liposomes loaded with nutraceuticals (Table 4.3).

### 4.4.2 Polymeric colloidal delivery systems

Biopolymers, mainly proteins and polysaccharides, are food-grade materials used extensively in the encapsulation of biocomponents and food ingredients. The formation of polymeric colloidal particles is based on two main physicochemical processes, namely: associative processes, in which attractive forces are involved (e.g., electrostatic, wan der Waals, hydrogen bonds or hydrophobic interactions) and chemical bonds and segregating processes, in which repulsive forces between polymers are involved (e.g., electrostatic repulsions, steric exclusion). Depending on the mechanisms of formation, polymeric particles can be of several types (Fig. 4.3): polymeric micelles, polymeric particles formed from a single polymer and polymeric particles formed from associated polymers.

#### *4.4.2.1 Polymeric micelles*

Some polymers have in their structure hydrophobic monomer units or chains of hydrophobic monomers linked to other hydrophilic monomer residues or chains of hydrophilic monomers. These polymers behave like amphiphilic molecules having the ability to self-assemble into nanometer-sized aggregates, called polymeric micelles, in which various biocomponents can be trapped. Compared to the micelles of molecular surfactants, polymeric micelles are more stable, provide better protection of bioactive compounds, and have higher encapsulation efficiency (Dima & Dima, 2016).

Among the proteins that form micelles, casein has been the most studied (de Kruif et al., 2012). Casein in mammalian milk is a mixture of four polypeptide fractions, with different content of amino acids, phosphate groups and carbohydrates, known as: $\alpha_{S1}$-, $\alpha_{S2}$-, $\beta$-, and $\kappa$-casein. Under certain conditions, caseins self-assemble in the amorphous stable aggregates, called casein micelles, with a radius between 50 and 500 nm and a mass between $1 \times 10^3$ and $3 \times 10^6$ kDa (Livney, 2010). The formation of

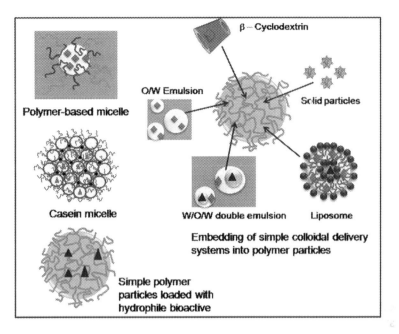

**FIG. 4.3**

Polymer-based colloidal delivery systems.

these aggregates is due to electrostatic and hydrophobic interactions, hydrogen bonds and van der Waals forces, which are manifested between polar and nonpolar amino acid sequences. The casein micelle has a complex structure. It contains approximately 94% casein molecules and 6% amorphous calcium phosphate, organized in clusters with a size of 3.5–5 nm, which binds the casein submicelles formed (Głab & Boratynski, 2017). The fractions $\alpha_{S1}$, $\alpha_{S2}$ and $\beta$ that form the submicelles, are inside the casein micelle, while κ-casein is on the outside, in the form of a brush that sterically stabilizes the micelle. The sequence of the 169 amino acids of κ-casein determines the formation of hydrophobic sequences, located at the beginning of the macromolecular chain and a hydrophilic sequence consisting of the terminal polar amino acids. Due to the amphiphilic structure, the κ-casein molecules (as of the other fractions) self-assemble in direct submicelles, with the hydrophobic inner volume and the negatively charged hydrophilic outer surface. Due to its gelling capacity, casein has been used as encapsulating material for encapsulating bioactive substances by various methods. Due to its gelling capacity, casein has been used to encapsulate bioactives by various methods. One method is to mix casein with bioactive compounds and gelling with gelling agents, such as: chymosin, glutaric aldehyde, transglutaminase and genipin (Głab & Boratynski, 2017).

Recently, NPs have been prepared by the method of reconstituting casein micelles. This method consists of three steps:

- destruction of casein micelles using enzymatic partial hydrolysis or by various techniques, such as high pressure and sonication or the addition of ethanol;
- solubilization of hydrophobic bioactive compounds in organic solvents such as ethanol;

- reassembly of micelles loaded with bioactive compounds, by releasing the solvent.
- Using this method some researchers encapsulated the fat-soluble vitamins $D_2$, $D_3$, $B_9$ (Levinson et al., 2016; Menéndez-Aguirre et al., 2014; Penalva et al., 2015), curcumin (Pan et al., 2014), polyunsaturated oils (Chen et al., 2017) and others.

### 4.4.2.2 Polymeric micro/nanoparticles from native polymers

An important property of biopolymers, described in Section 4.3.2.1 is gelling. This property has been extensively exploited by researchers for encapsulating biocomponents used in food fortification. Gelation of biopolymers can be achieved by physical or chemical methods, such as thermal gelation, ion gelation, osmotic gelation, and covalent gelation (Zhang et al., 2015).

*Thermal gelling* is based on the formation of gels by cooling or heating biopolymer solutions, below or above a certain temperature. For example, globular proteins, such as whey proteins, egg proteins, and soy proteins, form gels by heating due to hydrophobic interactions between nonpolar regions of protein chains that have undergone changes by thermal denaturation. Biopolymers with a fibrillar or helical structure, such as gelatin or agar, form reversible gels by cooling, due to the hydrogen bonds formed between certain regions of the polymer chains (Zhang et al., 2015). The "melting" temperature of the gel, the gelation temperature of the polymers and the $T_g$ are parameters that crucially influence the physico-chemical characteristics of the particles, as well as the behavior of the polymer particles during digestion and cellular uptake.

*Ion gelation* consists of the formation of a three-dimensional network between electrically charged polymer chains, by means of ions with electric charge opposite to that of the polymer. Polymers such as sodium alginate and pectins form gels in the presence of divalent ions of $Ca^{2+}$, $Zn^{2+}$, which provide the bond between two different polymer chains through carboxylate groups. The authors who prepared microparticles loaded with lipophilic bioactive compounds showed that the encapsulation efficiency, release rate and physical characteristics of microparticles depend on a number of parameters, such as: nature and concentration of biopolymer, nature and concentration of polyvalent ions, temperature, pH and ionic strength (Arroyo-Maya & McClements, 2015).

*Osmotic gelation* is based on the reduction of the excluded volume of the biopolymer in the presence of substances added to the biopolymer solution. For example, an aqueous solution of pectin, in an acidic medium, will gel by adding sugar in a higher concentration. The presence of sugar in the polymer solution creates an osmotic pressure that causes the elimination of sugar molecules near the biopolymer chain, causing a reduction in its excluded volume. Under these conditions, the chains of the biopolymer self-associate and form gels (Zhang et al., 2015).

*Gelation by covalent bonds* is based on the crosslinking of biopolymers (proteins, polysaccharides) either by heating, such as globular proteins, or by the use of substances such as glutaraldehyde, genipin, transglutaminase, which form covalent bonds with other functional groups of biopolymers (amino and hydroxyl groups).

### 4.4.2.3 Polymeric micro/nanoparticles from associated polymers

For the preparation of polymer micro/nanoparticles with high physico-chemical characteristics, intended to improve the nutritional and sensory properties of foods, some researchers used systems consisting of two or more polymers associated either by physical interactions (electrostatic, *van der Waals*, hydrogen bonds), or by chemical bonds. Polymers associated by physical interactions are also called *polyelectrolytic complexes*, and polymers associated by chemical bonds are called *conjugated polymers* or covalent complexes (Liu et al., 2017). The most used supramolecular polymeric structures applied to the preparation of food micro/nanoparticles are proteins and anionic polysaccharides.

*Protein-polysaccharide polyelectrolyte complexes* can be formed either by pH variation or temperature variation. The pH and ionic strength influence the net charge of proteins and the negative charge density of polysaccharides. At the isoelectric point (pI), proteins do not have a net charge (net charge is zero), but are regions located on the polymer chain with positive or negative effective charges that may be involved in their interaction with other polyelectrolytes. For example, at pH ≫ pI, the globular protein has a high net negative charge. Because of this, there is a segregating separation between the protein and an anionic polysaccharide, due to electrostatic repulsions between the polyanionic chains. When the concentration of one of the biopolymers increases above a certain value, phase separation occurs, in which each phase becomes richer in one of the biopolymers and poorer in the other biopolymer (Matalanis et al., 2011). At pH ≈ pI, the positive charges on certain segments of the protein chain interact with the anionic groups of the polysaccharide chain, forming a weak, unstable, water-soluble polyelectrolyte complex. At pH < pI, the number of positive groups ($-NH_3^+$) increases, causing an increase in the attractive electrostatic interactions between them and the anionic groups of polysaccharides. In this case, a phase separation (associative separation) takes place, also called complex coacervation. Coacervates are stable aggregates insoluble in water, which can be separated as micro/nanoparticles. Sometimes, by lowering the pH, electrostatic interactions between protein molecules and polysaccharides can lead to the formation of precipitates, which tend to stabilize. If the pH is reduced below the $pK_a$ values of the anionic groups of the polysaccharide chain, then the attractive interactions between the protein and polysaccharide molecules become very weak and the polyelectrolytic complex dissociates (Jones & McClements, 2011).

Polymeric micro/nanoparticles can also be formed by controlled heating of polymer complexes of globular proteins and ionic polysaccharides. The physico-chemical characteristics of the polymer particles formed (size, charge, morphology and behavior in the fluids of the GI tract) depend on the protein/polysaccharide ratio, the type of polysaccharide, pH, ionic strength and heating conditions. The polymer particles are formed when the native protein or the polyelectrolytic protein/polysaccharide complex is heated above the temperature of thermal denaturation of the globular protein, to a pH ≈ pI (pH ≈ 5.0–6.0). The preparation of polymer particles by heating the protein/polysaccharide system can be done differently depending on the order of heating of the components. One method consists in heating the native protein above the temperature of thermal degradation, at a certain pH, obtaining protein particles. During cooling, the polysaccharide is added over the protein particles. By reducing the pH to pI, the protein/polysaccharide polyelectrolyte complex is formed on the surface of the protein particles. In the second method, first the protein/polysaccharide polyelectrolyte complex solution obtained by adjusting the pH close to pI is heated. Heating takes place above the degradation temperature of the globular protein. The degraded protein cools, and the protein particles form micro/nanoparticles coated with polyelectrolyte complex. In practice, the second method is used more because more stable micro/nanoparticles are produced with higher encapsulation efficiency (Jones & McClements, 2011).

*Conjugated polymers* are other polymeric systems used in the preparation of micro/nanoparticles loaded with bioactive compounds. The formation of conjugated biopolymers can be achieved by chemical and enzymatic methods that produce covalent bonds between the functional groups of polymers (Liu et al., 2017). The most used chemical method of conjugating proteins with polysaccharides is the Maillard reaction. This reaction consists of the interaction between the carbonyl group of polysaccharides and the amino group of proteins, with the formation of an unstable Schiff base. By intramolecular isomerization (keto-enol tautomerism), the Schiff base is stabilized in an Amadori compound, which represents precisely the conjugated structure of proteins with polysaccharides. Experimentally, the Maillard reaction can be performed by heating, changing the pH or by applying a pulsating electric field to a mixture of proteins (milk proteins, egg proteins, meat proteins or plant proteins) and polysaccharides (dextran, chitosan, dextrins, pectins) (Guan et al., 2010). However, the preparation of micro- and nanocapsules loaded with nutraceuticals using conjugated biopolymers obtained by the Maillard reaction is restricted by altering the food color or by forming of compounds with toxicity risk (Liu et al., 2017). Proteins-polysaccharides conjugates can also be obtained by the action of enzymes, such as: transglutaminase, tyrosinase, laccase and peroxidase (Aljawish et al., 2015). Protein-polysaccharide conjugate systems show functional properties superior to native biopolymers, such as: solubility, thermal stability, resistance to pH, high emulsification and gelling capacity. Therefore, some researchers used conjugated polymeric systems to encapsulate nutraceuticals (Yao et al., 2015; Yi et al., 2014).

Polymeric micro/nanoparticles are widespread in the CDSs market with applications in the food and pharmaceutical industry. This, because they have a lower cost price than other CDSs, are more stable under storage conditions, are compatible with food matrices, provide good protection of encapsulated biocomponents and increase the bioavailability of nutraceuticals through controlled release during digestion. Also, polymeric micro/nanoparticles are obtained by less expensive techniques, with easy-to-handle machines and high yields. In practice, simple, homogeneous polymer particles containing hydrophilic bioactive compounds or polymeric particles filled with other types of CDSs can be prepared (Fig. 4.3). Polymeric micro/nanoparticles used in food fortification are prepared either by simple laboratory methods such as: heat treatment, stirring, coacervation, solvent evaporation, precipitation, or by special techniques such as: extrusion, spray drying, cooling or freezing drying, fluidized bed drying, among others. Numerous articles and monographs have recently been published that describe in detail the techniques for preparing nutraceutical-loaded polymer micro/nanoparticles (Garti & McClements, 2012; Jafari, 2020; Jones & McClements, 2011; Zhang et al., 2015). Some authors have studied the use of polymer micro/nanoparticles-based CDSs in the functionalization of some foods, bread, food beverages, meat product, dairy products (Table 4.3).

## 4.5 Conclusion and future perspective

Foods are systems with a complex structure in which the main materials exist in the form of colloidal dispersions with particles of different sizes, such as globular structures of proteins with a size of 1–10 nm, lipid aggregates with a size of 50–100 nm (self-assembled lipid nanostructures and nanoemulsions) or linear structures of polysaccharides whose fibers are several nanometers thick. Most of these colloidal systems are naturally introduced into foods from raw materials used in food preparation (e.g., meat, milk, vegetables, fruits) and play a crucial role in the texture of food. Lately, the food industry, like other economic sectors, has fully capitalized on the results of nanotechnologies, developing new food nanomaterials that helped to improve the sensory attributes of food. People's growing concern for their health has led food scientists and food producers to develop new foods that meet both nutritional and sensory requirements and especially those related to the prevention or treatment of some diseases. The group of these innovative foods also includes the functional foods whose market has experienced a great growth in recent years. One of the strategies for the preparation of functional foods is the encapsulation of bioactive compounds in different colloidal systems and their incorporation into food matrices. The design of CDSs is an essential step in the manufacture of functional foods because it critically analyzes the bioactive compounds, encapsulating materials, physicochemical characteristics of colloidal particles loaded with bioactive and technologies for obtaining CDSs, according to nutritional, sensory and health attributes demanded by consumers. The characteristics of colloidal particles, such as shape, size, charge, lipophilicity, but also the loading capacity, retention and release of bioactives must be carefully monitored because they can affect both the stability of CDSs, food sensory attributes and bioavailability of encapsulated bioactive components. Improving the bioavailability of bioactives from food and drugs remain a permanent concern of food scientists, doctors and pharmacists. In this regard, the use of micro and nanoparticles as bioactive delivery systems involves some challenges that should be overcome, including: (i) the development of new GRAS materials; (ii) release control of the bioactives and establishment of optimal doses corresponding to the required therapeutic effects; (iii) evaluation of the behavior of nanoparticles in the food matrix and in the gastrointestinal tract; (iv) development of new models and harmonization of investigation protocols for "in vitro" study of the main processes involved in the bioavailability of bioactives, such as: bioaccessibility, transport of nanoparticles and bioactives through the epithelial layer of the intestine and colon, distribution of bioactives to tissues and organs, metabolism and excretion of bioactives.

Although in the more than three decades of research, there are a large number functionalized foods with bioactive-loaded NPs on the market that have proven to be effective, there is still a high level of consumer caution when choosing these foods. Therefore, it is necessary to expand studies on the toxicity of NPs, to adopt international legislation governing the use of NPs in food and last but not least to correctly inform consumers about the presence of NPs in food and the possible risk of toxicity.

## Acknowledgment

Last author acknowledges the Chinese Ministry of Science and Technology "the Belt and Road" Innovative Talent Exchange Foreign Expert Project (Grant Number DL2021003001).

## References

Abu-Diak, O. A., Andrews, G. P., & Jones, D. S. (2012). Fundamentals and applications of controlled release drug delivery. *Controlled Release Society*, 47–74.

Aditya, N. P., Aditya, S., Yang, H. J., Kim, H. W., Park, S. O., Lee, J., & Ko, S. (2015). Curcumin and catechin co-loaded water-in-oil-in-water emulsion and its beverage application. *Journal of Functional Foods*, *15*, 35–43. https://doi.org/10.1016/j.jff.2015.03.013.

Aditya, N. P., Espinosa, Y. G., & Norton, I. T. (2017). Encapsulation systems for the delivery of hydrophilic nutraceuticals: Food application. *Biotechnology Advances*, *35*(4), 450–457. https://doi.org/10.1016/j.biotechadv.2017.03.012.

Ahmed, K., Li, Y., McClements, D. J., & Xiao, H. (2012). Nanoemulsion- and emulsion-based delivery systems for curcumin: Encapsulation and release properties. *Food Chemistry*, *132*(2), 799–807. https://doi.org/10.1016/j.foodchem.2011.11.039.

Alai, M. S., Lin, W. J., & Pingale, S. S. (2015). Application of polymeric nanoparticles and micelles in insulin oral delivery. *Journal of Food and Drug Analysis*, *23*(3), 351–358. https://doi.org/10.1016/j.jfda.2015.01.007.

Alehosseini, E., Jafari, S. M., & Shahiri Tabarestani, H. (2021). Production of d-limonene-loaded Pickering emulsions stabilized by chitosan nanoparticles. *Food Chemistry*, *354*, 129591. https://doi.org/10.1016/j.foodchem.2021.129591.

Aljawish, A., Chevalot, I., Jasniewski, J., Scher, J., & Muniglia, L. (2015). Enzymatic synthesis of chitosan derivatives and their potential applications. *Journal of Molecular Catalysis B: Enzymatic*, *112*, 25–39. https://doi.org/10.1016/j.molcatb.2014.10.014.

Amiri-Rigi, A., & Abbasi, S. (2019). Extraction of lycopene using a lecithin-based olive oil microemulsion. *Food Chemistry*, *272*, 568–573. https://doi.org/10.1016/j.foodchem.2018.08.080.

Arihara, K. (2014). Functional foods. In *Encyclopedia of meat sciences* (pp. 32–36). Elsevier Inc. https://doi.org/10.1016/B978-0-12-384731-7.00172-0.

Arroyo-Maya, I. J., & McClements, D. J. (2015). Biopolymer nanoparticles as potential delivery systems for anthocyanins: Fabrication and properties. *Food Research International*, *69*, 1–8. https://doi.org/10.1016/j.foodres.2014.12.005.

Arvizo, R. R., Miranda, O. R., Thompson, M. A., Pabelick, C. M., Bhattacharya, R., David Robertson, J., Rotello, V. M., Prakash, Y. S., & Mukherjee, P. (2010). Effect of nanoparticle surface charge at the plasma membrane and beyond. *Nano Letters*, *10*(7), 2543–2548. https://doi.org/10.1021/nl101140t.

Assadpour, E., Dima, C., & Jafari, S. M. (2020). 1. Fundamentals of food nanotechnology. In S. M. Jafari (Ed.), *Handbook of food nanotechnology* (pp. 1–35). Academic Press.

Assadpour, E., & Jafari, S. M. (2019). An overview of biopolymer nanostructures for encapsulation of food ingredients. In *Biopolymer nanostructures for food encapsulation purposes* (pp. 1–35). Elsevier. https://doi.org/10.1016/B978-0-12-815663-6.00001-X.

Auriemma, G., Mencherini, T., Russo, P., Stigliani, M., Aquino, R. P., & Del Gaudio, P. (2013). Prilling for the development of multi-particulate colon drug delivery systems: Pectin vs. pectin-alginate beads. *Carbohydrate Polymers*, *92*(1), 367–373. https://doi.org/10.1016/j.carbpol.2012.09.056.

Bajpai, V. K., Kamle, M., Shukla, S., Mahato, D. K., Chandra, P., Hwang, S. K., Kumar, P., Huh, Y. S., & Han, Y. K. (2018). Prospects of using nanotechnology for food preservation, safety, and security. *Journal of Food and Drug Analysis*, *26*(4), 1201–1214. https://doi.org/10.1016/j.jfda.2018.06.011.

Behzadi, S., Serpooshan, V., Tao, W., Hamaly, M. A., Alkawareek, M. Y., Dreaden, E. C., Brown, D., Alkilany, A. M., Farokhzad, O. C., & Mahmoudi, M. (2017). Cellular uptake of nanoparticles: Journey inside the cell. *Chemical Society Reviews*, *46*(14), 4218–4244. https://doi.org/10.1039/c6cs00636a.

Betoret, E., Betoret, N., Vidal, D., & Fito, P. (2011). Functional foods development: Trends and technologies. *Trends in Food Science and Technology*, *22*(9), 498–508. https://doi.org/10.1016/j.tifs.2011.05.004.

Bhatia, N. K., Kishor, S., Katyal, N., Gogoi, P., Narang, P., & Deep, S. (2016). Effect of pH and temperature on conformational equilibria and aggregation behaviour of curcumin in aqueous binary mixtures of ethanol. *RSC Advances*, *6*(105), 103275–103288. https://doi.org/10.1039/c6ra24256a.

Bigliardi, B., & Galati, F. (2013). Innovation trends in the food industry: The case of functional foods. *Trends in Food Science and Technology*, *31*(2), 118–129. https://doi.org/10.1016/j.tifs.2013.03.006.

Brandelli, A. (2020). The interaction of nanostructured antimicrobials with biological systems: Cellular uptake, trafficking and potential toxicity. *Food Science and Human Wellness*, *9*(1), 8–20. https://doi.org/10.1016/j.fshw.2019.12.003.

Burgos-Díaz, C., Opazo-Navarrete, M., Soto-Añual, M., Leal-Calderón, F., & Bustamante, M. (2020). Food-grade Pickering emulsion as a novel astaxanthin encapsulation system for making powder-based products: Evaluation of astaxanthin stability during processing, storage, and its bioaccessibility. *Food Research International*, *134*, 109244. https://doi.org/10.1016/j.foodres.2020.109244.

Caddeo, C., Pucci, L., Gabriele, M., Carbone, C., Fernàndez-Busquets, X., Valenti, D., Pons, R., Vassallo, A., Fadda, A. M., & Manconi, M. (2018). Stability, biocompatibility and antioxidant activity of PEG-modified liposomes containing resveratrol. *International Journal of Pharmaceutics*, *538*(1–2), 40–47. https://doi.org/10.1016/j.ijpharm.2017.12.047.

Cerqueira, M. A., Pinheiro, A. C., Silva, H. D., Ramos, P. E., Azevedo, M. A., Flores-López, M. L., Rivera, M. C., Bourbon, A. I., Ramos, O. L., & Vicente, A. A. (2014). Design of bio-nanosystems for oral delivery of functional compounds. *Food Engineering Reviews, 6*(1–2), 1–19. https://doi.org/10.1007/s12393-013-9074-3.

Chan, A. W., & Neufeld, R. J. (2009). Modeling the controllable pH-responsive swelling and pore size of networked alginate based biomaterials. *Biomaterials, 30*(30), 6119–6129. https://doi.org/10.1016/j.biomaterials.2009.07.034.

Chaudhari, A., Pan, Y., & Nitin, N. (2015). Beverage emulsions: Comparison among nanoparticle stabilized emulsion with starch and surfactant stabilized emulsions. *Food Research International, 69*, 156–163. https://doi.org/10.1016/j.foodres.2014.12.030.

Chen, F., Liang, L., Zhang, Z., Deng, Z., Decker, E. A., & McClements, D. J. (2017). Inhibition of lipid oxidation in nanoemulsions and filled microgels fortified with omega-3 fatty acids using casein as a natural antioxidant. *Food Hydrocolloids, 63*, 240–248. https://doi.org/10.1016/j.foodhyd.2016.09.001.

Chen, J., Zheng, J., Decker, E. A., McClements, D. J., & Xiao, H. (2015). Improving nutraceutical bioavailability using mixed colloidal delivery systems: Lipid nanoparticles increase tangeretin bioaccessibility and absorption from tangeretin-loaded zein nanoparticles. *RSC Advances, 5*(90), 73892–73900. https://doi.org/10.1039/c5ra13503f.

Chen, L., Song, Z., Zhi, X., & Du, B. (2020). Photoinduced antimicrobial activity of curcumin-containing coatings: Molecular interaction, stability and potential application in food decontamination. *ACS Omega, 5*(48), 31044–31054. https://doi.org/10.1021/acsomega.0c04065.

Choi, S. J., & McClements, D. J. (2020). Nanoemulsions as delivery systems for lipophilic nutraceuticals: Strategies for improving their formulation, stability, functionality and bioavailability. *Food Science and Biotechnology, 29*(2), 149–168. https://doi.org/10.1007/s10068-019-00731-4.

Couto, R., Alvarez, V., & Temelli, F. (2017). Encapsulation of Vitamin B2 in solid lipid nanoparticles using supercritical CO2. *The Journal of Supercritical Fluids, 120*(2), 432–442. https://doi.org/10.1016/j.supflu.2016.05.036.

Crowe-White, K. M., & Francis, C. (2013). Position of the academy of nutrition and dietetics: Functional foods. *Journal of the American Academy of Nutrition and Dietetics, 113*(8), 1096–1103. https://doi.org/10.1016/j.jand.2013.06.002.

da Silva Malheiros, P., Daroit, D. J., & Brandelli, A. (2010). Food applications of liposome-encapsulated antimicrobial peptides. *Trends in Food Science and Technology, 21*(6), 284–292. https://doi.org/10.1016/j.tifs.2010.03.003.

da Silva Santos, V., Badan Ribeiro, A. P., & Andrade Santana, M. H. (2019). Solid lipid nanoparticles as carriers for lipophilic compounds for applications in foods. *Food Research International, 122*, 610–626. https://doi.org/10.1016/j.foodres.2019.01.032.

Das, S., Ng, W. K., & Tan, R. B. H. (2012). Are nanostructured lipid carriers (NLCs) better than solid lipid nanoparticles (SLNs): Development, characterizations and comparative evaluations of clotrimazole-loaded SLNs and NLCs? *European Journal of Pharmaceutical Sciences, 47*(1), 139–151. https://doi.org/10.1016/j.ejps.2012.05.010.

David, S., & Livney, Y. D. (2016). Potato protein based nanovehicles for health promoting hydrophobic bioactives in clear beverages. *Food Hydrocolloids, 57*, 229–235. https://doi.org/10.1016/j.foodhyd.2016.01.027.

de Carvalho, R. J., de Souza, G. T., Pagán, E., García-Gonzalo, D., Magnani, M., & Pagán, R. (2018). Nanoemulsions of Mentha piperita L. essential oil in combination with mild heat, pulsed electric fields (PEF) and high hydrostatic pressure (HHP) as an alternative to inactivate Escherichia coli O157: H7 in fruit juices. *Innovative Food Science and Emerging Technologies, 48*, 219–227. https://doi.org/10.1016/j.ifset.2018.07.004.

de Kruif, C. G., Huppertz, T., Urban, V. S., & Petukhov, A. V. (2012). Casein micelles and their internal structure. *Advances in Colloid and Interface Science, 171–172*, 36–52. https://doi.org/10.1016/j.cis.2012.01.002.

Dickinson, E. (2017). Biopolymer-based particles as stabilizing agents for emulsions and foams. *Food Hydrocolloids*, *68*, 219–231. https://doi.org/10.1016/j.foodhyd.2016.06.024.

Dima, C., Assadpour, E., Dima, S., & Jafari, S. M. (2020a). Bioavailability and bioaccessibility of food bioactive compounds; overview and assessment by in vitro methods. *Comprehensive Reviews in Food Science and Food Safety*, *19*(6), 2862–2884. https://doi.org/10.1111/1541-4337.12623.

Dima, C., Assadpour, E., Dima, S., & Jafari, S. M. (2020b). Nutraceutical nanodelivery; an insight into the bioaccessibility/bioavailability of different bioactive compounds loaded within nanocarriers. *Critical Reviews in Food Science and Nutrition*, *61*(18), 3031–3065. https://doi.org/10.1080/10408398.2020.1792409.

Dima, C., Assadpour, E., Dima, S., & Jafari, S. M. (2020c). Bioactive-loaded nanocarriers for functional foods: From designing to bioavailability. *Current Opinion in Food Science*, *33*, 21–29. https://doi.org/10.1016/j.cofs.2019.11.006.

Dima, C., Assadpour, E., Dima, S., & Jafari, S. M. (2020d). *Characterization and analysis of nanomaterials in foods* (pp. 577–653). Elsevier BV. https://doi.org/10.1016/b978-0-12-815866-1.00015-7.

Dima, C., Assadpour, E., Dima, S., & Jafari, S. M. (2020e). Bioavailability of nutraceuticals: Role of the food matrix, processing conditions, the gastrointestinal tract, and nanodelivery systems. *Comprehensive Reviews in Food Science and Food Safety*, *19*(3), 954–994. https://doi.org/10.1111/1541-4337.12547.

Dima, C., Cotarlet, M., Alexe, P., & Dima, S. (2014). Microencapsulation of essential oil of pimento [Pimenta dioica (L) Merr.] by chitosan/k-carrageenan complex coacervation method. *Innovative Food Science and Emerging Technologies*, *20*, 203–211. https://doi.org/10.1016/j.ifset.2013.12.020.

Dima, C., & Dima, S. (2018). Water-in-oil-in-water double emulsions loaded with chlorogenic acid: Release mechanisms and oxidative stability. *Journal of Microencapsulation*, *35*(6), 584–599. https://doi.org/10.1080/02652048.2018.1559246.

Dima, C., & Dima, S. (2020). Bioaccessibility study of calcium and vitamin D3 co-microencapsulated in water-in-oil-in-water double emulsions. *Food Chemistry*, *303*, 125416. https://doi.org/10.1016/j.foodchem.2019.125416.

Dima, S., & Dima, C. (2016). Protection of bioactive compounds. In C. Apetrei (Ed.), *Frontiers in bioactive compounds: Natural sources physicochemical characterization and applications* (pp. 255–301). Bentham Science.

Duong, V. A., Nguyen, T. T. L., & Maeng, H. J. (2020). Preparation of solid lipid nanoparticles and nanostructured lipid carriers for drug delivery and the effects of preparation parameters of solvent injection method. *Molecules*, *25*(20). https://doi.org/10.3390/molecules25204781.

Ebrahimi, B., Homayouni Rad, A., Ghanbarzadeh, B., Torbati, M., & Falcone, P. M. (2020). The emulsifying and foaming properties of Amuniacum gum (Dorema ammoniacum) in comparison with gum Arabic. *Food Science & Nutrition*, *8*(7), 3716–3730. https://doi.org/10.1002/fsn3.1658.

Eisinaite, V., Juraite, D., Schroën, K., & Leskauskaite, D. (2017). Food-grade double emulsions as effective fat replacers in meat systems. *Journal of Food Engineering*, *213*, 54–59. https://doi.org/10.1016/j.jfoodeng.2017.05.022.

Flores, F. P., & Kong, F. (2017). In vitro release kinetics of microencapsulated materials and the effect of the food matrix. *Annual Review of Food Science and Technology*, *8*, 237–259. https://doi.org/10.1146/annurev-food-030216-025720.

Foroozandeh, P., & Aziz, A. A. (2018). Insight into cellular uptake and intracellular trafficking of nanoparticles. *Nanoscale Research Letters*, *13*. https://doi.org/10.1186/s11671-018-2728-6.

Fröhlich, E. (2012). The role of surface charge in cellular uptake and cytotoxicity of medical nanoparticles. *International Journal of Nanomedicine*, *7*, 5577–5591. https://doi.org/10.2147/IJN.S36111.

García-Rodríguez, A., Vila, L., Cortés, C., Hernández, A., & Marcos, R. (2018). Effects of differently shaped TiO2NPs (nanospheres, nanorods and nanowires) on the in vitro model (Caco-2/HT29) of the intestinal barrier. *Particle and Fibre Toxicology*, *15*(1). https://doi.org/10.1186/s12989-018-0269-x.

Garti, N., & Aserin, A. (2012). Micelles and microemulsions as food ingredient and nutraceutical delivery systems. In *Encapsulation technologies and delivery systems for food ingredients and nutraceuticals* (pp. 211–251). Elsevier Ltd. https://doi.org/10.1016/B978-0-85709-124-6.50009-3.

Garti, N., & McClements, D. J. (2012). Encapsulation technologies and delivery systems for food ingredients and nutraceuticals. In *Encapsulation technologies and delivery systems for food ingredients and nutraceuticals* (pp. 1–612). Elsevier Ltd. https://doi.org/10.1533/9780857095909.

Ghaderi-Ghahfarokhi, M., Barzegar, M., Sahari, M. A., & Azizi, M. H. (2016). Nanoencapsulation approach to improve antimicrobial and antioxidant activity of thyme essential oil in beef burgers during refrigerated storage. *Food and Bioprocess Technology, 9*(7), 1187–1201. https://doi.org/10.1007/s11947-016-1708-z.

Ghorbanzade, T., Jafari, S. M., Akhavan, S., & Hadavi, R. (2017). Nano-encapsulation of fish oil in nanoliposomes and its application in fortification of yogurt. *Food Chemistry, 216*, 146–152. https://doi.org/10.1016/j.foodchem.2016.08.022.

Ghosh, S., & Rousseau, D. (2011). Fat crystals and water-in-oil emulsion stability. *Current Opinion in Colloid and Interface Science, 16*(5), 421–431. https://doi.org/10.1016/j.cocis.2011.06.006.

Głab, T. K., & Boratynski, J. (2017). Potential of casein as a carrier for biologically active agents. *Topics in Current Chemistry (Z), 375*.

Gordillo-Galeano, A., & Mora-Huertas, C. E. (2018). Solid lipid nanoparticles and nanostructured lipid carriers: A review emphasizing on particle structure and drug release. *European Journal of Pharmaceutics and Biopharmaceutics, 133*, 285–308. https://doi.org/10.1016/j.ejpb.2018.10.017.

Guan, Y. G., Lin, H., Han, Z., Wang, J., Yu, S. J., Zeng, X. A., Liu, Y. Y., Xu, C. H., & Sun, W. W. (2010). Effects of pulsed electric field treatment on a bovine serum albumin-dextran model system, a means of promoting the Maillard reaction. *Food Chemistry, 123*(2), 275–280. https://doi.org/10.1016/j.foodchem.2010.04.029.

Guneser, O., & Karagul Yuceer, Y. (2012). Effect of ultraviolet light on water- and fat-soluble vitamins in cow and goat milk. *Journal of Dairy Science, 95*(11), 6230–6241. https://doi.org/10.3168/jds.2011-5300.

Guo, J., Li, P., Kong, L., & Xu, B. (2020). Microencapsulation of curcumin by spray drying and freeze drying. *LWT, 132*, 109892. https://doi.org/10.1016/j.lwt.2020.109892.

Handford, C. E., Dean, M., Henchion, M., Spence, M., Elliott, C. T., & Campbell, K. (2014). Implications of nanotechnology for the agri-food industry: Opportunities, benefits and risks. *Trends in Food Science and Technology, 40*(2), 226–241. https://doi.org/10.1016/j.tifs.2014.09.007.

He, X., Deng, H., & Hwang, H. M. (2019). The current application of nanotechnology in food and agriculture. *Journal of Food and Drug Analysis, 27*(1), 1–21. https://doi.org/10.1016/j.jfda.2018.12.002.

Herzi, S., & Essafi, W. (2018). Different magnesium release profiles from W/O/W emulsions based on crystallized oils. *Journal of Colloid and Interface Science, 509*, 178–188. https://doi.org/10.1016/j.jcis.2017.08.089.

Honda, S., Ishida, R., Hidaka, K., & Masuda, T. (2019). Stability of polyphenols under alkaline conditions and the formation of a xanthine oxidase inhibitor from gallic acid in a solution at pH 7.4. *Food Science and Technology Research, 25*(1), 123–129. https://doi.org/10.3136/fstr.25.123.

Ioannou, I., Chekir, L., & Ghoul, M. (2020). Effect of heat treatment and light exposure on the antioxidant activity of flavonoids. *Processes, 8*(9), 1078. https://doi.org/10.3390/pr8091078.

Islam Shishir, M. R., Karim, N., Gowd, V., Zheng, X., & Chen, W. (2019). Liposomal delivery of natural product: A promising approach in health research. *Trends in Food Science and Technology, 85*, 177–200. https://doi.org/10.1016/j.tifs.2019.01.013.

Jafari, S. (2020). *Handbook of food nanotechnology. Applications and approaches*. Elsevier Science.

Jafari, S. M., He, Y., & Bhandari, B. (2007). Optimization of nano-emulsions production by microfluidization. *European Food Research and Technology, 225*(5–6), 733–741. https://doi.org/10.1007/s00217-006-0476-9.

Jafari, S. M., Vakili, S., & Dehnad, D. (2019). Production of a functional yogurt powder fortified with nanoliposomal vitamin d through spray drying. *Food and Bioprocess Technology, 12*(7), 1220–1231. https://doi.org/10.1007/s11947-019-02289-9.

Jeong, C., Kim, S., Lee, C., Cho, S., & Kim, S. B. (2020). Changes in the physical properties of calcium alginate gel beads under a wide range of gelation temperature conditions. *Foods, 9*(2). https://doi.org/10.3390/foods9020180.

Jones, O. G., & McClements, D. J. (2011). Recent progress in biopolymer nanoparticle and microparticle formation by heat-treating electrostatic protein-polysaccharide complexes. *Advances in Colloid and Interface Science, 167*(1–2), 49–62. https://doi.org/10.1016/j.cis.2010.10.006.

Julio, L. M., Copado, C. N., Diehl, B. W. K., Ixtaina, V. Y., & Tomás, M. C. (2018). Chia bilayer emulsions with modified sunflower lecithins and chitosan as delivery systems of omega-3 fatty acids. *LWT, 89*, 581–590. https://doi.org/10.1016/j.lwt.2017.11.044.

Katouzian, I., Faridi Esfanjani, A., Jafari, S. M., & Akhavan, S. (2017). Formulation and application of a new generation of lipid nano-carriers for the food bioactive ingredients. *Trends in Food Science and Technology, 68*, 14–25. https://doi.org/10.1016/j.tifs.2017.07.017.

Kfoury, M., Hădărugă, N. G., Hădărugă, D. I., & Fourmentin, S. (2016). *Cyclodextrins as encapsulation material for flavors and aroma* (pp. 127–192). Elsevier BV. https://doi.org/10.1016/b978-0-12-804307-3.00004-1.

Kharat, M., & McClements, D. J. (2019). Recent advances in colloidal delivery systems for nutraceuticals: A case study—Delivery by Design of curcumin. *Journal of Colloid and Interface Science, 557*, 506–518. https://doi.org/10.1016/j.jcis.2019.09.045.

Komaiko, J., & McClements, D. J. (2015). Low-energy formation of edible nanoemulsions by spontaneous emulsification: Factors influencing particle size. *Journal of Food Engineering, 146*, 122–128. https://doi.org/10.1016/j.jfoodeng.2014.09.003.

Kumar, R., Kaur, K., Uppal, S., & Mehta, S. K. (2017). Ultrasound processed nanoemulsion: A comparative approach between resveratrol and resveratrol cyclodextrin inclusion complex to study its binding interactions, antioxidant activity and UV light stability. *Ultrasonics Sonochemistry, 37*, 478–489. https://doi.org/10.1016/j.ultsonch.2017.02.004.

Lacatusu, I., Mitrea, E., Badea, N., Stan, R., Oprea, O., & Meghea, A. (2013). Lipid nanoparticles based on omega-3 fatty acids as effective carriers for lutein delivery. Preparation and in vitro characterization studies. *Journal of Functional Foods, 5*(3), 1260–1269. https://doi.org/10.1016/j.jff.2013.04.010.

Lai, W. F., Wong, W. T., & Rogach, A. L. (2020). Molecular design of layer-by-layer functionalized liposomes for oral drug delivery. *ACS Applied Materials & Interfaces, 12*(39), 43341–43351. https://doi.org/10.1021/acsami.0c13504.

Lam, R. S. H., & Nickerson, M. T. (2013). Food proteins: A review on their emulsifying properties using a structure-function approach. *Food Chemistry, 141*(2), 975–984. https://doi.org/10.1016/j.foodchem.2013.04.038.

Le Meste, M., Champion, D., Roudaut, G., Blond, G., & Simatos, D. (2002). Glass transition and food technology: A critical appraisal. *Journal of Food Science, 67*(7), 2444–2458. https://doi.org/10.1111/j.1365-2621.2002.tb08758.x.

Lee, S. H., Bajracharya, R., Min, J. Y., Han, J. W., Park, B. J., & Han, H. K. (2020). Strategic approaches for colon targeted drug delivery: An overview of recent advancements. *Pharmaceutics, 12*(1). https://doi.org/10.3390/pharmaceutics12010068.

Levinson, Y., Ish-Shalom, S., Segal, E., & Livney, Y. D. (2016). Bioavailability, rheology and sensory evaluation of fat-free yogurt enriched with VD3 encapsulated in re-assembled casein micelles. *Food & Function, 7*(3), 1477–1482. https://doi.org/10.1039/c5fo01111f.

Lewis, C. A., Jackson, M. C., & Bailey, J. R. (2019). Understanding medical foods under FDA regulations. In *Nutraceutical and functional food regulations in the united states and around the world* (pp. 203–213). Elsevier. https://doi.org/10.1016/B978-0-12-816467-9.00015-0.

Li, Q., Shi, J., Liu, L., McClements, D. J., Duan, M., Chen, X., & Liu, J. (2021). Encapsulation of fruit peel proanthocyanidins in biopolymer microgels: Relationship between structural characteristics and encapsulation/release properties. *Food Hydrocolloids, 117*. https://doi.org/10.1016/j.foodhyd.2021.106693.

Li, S., & Malmstadt, N. (2013). Deformation and poration of lipid bilayer membranes by cationic nanoparticles. *Soft Matter, 9*(20), 4969–4976. https://doi.org/10.1039/c3sm27578g.

Li, Z., Jiang, H., Xu, C., & Gu, L. (2015). A review: Using nanoparticles to enhance absorption and bioavailability of phenolic phytochemicals. *Food Hydrocolloids, 43*, 153–164. https://doi.org/10.1016/j.foodhyd.2014.05.010.

Lipinski, C. A. (2004). Lead- and drug-like compounds: The rule-of-five revolution. *Drug Discovery Today: Technologies, 1*(4), 337–341. https://doi.org/10.1016/j.ddtec.2004.11.007.

Liu, C. H., & Wu, C. T. (2010). Optimization of nanostructured lipid carriers for lutein delivery. *Colloids and Surfaces A: Physicochemical and Engineering Aspects, 353*(2–3), 149–156. https://doi.org/10.1016/j.colsurfa.2009.11.006.

Liu, F., Ma, C., Gao, Y., & McClements, D. J. (2017). Food-grade covalent complexes and their application as nutraceutical delivery systems: A review. *Comprehensive Reviews in Food Science and Food Safety, 16*(1), 76–95. https://doi.org/10.1111/1541-4337.12229.

Liu, W., Chen, X. D., Cheng, Z., & Selomulya, C. (2016). On enhancing the solubility of curcumin by microencapsulation in whey protein isolate via spray drying. *Journal of Food Engineering, 169*, 189–195. https://doi.org/10.1016/j.jfoodeng.2015.08.034.

Liu, W., Ye, A., & Singh, H. (2015). Chapter 8. Progress in applications of liposomes in food systems. In L. M. C. Sagis (Ed.), *Microencapsulation and microspheres for food applications* (pp. 151–170). Academic Press. https://doi.org/10.1016/B978-0-12-800350-3.00025-X.

Liu, X., Wang, P., Zou, Y. X., Luo, Z. G., & Tamer, T. M. (2020). Co-encapsulation of Vitamin C and β-Carotene in liposomes: Storage stability, antioxidant activity, and in vitro gastrointestinal digestion. *Food Research International, 136*. https://doi.org/10.1016/j.foodres.2020.109587.

Livney, Y. D. (2010). Milk proteins as vehicles for bioactives. *Current Opinion in Colloid and Interface Science, 15*(1–2), 73–83. https://doi.org/10.1016/j.cocis.2009.11.002.

Lizardi-Mendoza, J., Argüelles Monal, W. M., & Goycoolea Valencia, F. M. (2016). Chemical characteristics and functional properties of chitosan. In *Chitosan in the preservation of agricultural commodities* (pp. 3–31). Elsevier Inc. https://doi.org/10.1016/B978-0-12-802735-6.00001-X.

Lucas, J., Ralaivao, M., Estevinho, B. N., & Rocha, F. (2020). A new approach for the microencapsulation of curcumin by a spray drying method, in order to value food products. *Powder Technology, 362*, 428–435. https://doi.org/10.1016/j.powtec.2019.11.095.

Luo, X., Zhou, Y., Bai, L., Liu, F., Deng, Y., & McClements, D. J. (2017). Fabrication of β-carotene nanoemulsion-based delivery systems using dual-channel microfluidization: Physical and chemical stability. *Journal of Colloid and Interface Science, 490*, 328–335. https://doi.org/10.1016/j.jcis.2016.11.057.

Maderuelo, C., Zarzuelo, A., & Lanao, J. M. (2011). Critical factors in the release of drugs from sustained release hydrophilic matrices. *Journal of Controlled Release, 154*(1), 2–19. https://doi.org/10.1016/j.jconrel.2011.04.002.

Malekjani, N., & Jafari, S. M. (2021). Modeling the release of food bioactive ingredients from carriers/nanocarriers by the empirical, semiempirical, and mechanistic models. *Comprehensive Reviews in Food Science and Food Safety, 20*(1), 3–47. https://doi.org/10.1111/1541-4337.12660.

Mansoor, S., Kondiah, P. P. D., Choonara, Y. E., & Pillay, V. (2019). Polymer-based nanoparticle strategies for insulin delivery. *Polymers, 11*(9). https://doi.org/10.3390/polym11091380.

Marsanasco, M., Márquez, A. L., Wagner, J. R., del Alonso, S. V., & Chiaramoni, N. S. (2011). Liposomes as vehicles for vitamins E and C: An alternative to fortify orange juice and offer vitamin C protection after heat treatment. *Food Research International, 44*(9), 3039–3046. https://doi.org/10.1016/j.foodres.2011.07.025.

Martínez-Huélamo, M., Tulipani, S., Estruch, R., Escribano, E., Illán, M., Corella, D., & Lamuela-Raventós, R. M. (2015). The tomato sauce making process affects the bioaccessibility and bioavailability of tomato phenolics: A pharmacokinetic study. *Food Chemistry, 173*, 864–872. https://doi.org/10.1016/j.foodchem.2014.09.156.

Matalanis, A., Jones, O. G., & McClements, D. J. (2011). Structured biopolymer-based delivery systems for encapsulation, protection, and release of lipophilic compounds. *Food Hydrocolloids, 25*(8), 1865–1880. https://doi.org/10.1016/j.foodhyd.2011.04.014.

Matos, M., Gutiérrez, G., Coca, J., & Pazos, C. (2014). Preparation of water-in-oil-in-water (W1/O/W2) double emulsions containing trans-resveratrol. *Colloids and Surfaces A: Physicochemical and Engineering Aspects*, *442*, 69–79. https://doi.org/10.1016/j.colsurfa.2013.05.065.

McClements, D. J. (2015). Encapsulation, protection, and release of hydrophilic active components: Potential and limitations of colloidal delivery systems. *Advances in Colloid and Interface Science*, *219*, 27–53. https://doi.org/10.1016/j.cis.2015.02.002.

McClements, D. J. (2016). *Food emulsions principles, practices, and techniques*. 3rd ed.

McClements, D. J. (2017a). The future of food colloids: Next-generation nanoparticle delivery systems. *Current Opinion in Colloid and Interface Science*, *28*, 7–14. https://doi.org/10.1016/j.cocis.2016.12.002.

McClements, D. J. (2017b). Designing biopolymer microgels to encapsulate, protect and deliver bioactive components: Physicochemical aspects. *Advances in Colloid and Interface Science*, *240*, 31–59. https://doi.org/10.1016/j.cis.2016.12.005.

McClements, D. J. (2018a). Delivery by design (DbD): A standardized approach to the development of efficacious nanoparticle- and microparticle-based delivery systems. *Comprehensive Reviews in Food Science and Food Safety*, *17*(1), 200–219. https://doi.org/10.1111/1541-4337.12313.

McClements, D. J. (2018b). Encapsulation, protection, and delivery of bioactive proteins and peptides using nanoparticle and microparticle systems: A review. *Advances in Colloid and Interface Science*, *253*, 1–22. https://doi.org/10.1016/j.cis.2018.02.002.

McClements, D. J., & Gumus, C. E. (2016). Natural emulsifiers—Biosurfactants, phospholipids, biopolymers, and colloidal particles: Molecular and physicochemical basis of functional performance. *Advances in Colloid and Interface Science*, *234*, 3–26. https://doi.org/10.1016/j.cis.2016.03.002.

McClements, D. J., & Rao, J. (2011). Food-grade nanoemulsions: Formulation, fabrication, properties, performance, biological fate, and potential toxicity. *Critical Reviews in Food Science and Nutrition*, *51*(4), 285–330. https://doi.org/10.1080/10408398.2011.559558.

McClements, D. J., & Xiao, H. (2014). Excipient foods: Designing food matrices that improve the oral bioavailability of pharmaceuticals and nutraceuticals. *Food & Function*, *5*(7), 1320–1333. https://doi.org/10.1039/c4fo00100a.

McClements, D. J., & Xiao, H. (2017). Is nano safe in foods? Establishing the factors impacting the gastrointestinal fate and toxicity of organic and inorganic food-grade nanoparticles. *npj Science of Food*. https://doi.org/10.1038/s41538-017-0005-1.

Meikle, T. G., Zabara, A., Waddington, L. J., Separovic, F., Drummond, C. J., & Conn, C. E. (2017). Incorporation of antimicrobial peptides in nanostructured lipid membrane mimetic bilayer cubosomes. *Colloids and Surfaces B: Biointerfaces*, *152*, 143–151. https://doi.org/10.1016/j.colsurfb.2017.01.004.

Menéndez-Aguirre, O., Kessler, A., Stuetz, W., Grune, T., Weiss, J., & Hinrichs, J. (2014). Increased loading of vitamin D2 in reassembled casein micelles with temperature-modulated high pressure treatment. *Food Research International*, *64*, 74–80. https://doi.org/10.1016/j.foodres.2014.06.010.

Mierczyńska, J., Cybulska, J., Sołowiej, B., & Zdunek, A. (2015). Effect of $Ca^{2+}$, $Fe^{2+}$ and $Mg^{2+}$ on rheological properties of new food matrix made of modified cell wall polysaccharides from apple. *Carbohydrate Polymers*, *133*, 547–555. https://doi.org/10.1016/j.carbpol.2015.07.046.

Mohamed, J. M., Alqahtani, A., Ahmad, F., Krishnaraju, V., & Kalpana, K. (2021). Pectin co-functionalized dual layered solid lipid nanoparticle made by soluble curcumin for the targeted potential treatment of colorectal cancer. *Carbohydrate Polymers*, *252*. https://doi.org/10.1016/j.carbpol.2020.117180.

Mohammadi, M., Pezeshki, A., Abbasi, M. M., Ghanbarzadeh, B., & Hamishehkar, H. (2017). Vitamin D3-loaded nanostructured lipid carriers as a potential approach for fortifying food beverages; in vitro and in vivo evaluation. *Advanced Pharmaceutical Bulletin*, *7*(1), 61–71. https://doi.org/10.15171/apb.2017.008.

Mozafari, M. R., Johnson, C., Hatziantoniou, S., & Demetzos, C. (2008). Nanoliposomes and their applications in food nanotechnology. *Journal of Liposome Research*, *18*(4), 309–327. https://doi.org/10.1080/08982100802465941.

Müller, R. H., Radtke, M., & Wissing, S. A. (2002). Nanostructured lipid matrices for improved microencapsulation of drugs. *International Journal of Pharmaceutics*, *242*(1–2), 121–128. https://doi.org/10.1016/S0378-5173(02)00180-1.

Muschiolik, G., & Dickinson, E. (2017). Double emulsions relevant to food systems: Preparation, stability, and applications. *Comprehensive Reviews in Food Science and Food Safety*, *16*(3), 532–555. https://doi.org/10.1111/1541-4337.12261.

Nagao, A., Kotake-Nara, E., & Hase, M. (2013). Effects of fats and oils on the bioaccessibility of carotenoids and vitamin e in vegetables. *Bioscience, Biotechnology, and Biochemistry*, *77*(5), 1055–1060. https://doi.org/10.1271/bbb.130025.

Neves, M. I. L., Desobry-Banon, S., Perrone, I. T., Desobry, S., & Petit, J. (2019). Encapsulation of curcumin in milk powders by spray-drying: Physicochemistry, rehydration properties, and stability during storage. *Powder Technology*, *345*, 601–607. https://doi.org/10.1016/j.powtec.2019.01.049.

Noello, C., Carvalho, A. G. S., Silva, V. M., & Hubinger, M. D. (2016). Spray dried microparticles of chia oil using emulsion stabilized by whey protein concentrate and pectin by electrostatic deposition. *Food Research International*, *89*, 549–557. https://doi.org/10.1016/j.foodres.2016.09.003.

Oxley, J. D. (2012). Spray cooling and spray chilling for food ingredient and nutraceutical encapsulation. In *Encapsulation technologies and delivery systems for food ingredients and nutraceuticals* (pp. 110–130). Elsevier Ltd. https://doi.org/10.1016/B978-0-85709-124-6.50005-6.

Ozturk, B. (2017). Nanoemulsions for food fortification with lipophilic vitamins: Production challenges, stability, and bioavailability. *European Journal of Lipid Science and Technology*, *119*.

Pabast, M., Shariatifar, N., Beikzadeh, S., & Jahed, G. (2018). Effects of chitosan coatings incorporating with free or nano-encapsulated Satureja plant essential oil on quality characteristics of lamb meat. *Food Control*, *91*, 185–192. https://doi.org/10.1016/j.foodcont.2018.03.047.

Pan, K., Luo, Y., Gan, Y., Baek, S. J., & Zhong, Q. (2014). PH-driven encapsulation of curcumin in self-assembled casein nanoparticles for enhanced dispersibility and bioactivity. *Soft Matter*, *10*(35), 6820–6830. https://doi.org/10.1039/c4sm00239c.

Pandita, D., Kumar, S., Poonia, N., & Lather, V. (2014). Solid lipid nanoparticles enhance oral bioavailability of resveratrol, a natural polyphenol. *Food Research International*, *62*, 1165–1174. https://doi.org/10.1016/j.foodres.2014.05.059.

Patel, A. R., Heussen, P. C. M., Hazekamp, J., Drost, E., & Velikov, K. P. (2012). Quercetin loaded biopolymeric colloidal particles prepared by simultaneous precipitation of quercetin with hydrophobic protein in aqueous medium. *Food Chemistry*, *133*(2), 423–429. https://doi.org/10.1016/j.foodchem.2012.01.054.

Peleg, M., Normand, M. D., & Corradini, M. G. (2012). The Arrhenius equation revisited. *Critical Reviews in Food Science and Nutrition*, *52*(9), 830–851. https://doi.org/10.1080/10408398.2012.667460.

Penalva, R., Esparza, I., Agüeros, M., Gonzalez-Navarro, C. J., Gonzalez-Ferrero, C., & Irache, J. M. (2015). Casein nanoparticles as carriers for the oral delivery of folic acid. *Food Hydrocolloids*, *44*, 399–406. https://doi.org/10.1016/j.foodhyd.2014.10.004.

Perry, S. L., & McClements, D. J. (2020). Recent advances in encapsulation, protection, and oral delivery of bioactive proteins and peptides using colloidal systems. *Molecules*, *25*(5). https://doi.org/10.3390/molecules25051161.

Piorkowski, D. T., & McClements, D. J. (2014). Beverage emulsions: Recent developments in formulation, production, and applications. *Food Hydrocolloids*, *42*, 5–41. https://doi.org/10.1016/j.foodhyd.2013.07.009.

Pramanik, S. K., Losada-Pérez, P., Reekmans, G., Carleer, R., Olieslaeger, M., Vanderzande, D., et al. (2017). Physicochemical characterizations of functional hybrid liposomal nanocarriers formed using photosensitive lipids. *Scientific Reports*, *7*.

Pujara, N., Jambhrunkar, S., Wong, K. Y., McGuckin, M., & Popat, A. (2017). Enhanced colloidal stability, solubility and rapid dissolution of resveratrol by nanocomplexation with soy protein isolate. *Journal of Colloid and Interface Science*, *488*, 303–308. https://doi.org/10.1016/j.jcis.2016.11.015.

Q8 (R2) Pharmaceutical Development. (2009). *International conference on harmonisation—quality (ICH). Final guidance.*

Qian, C., Decker, E. A., Xiao, H., & McClements, D. J. (2012). Nanoemulsion delivery systems: Influence of carrier oil on β-carotene bioaccessibility. *Food Chemistry, 135*(3), 1440–1447. https://doi.org/10.1016/j.foodchem.2012.06.047.

Qu, W., Zuo, W., Li, N., Hou, Y., Song, Z., Gou, G., & Yang, J. (2017). Design of multifunctional liposome-quantum dot hybrid nanocarriers and their biomedical application. *Journal of Drug Targeting, 25*(8), 661–672. https://doi.org/10.1080/1061186X.2017.1323334.

Rachmawati, H., Shaal, L. A., Müller, R. H., & Keck, C. M. (2013). Development of curcumin nanocrystal: Physical aspects. *Journal of Pharmaceutical Sciences, 102*(1), 204–214. https://doi.org/10.1002/jps.23335.

Ramesan, R. M., & Sharma, C. P. (2014). Recent advances in the oral delivery of insulin. *Recent Patents on Drug Delivery & Formulation, 8*(2), 155–159. https://doi.org/10.2174/1872211308666140527143407.

Rana, S., Arora, S., Gupta, C., & Kapila, S. (2019). Effect of sodium caseinate and vitamin A complexation on bioaccessibility and bioavailability of vitamin A in Caco-2 cells. *Food Research International, 121*, 910–918. https://doi.org/10.1016/j.foodres.2019.01.019.

Razavi, S. M. A., & Irani, M. (2019). Rheology of food gum. In *Reference series in phytochemistry* (pp. 1959–1985). Springer Science and Business Media B.V. https://doi.org/10.1007/978-3-319-78030-6_20.

Ribeiro, A. P. B., Masuchi, M. H., Miyasaki, E. K., Domingues, M. A. F., Stroppa, V. L. Z., de Oliveira, G. M., & Kieckbusch, T. G. (2015). Crystallization modifiers in lipid systems. *Journal of Food Science and Technology, 52*(7), 3925–3946. https://doi.org/10.1007/s13197-014-1587-0.

Robert, P., Zamorano, M., González, E., Silva-Weiss, A., Cofrades, S., & Giménez, B. (2019). Double emulsions with olive leaves extract as fat replacers in meat systems with high oxidative stability. *Food Research International, 120*, 904–912. https://doi.org/10.1016/j.foodres.2018.12.014.

Roohinejad, S., Oey, I., Everett, D. W., & Greiner, R. (2018). Microemulsions. In *Emulsion-based systems for delivery of food active compounds: Formation, application, health and safety* (pp. 231–262). Wiley. https://doi.org/10.1002/9781119247159.ch9.

Rostamabadi, H., Falsafi, S. R., & Jafari, S. M. (2019). Nanoencapsulation of carotenoids within lipid-based nanocarriers. *Journal of Controlled Release, 298*, 38–67. https://doi.org/10.1016/j.jconrel.2019.02.005.

Sala, M., Miladi, K., Agusti, G., Elaissari, A., & Fessi, H. (2017). Preparation of liposomes: A comparative study between the double solvent displacement and the conventional ethanol injection—From laboratory scale to large scale. *Colloids and Surfaces A: Physicochemical and Engineering Aspects, 524*, 71–78. https://doi.org/10.1016/j.colsurfa.2017.02.084.

Salminen, H., Helgason, T., Kristinsson, B., Kristbergsson, K., & Weiss, J. (2013). Formation of solid shell nanoparticles with liquid ω-3 fatty acid core. *Food Chemistry, 141*(3), 2934–2943. https://doi.org/10.1016/j.foodchem.2013.05.120.

Salvia-Trujillo, L., Qian, C., Martín-Belloso, O., & McClements, D. J. (2013). Modulating β-carotene bioaccessibility by controlling oil composition and concentration in edible nanoemulsions. *Food Chemistry, 139*(1–4), 878–884. https://doi.org/10.1016/j.foodchem.2013.02.024.

Salvia-Trujillo, L., Verkempinck, S. H. E., Sun, L., Van Loey, A. M., Grauwet, T., & Hendrickx, M. E. (2017). Lipid digestion, micelle formation and carotenoid bioaccessibility kinetics: Influence of emulsion droplet size. *Food Chemistry, 229*, 653–662. https://doi.org/10.1016/j.foodchem.2017.02.146.

Sarabandi, K., Jafari, S. M., Mohammadi, M., Akbarbaglu, Z., Pezeshki, A., & Khakbaz Heshmati, M. (2019). Production of reconstitutable nanoliposomes loaded with flaxseed protein hydrolysates: Stability and characterization. *Food Hydrocolloids, 96*, 442–450. https://doi.org/10.1016/j.foodhyd.2019.05.047.

Sarkar, P., Bhunia, A. K., & Yao, Y. (2017). Impact of starch-based emulsions on the antibacterial efficacies of nisin and thymol in cantaloupe juice. *Food Chemistry, 217*, 155–162. https://doi.org/10.1016/j.foodchem.2016.08.071.

Sedaghat Doost, A., Nikbakht Nasrabadi, M., Kassozi, V., Nakisozi, H., & Van der Meeren, P. (2020). Recent advances in food colloidal delivery systems for essential oils and their main components. *Trends in Food Science and Technology*, *99*, 474–486. https://doi.org/10.1016/j.tifs.2020.03.037.

Sekhon, B. S. (2010). Food nanotechnology—An overview. *Nanotechnology, Science and Applications*, *3*, 1–15. https://doi.org/10.2147/NSA.S8677.

Shi, A., Feng, X., Wang, Q., & Adhikari, B. (2020). Pickering and high internal phase Pickering emulsions stabilized by protein-based particles: A review of synthesis, application and prospective. *Food Hydrocolloids*, *109*, 106117. https://doi.org/10.1016/j.foodhyd.2020.106117.

Shukla, S., Haldorai, Y., Hwang, S. K., Bajpai, V. K., Huh, Y. S., & Han, Y. K. (2017). Current demands for food-approved liposome nanoparticles in food and safety sector. *Frontiers in Microbiology*, *8*. https://doi.org/10.3389/fmicb.2017.02398.

Siepmann, J., Siegel, R. A., & Siepmann, F. (2012). Fundamentals and applications of controlled release drug delivery. *Controlled Release Society*, 127–152.

Siepmann, J., & Siepmann, F. (2012). Modeling of diffusion controlled drug delivery. *Journal of Controlled Release*, *161*(2), 351–362. https://doi.org/10.1016/j.jconrel.2011.10.006.

Singh, V., & Chaudhary, A. K. (2010). Development and characterization of Rosiglitazone loaded gelatin nanoparticles using two step desolvation method. *International Journal of Pharmaceutical Sciences Review and Research*, *5*(1), 100–103. http://globalresearchonline.net/journalcontents/volume5issue1/Article-015.pdf.

Sobel, R., Versic, R., & Gaonkar, A. G. (2014). Introduction to microencapsulation and controlled delivery in foods. In A. G. Gaonkar, N. Vasisht, A. R. Khare, & R. Sobel (Eds.), *Microencapsulation in the food industry* (pp. 3–12). Elsevier Inc. https://doi.org/10.1016/B978-0-12-404568-2.00001-7.

Soottitantawat, A., Partanen, R., Neoh, T. L., & Yoshii, H. (2015). Encapsulation of hydrophilic and hydrophobic flavors by spray drying. *Japan Journal of Food Engineering*, *16*(1), 37–52. https://doi.org/10.11301/jsfe.16.37.

Stratulat, I., Britten, M., Salmieri, S., Fustier, P., St-Gelais, D., Champagne, C. P., & Lacroix, M. (2014). Enrichment of cheese with bioactive lipophilic compounds. *Journal of Functional Foods*, *6*(1), 48–59. https://doi.org/10.1016/j.jff.2013.11.023.

Szente, L., & Fenyvesi, É. (2018). Cyclodextrin-enabled polymer composites for packaging. *Molecules*, *23*(7), 1556. https://doi.org/10.3390/molecules23071556.

Tamjidi, F., Shahedi, M., Varshosaz, J., & Nasirpour, A. (2013). Nanostructured lipid carriers (NLC): A potential delivery system for bioactive food molecules. *Innovative Food Science & Emerging Technologies*, *19*, 29–43. https://doi.org/10.1016/j.ifset.2013.03.002.

Tavakoli, H., Hosseini, O., Jafari, S. M., & Katouzian, I. (2018). Evaluation of physicochemical and antioxidant properties of yogurt enriched by olive leaf phenolics within nanoliposomes. *Journal of Agricultural and Food Chemistry*, *66*(35), 9231–9240. https://doi.org/10.1021/acs.jafc.8b02759.

Thomopoulos, R., Baudrit, C., Boukhelifa, N., Boutrou, R., Buche, P., Guichard, E., Guillard, V., Lutton, E., Mirade, P. S., Ndiaye, A., Perrot, N., Taillandier, F., Thomas-Danguin, T., & Tonda, A. (2019). Multi-criteria reverse engineering for food: Genesis and ongoing advances. *Food Engineering Reviews*, *11*(1), 44–60. https://doi.org/10.1007/s12393-018-9186-x.

Uchiyama, H., Chae, J., Kadota, K., & Tozuka, Y. (2019). Formation of food grade microemulsion with rice glycosphingolipids to enhance the oral absorption of Coenzyme Q10. *Foods*, *8*(10), 502. https://doi.org/10.3390/foods8100502.

Wadher, K., Kakde, R., & Umekar, M. (2011). Formulation and evaluation of a sustained-release tablets of metformin hydrochloride using hydrophilic synthetic and hydrophobic natural polymers. *Indian Journal of Pharmaceutical Sciences*, *73*(2), 208–215. https://doi.org/10.4103/0250-474X.91579.

Walstra, P., Geurts, T. J., Noomen, A., Jellema, A., & van Boekel, M. A. J. S. (1999). *Dairy technology: Principles of milk. Properties and processes*. Marcel Dekker.

Wang, J., Yadav, V., Smart, A. L., Tajiri, S., & Basit, A. W. (2015). Toward oral delivery of biopharmaceuticals: An assessment of the gastrointestinal stability of 17 peptide drugs. *Molecular Pharmaceutics*, *12*(3), 966–973. https://doi.org/10.1021/mp500809f.

Wang, Y., Jia, J., Shao, J., Shu, X., Ren, X., Wu, B., & Yan, Z. (2018). Preservative effects of allicin microcapsules on daily foods. *LWT*, *98*, 225–230. https://doi.org/10.1016/j.lwt.2018.08.043.

Wang, Y. F., Shao, J. J., Zhou, C. H., Zhang, D. L., Bie, X. M., Lv, F. X., Zhang, C., & Lu, Z. X. (2012). Food preservation effects of curcumin microcapsules. *Food Control*, *27*(1), 113–117. https://doi.org/10.1016/j.foodcont.2012.03.008.

Wechtersbach, L., Poklar Ulrih, N., & Cigić, B. (2012). Liposomal stabilization of ascorbic acid in model systems and in food matrices. *LWT - Food Science and Technology*, *45*(1), 43–49. https://doi.org/10.1016/j.lwt.2011.07.025.

Wu, C., Li, L., Zhong, Q., Cai, R., Wang, P., Xu, X., Zhou, G., Han, M., Liu, Q., Hu, T., & Yin, T. (2019). Myofibrillar protein-curcumin nanocomplexes prepared at different ionic strengths to improve oxidative stability of marinated chicken meat products. *LWT*, *99*, 69–76. https://doi.org/10.1016/j.lwt.2018.09.024.

Yao, M., McClements, D. J., & Xiao, H. (2015). Improving oral bioavailability of nutraceuticals by engineered nanoparticle-based delivery systems. *Current Opinion in Food Science*, *2*, 14–19. https://doi.org/10.1016/j.cofs.2014.12.005.

Yi, J., Lam, T. I., Yokoyama, W., Cheng, L. W., & Zhong, F. (2014). Controlled release of β-carotene in β-lactoglobulin-dextran-conjugated nanoparticles in vitro digestion and transport with caco-2 monolayers. *Journal of Agricultural and Food Chemistry*, *62*(35), 8900–8907. https://doi.org/10.1021/jf502639k.

Zandi, M., Dardmeh, N., Pirsa, S., & Almasi, H. (2017). Identification of cardamom encapsulated alginate-whey protein concentrates microcapsule release kinetics and mechanism during storage, stew process and oral consumption. *Journal of Food Process Engineering*, *40*(1). https://doi.org/10.1111/jfpe.12314.

Zanotto, E. D., & Mauro, J. C. (2017). The glassy state of matter: Its definition and ultimate fate. *Journal of Non-Crystalline Solids*, *471*, 490–495. https://doi.org/10.1016/j.jnoncrysol.2017.05.019.

Zeng, M., Zhang, X., Qi, C., & Zhang, X. M. (2012). Microstructure-stability relations studies of porous chitosan microspheres supported palladium catalysts. *International Journal of Biological Macromolecules*, *51*(5), 730–737. https://doi.org/10.1016/j.ijbiomac.2012.07.017.

Zhang, R., Zhang, Z., Kumosani, T., Khoja, S., Abualnaja, K. O., & McClements, D. J. (2016). Encapsulation of β-carotene in nanoemulsion-based delivery systems formed by spontaneous emulsification: Influence of lipid composition on stability and bioaccessibility. *Food Biophysics*, *11*(2), 154–164. https://doi.org/10.1007/s11483-016-9426-7.

Zhang, Z., Zhang, R., Chen, L., Tong, Q., & McClements, D. J. (2015). Designing hydrogel particles for controlled or targeted release of lipophilic bioactive agents in the gastrointestinal tract. *European Polymer Journal*, *72*, 698–716. https://doi.org/10.1016/j.eurpolymj.2015.01.013.

Zhang, Z. P., Zhang, R. J., Zou, L. Q., & McClements, D. J. (2016). Tailoring lipid digestion profiles using combined delivery systems: Mixtures of nanoemulsions and filled hydrogel beads. *RSC Advances*, *6*.

Zheng, B., & McClements, D. J. (2020). Formulation of more efficacious curcumin delivery systems using colloid science: Enhanced solubility, stability, and bioavailability. *Molecules*, *25*(12). https://doi.org/10.3390/molecules25122791.

Zheng, B., Peng, S., Zhang, X., & McClements, D. J. (2018). Impact of delivery system type on curcumin bioaccessibility: Comparison of curcumin-loaded nanoemulsions with commercial curcumin supplements. *Journal of Agricultural and Food Chemistry*, *66*(41), 10816–10826. https://doi.org/10.1021/acs.jafc.8b03174.

Zhu, J., Zhuang, P., Luan, L., Sun, Q., & Cao, F. (2015). Preparation and characterization of novel nanocarriers containing krill oil for food application. *Journal of Functional Foods*, *19*, 902–912. https://doi.org/10.1016/j.jff.2015.06.017.

Zhu, Q., Zhao, L., Zhang, H., Saito, M., & Yin, L. (2017). Impact of the release rate of magnesium ions in multiple emulsions (water-in-oil-in-water) containing BSA on the resulting physical properties and microstructure of soy protein gel. *Food Chemistry*, *220*, 452–459. https://doi.org/10.1016/j.foodchem.2016.10.016.

Zou, L., Zheng, B., Zhang, R., Zhang, Z., Liu, W., Liu, C., Xiao, H., & McClements, D. J. (2016). Enhancing the bioaccessibility of hydrophobic bioactive agents using mixed colloidal dispersions: Curcumin-loaded zein nanoparticles plus digestible lipid nanoparticles. *Food Research International*, *81*, 74–82. https://doi.org/10.1016/j.foodres.2015.12.035.

## Further reading

Taheri, A., & Jafari, S. M. (2019). Gum-based nanocarriers for the protection and delivery of food bioactive compounds. *Advances in Colloid and Interface Science*, *269*, 277–295. https://doi.org/10.1016/j.cis.2019.04.009.

Yousefi, M., & Jafari, S. M. (2019). Recent advances in application of different hydrocolloids in dairy products to improve their techno-functional properties. *Trends in Food Science and Technology*, *88*, 468–483. https://doi.org/10.1016/j.tifs.2019.04.015.

**CHAPTER**

# How food structure influences the physical, sensorial, and nutritional quality of food products

## 5

Meliza Lindsay Rojas[a], Mirian T.K. Kubo[b,c], Maria Elisa Caetano-Silva[d], Gisandro Reis Carvalho[e], and Pedro E.D. Augusto[e,f,g]

[a]*Dirección de Investigación y Desarrollo, Universidad Privada del Norte (UPN), Trujillo, Peru,* [b]*Université de Technologie de Compiègne, UMR CNRS 7025, Enzyme and Cell Engineering Laboratory, Compiègne, France,* [c]*Institute of Biosciences, Humanities and Exact Sciences (IBILCE), Department of Food Engineering and Technology (DETA), São Paulo State University (UNESP), São Paulo, Brazil,* [d]*College of Agricultural, Consumer & Environmental Sciences, University of Illinois at Urbana-Champaign, Champaign, IL, United States,* [e]*Department of Agri-food Industry, Food and Nutrition (LAN), Luiz de Queiroz College of Agriculture (ESALQ), University of São Paulo (USP), São Paulo, Brazil,* [f]*Food and Nutrition Research Center (NAPAN), University of São Paulo (USP), São Paulo, Brazil,* [g]*Université Paris-Saclay, CentraleSupélec, Laboratoire de Génie des Procédés et Matériaux, SFR Condorcet FR CNRS 3417, Centre Européen de Biotechnologie et de Bioéconomie (CEBB), Pomacle, France*

## 5.1 Introduction

Foods have a central role in our lives, affecting not only our nutrition but also well-being, social behavior and other aspects of human society, including economical aspects and our relation with the environment (Augusto, 2020).

It is interesting to notice although some correlations are well established, we are still far from completely understand some others. For example, the relationship between composition and nutrition is frequently studied and known, however, how the structure affects the interaction between the food product and the human body is still a topic of discussion. Similarly, microbial thermo-resistance can vary widely with food composition, but the knowledge of how food structure affects it is still scarce. On the other hand, the relationship between structure and physical properties is better understood, although it is still a challenge to predict how food structure affects the process conditions and the energy needed.

## Chapter 5 Food structure influences

This chapter discusses the relationships among food structure, processing and properties. Different conventional and emerging technologies are detailed (as listed in Table 5.1), for food products from different origins (animal and vegetal), structuration (top-down and bottom-up, Fig. 5.2), state (solid, fluids, dispersions) and scale (molecules, cells, tissues and live organisms). The impact of structure modification is discussed on physical properties, sensorial aspects, nutritional quality and health aspects.

A clear understanding of the above mentioned relationship can help advance processes designed to tailor food structure, achieving different functionalities.

**Table 5.1** Main structural modifications obtained through food processing by applying conventional or emerging technologies (*) to obtain new products.

| Initial product/ constituents | Process/ technology | Structure modification | Products |
|---|---|---|---|
| Meat, grains, fruits, vegetables | TP, OD, D, F, MW, PEF, US, HHP | – Tissue softening/ compaction<br>– Channel formation<br>– Cell shrinkage, breaking and debonding<br>– Intercellular porosity changes<br>– Cell wall and membrane disruption<br>– Intracellular organelles distribution/disruption<br>– Compound structure modification | Ready to eat foods, snacks, canned foods, frozen, minimally processed, pre-treated products for further process |
| | M, G | – Tissue disintegration<br>– Cell disruption<br>– Size reduction and shape changes | Flour, powder, puree, pulp, juices |
| Puree, pulp, juices, milk | Hm, Cm, SD, TP HPH, US | – Particle size, distribution, and morphology modification<br>– Micelle formation/ disruption<br>– Cell wall and macromolecules disruption | Homogenized products, suspensions, beverages |
| Flour, powders, ground products | M, Mx, TP, E | – Particle/compounds interactions<br>– Compact (dense) or expanded (open) structures<br>– Structure modification of compounds | Sausages, extruded foods, restructured foods, pastries, bakery products |

Table 5.1 Main structural modifications obtained through food processing by applying conventional or emerging technologies (*) to obtain new products—cont'd

| Initial product/constituents | Process/technology | Structure modification | Products |
|---|---|---|---|
| Proteins | TP, E, F, Eh, I, US, HPH, HHP, PEF, CP | – Polypeptide chain breakdown, aggregation, cross-linking,<br>– Globular conformation changes (Unfolding, dissociation)<br>– Interaction changes | Emulsions, gel-like structures, texturized protein, peptides, protein hydrolysates |
| Polysaccharides | TP, E, F, Eh, I, US, HHP, HPH, PEF, Oz, CP | – Breakdown of intermolecular bonds,<br>– Branched/linear chain dissociation<br>– Bross-linked polymers<br>– Chain rearrangement | Hydrogel structures, hydrolysates, modified starch |
| Lipids-proteins-polysaccharides | Hm, Cm, Mh, TP, US, HPH | – Drops/particle size reduction<br>– Layer or multilayer structuring<br>– Interfacial architecture modification | Diverse emulsions, coated drops, hydrogel particles |

(*) *Conventional: TP=Thermal process; F=Freezing; OD=Osmotic dehydration; D=Drying; SD=Spray drying; E=Extrusion; F=Fermentation, Eh=Enzymatic hydrolysis; Mx=Mixing; M=Milling; G=Grinding; Hm=High-shear mixer; Cm=Colloid mill; Mh=Membrane homogenization. Emerging: MW=Microwave; US=Ultrasound; I=Irradiation; PEF=Pulsed electric fields; HHP=High hydrostatic pressure, HPH=High pressure homogenization; Oz=Ozone; CP=Cold plasma.*

## 5.2 Effect of food processing on food structure: Conventional and emerging technologies in food processing

Food processing has an inherent effect in modifying the original structure of raw materials or constituents, giving rise to new products. The different processes and technologies applied can be aimed at preserving, transforming, or creating new structures. However, whatever the goal, the modification of the original matrix structure is almost unavoidable. In Fig. 5.1, the existing interrelation among process-structure-properties is outlined.

Those methods to create new food structures could be classified according to two approaches: top-down and bottom-up (Aguilera, 2005; Lesmes & Mcclements, 2009; Velikov & Pelan, 2008). The top-down approach consists of the creation of new products by breaking the initial macroscopic material into smaller particles. In contrast, the bottom-up approach consists of the creation of new structures by using and assembling appropriate ingredients (such as molecules, particles) under controlled processes. This approach is commonly explored to produce engineered structures with desirable properties such as hydrogels or colloidal dispersions.

Therefore, through food processing, there are a series of structure modifications occurring at different scales, which are related to the matrix of the original product. According to Ubbink et al. (2008),

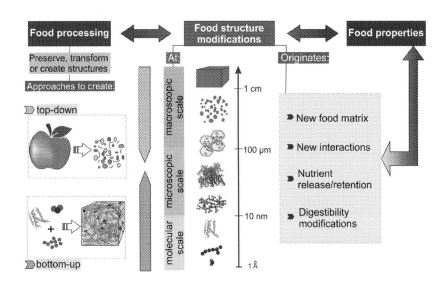

**FIG. 5.1**

Representation of the interrelation of process-structure-properties of food.

on one extreme, the product is macroscopic, and on the other, it is composed of molecules and atoms characterized by molecular length scales. The macroscopic scale (>100 μm) is composed of features that are perceived for human senses, while the microscopic scale (from ~10 nm to ~100 μm) is composed of aggregated molecules, colloids, granules and cells, among others. Finally, the molecular scale (from ~1 Å to ~10 nm) is composed of vitamins, flavor components, water, minerals, among others (Janowicz & Lenart, 2018; Ubbink et al., 2008). The structural elements that contribute to food identity and quality are microscopic, where interventions at the microscopic level are necessary to create new products (Aguilera, 2005). Therefore, the constituents or compounds and the dominant scale before processing will contribute to define the type of process or technology to be applied. The structure modifications, as a consequence of processing, will result in modifications on compounds (quantity and quality) and a new matrix with new interactions, in another dominant scale, which will define the obtained products and their properties.

Table 5.1 summarizes the initial products or typical constituents that make up the original matrix, the processes or technologies (examples of conventional or emerging) applied, the main modifications in the structure that occur during processing and examples of obtained products. To describe structural modifications, Table 5.1 provides different examples, comprising solid foods (of animal and plant origin), fluid systems (such as powders, pulp, juices) and the macromolecular constituents (proteins, polysaccharides, and lipids) that play an important role in forming the structure of processed foods (Aguilera et al., 2000). In addition, it is important to mention that the processing effects on food structure are difficult to predict or generalize due to the complexity of the food matrix. Therefore, the structure modifications summarized in Table 5.1 are those that commonly have been reported; however, these modifications will be specific for each raw material or compound, type of process and processing conditions.

### 5.2.1 Structural modifications in solid foods

In solid food structures, modifications at macrostructure and microstructure levels occur. For example, freezing causes modifications in the adhesion of adjacent cells, intercellular space enlargement and cell wall disruption (Li et al., 2018). Acidification replaces the air in the intercellular spaces with the acid solution, which induces fluid incorporation by the cells, increasing turgidity (Miano et al., 2017). On the other hand, tissue compaction or softening, cell shrinkage, breaking, plasmolysis, debonding and intercellular porosity changes occur during conventional thermal processes (blanching, pasteurization, cooking, roasting or frying) (Bouchon & Aguilera, 2001; Sensoy, 2014) and dehydration processes such as convective drying and osmotic dehydration (Aguilera et al., 2003; Mayor et al., 2008; Ramos et al., 2003).

Throughout processing with emerging technologies, modifications such as disruption or electroporation in the cell walls and membranes that modify the cell permeability were reported during HHP processing in vegetable and meat structure (Janowicz & Lenart, 2018; Jia et al., 2021; Oey et al., 2008) and during electric fields processing (Alam et al., 2018; Wang et al., 2018). In addition, erosion of sample surface, changes in the dispersion and disruption of intracellular plastids (as chromoplasts, amyloplast or elaioplast) and microchannel formation were reported during high-power ultrasound (US) processing in vegetable and meat tissue structures, as a consequence of acoustic cavitation and the sponge effect phenomena (Chen et al., 2020; Miano et al., 2018; Petigny et al., 2013; Rajewska & Mierzwa, 2017; Rojas & Augusto, 2018; Santos et al., 2021; Schössler et al., 2012). In the case of HHP (high hydrostatic pressure) and US (ultrasound) processes, an initial porous structure intensifies the obtained structure modifications (Aganovic et al., 2021; Janowicz & Lenart, 2018; Ozuna et al., 2014). The controlled structure modification during emerging technologies processing of solid foods was used to improve mass transfer phenomena. It is important to highlight that during the processing of solid products, modifications in the structure of their compounds could also occur, such as proteins and polysaccharides, which will be described in following sections.

Regarding the top-down approach, mechanical processes such as pulping (Gautam et al., 2020) and grinding (Huang et al., 2020) are used to create new particulate structures from initial macroscopic solid foods. These processes disintegrate the original tissues, disrupt cells, reduce the size from macro to microscale and change the shape and morphology of particles. As a result, products such as food powders, flour, or pulps are obtained, which could be used as constituents for further processes.

### 5.2.2 Structural modifications in particulate foods

Liquid foods such as milk and juices are made up of a dispersed phase (droplets, micelles, pieces of tissues, cells, cell fragments, insoluble polymers) and a continuous phase (water and solutes such as sugars, salts, and acids). In these systems, modifications in particle/droplet size, compound dissociation from micelles, distribution and shape, cell wall, membrane and organelles disruption occur during thermal processing (Dalgleish & Corredig, 2012; Moelants, Cardinaels, Van Buggenhout, et al., 2014), colloidal milling, homogenization or spray drying processes (Lesmes & Mcclements, 2009). The mentioned modifications during the conventional mechanical processes, also occur by using emerging technologies, such as PEF (pulsed electric fields) (Gabrić et al., 2018), US (Atalar et al., 2020; Gaikwad & Pandit, 2008) and HPH (high pressure homogenization) (Bot et al., 2017; Kubo et al., 2013; Moelants, Cardinaels, Jolie, et al., 2014). However, different structure modification steps can occur, such as the interesting mechanism described by Rojas, Leite, et al. (2016) during the US processing of juices, from the intact tissue to cell and macromolecules disruption, which triggered different mechanisms with a complex behavior. For instance, the disruption of cells is initiated with the breakage of organelles, releasing different compounds into the intracellular medium. It is followed by cell damage and release of intracellular content to the juice serum. Finally, the whole cells are disrupted, resulting in particle size reduction and dispersion of constituents. Those structural modifications affected the juice consistency, which was increased (which is interesting from a sensorial point of view, once pulp sedimentation is avoided and mouthfeel is enhanced without the need of adding other ingredients) or decreased (which is interesting from an industrial point of view, reducing friction and energy dissipation during processing), depending on processing conditions. In addition, during HPH, disruption occurs preferentially in larger particles and clusters of cells, being the smaller ones less susceptible to subsequent disruptions (Augusto et al., 2012b; Kubo et al., 2013; Tan & Kerr, 2015). As a consequence of structural modifications, new interactions are promoted within and between phases, which influence the properties of the obtained products.

On the other hand, initial solid particulates such as powers or ground products are usually subjected to processes such as thermal processes, mixing or extrusion. During processing, new network interactions occur among particles and compounds, successive structure modifications in compounds (e.g., starch and proteins), structure compaction or expansion, resulting in products such as bread, snacks, pasta (Petitot et al., 2009; Puerta et al., 2021). The suspensions prepared from these solid particulates can also be processed through the technologies mentioned above for the liquid particulate systems, obtaining similar structure modifications.

### 5.2.3 Structural modifications in food macromolecules

The most important macromolecules that compose the structure of foods are proteins, polysaccharides, and lipids. In proteins, restructuring of aggregates, unfolding and dissociation that modify their globular conformation and interactions, breakdown in polypeptide chains, and cross-linking among polypeptide chains or other compounds could occur during processing. On the one hand, modifications in the quaternary, tertiary and secondary structure of proteins without impact in the primary structure were reported through processes such as HHP, irradiation, PEF, US, cold plasma and mild thermal processing (Aganovic et al., 2021; Balny et al., 2002; Huang et al., 2017; Jia et al., 2021; Knorr et al., 2011; O'Sullivan et al., 2016; Rahaman et al., 2016; Vanga, Wang, Orsat, & Raghavan, 2020; Zhang et al., 2021). It is important to highlight that the emerging technologies show similar but not identical effects in proteins to those obtained with conventional thermal processes. In addition, the thermal process also promotes the interaction of proteins with other ingredients via Maillard reaction, which forms aggregates and new compounds. On the other hand, for modifications in the primary structure of proteins, more efficient methods such as fermentation or enzymatic hydrolysis are necessary to obtain peptides or protein hydrolysates (Udenigwe & Aluko, 2012). The study of structural modifications in proteins is especially important since processing, apart from the effects on physical or functional properties, also impacts protein allergenicity (Mills et al., 2009; Rahaman et al., 2016; Zhang et al., 2021), as well as it can be an interesting route to obtain bioactive peptides (Caetano-Silva et al., 2021; Daroit & Brandelli, 2021). Additionally, in enzymes, such as hydrolases and oxidases enzymes conformational structure modifications occur, which lead to enzyme inactivation during conventional thermal processing (Moelants, Cardinaels, Van Buggenhout, et al., 2014) or ohmic heating

(Makroo et al., 2020). In contrast, during US or HPH processing, depending on the applied conditions, conformational modification can modulate the enzyme activity (i.e., activation, stabilization or inactivation) and modify the enzyme interaction with some compounds (Dos Santos Aguilar et al., 2018; Rodríguez et al., 2017; Rojas, Trevilin, et al., 2017; Rojas, Trevilin, & Augusto, 2016; Soares et al., 2020). Generally, the application of mechanical emerging technologies also requires the application of mild heat treatments to achieve the desired levels of enzyme inactivation.

Regarding polysaccharides, pectin is a component found in the cell wall of fruits and vegetables, which undergoes modifications such as depolymerization during conventional thermal processing (Moelants, Cardinaels, Van Buggenhout, et al., 2014), but also structure modifications with emerging technologies application were reported. The cavitation effects of high-power US causes the breakdown of the structure and reduce the average molecular weight of pectin suspensions and other hydrocolloids such as guar and xanthan (Seshadri et al., 2003; Tiwari et al., 2010), also soluble polysaccharides can be cleaved during US processing due to both physical and sonochemical effects (Gogate & Prajapat, 2015). In addition, disruption of large molecular weight aggregates, modifications in the molecular weight distribution and reduction in average molecular weight of pectin were obtained during HPH processing (Corredig & Wicker, 2001). The structural modifications of pectin are particularly important since they impact the rheological properties of the food system. Furthermore, among the most important polysaccharides, starch has been highly studied due to its numerous applications and commercial importance. In starch, the thermal process (above gelatinization temperature), results in simultaneous loss of granular, lamellar, crystalline and double-helical order (Nayak et al., 2014). In addition, following the bottom-up approach, heat induces starch-protein interactions at the molecular scale (chain assembly from molecular to microstructure scale) forming larger-scale structures (Wang et al., 2021). The PEF application induces surface damages in starch granules and partial loss of crystalline structure (Knorr et al., 2011; Maniglia et al., 2020). However, with US processing, unclear results have been reported, while some studies reported changes in the structure such as cracks or pores in the granules with changes in crystalline and amorphous structure (Téllez-Morales et al., 2020; Zhu, 2015), other studies have reported minor modifications in the starch structure (Castanha et al., 2019), Further studies are necessary to corroborate this. In addition, with HPH (Apostolidis & Mandala, 2020; Che et al., 2007; Qiu et al., 2014) or HHP (Han et al., 2020; Kaur et al., 2019; Pei-Ling et al., 2010), especially in A-type and C-type starches, partial gelatinization, swelling and/or aggregation, changes on the particle size and distribution, modification of the interaction between amorphous and crystalline components were reported. Furthermore, the ozone application causes the increase of carbonyl and carboxyl groups content and depolymerization of both amylose and amylopectin molecules, modifying the molecular size and their interactions (Castanha et al., 2017; Çatal & İbanoğlu, 2012; Lima et al., 2020; Lima et al., 2021). During cold plasma application, modification of starch granule surface, and the interaction between reactive species of plasma and side chains of starch could result in crosslinking, depolymerization, crystallinity changes, and the inclusion of functional groups (Han et al., 2020; Thirumdas et al., 2017). More details about the emerging technologies application in starches can be found in the Maniglia et al. (2021) review.

Lipids can be found in the form of crystals in certain food products and temperatures, whose structure is especially sensitive to thermal processes that induce modification in the crystal structure, crystal network and crystal shape (Aguilera et al., 2000; Ubbink et al., 2008). However, in most processed foods, lipids are stabilized in the form of emulsions (e.g., oil-in-water, water-in-oil, multiple emulsions and multilayered emulsions) (Gallier & Singh, 2012). Lipids are also used to produce particulate delivery systems following the bottom-up and/or top-down approaches (Lesmes & Mcclements, 2009). These systems could be prepared by using HPH, high-shear mixers, US, membrane homogenization, microfluidization or colloid mill, resulting in different droplet sizes from 10 nm up to 100 μm, with one or multiple layers. In addition, emulsion processing requires the presence of food emulsifiers such as proteins, phospholipids, or other polymers (Gallier & Singh, 2012). Therefore, proteins, polysaccharides and lipids are used as ingredients following the bottom-up or top-down approaches of processing to create new structures in colloidal state such as emulsions, gels, foams and dispersions, which can be used as end products in themselves or used as ingredients for more complex products (Cao & Mezzenga, 2020; Dalgleish, 2006; Dickinson, 2015; Lam & Nickerson, 2013).

In conclusion, processing modifies the structure of the original food matrix (solid foods, particulate foods, or food constituents) at different scales. The level of structure modification depends on the initial structure, composition and type of food matrix, and processing conditions such as energy supplied to the system, power, temperature, pH, and processing time. The structure modifications at macro and micro scales are typically achieved by applying mechanical or mild physical processes. Therefore, the primary structure of food constituents is not significantly modified with these processes, without direct effects on vitamins or secondary metabolites. However, modifications at a molecular scale are achieved by applying severe physical (including thermal processes) or chemical processes. Consequently, the modifications at different levels of the structure during processing will impact physical and sensorial properties but also can generate direct or indirect effects on the quantity and quality of compounds. These aspects will be described below.

## 5.3 Structure modification and impact on physical properties, sensorial aspects and nutritional quality

### 5.3.1 Structure modification and impact on physical properties

Food processing involves several unit operations and processes, such as heat, mass and momentum transfer, resulting in various physical, biochemical, sensorial and nutritional changes. Many of these changes are related to the structure modification of the food matrix.

The relationship between structure and properties is quite evident and may be associated with different physical aspects. Food structure provides information on different properties, such as mechanical, rheological, dielectric and thermophysical properties. Depending on the processing technology and process parameters as well as the type of food matrix and its composition, the structure modification can cause either negative or positive effects on food properties.

Food drying provides good examples to illustrate how the various macro and microstructural changes resulting from this processing reflect on the properties of the dried solid food materials. After the migration of water, the structure of the dried food may become porous, shrunken, stiffer, glassy and brittle, which in turn affect its mechanical and mass-volume-area-related properties. During drying, foods can undergo deformation, changing their shape and size/volume. Generally, a shrinkage of the food material is observed, because of the collapse of cells during water vaporization. Shrinkage extension is very dependent on the type of food, the initial moisture content, the drying method and the applied parameters, in special temperature (Koç et al., 2008; Pacheco-Aguirre et al., 2015; Panyawong & Devahastin, 2007). In fact, the use of novel and/or combinations of drying techniques have been applied as an approach to control the shrinkage of dehydrated foods. For example, the combination of US and conventional drying methods can lead to the acceleration of the drying rate and a decrease of shrinkage phenomenon (Oikonomopoulou & Krokida, 2013). The use of microwave-convective and infrared techniques instead of conventional convective drying methods results in dried apple slices with lower shrinkage values and greater porosity (Witrowa-Rajchert & Rząca, 2009).

The formation of cavities and the porous structure is also commonly observed during drying and extrusion processes. Porosity can be defined as the ratio of total enclosed air space or void space to the total volume of food material. Together, porosity and shrinkage can directly affect physical properties, such as the true and apparent densities, and further aspects, such as rehydration rate and diffusion coefficient (Oikonomopoulou & Krokida, 2013). Besides, variation of density, porosity and degree of shrinkage can be correlated with mechanical properties, particularly hardness or firmness (Funebo et al., 2000). For example, the hardness of extruded cereals was influenced by porosity: products with higher porosity presented higher aeration, lower hardness and higher number of mechanical fractures (Chanvrier et al., 2014).

Moreover, structure modification can also have an impact on thermophysical (or transport) properties, such as thermal conductivity and thermal diffusivity. In fact, the thermal properties of foods are determined not only by their temperature and chemical composition (e.g., water, protein, carbohydrate, fat, ash content) but also by their structure. Therefore, factors such as cell structure, air trapped between cells, void/air space fraction, pore size and distribution must be considered, especially in the case of highly porous foods such as dehydrated and extruded products. In this way, the thermal properties can be calculated based on their constituents as well as air. For that, well-established equations as those described by Choi and Okos (1986)) can be used—although further improvement is available, as in the work of (Phinney et al., 2017). For example, the thermal conductivity of foodstuffs has intermediate values between two extremes: water conductivity (highest) and air conductivity (lowest). As a result, dried porous foods are poorly conductive materials due to the low moisture content and the presence of air spaces—actually, in this case, an apparent or effective thermal conductivity is measured, assuming that heat is transferred through the solid and void spaces (Sahin & Sumnu, 2006).

Changes in food structure also influence the dielectric properties, which determine the distribution of electromagnetic energy during dielectric heating and are relevant properties to radiofrequency and microwave heating. Besides being very dependent on the electric field frequency, temperature and chemical composition, dielectric properties are also influenced by the density and the structure of food materials (İçier & Baysal, 2004; Venkatesh & Raghavan, 2004). Dielectric properties are affected by the amount of mass interacting with the electromagnetic field (İçier & Baysal, 2004). Since air presents low dielectric properties—generally, the air is considered transparent to microwaves—structural factors such as density and porosity have an impact on the food dielectric properties. For example, highly porous materials present lower dielectric properties due to the entrapped air (Feng et al., 2012). Similarly, in particulate materials, bulk density has also proven to be an important effect on dielectric constant and loss factor (Venkatesh & Raghavan, 2004).

Additionally, the rheological properties, which are of great importance in the design of manufacturing machinery and unit operations, quality control, and sensory acceptance, are closely related to food structure. These properties are associated with the deformation and flow of materials subjected to mechanical forces. In general, food materials present a viscoelastic behavior. Although rheological responses are measured at a macroscopic level, they are very affected by changes at a microscopic level (Rao, 2007). For example, the time-dependent rheological behavior of foods is based on the internal structural changes and the balance between structural breakdown due to shear and reorganization due to particle attractive forces. In this way, homogenized products, in which the size of particles was modified and the interaction forces were increased, present the formation of networks and particle aggregates, leading to an increase of thixotropy, as observed in tomato juice after HPH (Augusto et al., 2012b). By the way, by reducing the diameter and size distribution of suspended particles, HPH not only increased thixotropic behavior but also increased the tomato juice consistency (Augusto et al., 2012b) and the elastic and viscous behaviors (Augusto, Ibarz, & Cristianini, 2013). This further led to a reduction of sedimentation and phase separation, improving the physical stability of the tomato juice during storage (Kubo et al., 2013). However, it is important to note that the rheological response depends on the type of food and its matrix characteristics and composition. For instance, an increase of consistency after HPH was reported for tomato, apricot and mango juices (Patrignani et al., 2009; Zhou et al., 2017), however a decrease of (apparent) viscosity was observed in cashew apple, banana, orange and pineapple products (Calligaris et al., 2012; Leite et al., 2014; Leite et al., 2015; Silva et al., 2010).

In addition to HPH, US processing also provides great examples of the influence of structural modification on rheological properties. As mentioned, US can change the food structure by causing cell damage and disruption, the release of intracellular content, and particle size reduction. Similarly to HPH, these US-related changes can influence the forces of interaction (particle-particle and particle-serum). For example, the apparent viscosity of peach juice increased, then decreased, and finally increased again along US processing time, including changes in the yield stress. This behavior and the possible mechanisms involved were thoroughly described by Rojas, Leite, et al. (2016). Further information on the relationship between food structure and rheological properties of plant-tissue-based food suspensions can be found in a review by Moelants, Cardinaels, Van Buggenhout, et al. (2014).

### 5.3.2 Structure modification and impact on sensorial aspects

The food composition and its structure can be an important factor affecting the sensorial aspects. For example, more marbled beef with streaky fat presents greater softness and juiciness than beef with a superficial fat layer or even more than a low-fat beef. Therefore, the quantity and the way the fat is distributed into beef affects its sensory perception. This occurs in different types of food products and the more complex the structure more modifications can influence sensorial aspects.

As an illustrative example, the egg white can be considered (Fig. 5.2). This egg component is naturally in a liquid state and presents in its composition mostly water and proteins, like albumin. The mechanical agitation promotes the air incorporation, protein denaturation and formation of bubbles. With the agitation progress, the bubbles become smaller and form a network with the proteins changing the structure and generating a foamy aspect.

This modification in egg white promoted by the agitation also results in sensorial changes, which are visually apparent, like color alteration and also textural changes.

Therefore, both conventional and emerging technologies of food processing result in structural modification (as described in Table 5.1) and, consequently, sensorial changes in the product.

Color is an important sensorial aspect because it is one of the first attributes evaluated in a product. For juices and purees, it is desirable that the color of processed products remains similar to that of fresh products; additionally, that the color does not change during the storage period (Kubo et al., 2021). So

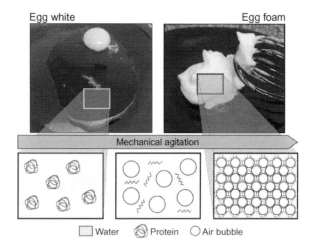

**FIG. 5.2**

Modification in egg white promoted by mechanical agitation.

that, color can also be considered an important indicator of quality and nutritional losses of liquid foods during processing and/or subsequent storage (Rojas, Miano, & Augusto, 2017).

Even so, some technologies can improve the color characteristics. Calligaris et al. (2012) obtained brighter and lighter banana juice applying HPH treatments from 150 to 400 MPa when compared with raw juice. This effect was attributed to the ability of small size particles generated by HPH scatter light. When applied in mango juice, at 190 MPa, 1 and 3 passes at an inlet temperature of 60 °C, the HPH treatment resulted in no visible difference when compared with control samples (Guan et al., 2016).

Regarding US application, this technology can enhance the color stabilization of many juices during storage, as reported by Costa et al. (2013) for pineapple juice and by Rojas, Leite, et al. (2016) for peach juice. Additionally, the US application can also increase the lightness ($L^*$) parameter of color, as reported by Aadil et al. (2015) for grapefruit juice, by Abid et al. (2014) for apple juice, and Rawson et al. (2011) for watermelon juice. This increase in lightness can be attributed to the higher reflection due to small particles obtained after the US application. The US also can affect the color of solid products submitted to processes like drying and, in these cases, color is a decisive factor to consider when judging the quality of the product (Musielak et al., 2016). Positive effects of US application on the color of dried products were also reported for carrot (Kroehnke et al., 2015) and green pepper (Szadzińska et al., 2015). The mitigation of color degradation or even the improvement in color parameters because of US application can be attributed to the acceleration that this technology promotes in the drying process. The cavitation, sponge effect and microchannels formation in the structure of the product enhances the drying process avoiding degradation.

In relation to PEF, the application of this technology also has been presented interesting results considering the product color maintenance. The PEF application promotes structural changes in the products, mainly due to electroporation, enhancing the mass transfer during the drying process (De Vito et al., 2008). Liu et al. (2020) reported less modification in overall color ($\Delta E$) with the application of PEF before vacuum-drying of carrot when compared with untreated samples. Similar results were found during treatment of bovine *logissimus thoracis* muscles with PEF, where the color parameters ($L^*$, $a^*$ and $b^*$) were maintained (Chian et al., 2019).

The microwave also is a technology highly employed in food processing. Its application in drying can promote a large reduction in processing time, however, due to its thermal action, large modification in quality parameters can occur, mainly in the color of the products. Consequently, the application of MW (microwave) needs to be well-balanced in terms of improving the process without excessively damaging the product. Usually, the combination of other technologies such as US, PEF, vacuum and others, can help to reach these goals. It is common to found in literature works with microwave applications obtaining high-quality products. Regarding color parameters, (Monteiro et al., 2020) related a reduction in overall color ($\Delta E$) change during MW-vacuum drying of sweet potato chips when compared with blanched fresh samples. Additionally, the authors obtained chips with uniform color and the absence of burned parts. Monteiro et al. (2018) also obtained lower overall color ($\Delta E$) change in pumpkin using the microwave multi-flash drying.

The texture is also an important sensorial parameter that determines the acceptance of the product by the consumers. Structural changes, including the promoted by the application of different technologies during the processing, will be determinant to the texture characteristics of the final product. Different textures parameters can be relevant and evaluated depending on the product taken into account. For liquid and semi-solid products like juices, purees, creams, and others, the viscosity is an important parameter that can be correlated with the consistency and mouthfeel which, in turn, are valuable attributes to quality and consumer acceptance (Kubo et al., 2019). The effect of different technologies applied on the processing of liquids and semi-solids depends on the structure of each product, for example, the conditions and effects on carrot processing are different on tomato processing, because they present distinct structures.

For some types of products, such as tomato juice (Augusto et al., 2012a; Augusto et al., 2012b; Augusto, Falguera, et al., 2013), tomato-based products (Colle et al., 2010; Lopez-Sanchez et al., 2011) and mango juice (Zhou et al., 2017), the HPH application increased the consistency, viscous and elastic behaviors. These results are obtained due to the reduction in suspended particle dimension and distribution and consequent higher interaction between the particles. Considering the technological point of view, it is an interesting result with HPH application to reduce pulp sedimentation while allowing for greater consistency in the product without the addition of stabilizers/hydrocolloids. On the other hand, as described for products like carrot, broccoli, cashew apple, banana, orange and pineapple, the application of HPH resulted in the reduction of the apparent viscosity (Calligaris et al., 2012; Leite et al., 2014; Leite et al., 2015; Lopez-Sanchez et al., 2011; Silva et al., 2010).

US processing of fluid products presents similar structural effects when compared with the HPH. US can promote the reduction in particle size dispersed in the product, although its effects on juices rheology can present complex behavior. Depending on the energy applied by the ultrasound, the consistency of the product can be modified permanently or temporarily (Soria & Villamiel, 2010). Augusto et al. (2012a) and Rojas, Leite, et al. (2016) presented possible approaches in relation to structural changes promoted by ultrasound processing that causes the increase/decrease in juice consistency correlating the increase in juice consistency with an increase in the interaction forces between small particles, the release of intracellular compounds and increase in dispersibility. On the other hand, the reduction in consistency can be related to the decrease in dispersibility and reduction in polysaccharides size. The rheological changes due to structural modifications also affected pulp sedimentation, being the US processing an alternative to reduce it—such as demonstrated in peach (Rojas, Leite, et al., 2016) and guava (Campoli et al., 2018) juices.

The US application is also interesting in the point of view of texture when applied in solid food products. The application can be made as a pre-treatment to other subsequent processes or even assisting the process itself. US was able to modify the texture of fruit and vegetable, meat, and dairy products (Terefe et al., 2016), being a function of several factors, such as composition, the porosity of the product and the power and intensity of US (Rodríguez et al., 2018). Regarding fruits and vegetables, low-intensity ultrasound pre-treatments at low to medium frequencies (40–400 kHz) showed improvement in the firmness of potato, carrot, and pear tissue (Terefe et al., 2016), which can be interesting to the production of snacks. Significant structural changes were observed in a less porous material like the one from eggplant and apple. On the contrary, more compact products, such as potato and cassava, presented smaller modifications in structure.

PEF also can be able to modify the texture of food products. Jin et al. (2017) demonstrated that the application of PEF in whole blueberry fruits led to softer texture in comparison to untreated samples. This softening may be due to cell membrane breakdown (Sungsinchai et al., 2019). Llavata et al., 2020) also reported the softening effect of PEF for several applications.

The MW also impacts the structure and texture of food products. When applied to drying food, particularly added by vacuum, the microwave can produce more crispy products. Due to the particular heating mechanism of microwaves, which heats the product from inside, the water vaporizes before leaving the product, increasing the pressure inside the product. This phenomenon promotes the formation of pores inside the product and these pores can give the crispy aspect of the product, desirable in some snacks. Depending on the product and its structure, the pressure increase can result in a puffing effect (commonly observed in grains). For example, (Monteiro et al., 2016) obtained crispy bananas with high porosity in the MW multi-flash drying. The MW vacuum drying was also effective to obtain oil-free sweet potato chips, as described by Monteiro et al. (2020), resulting in a highly porous product.

Other sensorial aspects like aroma and flavor can also be affected by modifications in the structure. As before mentioned, utilization of technologies like HPH, US and PEF can promote cell disruption in the product and consequent release of compounds responsible for aroma and flavor. The harmful effect in aroma and flavor will depend on the conditions of application, the product being processed and the exposure to oxygen and light, since there is literature reporting the maintenance of aroma and flavor even with the application of technologies like PEF (Aguillar-Rosas et al., 2007), US (Singla & Sit, 2021) and HPH (Oey et al., 2008).

To summarize, several technologies can be applied in order to promote structural modifications in food products, affecting the sensorial aspects, such as color, texture, aroma, and flavor. The modifications in sensorial attributes will depend on the product, the technology or combination of technologies utilized, and the processing conditions. The combination of these factors can enhance the sensorial attributes, or making them worse. Therefore, it is important to study different technologies and conditions to provide data regarding the modification of the structure of food products and the consequences on the sensorial aspects.

### 5.3.3 Structure modification and impact on nutritional quality and health aspects

Besides the physical properties and sensorial aspects, the food structure can greatly affect the nutritional quality and health aspects of food products, which means it plays an essential role in how the nutrients and bioactive compounds are absorbed and thus used by the organism. Noteworthy is that the nutrient content of a food product does not reflect the nutrient absorption and the metabolic pathways used by the Human organism, since several intrinsic and extrinsic factors influence the absorption rate. In addition, the structural modification of tissues, cells, and molecules promoted by food processing can dramatically affect how the food products will be digested and the nutrients or bioactive compounds will be available to be absorbed, resulting in different food digestibility and bioaccessibility (Verkempinck et al., 2020). Therefore, foods containing the same nutrients and bioactive compounds can present different nutritional qualities and differ on the health effects they present.

Nutrients are compounds necessary in metabolic pathways, being essential for the maintenance and growth of the organism. On the other hand, bioactive compounds are defined as compounds that, in small quantities, can present health benefits regardless its nutritional value, i.e., being or not part of an essential metabolic pathway. An example can be the phenolic compounds, which are non-nutrients that can present several health benefits. The food industry has been focusing on developing processing technologies that somehow can be beneficial for both the nutritional value, related to how much a nutrient will be absorbed and effectively used by the organism and the bioactivity, related to health effects beyond the nutritional aspect.

It must be considered that to present their bioactivity, the compounds must be soluble in intestinal pH (6.0 to 7.4) and absorbable by the human body. These two processes (solubility after gastrointestinal digestion and absorption by the intestinal cells) are related to two terms often used when evaluating the bioactivity of food compounds: bioaccessibility and bioavailability. It is of utmost importance to consider the capacity of a compound to be released from the food matrix and become available to be absorbed (bioaccessibility), which is highly influenced by the food structure, and its capacity to be absorbed by the enterocytes and be available to be transported to the target tissue to be used by the organism (bioavailability).

As previously mentioned, food products can be processed by different technologies, and the different conditions to which those products are submitted play a crucial role in determining the structure of both food matrix and molecules. Such conditions can promote partial degradation, exposure, or entrapment of molecules, affecting their interaction with other molecules during gastrointestinal digestion. Food processing can promote cell damage and disruption, with the consequent release of intracellular compounds (Rojas, Leite, et al., 2016), consequently altering their stability and accessibility. Once outside the organelle or cell structure, the compounds can be more prone to degradation; however, it can also facilitate their absorption and consequent use by the organism since it can increase the likelihood to enter the enterocytes membrane (Rojas et al., 2021). For example, bioactive compounds can be entrapped into the food matrix, which hinders their solubility, and thus bioaccessibility, as well as impact their bioactivity. When released from the food matrix, such compounds are more likely to be available for absorption. In fact, the bioaccessibility and bioavailability are significantly affected by some important characteristics of bioactive compounds, such as molecular mass (MM), solubility under different conditions, hydrophobicity, and isomer configuration (Barba et al., 2017). For instance, the structure of molecules with low MM is less likely to be affected by food processing; however, it can influence how these small molecules complex with other compounds, significantly affect their absorption. The solubility of molecules previously intracellular located can be affected upon cell-wall rupture since the new environment after release will be completely different from the intracellular one. Lipophilic compounds, such as carotenoids, have a different absorption mechanism than hydrophilic compounds, such as polyphenols and hydrophilic vitamins. The former ones must be released from the food matrix and incorporated into micelles to be absorbed in the small intestine, while the latter ones do not depend on micelle incorporation (Barba et al., 2017).

It is worthy of highlighting that the changes in food structure promoted by food processing do not increase the amount of bioactive compounds, but they markedly affect the accessibility of such compounds after being released from the food matrix. The access of gastrointestinal enzymes, the solubility of the compounds through digestion, the capability to be transported across the enterocyte membrane and to reach the target tissue, can all be influenced, even if the amount of given compound remains the same or even if the compound is partially degraded (Rojas et al., 2021). Thus, the measurement of bioactive compounds using different methods can vary depending on the accessibility of the compounds to be measured.

Consequently, food processing has been described as a strategy to manage digestive barriers and improve both the digestibility and bioaccessibility of compounds (Verkempinck et al., 2020). For example, although the content of some bioactive compounds can be decreased during processing due to degradation, their bioaccessibility can be increased due to the rupture of the food matrix, which facilitates the compound release. In the case of carotenoids, this process can also facilitate their solubilization in mixed micelles, aiding their absorption (Barba et al., 2017).

HPH is a valuable tool to improve the nutritional quality of food products, especially regarding bioactive compounds, which are, in general, sensitive to thermal processing (Kubo et al., 2021). HPH can affect the food microstructure by reducing the particle size distribution and inducing cell-wall rupture, thus influencing the bioaccessibility of several compounds. Fernández-Jalao et al. (2017) reported that high-pressure processing combined with freeze-drying improved the extractability of bioactive compounds in onions such as flavonols, hence improving their bioactivity after gastrointestinal digestion. The reduction in size distribution due to the complete disruption of individual plant cells is reported to cause an increase in the release of intracellular compounds, such as proteins and polyphenols, and a consequent increase in more than 20% of the antioxidant activity of tomato peels, byproducts of tomato processing, treated by HPH (Jurić et al., 2019).

In the meantime, the US processing of guava juice promoted lycopene release, resulting in accessibility more than twofold higher after processing (Campoli et al., 2018). In tomato juice, US treatment was reported to promote the isomerization of all-trans lycopene and cell breakage, resulting in a bioaccessibility increase of up to 1.76-fold (Zhang et al., 2019).

Certain bioactivities can be affected by US treatment, such as antioxidants and anti-inflammatory compounds. De Souza Carvalho et al. (2020) reported an increase in the antioxidant activity of Amazon fruits açaí and buriti after juice treatment with high-power US processing due to the release of aromatic compounds, such as anthocyanins and carotenoids. The US pre-treatment of soybeans has been reported to increase the antioxidant and anti-inflammatory activities, evidenced by the reduction of nuclear factor kappa B (NF-κB) activation in macrophages. Interestingly, the capability of isoflavones in reducing the NF-κB activation indicates the antioxidant molecules have entered the macrophages, suggesting the US treatment led to the complete release of isoflavones molecules from the soybean matrix (Falcão et al., 2018).

PEF is another non-thermal technology that can be applied to increase the bioaccessibility of bioactive compounds. The bioaccessibility of phenolic and carotenoid compounds of PEF-treated oil-added carrot puree increased by threefold when compared to thermal treatment, due to the release of the bioactive compounds from the matrix during digestion. The enhancement effect was attributed to the microstructural changes characterized by smaller particle sizes after PEF treatment, although further studies are necessary to explain better the relationship between PEF-induced changes and bioaccessibility (López-Gámez et al., 2021).

The mechanism by which PEF treatment impacts the bioaccessibility of bioactive compounds seems to be related to the electroporation phenomenon, a direct result of electrical breakdown. The electroporation theory suggests that exposure to an electric field involves membrane compression and pore formation. The perforation of the cytoplasmic membrane would lead to the leakage of the cell content, possibly facilitating access to the internal components and their absorption. Moreover, rheological changes of the liquids induced by PEF treatments could also be related to the increase in enzymatic hydrolysis and increased bioaccessibility of carotenoids (Barba et al., 2017).

The interactions of bioactive compounds with compounds that commonly influence absorption during gastrointestinal digestion, such as fiber and fats, can be dramatically influenced by food processing. HPP may affect the location of tocopherols in the food matrix, making these lipophilic compounds more accessible for incorporation into micelles to be absorbed (Cilla et al., 2012). US has been reported to promote the rupture of carotenoid-protein binding in mango and papaya juices, leading to the release of bound carotenoids and consequent easier extractability of such compounds (Buniowska et al., 2017).

Dietary fiber can entrap compounds in its structure and act as a hindrance to the bioaccessibility of such compounds. Changing the structure of fiber in the food can result in the release of bioactive compounds and increase their bioaccessibility. US applied in peel and paste of mango resulted in the release of carotenoids such as β-carotene, lutein, and β-cryptoxanthin, and their consequent higher bioaccessibility (Mercado-Mercado, Montalvo-González, González-Aguilar, Alvarez-Parrilla, & Sáyago-Ayerdi, 2018). However, it must be highlighted that the release of nutrients and bioactive compounds upon cell disruption can also be undesirable since it can increase oxidation by exposing sensitive compounds (Kubo et al., 2021).

The structural changes promoted by food processing depend not only on the compounds involved and applied technology but also on their food matrix. Kubo et al. (2021) reviewed different studies from the last decade on the HPH impact on food structure and reported controversial results regarding the compounds bioaccessibility: either an increase of the in vitro bioaccessibility of α- and β-carotene from carrot due to the cell disruption by HPH, or a decrease of the in vitro bioaccessibility of lycopene from tomato due to the formation of a fiber network that may entrap the bioactive compounds. A similar observation was described by Rojas et al. (2021)) for US processing. The US treatment can also lead to forming a new network based on hydrogen bonding and hydrophobic interactions among saponified pectin molecules (Rojas et al., 2021). For instance, ultrasound decreased the degree of pectin esterification in tomato pulp, promoting the formation of a strong network that led to a decrease in lycopene bioaccessibility (Anese et al., 2013).

A crucial impact of food processing on food structure is the conformational changes on protein molecules that can lead to different access of gastrointestinal enzymes, resulting in different protein absorption. The bioavailability of proteins (rice, oat, corn, and soy) has been reported to be increased by up to 18.7% upon US treatment due to changes in their secondary structure, resulting in greater accessibility by the digestive enzymes and a consequent increase in absorption of the molecules with lower MM (peptides with MM of 200–1000 Da) (Yuanqing et al., 2020). Also, the interaction of proteins and peptides with other molecules can have a significant impact. For instance, the digestibility of pea protein added to apple puree was lower when added to carrot puree, due to the protein binding to apple procyanidins, reducing the accessibility of proteolytic enzymes during gastrointestinal digestion. However, when HPH-treated, both vegetable purees enriched with pea protein showed no matrix effect on protein digestibility (Laguna et al., 2017). Furthermore, (Xue et al., 2020) reported the positive effect of HPH on the in vitro digestibility of gel-type meat products, being the myofibrillar the major contributors to the high pressure-induced changes.

The digestibility of plant-based protein can be greatly impacted by the presence of anti-nutritional factors, such as phytates, tannins, trypsin inhibitors, and lectins, leading to lower digestibility of plant-based proteins when compared to animal proteins. Protease inhibitors can hinder the action of digestive enzymes in the small intestine, reducing protein digestibility and, consequently, absorption. In the meanwhile, polyphenols might complex gastrointestinal enzymes, decreasing their activity and leading to a lower protein digestibility (Sá et al., 2020). Recent studies have shown that changes in food structure can lead to the inactivation of anti-nutritional factors and a consequent increase in plant protein digestibility (Alsalman & Ramaswamy, 2020; Joye, 2019; Vanga, Wang, & Raghavan, 2020). For instance, HP treatment combined with pre-soaking of chickpeas reduced fivefold the tannin and phytic acid content compared to raw chickpeas (Alsalman & Ramaswamy, 2020). Ultrasound or microwave processing reduced up to 52% the trypsin inhibitor activity of soymilk, significantly improving the soy protein digestibility (Vanga, Wang, & Raghavan, 2020). The increasing worldwide demand for plant-based protein with higher digestibility has driven the food industry to look for a novel or combining existing processing techniques to address protein digestibility besides the changes in functionality. However, despite the promising results, the enhancement of protein digestibility of a plant source remains a scientific and technological challenge for the food industry.

Besides protein digestibility, HPH has been reported to potentially decrease protein allergenicity due to its impact on the structure. Considering that the allergenic potential of proteins is mainly related to the secondary and tertiary structures of allergens, HPH can have an impact on the epitopes involved in allergic reactions and reduce the allergenicity of several proteins (Pottier et al., 2017).

Protein structure is also affected by processes such as the cheese-making process, where proteolysis influence not only the final product structure but also its bioactivity due to the release of bioactive compounds such as bioactive peptides. These small protein fragments are reported to exhibit a range of different bioactivities such as antihypertensive, antioxidant, anti-inflammatory, immunomodulatory (Santiago-López et al., 2018). Additional processing applied as pre-treatment for cheese production can also result in significant changes in food structure that impact both the nutritional aspect, improving protein digestibility, and the bioactive capacity of the final product. Munir et al. (2020) reported the positive effect of sonication, microwaves, and high-pressure processing on ACE-inhibitory activity and antioxidant potential of Cheddar cheese during ripening.

In summary, nonconventional or emerging technologies have been proven to be valuable tools to enhance the nutritional and bioactive properties of food products, in addition to their well-known effect on technological and functional properties. However, the food structure changes promoted by such processes and the consequent impact on bioaccessibility, bioavailability, and bioactivity of nutrients and bioactive compounds are highly dependent on the composition of the food matrix, processing conditions applied, and type of compound. Therefore, it is of utmost importance to design specific parameters for the process of each food matrix and their impact on the nutritional value of the final product.

Furthermore, considering the bioactive compounds and their evaluation, the use of static in vitro protocols (Brodkorb et al., 2019) used for most of the studies do not consider the relevant rheological changes after processing and during digestion, that potentially affect the fluid dynamics of digestion (Ferrua et al., 2011; Ferrua & Singh, 2015) and, probably, the nutrient bioaccessibility. Thus, the lack of standardization of in vitro digestion protocols throughout the literature still difficult the comparison of absolute values and conclusions on the expected effect of each structure modification on the absorption of such compounds.

## 5.4 Conclusion and future perspectives

The processing affects food structure and, consequently, food properties. This chapter discussed how food structure influences physical, sensorial, nutritional quality and health aspects of food products. The effect of both conventional and emerging technologies of food processing is detailed.

Although recent efforts have expanded the knowledge about the relationships among food structure, processing and properties, it is still a challenge to predict the exact conditions to achieve the different functionalities. Progress can be seen in establishing it for bottom-up structuring of food products, with a focus on personalization, while the top-down strategies are highly dependent on experimental conditions and must be better developed. A better understanding of this topic is still necessary to tailor structures using different technologies to obtain products with standardized quality attributes despite raw material variability and process condition, and enhanced nutrient availability. Moreover, the development of appropriate strategies at an industrial scale is still necessary for most of the recent approaches.

Food structure engineering and design can be a cleaver approach to improve nutrition, health and well-being. However, further studies are still necessary, with multidisciplinary teams, to pave this road and achieve the ultimate goals. We hope this chapter can be useful for the next steps.

## References

Aadil, R. M., Zeng, X.-A., Abbasi, A. M., Khan, M. S., Khalid, S., Jabbar, S., & Abid, M. (2015). Influence of power ultrasound on the quality parameters of grapefruit juice during storage. *Science Letters*, *3*, 6–12.

Abid, M., Jabbar, S., Hu, B., Hashim, M. M., Wu, T., Lei, S., Khan, M. A., & Zeng, X. (2014). Thermosonication as a potential quality enhancement technique of apple juice. *Ultrasonics Sonochemistry*, *21*, 984–990.

Aganovic, K., Hertel, C., Vogel, R. F., Johne, R., Schlüter, O., Schwarzenbolz, U., Jäger, H., Holzhauser, T., Bergmair, J., Roth, A., Sevenich, R., Bandick, N., Kulling, S. E., Knorr, D., Engel, K.-H., & Heinz, V. (2021). Aspects of high hydrostatic pressure food processing: Perspectives on technology and food safety. *Comprehensive Reviews in Food Science and Food Safety*, *20*.

Aguilera, J. M. (2005). Why food microstructure? *Journal of Food Engineering*, *67*, 3–11.

Aguilera, J. M., Chiralt, A., & Fito, P. (2003). Food dehydration and product structure. *Trends in Food Science & Technology*, *14*, 432–437.

Aguilera, J. M., Stanley, D. W., & Baker, K. W. (2000). New dimensions in microstructure of food products. *Trends in Food Science & Technology*, *11*, 3–9.

Aguillar-Rosas, S., Ballinas-Casarrubias, M., Nevarez-Moorillon, G., Martin-Belloso, O., & Ortega-Rivas, E. (2007). Thermal and pulsed electric fields pasteurization of apple juice: Effects on physicochemical properties and flavour compounds. *Journal of Food Engineering*, *83*, 41–46.

# References

Alam, M. R., Lyng, J. G., Frontuto, D., Marra, F., & Cinquanta, L. (2018). Effect of pulsed electric field pretreatment on drying kinetics, color, and texture of parsnip and carrot. *Journal of Food Science, 83*, 2159–2166.

Alsalman, F. B., & Ramaswamy, H. (2020). Reduction in soaking time and anti-nutritional factors by high pressure processing of chickpeas. *Journal of Food Science and Technology*, 1–14.

Anese, M., Mirolo, G., Beraldo, P., & Lippe, G. (2013). Effect of ultrasound treatments of tomato pulp on microstructure and lycopene in vitro bioaccessibility. *Food Chemistry, 136*, 458–463.

Apostolidis, E., & Mandala, I. (2020). Modification of resistant starch nanoparticles using high-pressure homogenization treatment. *Food Hydrocolloids, 103*, 105677.

Atalar, I., Saricaoglu, F. T., Odabas, H. I., Yilmaz, V. A., & Gul, O. (2020). Effect of ultrasonication treatment on structural, physicochemical and bioactive properties of pasteurized Rosehip (Rosa canina L.) nectar. *LWT, 118*, 108850.

Augusto, P. E., Ibarz, A., & Cristianini, M. (2013). Effect of high pressure homogenization (HPH) on the rheological properties of tomato juice: Viscoelastic properties and the Cox-Merz rule. *Journal of Food Engineering, 114*, 57–63.

Augusto, P. E. D. (2020). Challenges, trends and opportunities in food processing. *Current Opinion in Food Science, 35*, 72–78.

Augusto, P. E. D., Falguera, V., Cristianini, M., & Ibarz, A. (2013). Viscoelastic properties of tomato juice: Applicability of the Cox-Merz rule. *Food and Bioprocess Technology, 6*, 839–843.

Augusto, P. E. D., Ibarz, A., & Cristianini, M. (2012a). Effect of high pressure homogenization (HPH) on the rheological properties of a fruit juice serum model. *Journal of Food Engineering, 111*, 474–477.

Augusto, P. E. D., Ibarz, A., & Cristianini, M. (2012b). Effect of high pressure homogenization (HPH) on the rheological properties of tomato juice: Time-dependent and steady-state shear. *Journal of Food Engineering, 111*, 570–579.

Balny, C., Masson, P., & Heremans, K. (2002). High pressure effects on biological macromolecules: From structural changes to alteration of cellular processes. *Biochimica et Biophysica Acta, 1595*, 3–10.

Barba, F. J., Mariutti, L. R. B., Bragagnolo, N., Mercadante, A. Z., Barbosa-Cánovas, G. V., & Orlien, V. (2017). Bioaccessibility of bioactive compounds from fruits and vegetables after thermal and nonthermal processing. *Trends in Food Science & Technology, 67*, 195–206.

Bot, F., Calligaris, S., Cortella, G., Nocera, F., Peressini, D., & Anese, M. (2017). Effect of high pressure homogenization and high power ultrasound on some physical properties of tomato juices with different concentration levels. *Journal of Food Engineering, 213*, 10–17.

Bouchon, P., & Aguilera, J. M. (2001). Microstructural analysis of frying potatoes. *International Journal of Food Science & Technology, 36*, 669–676.

Brodkorb, A., Egger, L., Alminger, M., Alvito, P., Assunção, R., Ballance, S., Bohn, T., Bourlieu-Lacanal, C., Boutrou, R., Carrière, F., Clemente, A., Corredig, M., Dupont, D., Dufour, C., Edwards, C., Golding, M., Karakaya, S., Kirkhus, B., Le Feunteun, S., … Recio, I. (2019). Infogest static in vitro simulation of gastrointestinal food digestion. *Nature Protocols, 14*, 991–1014.

Buniowska, M., Carbonell-Capella, J. M., Frigola, A., & Esteve, M. J. (2017). Bioaccessibility of bioactive compounds after non-thermal processing of an exotic fruit juice blend sweetened with Stevia rebaudiana. *Food Chemistry, 221*, 1834–1842.

Caetano-Silva, M. E., Netto, F. M., Bertoldo-Pacheco, M. T., Alegría, A., & Cilla, A. (2021). Peptide-metal complexes: Obtention and role in increasing bioavailability and decreasing the pro-oxidant effect of minerals. *Critical Reviews in Food Science and Nutrition, 61*, 1470–1489.

Calligaris, S., Foschia, M., Bartolomeoli, I., Maifreni, M., & Manzocco, L. (2012). Study on the applicability of high-pressure homogenization for the production of banana juices. *LWT - Food Science and Technology, 45*, 117–121.

Campoli, S. S., Rojas, M. L., Do Amaral, J. E. P. G., Canniatti-Brazaca, S. G., & Augusto, P. E. D. (2018). Ultrasound processing of guava juice: Effect on structure, physical properties and lycopene in vitro accessibility. *Food Chemistry, 268*, 594–601.

Cao, Y., & Mezzenga, R. (2020). Design principles of food gels. *Nature Food, 1*, 106–118.

Castanha, N., Lima, D. C., Matta Junior, M. D., Campanella, O. H., & Augusto, P. E. D. (2019). Combining ozone and ultrasound technologies to modify maize starch. *International Journal of Biological Macromolecules, 139*, 63–74.

Castanha, N., Matta Junior, M. D. D., & Augusto, P. E. D. (2017). Potato starch modification using the ozone technology. *Food Hydrocolloids, 66*, 343–356.

Çatal, H., & İbanoğlu, Ş. (2012). Ozonation of corn and potato starch in aqueous solution: Effects on the thermal, pasting and structural properties. *International Journal of Food Science & Technology, 47*, 1958–1963.

Chanvrier, H., Jakubczyk, E., Gondek, E., & Gumy, J.-C. (2014). Insights into the texture of extruded cereals: Structure and acoustic properties. *Innovative Food Science & Emerging Technologies, 24*, 61–68.

Che, L., Li, D., Wang, L., Özkan, N., Chen, X. D., & Mao, Z. (2007). Effect of high-pressure homogenization on the structure of Cassava starch. *International Journal of Food Properties, 10*, 911–922.

Chen, F., Zhang, M., & Yang, C.-H. (2020). Application of ultrasound technology in processing of ready-to-eat fresh food: A review. *Ultrasonics Sonochemistry, 63*, 104953.

Chian, F., Kaur, L., Oey, I., Astruc, T., Hodgkinson, S., & Boland, M. (2019). Effect of pulsed electric fields (PEF) on the ultrastructure and in vitro protein digestibility of bovine Longissimus thoracis. *LWT-Food Science and Technology, 103*, 253–259.

Choi, Y., & Okos, M. (1986). Effect of temperature and composition on the thermal properties of foods. In M. Le Maguer, & P. Jelen (Eds.), *Food engineering and process applications: Transport phenomena*. London: Elsevier Applied Science Publishers.

Cilla, A., Alegría, A., De Ancos, B., Sánchez-Moreno, C., Cano, M. P., Plaza, L., Clemente, G., Lagarda, M. J., & Barberá, R. (2012). Bioaccessibility of tocopherols, carotenoids, and ascorbic acid from milk- and soy-based fruit beverages: Influence of food matrix and processing. *Journal of Agricultural and Food Chemistry, 60*, 7282–7290.

Colle, I., Van Buggenhout, S., Van Loey, A., & Hendrickx, M. (2010). High pressure homogenization followed by thermal processing of tomato pulp: Influence on microstructure and lycopene in vitro bioaccessibility. *Food Research International, 43*, 2193–2200.

Corredig, M., & Wicker, L. (2001). Changes in the molecular weight distribution of three commercial pectins after valve homogenization. *Food Hydrocolloids, 15*, 17–23.

Costa, M. G. M., Fonteles, T. V., De Jesus, A. L. T., Almeida, F. D. L., De Miranda, M. R. A., Fernandes, F. A. N., & Rodrigues, S. (2013). High-intensity ultrasound processing of pineapple juice. *Food and Bioprocess Technology, 6*, 997–1006.

Dalgleish, D. G. (2006). Food emulsions—Their structures and structure-forming properties. *Food Hydrocolloids, 20*, 415–422.

Dalgleish, D. G., & Corredig, M. (2012). The structure of the casein micelle of milk and its changes during processing. *Annual Review of Food Science and Technology, 3*, 449–467.

Daroit, D. J., & Brandelli, A. (2021). In vivo bioactivities of food protein-derived peptides—A current review. *Current Opinion in Food Science, 39*, 120–129.

De Souza Carvalho, L. M., Lemos, M. C. M., Sanches, E. A., Da Silva, L. S., De Araújo Bezerra, J., Aguiar, J. P. L., das Chagas Do Amaral Souza, F., Alves Filho, E. G., & Campelo, P. H. (2020). Improvement of the bioaccessibility of bioactive compounds from amazon fruits treated using high energy ultrasound. *Ultrasonics Sonochemistry, 67*, 105148.

De Vito, F., Ferrari, G., Lebovka, I., Shynkaryk, N. V., & Vorobiev, E. (2008). Pulse duration and efficiency of soft cellular tissue disintegration by pulsed electric fields. *Food and Bioprocess Technology, 1*, 307–313.

Dickinson, E. (2015). Colloids in food: Ingredients, structure, and stability. *Annual Review of Food Science and Technology, 6*, 211–233.

Dos Santos Aguilar, J. G., Cristianini, M., & Sato, H. H. (2018). Modification of enzymes by use of high-pressure homogenization. *Food Research International, 109*, 120–125.

Falcão, H. G., Handa, C. L., Silva, M. B. R., De Camargo, A. C., Shahidi, F., Kurozawa, L. E., & Ida, E. I. (2018). Soybean ultrasound pre-treatment prior to soaking affects β-glucosidase activity, isoflavone profile and soaking time. *Food Chemistry, 269*, 404–412.

Feng, H., Yin, Y., & Tang, J. (2012). Microwave drying of food and agricultural materials: Basics and heat and mass transfer modeling. *Food Engineering Reviews, 4*, 89–106.

Fernández-Jalao, I., Sánchez-Moreno, C., & De Ancos, B. (2017). Influence of food matrix and high-pressure processing on onion flavonols and antioxidant activity during gastrointestinal digestion. *Journal of Food Engineering, 213*, 60–68.

Ferrua, M. J., Kong, F., & Singh, R. P. (2011). Computational modeling of gastric digestion and the role of food material properties. *Trends in Food Science & Technology, 22*, 480–491.

Ferrua, M. J., & Singh, R. P. (2015). Computational modelling of gastric digestion: Current challenges and future directions. *Current Opinion in Food Science, 4*, 116–123.

Funebo, T., Ahrné, L. L. A., Kidman, S., Langton, M., & Skjöldebrand, C. (2000). Microwave heat treatment of apple before air dehydration—Effects on physical properties and microstructure. *Journal of Food Engineering, 46*, 173–182.

Gabrić, D., Barba, F., Roohinejad, S., Gharibzahedi, S. M. T., Radojčin, M., Putnik, P., & Bursać Kovačević, D. (2018). Pulsed electric fields as an alternative to thermal processing for preservation of nutritive and physicochemical properties of beverages: A review. *Journal of Food Process Engineering, 41*, E12638.

Gaikwad, S. G., & Pandit, A. B. (2008). Ultrasound emulsification: Effect of ultrasonic and physicochemical properties on dispersed phase volume and droplet size. *Ultrasonics Sonochemistry, 15*, 554–563.

Gallier, S., & Singh, H. (2012). The physical and chemical structure of lipids in relation to digestion and absorption. *Lipid Technology, 24*, 271–273.

Gautam, A., Dhiman, A. K., Attri, S., & Kathuria, D. (2020). Optimization of pulping method for extraction of pulp from ripe persimmon (Diospyros kaki L.) and its stability during storage. *Journal of Applied and Natural Science, 12*, 618–627.

Gogate, P. R., & Prajapat, A. L. (2015). Depolymerization using sonochemical reactors: A critical review. *Ultrasonics Sonochemistry, 27*, 480–494.

Guan, Y., Zhou, L., Bi, J., Yi, J., Liu, X., Chen, Q., Wu, X., & Zhou, M. (2016). Change of microbial and quality attributes of mango juice treated by high pressure homogenization combined with moderate inlet temperatures during storage. *Innovative Food Science & Emerging Technologies, 36*, 320–329.

Han, Z., Shi, R., & Sun, D.-W. (2020). Effects of novel physical processing techniques on the multi-structures of starch. *Trends in Food Science & Technology, 97*, 126–135.

Huang, L., Ding, X., Dai, C., & Ma, H. (2017). Changes in the structure and dissociation of soybean protein isolate induced by ultrasound-assisted acid pretreatment. *Food Chemistry, 232*, 727–732.

Huang, X., Liang, K.-H., Liu, Q., Qiu, J., Wang, J., & Zhu, H. (2020). Superfine grinding affects physicochemical, thermal and structural properties of Moringa Oleifera leaf powders. *Industrial Crops and Products, 151*, 112472.

İçier, F., & Baysal, T. (2004). Dielectrical properties of food materials—1: Factors affecting and industrial uses. *Critical Reviews in Food Science and Nutrition, 44*, 465–471.

Janowicz, M., & Lenart, A. (2018). The impact of high pressure and drying processing on internal structure and quality of fruit. *European Food Research and Technology, 244*, 1329–1340.

Jia, G., Orlien, V., Liu, H., & Sun, A. (2021). Effect of high pressure processing of pork (Longissimus dorsi) on changes of protein structure and water loss during frozen storage. *LWT, 135*, 110084.

Jin, T. Z., Yu, Y., & Gurtler, J. B. (2017). Effects of pulsed electric field processing on microbial survival, quality change and nutritional characteristics of blueberries. *LWT, 77*, 517–524.

Joye, I. (2019). Protein digestibility of cereal products. *Food*, *8*.

Jurić, S., Ferrari, G., Velikov, K. P., & Donsì, F. (2019). High-pressure homogenization treatment to recover bioactive compounds from tomato peels. *Journal of Food Engineering*, *262*, 170–180.

Kaur, M., Punia, S., Sandhu, K. S., & Ahmed, J. (2019). Impact of high pressure processing on the rheological, thermal and morphological characteristics of mango kernel starch. *International Journal of Biological Macromolecules*, *140*, 149–155.

Knorr, D., Froehling, A., Jaeger, H., Reineke, K., Schlueter, O., & Schoessler, K. (2011). Emerging technologies in food processing. In M. P. Doyle, & T. R. Klaenhammer (Eds.), *Vol. 2. Annual review of food science and technology*. Annual Reviews: Palo Alto.

Koç, B., Eren, I., & Kaymak Ertekin, F. (2008). Modelling bulk density, porosity and shrinkage of quince during drying: The effect of drying method. *Journal of Food Engineering*, *85*, 340–349.

Kroehnke, J., Radziejewska-Kubzdela, E., Musielak, G., Stasiak, M., & Bieganska-Marecik, R. (2015). Ultrasonic-assisted and microwave-assisted convective drying of carrot: Drying kinetics and quality analysis. In *Vol. 2015. Proceedings of 5th European drying conference, Budapest, Hungray* (pp. 21–23).

Kubo, M. T. K., Augusto, P. E. D., & Cristianini, M. (2013). Effect of high pressure homogenization (HPH) on the physical stability of tomato juice. *Food Research International*, *51*, 170–179.

Kubo, M. T. K., Rojas, M. L., Miano, A. C., & Augusto, P. E. D. (2019). Chapter 1. Rheological properties of tomato products. In *Tomato chemistry, industrial processing and product development* The Royal Society Of Chemistry.

Kubo, M. T. K., Tribst, A. A. L., & Augusto, P. E. D. (2021). 3.19. High pressure homogenization in fruit and vegetable juice and puree processing: effects on quality, stability and phytochemical profile. In K. Knoerzer, & K. Muthukumarappan (Eds.), *Innovative food processing technologies*. Oxford: Elsevier.

Laguna, L., Picouet, P., Guàrdia, M. D., Renard, C. M. G. C., & Sarkar, A. (2017). In vitro gastrointestinal digestion of pea protein isolate as a function of pH, food matrices, autoclaving, high-pressure and re-heat treatments. *LWT*, *84*, 511–519.

Lam, R. S. H., & Nickerson, M. T. (2013). Food proteins: A review on their emulsifying properties using a structure-function approach. *Food Chemistry*, *141*, 975–984.

Leite, T. S., Augusto, P. E. D., & Cristianini, M. (2014). The use of high pressure homogenization (HPH) to reduce consistency of concentrated orange juice (COJ). *Innovative Food Science & Emerging Technologies*, *26*, 124–133.

Leite, T. S., Augusto, P. E. D., & Cristianini, M. (2015). Using high pressure homogenization (HPH) to change the physical properties of cashew apple juice. *Food Biophysics*, *10*, 169–180.

Lesmes, U., & Mcclements, D. J. (2009). Structure-function relationships to guide rational design and fabrication of particulate food delivery systems. *Trends in Food Science & Technology*, *20*, 448–457.

Li, D., Zhu, Z., & Sun, D.-W. (2018). Effects of freezing on cell structure of fresh cellular food materials: A review. *Trends in Food Science & Technology*, *75*, 46–55.

Lima, D. C., Castanha, N., Maniglia, B. C., Matta Junior, M. D., La Fuente, C. I. A., & Augusto, P. E. D. (2021). Ozone processing of cassava starch. *Ozone: Science & Engineering*, *43*, 60–77.

Lima, D. C., Villar, J., Castanha, N., Maniglia, B. C., Matta Junior, M. D., & Duarte Augusto, P. E. (2020). Ozone modification of arracacha starch: Effect on structure and functional properties. *Food Hydrocolloids*, *108*, 106066.

Liu, C., Pirozzi, A., Ferrari, G., Vorobiev, E., & Grimi, N. (2020). Impact of pulsed electric fields on vacuum drying kinetics and physicochemical properties of carrot. *Food Research International*, *137*, 109658.

Llavata, B., García-Pérez, J. V., Simal, S., & Cárcel, J. A. (2020). Innovative pre-treatments to enhance food drying: A current review. *Current Opinion in Food Science*, *35*, 20–26.

López-Gámez, G., Elez-Martínez, P., Martín-Belloso, O., & Soliva-Fortuny, R. (2021). Pulsed electric field treatment strategies to increase bioaccessibility of phenolic and carotenoid compounds in oil-added carrot purees. *Food Chemistry*, *364*, 130377.

Lopez-Sanchez, P., Nijsse, J., Blonk, H. C., Bialek, L., Schumm, S., & Langton, M. (2011). Effect Of mechanical and thermal treatments on the microstructure and rheological properties of carrot, broccoli and tomato dispersions. *Journal of the Science of Food and Agriculture*, *91*, 207–217.

Makroo, H. A., Rastogi, N. K., & Srivastava, B. (2020). Ohmic heating assisted inactivation of enzymes and microorganisms in foods: A review. *Trends in Food Science & Technology*, *97*, 451–465.

Maniglia, B. C., Castanha, N., Le-Bail, P., Le-Bail, A., & Augusto, P. E. D. (2020). Starch modification through environmentally friendly alternatives: A review. *Critical Reviews in Food Science and Nutrition*, 1–24.

Maniglia, B. C., Castanha, N., Rojas, M. L., & Augusto, P. E. D. (2021). Emerging technologies to enhance starch performance. *Current Opinion in Food Science*, *37*, 26–36.

Mayor, L., Pissarra, J., & Sereno, A. M. (2008). Microstructural changes during osmotic dehydration of parenchymatic pumpkin tissue. *Journal of Food Engineering*, *85*, 326–339.

Mercado-Mercado, G., Montalvo-González, E., González-Aguilar, G. A., Alvarez-Parrilla, E., & Sáyago-Ayerdi, S. G. (2018). Ultrasound-assisted extraction of carotenoids from mango (*Mangifera indica* L. 'Ataulfo') by-products on in vitro bioaccessibility. *Food Bioscience*, *21*, 125–131.

Miano, A. C., Da Costa Pereira, J., Miatelo, B., & Augusto, P. E. D. (2017). Ultrasound assisted acidification of model foods: Kinetics and impact on structure and viscoelastic properties. *Food Research International*, *100*, 468–476.

Miano, A. C., Rojas, M. L., & Augusto, P. E. D. (2018). Structural changes caused by ultrasound pretreatment: Direct and indirect demonstration in potato cylinders. *Ultrasonics Sonochemistry*, *52*.

Mills, E. N. C., Sancho, A. I., Rigby, N. M., Jenkins, J. A., & Mackie, A. R. (2009). Impact of food processing on the structural and allergenic properties of food allergens. *Molecular Nutrition & Food Research*, *53*, 963–969.

Moelants, K. R., Cardinaels, R., Jolie, R. P., Verrijssen, T. A., Van Buggenhout, S., Van Loey, A. M., Moldenaers, P., & Hendrickx, M. E. (2014). Rheology of concentrated tomato-derived suspensions: Effects of particle characteristics. *Food and Bioprocess Technology*, *7*, 248–264.

Moelants, K. R. N., Cardinaels, R., Van Buggenhout, S., Van Loey, A. M., Moldenaers, P., & Hendrickx, M. E. (2014). A review on the relationships between processing, food structure, and rheological properties of plant-tissue-based food suspensions. *Comprehensive Reviews in Food Science and Food Safety*, *13*, 241–260.

Monteiro, R. L., Carciofi, B. A. M., & Laurindo, J. B. (2016). A microwave multi-flash drying process for producing crispy bananas. *Journal of Food Engineering*, *178*, 1–11.

Monteiro, R. L., De Moraes, J. O., Domingos, J. D., Carciofi, B. A. M., & Laurindo, J. B. (2020). Evolution of the physicochemical properties of oil-free sweet potato chips during microwave vacuum drying. *Innovative Food Science & Emerging Technologies*, *63*, 102317.

Monteiro, R. L., Link, J. V., Tribuzi, G., Carciofi, B. A. M., & Laurindo, J. B. (2018). Effect of multi-flash drying and microwave vacuum drying on the microstructure and texture of pumpkin slices. *LWT*, *96*, 612–619.

Munir, M., Nadeem, M., Mahmood Qureshi, T., Gamlath, C. J., Martin, G. J. O., Hemar, Y., & Ashokkumar, M. (2020). Effect of sonication, microwaves and high-pressure processing on ace-inhibitory activity and antioxidant potential of cheddar cheese during ripening. *Ultrasonics Sonochemistry*, *67*, 105140.

Musielak, G., Mierzwa, D., & Kroehnke, J. (2016). Food drying enhancement by ultrasound—A review. *Trends in Food Science & Technology*, *56*, 126–141.

Nayak, B., De Berrios, J., & Tang, J. (2014). Impact of food processing on the glycemic index (GI) of potato products. *Food Research International*, *56*, 35–46.

Oey, I., Lille, M., Van Loey, A., & Hendrickx, M. (2008). Effect of high-pressure processing on color, texture and flavour of fruit- and vegetable-based food products: A review. *Trends in Food Science & Technology*, *19*, 320–328.

Oikonomopoulou, V. P., & Krokida, M. K. (2013). Novel aspects of formation of food structure during drying. *Drying Technology, 31*, 990–1007.

O'Sullivan, J., Murray, B., Flynn, C., & Norton, I. (2016). The effect of ultrasound treatment on the structural, physical and emulsifying properties of animal and vegetable proteins. *Food Hydrocolloids, 53*, 141–154.

Ozuna, C., Gómez Álvarez-Arenas, T., Riera, E., Cárcel, J. A., & Garcia-Perez, J. V. (2014). Influence of material structure on air-borne ultrasonic application in drying. *Ultrasonics Sonochemistry, 21*, 1235–1243.

Pacheco-Aguirre, F. M., García-Alvarado, M. A., Corona-Jiménez, E., Ruiz-Espinosa, H., Cortés-Zavaleta, O., & Ruiz-López, I. I. (2015). Drying modeling in products undergoing simultaneous size reduction and shape change: Appraisal of deformation effect on water diffusivity. *Journal of Food Engineering, 164*, 30–39.

Panyawong, S., & Devahastin, S. (2007). Determination of deformation of a food product undergoing different drying methods and conditions via evolution of a shape factor. *Journal of Food Engineering, 78*, 151–161.

Patrignani, F., Vannini, L., Kamdem, S. L. S., Lanciotti, R., & Guerzoni, M. E. (2009). Effect of high pressure homogenization on saccharomyces cerevisiae inactivation and physico-chemical features in apricot and carrot juices. *International Journal of Food Microbiology, 136*, 26–31.

Pei-Ling, L., Xiao-Song, H., & Qun, S. (2010). Effect of high hydrostatic pressure on starches: A review. *Starch, 62*, 615–628.

Petigny, L., Périno-Issartier, S., Wajsman, J., & Chemat, F. (2013). Batch and continuous ultrasound assisted extraction of boldo leaves (Peumus boldus Mol.). *International Journal of Molecular Sciences, 14*, 5750–5764.

Petitot, M., Abecassis, J., & Micard, V. (2009). Structuring of pasta components during processing: Impact on starch and protein digestibility and allergenicity. *Trends in Food Science & Technology, 20*, 521–532.

Phinney, D. M., Frelka, J. C., & Heldman, D. R. (2017). Composition-based prediction of temperature-dependent thermophysical food properties: Reevaluating component groups and prediction models. *Journal of Food Science, 82*, 6–15.

Pottier, L., Villamonte, G., & De Lamballerie, M. (2017). Applications of high pressure for healthier foods. *Current Opinion in Food Science, 16*, 21–27.

Puerta, P., Garzón, R., Rosell, C. M., Fiszman, S., Laguna, L., & Tárrega, A. (2021). Modifying gluten-free bread's structure using different baking conditions: Impact on oral processing and texture perception. *LWT, 140*, 110718.

Qiu, S., Li, Y., Chen, H., Liu, Y., & Yin, L. (2014). Effects of high-pressure homogenization on thermal and electrical properties of wheat starch. *Journal of Food Engineering, 128*, 53–59.

Rahaman, T., Vasiljevic, T., & Ramchandran, L. (2016). Effect of processing on conformational changes of food proteins related to allergenicity. *Trends in Food Science & Technology, 49*, 24–34.

Rajewska, K., & Mierzwa, D. (2017). Influence of ultrasound on the microstructure of plant tissue. *Innovative Food Science & Emerging Technologies, 43*, 117–129.

Ramos, I. N., Brandão, T. R. S., & Silva, C. L. M. (2003). Structural changes during air drying of fruits and vegetables. *Food Science and Technology International, 9*, 201–206.

Rao, M. A. (2007). Influence of food microstructure on food rheology. In D. J. Mcclements (Ed.), *Understanding and controlling the microstructure of complex foods* Woodhead Publishing.

Rawson, A., Tiwari, B. K., Patras, A., Brunton, N., Brennan, C., Cullen, P. J., & O'donnell, C. (2011). Effect of thermosonication on bioactive compounds in watermelon juice. *Food Research International, 44*, 1168–1173.

Rodríguez, Ó., Eim, V., Rosselló, C., Femenia, A., Cárcel, J. A., & Simal, S. (2018). Application of power ultrasound on the convective drying of fruits and vegetables: Effects on quality. *Journal of the Science of Food and Agriculture, 98*, 1660–1673.

Rodríguez, Ó., Gomes, W., Rodrigues, S., & Fernandes, F. A. N. (2017). Effect of acoustically assisted treatments on vitamins, antioxidant activity, organic acids and drying kinetics of pineapple. *Ultrasonics Sonochemistry, 35*, 92–102.

Rojas, M. L., & Augusto, P. E. D. (2018). Ethanol and ultrasound pre-treatments to improve infrared drying of potato slices. *Innovative Food Science & Emerging Technologies, 49*, 65–75.

Rojas, M. L., Kubo, M. T. K., Caetano-Silva, M. E., & Augusto, P. E. D. (2021). Ultrasound processing of fruits and vegetables, structural modification and impact on nutrient and bioactive compounds: A review. *International Journal of Food Science & Technology, 56*.

Rojas, M. L., Leite, T. S., Cristianini, M., Alvim, I. D., & Augusto, P. E. D. (2016). Peach juice processed by the ultrasound technology: Changes in its microstructure improve its physical properties and stability. *Food Research International, 82*, 22–33.

Rojas, M. L., Miano, A. C., & Augusto, P. E. D. (2017). Chapter 7. Ultrasound processing of fruit and vegetable juices. In D. Bermudez-Aguirre (Ed.), *Ultrasound: Advances for food processing and preservation* Academic Press.

Rojas, M. L., Trevilin, J. H., & Augusto, P. E. D. (2016). The ultrasound technology for modifying enzyme activity. *Scientia Agropecuaria, 7*, 5.

Rojas, M. L., Trevilin, J. H., Funcia, E. D. S., Gut, J. A. W., & Augusto, P. E. D. (2017). Using ultrasound technology for the inactivation and thermal sensitization of peroxidase in green coconut water. *Ultrasonics Sonochemistry, 36*, 173–181.

Sá, A. G. A., Moreno, Y. M. F., & Carciofi, B. A. M. (2020). Food processing for the improvement of plant proteins digestibility. *Critical Reviews in Food Science and Nutrition, 60*, 3367–3386.

Sahin, S., & Sumnu, S. G. (2006). Thermal properties of foods. In S. Sahin, & S. G. Sumnu (Eds.), *Physical properties of foods*. New York, NY: Springer.

Santiago-López, L., Aguilar-Toalá, J. E., Hernández-Mendoza, A., Vallejo-Cordoba, B., Liceaga, A. M., & González-Córdova, A. F. (2018). Invited review: Bioactive compounds produced during cheese ripening and health effects associated with aged cheese consumption. *Journal of Dairy Science, 101*, 3742–3757.

Santos, K. C., Guedes, J. S., Rojas, M. L., Carvalho, G. R., & Augusto, P. E. D. (2021). Enhancing carrot convective drying by combining ethanol and ultrasound as pre-treatments: Effect on product structure, quality, energy consumption, drying and rehydration kinetics. *Ultrasonics Sonochemistry, 70*, 105304.

Schössler, K., Thomas, T., & Knorr, D. (2012). Modification of cell structure and mass transfer in potato tissue by contact ultrasound. *Food Research International, 49*, 425–431.

Sensoy, I. (2014). A review on the relationship between food structure, processing, and bioavailability. *Critical Reviews in Food Science and Nutrition, 54*, 902–909.

Seshadri, R., Weiss, J., Hulbert, G. J., & Mount, J. (2003). Ultrasonic processing influences rheological and optical properties of high-methoxyl pectin dispersions. *Food Hydrocolloids, 17*, 191–197.

Silva, V. M., Sato, A. C. K., Barbosa, G., Dacanal, G., Ciro-Velásquez, H. J., & Cunha, R. L. (2010). The effect of homogenisation on the stability of pineapple pulp. *International Journal of Food Science & Technology, 45*, 2127–2133.

Singla, M., & Sit, N. (2021). Application of ultrasound in combination with other technologies in food processing: A review. *Ultrasonics Sonochemistry, 73*.

Soares, A. D. S., Leite Júnior, B. R. D. C., Tribst, A. A. L., Augusto, P. E. D., & Ramos, A. M. (2020). Effect of ultrasound on goat cream hydrolysis by lipase: Evaluation on enzyme, substrate and assisted reaction. *LWT, 130*, 109636.

Soria, A. C., & Villamiel, M. (2010). Effect of ultrasound on the technological properties and bioactivity of food: A review. *Trends in Food Science & Technology, 21*, 323–331.

Sungsinchai, S., Niamnuy, C., Wattanapan, P., Charoenchaitrakool, M., & Devahastin, S. (2019). Texture modification technologies and their opportunities for the production of dysphagia foods: A review. *Comprehensive Reviews in Food Science and Food Safety, 18*, 1898–1912.

Szadzińska, J., Łechtańska, J., & Kowalski, S. J. (2015). Microwave and ultrasonic assisted convective drying of green pepper: Drying kinetics and quality. In *Proceedings of the 5th European drying conference (Eurodrying'2015), Budapest, Hungary* (pp. 391–398).

Tan, J., & Kerr, W. (2015). Rheological properties and microstructure of tomato puree subject to continuous high pressure homogenization. *Journal of Food Engineering, 166*, 45–54.

Téllez-Morales, J. A., Hernández-Santo, B., & Rodríguez-Miranda, J. (2020). Effect of ultrasound on the techno-functional properties of food components/ingredients: A review. *Ultrasonics Sonochemistry*, *61*, 104787.

Terefe, N. S., Sikes, A. L., & Juliano, P. (2016). 8. Ultrasound for structural modification of food products. In K. Knoerzer, P. Juliano, & G. Smithers (Eds.), *Innovative food processing technologies* Woodhead Publishing.

Thirumdas, R., Kadam, D., & Annapure, U. (2017). Cold plasma: An alternative technology for the starch modification. *Food Biophysics*, *12*, 129–139.

Tiwari, B. K., Muthukumarappan, K., O'Donnell, C. P., & Cullen, P. J. (2010). Rheological properties of sonicated guar, xanthan and pectin dispersions. *International Journal of Food Properties*, *13*, 223–233.

Ubbink, J., Burbidge, A., & Mezzenga, R. (2008). Food structure and functionality: A soft matter perspective. *Soft Matter*, *4*, 1569–1581.

Udenigwe, C. C., & Aluko, R. E. (2012). Food protein-derived bioactive peptides: Production, processing, and potential health benefits. *Journal of Food Science*, *77*, R11–R24.

Vanga, S. K., Wang, J., Orsat, V., & Raghavan, V. (2020). Effect of pulsed ultrasound, a green food processing technique, on the secondary structure and in-vitro digestibility of almond milk protein. *Food Research International*, *137*, 109523.

Vanga, S. K., Wang, J., & Raghavan, V. (2020). Effect of ultrasound and microwave processing on the structure, in-vitro digestibility and trypsin inhibitor activity of soymilk proteins. *LWT*, *131*, 109708.

Velikov, K. P., & Pelan, E. (2008). Colloidal delivery systems for micronutrients and nutraceuticals. *Soft Matter*, *4*, 1964–1980.

Venkatesh, M. S., & Raghavan, G. S. V. (2004). An overview of microwave processing and dielectric properties of agri-food materials. *Biosystems Engineering*, *88*, 1–18.

Verkempinck, S., Pallares Pallares, A., Hendrickx, M., & Grauwet, T. (2020). Processing as a tool to manage digestive barriers in plant-based foods: Recent advances. *Current Opinion in Food Science*, *35*, 1–9.

Wang, J., Zhao, S., Min, G., Qiao, D., Zhang, B., Niu, M., Jia, C., Xu, Y., & Lin, Q. (2021). Starch-protein interplay varies the multi-scale structures of starch undergoing thermal processing. *International Journal of Biological Macromolecules*, *175*, 179–187.

Wang, Q., Li, Y., Sun, D.-W., & Zhu, Z. (2018). Enhancing food processing by pulsed and high voltage electric fields: Principles and applications. *Critical Reviews in Food Science and Nutrition*, *58*, 2285–2298.

Witrowa-Rajchert, D., & Rząca, M. (2009). Effect of drying method on the microstructure and physical properties of dried apples. *Drying Technology*, *27*, 903–909.

Xue, S., Wang, C., Kim, Y. H. B., Bian, G., Han, M., Xu, X., & Zhou, G. (2020). Application of high-pressure treatment improves the in vitro protein digestibility of gel-based meat product. *Food Chemistry*, *306*, 125602.

Yuanqing, H., Min, C., Lingling, S., Quancai, S., Pengyao, Y., Rui, G., Sijia, W., Yuqing, D., Haihui, Z., & Haile, M. (2020). Ultrasound pretreatment increases the bioavailability of dietary proteins by dissociating protein structure and composition. *Food Biophysics*, *15*.

Zhang, W., Yu, Y., Xie, F., Gu, X., Wu, J., & Wang, Z. (2019). High pressure homogenization versus ultrasound treatment of tomato juice: Effects on stability and in vitro bioaccessibility of carotenoids. *LWT*, *116*, 108597.

Zhang, Y., Dong, L., Zhang, J., Shi, J., Wang, Y., & Wang, S. (2021). Adverse effects of thermal food processing on the structural, nutritional, and biological properties of proteins. *Annual Review of Food Science and Technology*, *12*, 259–286.

Zhou, L., Guan, Y., Bi, J., Liu, X., Yi, J., Chen, Q., Wu, X., & Zhou, M. (2017). Change of the rheological properties of mango juice by high pressure homogenization. *LWT - Food Science and Technology*, *82*, 121–130.

Zhu, F. (2015). Impact of ultrasound on structure, physicochemical properties, modifications, and applications of starch. *Trends in Food Science & Technology*, *43*, 1–17.

# CHAPTER 6

# Structure design for gastronomy applications

Alessandra Massa[a], Juan-Carlos Arboleya[a,b], Fabiola Castillo[a], and Eneko Axpe[a,c,d]

[a]Basque Culinary Center, Mondragon Unibertsitatea, Donostia-San Sebastián, Gipuzkoa, Spain, [b]BCC Innovation, Technological Center in Gastronomy, Donostia-San Sebastián, Gipuzkoa, Spain, [c]Azti, Parque Tecnológico de Bizkaia, Derio, Bizkaia, Spain, [d]Ikerbasque, Basque Foundation for Science, Bilbao, Bizkaia, Spain

## 6.1 Introduction

Scientific knowledge applied into the kitchen enables a practical way to optimize chefs' creativity to encourage rational approaches and foster innovative food product designs. By interacting closely, chefs and food scientists can contribute to develop healthy and delicious food products.

This relationship is also reciprocal, and while scientists can bring a great benefit to the cooks, scientists can also take advantage of their expertise, craft and invention from chefs. Chefs are incessantly generating innovative ideas, some of which are highly thought-inspiring from a scientific and industrial perspective. Translational teams consisting of food scientists, cooks and nutrition experts will continue to define consumer needs and outline new ideas to re-design food products that meet consumer expectations.

For most of human history, great achievements of scientific works developed by chemists, physicists, and biologists were thought to be intrinsically disconnected from those of the great chefs and their teams. Recently, the existing gap between gastronomic and scientific interests has remarkably diminished. Precursors contributing to this change are, probably, *haute cuisine* (high cuisine) and the preliminary work developed by Francois La Varenne, in the 16th century, and by Antonin Carême, in the 17th century. These two French chefs played a pioneering role raising questions which pushed the kitchen beyond conventional scenarios, thus discovering brand-new aspects of it at two influential times in French history (Beaugé, 2012). This call to investigate and create in the kitchen has linked two worlds that otherwise may have continued looking at one another from afar.

In the past century, this coexistence has grown to research the way in which food products affect our bodies and enrich the sensory and aesthetic experience of eating. The effects of this investigation are not just restricted to the functionality of food, but also include how to make this functionality work in the frame of a delicious and healthy snack or dish. As a matter of fact, structuring food, while keeping it nutritive and tasty, is a complex task faced continuously by chefs and food scientists. Food scientists, nutritionists, chefs and the food-eating population need to work closely together with the aim of finding solutions that improve our physical health and, at the same time, be integrated in gastronomy.

Within the last 20 years, gastronomic players around the world have developed a powerful approach to different fields of science such as food technology, food engineering, nutrition and medicine among others. This cooperation between restaurants and food research institutions has been growing exponentially. Examples include the interactions between *Mugaritz Restaurant* and *Azti* (Spain), *Noma Restaurant* and *Nordic Food Lab* (Denmark), *The Fat Duck Restaurant* and *University of Reading* (UK) or *Boragó Restaurant* and *Pontifical Catholic University of Chile* (Chile), resulting in new research areas and interesting new foodstuffs. Even with this satisfactory context, no full transdisciplinary interactions have been developed so far to fuse gastronomy and science within a rigorous scientific approach. Discussion and further consensus on methodologies and ultimate goals are fundamental to set standards and join efforts to cooperatively build Gastronomic Sciences and its different branches.

## 6.2 Interaction between science and gastronomy: A good match for food product design

Some cooks have already carried out a way to cook that integrates a basic understanding of science and technology, developing dishes that go beyond standard ideas of what food should be (Vega & Ubbink, 2008). Chefs are also finding new ways of inspiration to make innovative dishes. In fact, their creative ability to generate innovative food is well recognized by technologists (van der Linden et al., 2008). As a result, it seems fundamental to include chefs into translational teams to develop innovation (Linnemann & van Boekel, 2007) and to explore alternative ways to design food products. New food product design could then be used to unravel ways to scale up new culinary concepts to develop consistent and high-quality foods for the general public.

In addition to appearance, these are the aspects considered when composing a dish: flavor, aroma and mouthfeel. These three elements inherently focus on the chemical and physico-chemical characteristics of the dish, such as the macromolecules that make it up: fats, proteins and carbohydrates. Whereas what is also significant, and less understood, are the food properties inherent to the structure itself. Most food is categorized as soft matter; a physical state considered to be between a fluid liquid and an ideal crystalline solid. Understanding the processes which occur inside the product, rather than the processes necessary to create a food product are of utmost importance. This is exactly where technical and engineering knowledge of structural food components play a primary role. For instance, liquid dairy cream and whipped cream are products sharing the same chemical composition, but different structures, uses, and tastes. According to this, knowing the chemical composition of a product is not the only aspect that gives you an insight into how to use or consume a new food product.

Basic knowledge of food colloids applied to the kitchen can be significantly beneficial for the gastronomy sector. Colloids are mixtures of two (or more) immiscible substances in form of a foam, an emulsion, a gel or a sol. The fundamental part of a food colloid is the dispersed phase, such as bubbles or droplets, which is surrounded by another immiscible or continuous phase which is often supported by a surfactant to keep the interfacial layer intact (Dickinson, 2015). Surfactants are amphiphilic molecules (proteins, polysaccharides or phospholipids) that break up the intermolecular forces, keeping interfacial surface tension lower than normal. This drives down the droplet or bubble's tendency to coalesce and keeps the colloid stable (Friberg et al., 2004).

In addition, fresh ideas are constantly coming up to make more stable foams and gels with numerous materials that can be found in the back-of-house. For instance, the usage of enzymes to hydrolyze eggs has led to an entirely new functionality of an ever-present product. Garcés-Rimón et al. (2016) used aminopeptidases in different conditions with a variety of ingredients to bring smooth, creamy and rigid textures to egg white proteins. This collaboration between science and gastronomy showed chefs' desire to improve the dishes and bring science into the dining room such as Mario Sandoval at Restaurante Coque in Madrid, Spain. On the one hand, these new textures are sensational and have fully entered into the world of *haute cuisine*, but on the other hand, they can also be very beneficial for elderly people with chewing difficulties as these dishes are easier to consume while enjoying the same flavor. These kind of enzymatic hydrolysis processes can be used in different protein media hosts to develop diverse textures or to use traditional proteins in new ways (Table 6.2).

Knowledge on food colloids can also help to re-design traditional products using their base components. In ice cream, a standard base consists of cream, milk, sugars and a stabilizer to form a water, lipid and air colloid (Corvitto, 2011). A stabilizer, in this case, is able to reduce ice crystal growth to keep a creamy texture of the ice cream while it is being stirred and cooled. By knowing the part that each of these ingredients play in the recipe, it is possible to reduce some ingredients (cream or milk) to basic, functional components and create a new type of ice cream. Fiol et al. (2017) introduced a new family of ice creams, in which milk was substituted for sodium caseinate, lactose and other kind of fats. These new formulations seemed to improve the final flavor perception and they showed a higher consumer acceptance than the traditional ones. From a culinary point of view, a good and deep understanding on the colloidal behavior of different food structures opens new possibilities to create novel textures.

Among the traditional dishes in Basque Cuisine we find *mamia* or curd. This ancient product is made by immersing rennet in milk and in the past, this rennet came from the stomach of lactating calves or sheep. It contains chymosin that causes casein coagulation until curd's characteristic texture is finally achieved. In our quest for ancestral recipes, through the search for old recipes and interviews with people closely linked to popular cuisine, it was found that nettles were used to make *mamia*. The ancient recipe was written as presented below.

## 6.2.1 Mammia (curd)

### 6.2.1.1 Ingredients
- Sheep's milk
- Salt
- Nettles
- Lamb rennet

### 6.2.1.2 Method
Heat the sheep's milk and salt to taste. When cooled, put the cream in a cup (katillu). Put plenty of nettles in a colander and strain the milk to remove impurities. When the milk is warm, add a small amount of lamb rennet mixed with a little water and strain through a cloth. Stir the milk and quickly share it out in cups, in a clay pan or in a cooking pot. Also made of clay. When the milk has set, spread the cream that was removed at the beginning over it. It can be drizzled with honey or sprinkled with sugar.

This exciting discovery led to further investigation about the traditional uses of the nettle. This plant is considered a medicinal herb and a highly nutritious food. Used as an infusion to purify the blood, it was also administered as a treatment for arthritis and anemia (Letcher, 2010). This highly-valued herb had multiple uses. Young leaves were sautéed for soups, creams and sauces; it was also dehydrated to

**FIG. 6.1**

Nettle Cheese inspired in the Basque *mamia*.

make infusions as a digestive remedy (Irving, 2009); and it was also simply treated for other preparations, as if it were spinaches, only that the nettle is much tastier.

These species abounded in the area and as this plant has a large number of stinging hairs which serve to remove impurities, when filtering the milk, nettle was used to obtain *mamia*. All evidence seems to indicate that nettle could also be directly related to the coagulation of the milk. On analyzing these data, curd was elaborated in the context of Basque *mamia*, using only fresh nettle leaves to produce fresh cheese, curdled and flavored with nettles (Fig. 6.1) (Fiol et al., 2016).

## 6.3 Food colloids in gastronomy

The term colloid has its roots in the ancient Greek word κόλλα, which means glue, and the word-forming element *-oid* meaning *like*. Thus, the word *colloid*, coined by the Scottish chemist Thomas Graham (Graham, 1861), would basically mean "like a glue". Nowadays, a colloid is defined as a mixture of insoluble particles dispersed in a medium. Thus, a colloid is a combination of two phases that are immiscible. In colloids, suspended particles form the dispersed phase and do not settle, and a matrix that holds the particles forms the continuous phase. Food colloids have shaped gastronomies around the world. Numerous food products and dishes that are consumed everyday in every continent are colloids (natural and artificial) such as milk, ice cream, mayonnaise, etc.

### 6.3.1 Types of food colloids

Colloids have unique structures, and provide different textures to food products and dishes that impact the mouthfeel of consumers and diners. The types of food colloids include sols, gels, emulsions and foams (Dickinson & Miller, 2001). As a curious note, the now legendary dish *Menestra en texturas* (texture panache, Fig. 6.2), created by Ferrán Adrià with his team at El Bulli in 1994, included mousses, slushes, gelatins, sorbets, etc., making it a dish that contained all types of food colloids: sols, gels, emulsions and foams.

### 6.3 Food colloids in gastronomy    143

**FIG. 6.2**

*Menestra en texturas* (texture panache) by Ferrán Adrià.

*Copyright elBulli.com.*

#### 6.3.1.1 Sol
A sol is a colloidal suspension with solid particles in a liquid. Cells and proteins make blood a natural sol. Also, a sauce thickened with flour, is a colloid in which long starch molecules are dispersed in a liquid. Commonly used sols in the kitchen include soups, jams, gravy and ketchup. The viscosity of the sol depends on the concentration of solid particles, temperature, etc. Thickening agents include starch, xanthan, guar gum, locust bean gum, gum karaya, gum tragacanth, gum Arabic, etc.

#### 6.3.1.2 Gel
A gel is a solid matrix that holds a liquid. For instance, tapioca is a solid colloid, a gel, that consists of water molecules dispersed and immobilized in a network of cross-linked starch. The texture of gels (elastic/brittle, chewy/creamy) vary with the type of gelling agent, the concentration used, the pH, salt content, temperature, etc. Gelling agents include alginate, pectin, carrageenan, gelatin, gellan, agar and methycellulose.

#### 6.3.1.3 Emulsion
Emulsion is a colloid formed, typically, between two liquids. Mayonnaise is a liquid emulsion, made up of proteins and oil droplets dispersed in liquid (also known as oil in water, O/W, emulsions). The liquid with the higher surface tension forms small droplets or the dispersed phase. There also exist solid emulsions such as butter or margarine (also known as water in oil, W/O, emulsions). Emulsions can be temporary, semi-permanent or permanent.

#### 6.3.1.4 Foam
A foam is formed when gas bubbles are trapped in a liquid or solid. When a cook beats egg whites, the energy applied when whipping mixes air bubbles (gas) within the egg white (liquid) to form foams. The whipping motion forces denatures or coagulates the protein at the air-water interface and makes the foam stable. Although typically air is used to form colloidal dispersions for foams, other gases can be also used. For instance, the famous dessert named *Viaje a la Habana* (travel to Habana), served at El Celler del Can Roca, was cooked with a real Partagás serie D n° 4 habano cigar smoke, used

**144 Chapter 6** Structure design for gastronomy applications

**FIG. 6.3**

*Viaje a la Habana* (travel to Habana).

*Permission from Celler de Can Roca.*

to create a foam that is later introduced in a chocolate cylinder, giving the dish a cigar appearance, aroma and flavor (Fig. 6.3).

### 6.3.1.5 Aerosols

Aerosols contain small liquid or solid particles dispersed in a gas. Excepting spray oil, which is not consumed as an aerosol, but is used e.g. to spread the oil in a pan, the presence of food aerosols is very limited. This opens a window for creating new products and dishes in this colloid form, which would surely give the diner a new culinary experience.

Table 6.1 summarizes different types of food colloids.

**Table 6.1 Types of food colloids.**

| Colloid | Dispersed phase | Continuous matrix | Food examples |
| --- | --- | --- | --- |
| Sol | Solid | Liquid | Soups, jams, gravy, ketchup |
| Gel | Liquid | Solid | Tapioca, gelatin, agar |
| Emulsion | Liquid | Liquid | Mayonnaise |
| Solid emulsion | Liquid | Solid | Butter, margarine, milk chocolate |
| Foam | Gas | Liquid | Whisked egg white, whipped cream |
| Solid foam | Gas | Solid | Ice cream, meringue |

### 6.3.2 Developing and emerging applications of food colloids in gastronomy

#### 6.3.2.1 Hydrocolloids and plant-based products

Hydrocolloids are polysaccharides and proteins that are colloidally dispersible in liquids. Being effectively dispersible, hydrocolloids tune the rheology of liquids by raising the viscosity and/or inducing gelation (Phillips & Williams, 2009). Widely applied to improve shelf-life of food products and their texture attributes, hydrocolloids are typically used for thickening (soups, gravies, dressings, sauces) and gelling (jams, jellies, marmalades, low calorie gels). Hydrocolloids are also employed as stabilizing, emulsifying and coating agents.

When developing a plant-based product that aims to mimic its animal counterpart, texture is probably the most challenging attribute to replicate. The notably increasing demand of plant-based dairy and meat alternatives is boosting the use and development of hydrocolloids due to their texturizing properties. Hydrocolloids grew at a 7% a year from 2015 to 2019, according to Innova Market Insights. Hydrocolloids are used to improve the rheological properties (viscosity) in plant-based dairy and the mechanical properties (hardness, toughness) of plant-based meat products, aiming to improve their sensory properties and mouthfeel. Hydrocolloids used in plant-based beverages typically include carrageenan, high-acyl gellan gum and locust bean gum. Methylcellulose and starches are also widely used in plant-based meat products.

#### 6.3.2.2 Oleogels as fat replacements

Oleogels are gels in which the dispersed liquid phase is oil. Oleogels have potential applications as trans fat replacements (Marangoni & Garti, 2018). By controlling the mobility of the oil phase with a network, oleogels can provide solid-like properties, making them a healthier alternative to the excessive use of trans and saturated fats, decreasing the risk of obesity and cardiovascular diseases. The typical gelling agents used to form oleogels are non-triacylglycerolic oleogelators and lipid-based gelators (sunflower wax, rice bran wax, candelilla wax, fatty acids, fatty alcohols, monoglycerides).

#### 6.3.2.3 Oil bodies (oleosomes) in plant-based products

Oil bodies, also named oleosomes, are microscopic natural oil droplets that are abundant in seeds, nuts, and some fruits (Abdullah et al., 2020). In seeds, the content of oil bodies varies from 20 to 50 wt%. Their natural function is to store energy to be used during germination. Oil bodies are oil-in-water emulsions, with a unique phospholipid/proteinaceous surface which makes them hardly soluble and gives them remarkable stability against physical (high temperatures), mechanical and chemical stresses. Oil bodies are relatively easy to extract and possess a high yield recovery (Nikiforidis, 2019). Due to the cost-competitive advantage of plant-based emulsifiers, potential applications in the food industry include vegan products such as plant-based milk, plant-based dairy, sauces, tofu, etc.

#### 6.3.2.4 Nanoemulsions as delivery systems

Calcium in milk is kept in a porous colloid named casein. Millions of years of evolution have engineered casein to be a natural way to protect and deliver calcium at the nano-level. Recent advances in nanotechnology are also giving us tools to engineer nanoemulsions that are able to deliver nutrients with improved bioavailability, increase the shelf-life of products, or deliver color/flavor upon cooking/eating (McClements & Rao, 2011). Engineered nanoemulsions are currently studied in food science, especially in the functional foods area. Functional foods aim to deliver nutraceuticals and/or micronutrients. The potential applications of functional foods to prevent or overcome health issues are quite promising, as we humans, in general, prefer to consume foods over drugs. Oral processing and gastrointestinal digestion of food colloids like nanoemulsions are becoming a relevant research topic (Lu et al., 2021).

## 6.4 Designing new food microstructures by biotechnology processes in the kitchen

### 6.4.1 The role of fermentation: a revolutionary technology that always has been with us

Although fermented products have become of particular interest to the food industry in recent years, it is worth recalling that fermentation has been applied for millennia, and it is the oldest human technique applied to create new kinds of food. There are references to different fermented products worldwide as Kombucha, originally from Asia, ginger beer from Ireland, tepache from Mexico, and Kefir milk from the Caucasus mountains of central Asia (Marshall & Meja, 2011). The first references to products developed through fermentation techniques to preserve perishable food over time come from ancient Egypt (Mohamed, 2020).

Nowadays, the application of fermentation in the food industry is in constant evolution. Modern and ancient methods of fermentation jointly concurred to improve sensory properties and food preservation and protect the cultural heritage. New fermentation techniques (isolation of mutants with desired properties, adaptive laboratory evolution, genome shuffling, and genome editing) are introduced in the development process to enhance food quality and produce many desirable attributes such as taste, texture, and aroma (Johansen, 2018).

During the fermentation process, the production of enzymes can likely change the food microstructure. Carbohydrates, proteins, and lipids are the principal nutrients changed by enzymes along with the fermentation technique, giving them new chemical structure properties, new functional properties and improving organoleptic characteristics of foods, including a range of health benefits (Ruddle & Ishige, 2010). These changes by using enzymes may create the basis for the new food.

### 6.4.2 A key player in fermentation: enzymes

To define fermented food, we consider those that go through a microbiological process involving enzymatic action to change its (i) structure, (ii) organoleptic, and (iii) physico-chemical characteristics. Enzymes have always been necessary for food processing (juice industry; starch processing, brewing industry, meat processing, alcohol, wine production) (Chaudhary et al., 2015) and they are well defined as "*a product obtained from plants, animals or micro-organisms or products thereof including a product obtained by a fermentation process using micro-organisms: containing one or more enzymes capable of catalyzing a specific biochemical reaction; and added to food for a technological purpose at any stage of the manufacturing, processing, preparation, treatment, packaging, transport or storage of foods*".

Some clearly examples of a smart use of enzymes in the kitchens are the work done by chef Ángel León, who tenderizes crustacean skins with xylanases; or chef Mario Sandoval, who demonstrated that egg white can turn into a creamy material thanks to the action of an aminopeptidase from *Aspergillus oryzae* (Garcés-Rimón et al., 2016).

A remarkable culinary application is the addition of pectinase in cocktails, which improves carbonation by clarifying the juice. It can also alter the drink texture by removing solids and reducing its body

(Nighojkar et al., 2019). Another example of a gastronomic application in this field by chef Andoni Luis Aduriz from Mugaritz (Errentería, Spain), is the *Condensed Onion or Mugaritz Onion Soup* (Mugaritz, 2018), where the use of pectinase, cellulase and hemicellulase results in a puree type texture. Furthermore, Mugaritz also developed "*Albedo in lime, aperitif*", which uses pectinase as well as lime, in order to make the inside soft and lime in order to make the outside firm (Mugaritz, 2018).

It is curious how in one of the emblematic dishes of Mexican cuisine, *the taco al pastor*, a pineapple is placed on the surface of the meat during the cooking process. Surely, in addition to providing flavors and sugars necessary for the correct cooking of this meat, without knowing it, the pineapple provides a softening of the meat, due to the bromelain (Muñoz Murillo et al., 2019). Meat texture has been also studied by William Meyers, executive chef at the Stone Harbor Golf Club (Cape May, EE.UU.). By replacing flour by transglutaminase, he prepared gluten free scallops with prosciutto. In addition, David Arnold, Director of Technology at the French Culinary Institute in New York City designed Mokume-Gane fish, a dish with salmon and halibut glued with transglutaminase and then sliced; or the elaboration of low salt sausage (Wolinsky & Husted, 2015).

Today, numerous enzymes are being used by the industry in designing new food microstructures and indifferent food applications (Table 6.2).

**Table 6.2 Enzymes application for food changes.**

| Food application | Enzymes | Source | Type |
|---|---|---|---|
| Baking | *a*-Amylase | *Aspergillus oryzae* | Fungal enzymes |
| | *a*-Amylase | *Basil* spp. | Bacterial enzymes |
| | Amylase | Malted cereals | Plant enzymes |
| | Lipoxygenase | Soybean | Plant enzymes |
| | Lactase | *Aspergillus oryzae* | Fungal enzymes |
| | Lactase | *Saccharomyces fragilis* | Fungal enzymes |
| Dairy | Lactoperoxidase | Cheese whey, bovine colostrum | Animal enzymes |
| | Lipase | Candida | Yeast enzymes |
| | Lipase | Bovine/porcine pancreas | Animal enzymes |
| | Catalase | Bovine/porcine pancreas | Animal enzymes |
| Beer making | Glucanase | *Bacillus* spp. | Bactrial enzymes |
| | Glucanselus | Malted barley | Plant enzymes |
| Winemaking | Glucosidase | *Aspergillus flavus* | Fungal enzymes |
| | Glucanase | *Trichoderma harizanum* | Fungal enzymes |
| | Cellulase Hemicellulose Pectinase | *Aspergillus niger* | Fungal enzymes |
| | Hemicellulose | *Bacillus subtilis* | Bacterial enzymes |
| Meat tenderization | Bromelain | *Ananas comosus* | Plant enzymes |
| | Fiction | Fig latex | Plant enzymes |
| | Papain | *Carica papaya* | Plan enzymes |

*Adapted from Ackaah-Gyasi, N. A., Patel, P., Zhang, Y., & Simpson, B. K. (2015). Current and future uses of enzymes in food processing.* Improving and Tailoring Enzymes for Food Quality and Functionality, *103–122.*

### 6.4.3 Fermentation in the kitchen: relationship with sciences, and new food design

Science and gastronomy march together and fermentation is a good example of this multidisciplinary field. In this respect in recent years, in some Michelin restaurants around the world, such as Mugaritz (Errenteria, Spain), Alchemist (Copenhagen, Denmark), or Noma (Copenhagen, Denmark), the microbiological phenomenon of fermentation has become an important part of high end kitchens, often linked to a new concept of healthy food. Chefs who apply fermentation in their laboratory kitchens as a technique to generate interesting, unique, tasty, and novel dishes.

A clear example is the case of the Mugaritz dish called "*podedumbre noble*" (noble rottenness) (Fig. 6.4), explained in book Mugaritz. Puntos de Fuga, written by Aduriz (2019), whereby impregnating an apple in a bath of milk and *Penicillium roqueforti*, they obtained a dish that forms part of the gastronomic *avant-garde*. This dish uses the enzymatic action of the fungus growing on the apple, impregnated with milk and lactose, to develop an apple with cheese flavors. *Penicillium roqueforti* fungus is used as a starter for the culture of blue-veined cheeses (Danablu, Gorgonzola, Roque-fort, and Stilton). The proteolytic enzymes present in *Penicillium roqueforti* need lactose as a nutrient, to grow and give to the cheese the characteristic taste of blue cheese (Coton et al., 2022). Interpreting the needs of these enzymes, they understood that, by introducing the nutrients necessary for the growth of the fungus (in this case, milk and lactose) into the apple, they could have designed a new food and generated the growth of the fungus in a substrate different than milk, thus giving this product the characteristic flavor of blue cheese.

This technique has led to the development of innovative dishes and studies, that today form part of the scientific-gastronomic revolution that is part of the kitchens. Another great example of the fusion between science and gastronomy is the Noma restaurant.

**FIG. 6.4**

Penicillium roqueforti on apple (Mugaritz).

*Source: Puntos de fuga Permission from Mugaritz; photo made by Jose luis lopez de Zubiria.*

Noma was the first restaurant to publish a gastronomic book talking about fermentation; The Noma Guide of Fermentation, written by Renè Zepei and David Zilber. They set out an easy development methodology for applying fermentation techniques to the kitchen and bringing new flavors and characteristics into the kitchen. An exciting application presented in this book is acid-lactic fermentation. This fermentation applied to food, has the function to extend the shelf life of products over time and preserve it by just adding 2% salt in anaerobic conditions. At Noma they use lacto-fermented foods to give a touch of acidity and umami to the dishes (Redzepi & Zilber, 2018). It is important to note that the decrease of the pH below 4 through acid production, inhibits the growth of pathogenic microorganisms, which can cause food spoilage, food poisoning and disease, and antifungal, antiseptic activity (Admassie, 2018).

Another type of microorganism currently used in restaurants to develop products is the SCOOBY of Kombucha, a symbiotic culture of bacteria and yeast used to develop carbonated beverages commonly developed with the black tea base. Understanding the metabolic development of Kombucha SCOOBY it could be possible to create a huge scale of different flavors of kombucha fermented beverage. Yeasts enzymes present in the symbiotic culture of kombucha have the function to hydrolyze the sugars present in the water infusion substrate (sucrose into glucose and fructose). This produces ethanol by acetic acid bacteria, also converting glucose to gluconic acid, fructose to acetic acid, and producing cellulose that gives structure to SCOOBY (Malbaša et al., 2008).

The connection between science and cuisine, together with new product developments in industry, has resulted in this new interdisciplinary research carried out in restaurant laboratories.

### 6.4.4 New fermented food for diet and health

By eating food, we take the nutrients needed to perform our basic functions. However, eating inadequate amounts of nutrients can cause poor health (Objective, 2022). Recent studies are emerging on functional food and fermented food for food related diseases reduction. The presence of probiotics and prebiotics in fermented foods have a significant role in consumers' health. For this reason, the development of innovative fermented food products is increasing (Kailasapathy, n.d.). Various studies investigated the healthiness of these foods and their benefits to the human body (Marco et al., 2017).

Positive consequences of the intake of fermented food, include, thanks to the probiotic organism: (i) benefits to the intestinal tract; (ii) improve the immune system by synthesizing the bioavailability of nutrients; (iii) decreasing the allergy and intolerance to many products in susceptible individuals (e.g. lactose intolerance) (Hasan et al., 2014).

Humans consumed fermented food way before discovering fermentation. Spontaneous fermentation still brings to our diet products that allow our body to reach probiotic elements that help our microflora stay healthy. Colonic bacteria present in our stomachs can produce many compounds that have several potential effects on our gut physiology. Bacteria such as bifidobacteria and lactobacilli are the main compounds of acid, lactic fermentation bacteria, the fermentation technique used to extend the shelf-life of foods (Kailasapathy, n.d.).

Today, thanks to re-discovering this millenial technique, fermentation opens new opportunities to develop novel functional and healthy products to bring to market.

## 6.5 Structuring food for health and wellness

### 6.5.1 Introduction

The use of new culinary techniques, methods, ingredients, and technologies in the kitchen has allowed progress in the field of sustainability and nutritional improvements. The cooperation between science and gastronomy is surely the way for the food industry to improve products in order to give the consumer not only a nutritional benefit but also a sustainable way of living.

Studies related with the design of new products and aiming to improve health are important and necessary for the food industry. Issues such as the reduction of sugar and fat, the improvement of the protein composition or creating more satiating products through emulsions or any other physical chemical method are being relevant for our society that is now facing health and sustainable problems. In this regard, gastronomy has already shown a great potential for the design of healthy products.

### 6.5.2 Reduction or replacement of fat through emulsions, hydrogels, oleogels and oleofoams

The use of food emulsions such as emulsions, hydrogels, oleogels and oleofoams can be considered to create different reduced fat foods and reduce the caloric content of chocolate (Nguyen et al., 2017; Norton et al., 2013). For example, if cocoa butter is reduced, the overall viscosity would increase either in the liquid chocolate and solid chocolate (hardness is increased). It would also affect the quality of the product and its uses (Aidoo & Afoakwa, 2014). Cocoa butter can be partially replaced by similar products without changing sensorial properties and the product's quality. Limonene is a cyclic monoterpene that has been considered for use in reduced cocoa butter chocolate products since 1999 (Beckett, 2008). The addition of limonene changes the kinetic of cocoa butter crystallization as it dilutes the cocoa butter, causing fewer cocoa butter crystals to form and disrupting the crystals packing (Do et al., 2008). Another solution to reduce cocoa butter content was replacing it with water. A water-in-cocoa butter emulsion with 40% water (all water droplets below 5 μm diameter size) has been suggested to be comparable with normal milk chocolate (Prosapio & Norton, 2019).

The use of margarine processing equipment to add water in molten cocoa butter could be easily incorporated into chocolate processing plants creating form V cocoa butter crystals (Sullo et al., 2014). Starting from this *water-in* strategy, different emulsifiers were also used to stabilize even more water-in-oil emulsions. The emulsifier soybean lecithin caused flocculation of the water droplets, leading to an increase in chocolate viscosity (Sullo et al., 2014). The emulsifier polyglycerol polyricinoleate (PGPR) was found to provide the best rheological properties in liquid chocolate (Prosapio & Norton, 2019). This study shows the importance of knowing the kind of additives used in a recipe when using water as a fat replacer and also what kind of them works against, causing unstable cocoa butter crystallization.

An acceptable alternative chocolate product could be formulated by replacing up to 50% (v/v) cocoa butter with a sodium alginate-pectin-citric acid hydrogel. This strategy works well to the development of the desired form V cocoa butter polymorph. Another advantage of this strategy by using this hydrogel is that the resulting chocolate is more resistant to heat. In fact, at 50–100% (v/v) substitution the final product kept its original structure at temperatures up to 80 °C (Francis & Chidambaram, 2019).

Oleogels have also been considered as a fat replacer for cocoa butter. Oleogels can entrap liquids and oils by forming gels using ingredients such as ethyl cellulose, waxes and monoglycerides (Martins et al., 2018). As well as all the previous strategies, another one used to reduce calorific content of chocolate is air incorporation. The addition of air bubbles does not add any nutritional value to food products, but it lowers their calorific density and significantly affects rheology and texture (Campbell & Mougeot, 1999).

### 6.5.3 Reduction of fat in mayonnaise through different kinds of fat mimetics

Reduction of the amount and type of fat consumed has been an important topic for the last few years. A higher fat intake is linked to several chronical diseases such as obesity, cardiovascular diseases and cancer. Multiple studies have been carried out with the objective of reducing fat content of traditionally high-fat content products but keeping the original sensory properties. It should be healthy, but it must be tasty. One of these products could be the mayonnaise.

From a physical point of view, mayonnaise is an oil-in-water emulsion (O/W). In order to keep the same overall viscosity with a lower fat content, one strategy could be decreasing oil volume fraction and increasing the viscosity of the continuous phase. Some fat mimetics such as modified starch (Murphy, 1999), inulin (James, 1998), pectin (Pedersen & Christian, 1997), microcrystalline cellulose (Chouard, 2005; Grodzka et al., 2005) or carrageenan (Trueck, 1997) were generally used. Culinary research already made important contributions in this matter as chefs has been trying new alternatives with improved texture and sensory properties like cod gelatin or linseed (Arboleya et al., 2008).

### 6.5.4 How aerated food can help on the expected satiety

Satiety is a concept that can be subjective for each person. It has been reported that deficiencies in satiation show more connections to obesity and binge eating than actual satiety (Kissileff, 1995; Spiegel et al., 1989). There are many methods for the development of products related to this effect (satiation), which are based on lowering caloric density or modifying structure. For instance, the use of non-caloric ingredients, immobilizing high quantities of water and incorporating air as small dispersed bubbles (Zúñiga & Aguilera, 2008). Considering that satiation is partially governed by sensory factors (Blundell et al., 2010), the volume of the own food could have an important role, that is, aerated products could potentially achieve a reduction in the caloric density, and it could induce satiety by making novel gastronomic structures (Arboleya et al., 2014). When does the feeling of being full start? When you eat the food, or when you first look at the food? Expected satiety was studied by using different levels of aeration in a meringue. High aerated product was formed by mixing pasteurized egg white, methyl cellulose, amorphous silica and maltodextrin, which helped in the reinforcement of the foam structure. A low aerated product was made just by pasteurized egg white, methyl cellulose and maltodextrin (Fig. 6.5).

The study proved how expected satiety was considerably higher for the high aerated foams and an actual lower intake. This interaction between science and gastronomy was interesting not only for the design of new dishes in the restaurant but also to the food industry to design healthier products.

**FIG. 6.5**

Expected satiety in a highly aerated and low aerated meringue.

## 6.6 Conclusions

Interaction between science and gastronomy can provide seamless integration of knowledge in physics, chemistry and food technology into sensory perception and pleasure. Science is the tool to be used in order to control such stimuli and the sensation of pleasure while eating. Within the sensory world of the kitchen, science can serve as an instrument to observe, control and hypothesize what is already happening. This tool can be used alongside the extraordinary imagination of the chef to brilliantly integrate scientific, psychological and cultural aspects. This transdisciplinary perspective is appealing from a scientific point of view, and potentially influential in making a major contribution to more appropriate eating habits worldwide.

## References

Abdullah, Weiss, J., & Zhang, H. (2020). Recent advances in the composition, extraction and food applications of plant-derived oleosomes. *Trends in Food Science & Technology, 106*, 322–332.

Admassie, M. (2018). A review on food fermentation and the biotechnology of lactic acid bacteria. *World Journal of Food Science and Technology*, 2(1), 19.

Aduriz, A. L. (2019). *Mugaritz. Puntos de Fuga*. Planeta. In press.

Aidoo, E. O., & Afoakwa, K. D. (2014). Optimization of inulin and polydextrose mixtures as sucrose replacers during sugar-free chocolate manufacture—Rheological, microstructure and physical quality characteristics. *Journal of Food Engineering*, 126, 35–42.

Arboleya, J.-C., García-Quiroga, M., Lasa, D., Oliva, O., & Luis-Aduriz, A. (2014). Effect of highly aerated food on expected satiety. *International Journal of Gastronomy and Food Science*, 2(1), 14–21.

Arboleya, J. C., Olabarrieta, I., Luis-Aduriz, A., Lasa, D., Vergara, J., Sanmartin, E., … de Maranon, I. M. (2008). From the chef's mind to the dish: How scientific approaches facilitate the creative process. *Food Biophysics*, 3(2), 261–268.

Beaugé, B. (2012). On the idea of novelty in cuisine. *International Journal of Gastronomy and Food Science*, 1(1), 5–14.

Beckett, T. (2008). *The science of chocolate* (2nd ed.). Cambridge: The Royal Society of Chemistry.

Blundell, J., de Graaf, C., Hulshof, T., Jebb, S., Livingstone, B., Lluch, A., Mela, D., Salah, S., Schuring, E., Van Der Knaap, H., & Westerterp, M. (2010). Appetite control: Methodological aspects of the evaluation of foods. *Obesity Reviews*, 11(3), 251–270.

Campbell, G. M., & Mougeot, E. (1999). Creation and characterization of aerated food products. *Trends Food Science and Technology*, 10, 283–296.

Chaudhary, S., Sagar, S., Kumar, M., Sengar, R. S., & Tomar, A. (2015). The use of enzymes in food processing: A review. *South Asian Journal of Food Technology and Environment*, 1, Issue 4.

Chouard, G. (2005). 100% indulgence 0% guilt—Be in phase, bring connectivity to your low-fat mayonnaise. *Innovations in Food Technology*, 29, 98–100.

Coton, E., Jany, J.-L., & Coton, M. (2022). *Penicillium roqueforti*. Encyclopedia of Dairy Sciences: Elsevier.

Corvitto, A. (2011). *The secrets of ice cream: Ice cream without secrets* (2nd ed.). Spain: Sant Cugat de Valles, Vilbo.

Dickinson, E. (2015). Colloids in food: Ingredients, structure, and stability. *Annual Review in Food Science and Technology.*, 6, 211–233.

Dickinson, E., & Miller, R. (2001). *Food colloids: Fundamentals of formulation*.

Do, T. A. L., Vieira, J., Hargreaves, J. M., Wolf, B., & Mitchell, J. R. (2008). Impact of limonene on the physical properties of reduced fat chocolate. *Journal of American Oil Chemistry Society*, 85, 911–920.

Marshall, E., & Meja, D. (2011). *Traditional fermented food beverages for improved livelihoods*. FAO Diversification Booklet No. 21.

Friberg, S., Larsson, K., & Sjoblom, J. (2004). *Food emulsions*. New York, USA: CRC Press.

Fiol, C., Prado, D., Romero, C., Laburu, N., Mora, M., & Alava, J. I. (2017). Introduction of a new family of ice creams. *International Journal of Gastronomy and Food Science*, 7, 5–10.

Fiol, C., Prado, D., Mora, M., & Alava, J. I. (2016). Nettle cheese: Using nettle leaves (*Urtica dioica*) to coagulate milk in the fresh cheese making process. *International Journal of Gastronomy and Food Science*, 4, 19–24.

Francis, F. P., & Chidambaram, R. (2019). Hybrid hydrogel dispersed low fat and heat resistant chocolate. *Journal of Food Engineering*, 256, 9–17.

Garcés-Rimón, M., Sandoval, M., Molina, E., López-Fandiño, R., & Miguel, M. (2016). Egg protein hydrolysates: New culinary textures. *International Journal of Gastronomy and Food Science*, 3, 17–22.

Graham, T. (1861). Liquid diffusion applied to analysis. *Phylosophical Transactions*, 151, 183–197.

Grodzka, K., Maciejec, A., & Krygier, K. (2005). Attempts to apply microcrystalline cellulose as a fat replacer in low fat mayonnaise emulsions. *Zywnosc*, 12, 52–61.

Hasan, M. N., Sultan, M. Z., & Mar-E-Um, M. (2014). Significance of fermented food in nutrition and food science. *Journal of Scientific Research*, 6(2), 373–386.

Irving, M. (2009). *The forager handbook: A guide to the edible plants of Britain*. London: Ebury.

James, S. M. (1998). *Method for producing fat-free and low-fat viscous dressings using inulin*. United States of American Patent No. 5721004.

Johansen, E. (2018). Use of natural selection and evolution to DevelopNew starter cultures fermented foods. *Annual Review of Food Science and Technology, 9*, 411–439.

Kailasapathy, K. (n.d.). Fermented foods and beverages of the world.

Kissileff, H. R. (1995). Inhibitions of eating in humans: Assessment, mechanisms, and disturbances. *Appetite, 1995*, 225–299.

Letcher, K. (2010). *The complete guide to edible wild plants, mushrooms, fruits, and nuts: How to find, identify, and cook them* (2nd revised ed.). USA: Falcon Guides.

Linnemann, A. R., & van Boekel, M. A. J. S. (2007). *Food product design. An Integrated Approach*. Wageningen, The Netherlands: Wageningen Academic Publishers.

Lu, W., Nishinari, K., Phillips, G. O., & Fang, Y. (2021). Colloidal nutrition science to understand food-body interaction. *Trends in Food Science & Technology, 109*(2021), 352–364.

Malbaša, R., Lončar, E., & Djurić, M. (2008). Comparison of the products of Kombucha fermentation on sucrose and molasses. *Food Chemistry, 106*(3), 1039–1045.

Marangoni, A. G., & Garti, N. (2018). *Edible oleogels: Structure and health implications*. 978-0128102220.

Marco, M. L., Heeney, D., Binda, S., Cifelli, C. J., Cotter, P. D., Foligné, B., Gänzle, M., Kort, R., Pasin, G., Pihlanto, A., Smid, E. J., & Hutkins, R. (2017). Health benefits of fermented foods: Microbiota and beyond. *Current Opinion in Biotechnology, 44*, 94–102.

Martins, A. J., Vicente, A., Cunha, R. L., & Cerqueira, M. A. (2018). Edible oleogels: An opportunity for fat replacement in foods. *Food Functional, 9*, 758–773.

McClements, D. J., & Rao, J. (2011). Food-grade nanoemulsions: Formulation, fabrication, properties, performance, biological fate, and potential toxicity. *Critical Reviews in Food Science and Nutrition, 51*(4), 285–364.

Mohamed, E. (2020). The ancient Egyptian bread and fermentation. *Microbial Biosystems, 5*(1), 52–53.

Muñoz Murillo, J. P., Zambrano Vélez, M. I., Párraga Álava, R. C., & Verduga López, C. D. (2019). Uso de papaína y bromelina y su efecto en las características organolépticas y bromatológicas de chuletas de cerdo ahumadas. *RECUS. Revista Electrónica Cooperación Universidad Sociedad, 4*(2), 38. ISSN 2528-8075.

Murphy, P. (1999). Low fat developments with speciality starches. *Food Technology International*, 22–24.

Mugaritz. (2018). *Mugaritz*. Planeta Gastro, Spain: Puntos de Fuga.

Nighojkar, A., Patidar, M. K., & Nighojkar, S. (2019). Pectinases: Production and applications for fruit juice beverages. In A. M. Grumezescu, & A. M. Holban (Eds.), *Vol. 2. Processing and Sustainability of Beverages* (pp. 235–273). Woodhead Publishing.

Nguyen, P. T. M., Kravchuk, O., Bhandari, B., & Prakash, S. (2017). Effect of different hydrocolloids on texture, rheology, tribology and sensory perception of texture and mouthfeel of low-fat pot-set yogurt. *Food Hydrocolloids, 72*, 90–104.

Nikiforidis, C. V. (2019). Structure and functions of oleosomes (oil bodies). *Advances in Colloid and Interface Science, 274*, 102039.

Norton, J. E., Fryer, P. J., & Norton, I. T. (2013). Design of food structures for consumer acceptability. In *Formulation Engineering of Foods* (pp. 253–280). Chichester: John Wiley & Sons.

Objective, L. (2022). *1.3: What Are Nutrients?* (pp. 1–6).

Pedersen, A., & Christian, H. (1997). *No and low fat mayonnaise compositions*. United States of American Patent No. 5641533.

Phillips, G. O., & Williams, P. A. (2009). *Handbook of hydrocolloids*.

Prosapio, V., & Norton, I. T. (2019). Development of fat-reduced chocolate by using water-in-cocoa butter emulsions. *Journal of Food Engineering, 261*, 165–170.

Redzepi, R., & Zilber, D. (2018). *The Noma guide of fermentation*. Workman Publishing. In press.

Ruddle, K., & Ishige, N. (2010). On the Origins, Diffusion and Cultural Context of Fermented Fish Products in Southeast Asia. *Globalization, Food and Social Identities in the Asia Pacific Region, January, 2010*, 18.

Sullo, A., Arellano, M., & Norton, I. T. (2014). Formulation engineering of water in cocoa—Butter emulsion. *Journal of Food Engineering., 142*, 100–110.

Spiegel, T. A., Shrager, E. E., & Stellar, E. (1989). Responses of lean and bese subjects to preloads, deprivation, and palatability. *Appetite, 13*, 45–69.

Trueck, H. U. (1997). *A mayonnaise-like product*. European Patent Application No. EP 0768042A1.

van der Linden, E., McClements, D. J., & Ubbink, J. (2008). Molecular gastronomy: A food fad or an interface for science-based cooking? *Food Biophysics, 3*(2), 246–254.

Vega, C., & Ubbink, J. (2008). Molecular gastronomy: A food fad or science supporting innovative cuisine? *Trends in Food Science & Technology, 19*, 372–382.

Wolinsky, H., & Husted, K. (2015). Science for food: molecular biology contributes to the production and preparation of food. *EMBO Rep., 16*, 272–275.

Zúñiga, R. N., & Aguilera, J. M. (2008). Aerated food gels: fabrication and potential applications. *Trends of Food Science and Technology., 19*(4), 176–187.

# Development of healthy products

PART III

# CHAPTER 7

# Design of functional foods with targeted health functionality and nutrition by using microencapsulation technologies

Guilherme de Figueiredo Furtado, Juliana Domingues dos Santos Carvalho, Gabriela Feltre, and Miriam Dupas Hubinger

*Department of Food Engineering, Faculty of Food Engineering, University of Campinas, Campinas, SP, Brazil*

## 7.1 Introduction

With the growing concern with health, consumers are increasingly looking for food products with functional claims. Many foods have functional ingredients in their formulation that act improving their physicochemical properties, increasing their stability during storage, or increasing their nutritional value and their respective physiological benefits. However, these ingredients when added into foodstuffs may lose their functionality due to undesirable reactions and degradation during processing and storage. To overcome these challenges, encapsulation technology has been used as a good alternative (Rostamabadi et al., 2021; Schrooyen et al., 2001; Sobel et al., 2014). Moreover, these functional ingredients have different molecular characteristics that lead to differences in physicochemical properties (e.g., chemical stability, physical state, optical characteristics and solubility) and, consequently, different delivery systems are necessary to protect the various types of functional ingredients (Mcclements et al., 2009).

Functional foods have been one of the most intensively investigated and widely promoted areas in the food and nutrition sciences (Hasler, 2002; Martirosyan et al., 2021) and the use of encapsulation technology by the food industry involves many characteristic changes on the macroscale (e.g., texture, taste and color), which has also led to the development of many products (Kwak, 2014). In addition, with the development of this technology, it has been possible to obtain solutions to improve the bioavailability of many functional compounds (Chau et al., 2007; de Souza Simões et al., 2017; Gouin, 2004; Hosseini & Jafari, 2020). Some of the most common materials and methods used to design functional foods with targeted health functionality are described in this chapter.

## 7.2 Strategies of microencapsulation

Microencapsulation is a technique that creates an external membrane (wall material) over another material (core material) and it is used to mask undesirable flavors and aromas or to protect bioactives and volatile compounds from biochemical and thermal degradation (Desai & Park, 2005; Shishir et al., 2018).

Microencapsulation techniques have been widely used by the food industry and different processes have been developed so far. Various methods are used to produce microencapsulated food systems. These methods are divided into physical, chemical and physico-chemical methods (Jafari et al., 2008). The most common strategies of microencapsulation include spray drying, spray chilling and ionic gelation. The main characteristics, advantages, and disadvantages of these techniques are presented in Table 7.1.

Table 7.1 Characteristics of the main microencapsulation methods.

| Techniques | Characteristics | Advantages | Disadvantages | References |
|---|---|---|---|---|
| Spray drying | Water is removed by spraying the feed dispersion/emulsion into a heated atmosphere; the active core is usually hydrophobic and wall materials are hydrophilic. | It is easily applicable, easily scalable, and economical; Different wall materials sources can be used. | The high drying air temperatures may cause deterioration of thermo-sensitive compounds; It can produce fine powders which may need further processing such as agglomeration. | Gharsallaoui et al. (2007), Abbas et al. (2012), Selvamuthukumaran (2019) |
| Spray Chilling | Use of hydrophobic materials (vegetable oils and fats) as wall material; Application of cold air to solidify the wall material; Water-insoluble particles. | It does not use solvents in the formulation and production process; The processing time is short; Low-cost technique; Protection of hydrophilic actives. | A significant amount of active material located on the surface of particles; Unfavorable polymorphic change; Unanticipated expulsion of the active compound. | Ribeiro and Veloso (2021), Abbas et al. (2012) |
| Ionic Gelation | Occurs through the bonding between polysaccharide chains and ions of opposite charges; Diffusion of ions into the polysaccharide chains; Formation of a solid matrix. | It does not use organic solvents; it does not require high process temperature; it is ideal for encapsulating heat-sensitive compounds; Low cost. | Difficulty of scaling up. | Arriola et al. (2019), Naranjo-Durán et al. (2021), Burey et al. (2008) |

## 7.2 Strategies of microencapsulation

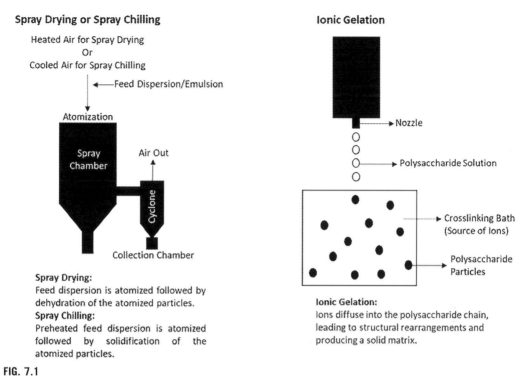

**FIG. 7.1**

Scheme of microencapsulation methodologies by spray drying, spray chilling and ionic gelation.

*Own authorship.*

The choice of the microencapsulation process must consider the physicochemical properties of the core and the wall materials, the expected release rate, processing steps, particle size and the desired applications of the food ingredients (Chen, 2009; de Souza Simões et al., 2017; Desai & Park, 2005; Estevinho et al., 2017).

Different systems can be obtained depending on the microencapsulation technique. The architecture of the microcapsules is classified based on the way the active core is distributed within the system: (a) mononucleated system when the active material is concentrated in the center of the matrix and surrounded by a continuous film denominated as wall material, and (b) multinucleated system when the core is uniformly dispersed throughout the matrix (Assadpour & Jafari, 2019; I Ré, 1998). The following sections briefly review the most common microencapsulation techniques. A scheme of these methodologies is also presented in Fig. 7.1.

### 7.2.1 Spray drying

Spray drying is a unit operation by which a liquid product is atomized (pulverized) in a hot air stream to obtain a powder. This technique is considered a dehydration process successfully employed for the drying of solutions, slurries and pastes; however, it can also be used as an encapsulation method when

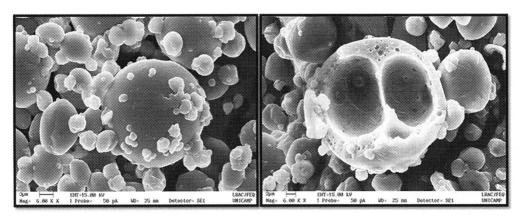

**FIG. 7.2**

Typical scanning electron microscopy (SEM) images of spray dried microparticles.

*Adapted from de Figueiredo Furtado, G., da Silva Carvalho, A. G. & Hubinger, M. D. (2021). Model infant formulas: Influence of types of whey proteins and oil composition on emulsion and powder properties.* Journal of Food Engineering, 292, *110256. https://doi.org/10.1016/j.jfoodeng.2020.110256.*

it entraps functional materials within a protective matrix, which is essentially inert to the encapsulated material (Gharsallaoui et al., 2007; I Ré, 1998). Moreover, spray drying is one of the most commonly used techniques of microencapsulation since it is easily applicable, easily scalable, and economical (Selvamuthukumaran, 2019).

The spray drying process is composed of different steps, such as preparation of the liquid product (which can be a solution, an emulsion, or a suspension), nebulization, formation of the droplet/air contact, evaporation of the water, separation of the dry product/humid air, and collection of the powder (Desai & Park, 2005; Estevinho et al., 2017; Gharsallaoui et al., 2007).

The spray dried particles are considered as matrix type, where the core exists as microparticles or microdroplets distributed within the dry solid matrix (Fig. 7.2). The spray drying process can produce very fine particles (10–50 μm) or large size ones (2–3 mm) depending on many factors, such as feeding material characteristics (composition and rheological properties), process parameters (type of atomizer) and external factors (humidity and temperature). These factors can also impact the drying performance and the quality of the final product (Gharsallaoui et al., 2007; I Ré, 1998). In addition, a successful spray drying process must produce powdered particles with maximum retention of the encapsulated compound and a minimal amount of the compound at the surface (Carneiro et al., 2013).

### 7.2.2 Spray chilling

Spray chilling is a promising physical method of microencapsulation that uses hydrophobic compounds as carrier agents. This procedure is mainly used in the microencapsulation of hydrophobic actives, but it has been studied to protect hydrophilic compounds. Some examples of compounds that can be used as a core material are oil matrices, vitamins, enzymes, probiotics, acidifying medications, and some aromas (Augustin & Sanguansri, 2008).

In this process, the lipid carrier agents are melted at a temperature above the system melting point. Subsequently, the active compound is incorporated and homogenized in this lipid system by dispersion or emulsification process. A peristaltic pump sends the molten matrix to the atomization nozzle, responsible for spraying the material inside the cooling chamber. The lipid droplets resulting from this process crystallize quickly on the first contact with cold air, resulting in solid lipid microparticles (SLMs). Then, SLMs are conducted through the air to a cyclone, in which they are collected into a collecting flask (Jaskulski et al., 2017; Oxley, 2012).

This method has the advantages of not using solvents in the formulation and the production process of the microparticles; the processing time is relatively short and is a technique that has a low cost (use of low temperature) and it is easy to scale up (Bertoni et al., 2018; Gouin, 2004). However, the spray chilling technique can have the following disadvantages: low retention of active compounds in the SLMs structure, polymorphic rearrangements, oxidation of the lipid carriers during the storage and, the hydrophobic character of the particles can make it difficult for some applications (Gouin, 2004; Okuro et al., 2013a).

SLMs obtained by this technique are classified according to the matrix type, in which the active compound is totally dispersed in the carrier agent. There is no solvent evaporation in this technique and the particles have different diameters (high polydispersity), spherical shape, and porous with roughness but without cracks (Fig. 7.3). SLMs diameters range from 50 to 500 µm (Bertoni et al., 2018). Some factors can influence the size distribution of SLMs, such as diameter nozzle, feed flow rate, pressure, cold air temperature, carrier material type (lipid or emulsion), the active compound (powder or solution), viscosity, among others.

**FIG. 7.3**

Typical scanning electron microscopy (SEM) images of lipid microparticles obtained by spray chilling.

*Adapted from Carvalho, J. D. dos S., Oriani, V. B., de Oliveira, G. M. & Hubinger, M. D. (2019). Characterization of ascorbic acid microencapsulated by the spray chilling technique using palm oil and fully hydrogenated palm oil. LWT, 101, 306–314. https://doi.org/10.1016/j.lwt.2018.11.043.*

In the spray chilling technique, the materials used as hydrophobic carriers include food-grade waxes, oils and fats. These components allow less diffusion of hydrophilic compounds (Gouin, 2004). Therefore, SLMs from this process are insoluble in water and provide physicochemical properties suitable for encapsulating hydrophobic and hydrophilic components. The active compounds may be present in crystals, dry particles, or water-in-oil emulsions (Abbas et al., 2012).

As a release mechanism, the active can be expelled to the external environment by the action of osmotic forces, diffusion, mechanical rupture, polymorphism, and, mainly, by the fusion of the constituent lipids of the matrix (temperature of the medium) (Gouin, 2004). However, the release kinetics can be improved by modifying the lipid crystals that are part of the wall material structure through combinations between the wall materials (Bertoni et al., 2018; Gouin, 2004).

### 7.2.3 Ionic gelation

Ionic gelation is a process in which gels are formed due to the bonding between chains of some polysaccharides (e.g., sodium alginate and pectin) and ions of opposite charges (e.g., calcium and potassium), resulting in a structural rearrangement and producing a solid matrix (Burey et al., 2008; Funami et al., 2009).

Gels can be formed with sodium alginate containing a high amount of G- segments (guluronic acid), that are able to interact with divalent cations (Grant et al., 1973). In relation to pectin, only that with a low degree of methoxylation can form gels in the presence of divalent cations without the addition of sugars (Dickinson, 2003). Ionic gelation method can produce particles with a wide range of size, from 10 µm to larger than 1 mm (da Silva Carvalho et al., 2019), depending on some factors, such as the type of equipment, diameter of the nozzle and operation conditions. Fig. 7.4 shows a typical structure of particles formed by ionic gelation. This technique presents some advantages, such as the fact that it is not necessary to use organic solvents and extreme conditions, making it less expensive when compared to other microencapsulation techniques. In addition, it can be used to encapsulate heat-sensitive compounds since the technique does not require high process temperatures (Arriola et al., 2019; Naranjo-Durán et al., 2021). Particles produced by ionic gelation can be insoluble in water and incorporated into high moisture products, such as yogurt (Carvalho et al., 2020; Comunian et al., 2017; de Moura et al., 2019; Silva, Cezarino, et al., 2018), cheese (Cardoso et al., 2020), vegan milks (Lopes et al., 2020) and goat ricotta (Lopes et al., 2021).

There are two types of ionic gelation: external and internal gelation. In both methods a source of ions is needed. External gelation is more common and occurs when the blend of polysaccharides and the compound come in contact with an external ion source, such as calcium chloride bath, at neutral pH (Davarcı et al., 2017). External gelation begins at the surface of the particle, where the ions react instantly, trapping water molecules in the network. However, the water is still free, allowing calcium ions to migrate by diffusion into the interior of the particle, favoring cross-linking (Helgerud et al., 2010; Smrdel et al., 2008). Internal gelation is based on the addition of ions to the polysaccharide solution and crosslinking occurs when there is a pH change and release of calcium ions. To make internal gelation happen, an insoluble salt (commonly $CaCO_3$) is added to the polymer solution, being insoluble. Then, an organic acid is added to reduce the pH, allowing the solubilization of the calcium ions, which are complexed with the polymer's carboxylic groups. The gel is formed as the acid penetrates the aqueous phase, releasing calcium ions (Funami et al., 2009; Holkem et al., 2017; Lupo et al., 2015; Naranjo-Durán et al., 2021; Poncelet et al., 1992). Internal gelation was applied to encapsulate probiotics

**FIG. 7.4**

Typical scanning electron microscopy (SEM) image of alginate microparticles obtained by ionic gelation.

*Own authorship, nonpublished.*

(Holkem et al., 2017; Martin et al., 2013; Pankasemsuk et al., 2016; Sánchez et al., 2017), polyphenolic compounds from dandelion (Belščak-Cvitanović et al., 2016), anthocyanins from grape skin (Zhang et al., 2020) and thyme essential oil (Benavides et al., 2016). The main difference between external and internal gelation is the reaction kinetics that influences the properties of the particles, such as size, structure, encapsulation efficiency and compound release (Lupo et al., 2015; Naranjo-Durán et al., 2021).

Usually, ionic gelation is applied in combination with other encapsulation techniques, such as extrusion (Arriola et al., 2016; Arriola et al., 2019; da Silva et al., 2021; Kim et al., 2017; Menin et al., 2018; Xu et al., 2019) and atomization (Cardoso et al., 2020; Feltre et al., 2020; Silva et al., 2019). Extrusion is most widely used and occurs by passing the liquid containing the polysaccharide and the active compound through a nozzle, dripping in a gelling bath. In this process, a simple syringe or more complex tools are used (Lupo et al., 2015). Emulsification is commonly used in combination with ionic gelation. It can be applied in some situations, such as previously to extrusion/atomization when it is desired to encapsulate a hydrophobic material (Feltre et al., 2020; Silva et al., 2020; Xu et al., 2019). The compound is incorporated into a polymeric solution through the emulsification process, and then, this emulsion is dripped/atomized in a crosslinking bath (Paques et al., 2014). Fig. 7.5 shows the internal structure of a particle formed by ionic gelation, where the oil droplets are encapsulated within the three-dimensional network formed from crosslinking between sodium alginate and calcium. Although the ionic gelation is a suitable method to encapsulate both hydrophobic and hydrophilic compounds, for hydrophilic materials, there are some limitations, because they are miscible with the polysaccharide solutions, making difficult the phase separation between core and wall materials.

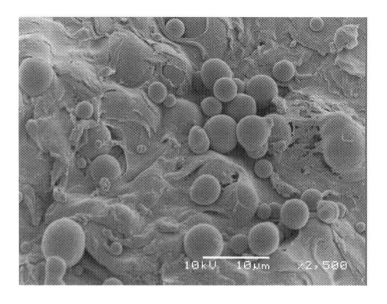

**FIG. 7.5**

Typical scanning electron microscopy (SEM) image of the internal structure of microparticles obtained by ionic gelation.

*Adapted from Feltre, G., Sartori, T., Silva, K. F. C., Dacanal, G. C., Menegalli, F. C. & Hubinger, M. D. (2020). Encapsulation of wheat germ oil in alginate-gelatinized corn starch beads: Physicochemical properties and tocopherols' stability.* Journal of Food Science, *85(7), 2124–2133. https://doi.org/10.1111/1750-3841.15316.*

Moreover, gel matrices are porous, which can lead to a partial loss of the encapsulated compound during the encapsulation process, storage and burst release effect (Arriola et al., 2016; Arriola et al., 2019; Chan et al., 2006; Kurozawa & Hubinger, 2017). In order to improve the retention of hydrophilic compounds inside the particles, studies have proposed some strategies. da Silva Carvalho et al. (2019) produced alginate beads with no core and placed them in a concentrated solution of anthocyanins from jussara extract, so that this compound was adsorbed by the beads. The process allowed to increase the stability of the anthocyanins and protected them from environmental conditions. de Moura et al. (2018) encapsulated anthocyanins from hibiscus extract, that is hydrophilic, using emulsification previously to the ionic gelation. First, an emulsion was produced with hibiscus extract, canola oil and a surfactant. Then, a double emulsion was obtained with pectin solution. From the double emulsion, particles were generated by ionic gelation by both atomization and extrusion methods.

## 7.3 Wall materials

Wall materials have the primary function of providing a protection system for the active compound against factors that may cause its degradation. Therefore, the choice of wall material will depend on the type of active compound to be encapsulated, the microencapsulation technique, and these agents' physical–chemical characteristics (Comunian & Favaro-Trindade, 2016).

Wall materials used for spray drying microencapsulation need to fulfill the following conditions: excellent protection barrier; high aqueous solubility; and good film forming property. In addition, these materials should have emulsifying ability to stabilize emulsions prior to spray drying (Furuta & Neoh, 2021; Gharsallaoui et al., 2007). The wall materials suitable for microencapsulation by spray drying for food applications are predominantly carbohydrates (e.g., carrageenan, carboxymethyl cellulose, chitosan, corn syrup, gum Arabic, inulin, maltodextrin, pectin and modified starches) and proteins (e.g., casein, gelatin, gluten, whey proteins and vegetable proteins from soy, pea, oat, rice and sunflower) (Coimbra et al., 2020; Zuidam & Nedović, 2010). New emerging materials, such as different Maillard reaction products (e.g., Maillard conjugates from sodium caseinate-corn starch hydrolysates (Consoli et al., 2018) and from spent brewer's yeast (Aguilera et al., 2008; Lee et al., 2017; Marson et al., 2020), or mixtures of two or more polymers (carbohydrates, proteins and/or synthetic emulsifiers) have also been used to enhance the encapsulation efficiency and powder characteristics (Coimbra et al., 2020; Labuschagne, 2018; Shamaei et al., 2017).

Regarding the spray chilling technique, hydrophobic compounds are used as wall material, usually lipids, and these can be of animal or plant origin. Among the fats of animal origin used as wall materials are lard, fish oil, butter, and milk fat. In contrast, soy, peanuts, canola, corn, cotton, palm and olive oils are obtained from plant sources and used as wall material (Vasisht et al., 2014).

Oils and fats are lipid molecules made up of approximately 98% triacylglycerols (TAGs); these molecules, in general, are soluble in organic solvents and insoluble in water. Fats and lipid systems with higher saturated fatty acid (FA) content are solid at room temperature (20–40 °C) and present an elevated melting point (MP). However, unsaturated FA in the *cis* form is naturally present in the liquid state, and their melting point is relatively low. *Trans* FAs are more linear than *cis* FA, leading to a more robust packaging between chains and higher melting points (Nichols & Sanderson, 2003).

The lipids exhibit a complex mixture of TAGs with different MP and can therefore fuse over a wide temperature range (Damodaran et al., 2007). Table 7.2 shows the melting point of some typical carrier agents. Examples of lipid carriers include fatty acids, alcohols, polyols, mono or diacylglycerol, fractionated or hydrogenated vegetable oils, waxes, and a mix of them (Abbas et al., 2012; Bertoni et al., 2018). The wide availability of these sources in nature promotes diverse materials for the coating of SLMs.

The melting point of wall materials of lipid origin must be determined, as well as their polymorphic behavior. Such factors will decide the particle's stability and active compound's early release during storage.

Polymorphism is when solid lipids present the same chemical composition but differ in their crystalline structures and molecular packaging. $\alpha$, $\beta'$, and $\beta$ are the crystalline forms available by the fatty acid esters present in lipids. These structures exhibit crystals of different sizes, hydrophilicity and densities, directly impacting the physical properties and encapsulation characteristics of SLMs. The more stable polymorphic forms tend to present a denser packaging of the molecules. This fact can influence the retention of the active material inside the SLMs (Gouin, 2004; Sato & Ueno, 2011).

Fats are monotropic, and their polymorphic transformation will always happen in the direction of the most stable crystal. Thermodynamic stability and the melting point of crystals increase in the following order: $\alpha < \beta' < \beta$ (Damodaran et al., 2007). Under normal conditions of subcooling, fats with a tendency to crystallize in the $\beta$ form include soybean, peanut, canola, corn, and olive oils. In contrast, palm and cotton oils, milk fat, and tallow usually crystallize in $\beta'$ form (Foubert et al., 2007; Ribeiro, Masuchi, et al., 2015).

Table 7.2 Melting point and applications of some lipid carriers.

| Lipid source | Melting point (°C) | Applications |
|---|---|---|
| Triglycerides | 55–73 | Bovine serum albumin by Di Sabatino et al. (2012); Insulin by Maschke et al. (2007) |
| Partial glycerides | 53–70 | Glibenclamide by Albertini et al. (2015); Econazole nitrate by Passerini et al. (2009) |
| Palmitic acid | 63 | Ginger oleoresin encapsulation by Oriani et al. (2016) |
| Stearic acid | 69 | Ascorbic acid encapsulation by Alvim et al. (2016); Glucose encapsulation by Ribeiro, Arellano, et al. (2012) |
| Fatty alcohols | 50–61 | Ibuprofen by Wong et al. (2015a, 2015b); Verapamil hydrochloride by Passerini et al. (2003) |
| Hydrogenated castor oil | 83–88 | Ibuprofen encapsulation by Vieira et al. (2013); Felodipine encapsulation by Savolainen et al. (2002) |
| Hydrogenated palm oil | 55–60 | Ascorbic acid encapsulation by Carvalho et al. (2019); Cinnamon bark oleoresin by Procopio et al. (2018); *Lactubacillus* encapsulation by Okuro et al. (2013b) |
| Hydrogenated soybean oil | 57–70 | Gallic acid encapsulation by Consoli et al. (2016) |
| Beeswax | 62–65 | Yerba mate extract encapsulation by Gris et al. (2021); Vitamin $D_3$ encapsulation by Paucar et al. (2016) |
| Carnauba wax | 82–86 | Urea encapsulation by Medeiros et al. (2019); Ammnonium bicarbonate encapsulation by Ding et al. (2018) |

*Adapted from Bertoni, S., Dolci, L. S., Albertini, B. & Passerini, N. (2018). Spray congealing: A versatile technology for advanced drug-delivery systems. Therapeutic Delivery, 9(11), 833–845. https://doi.org/10.4155/tde-2018-0049.*

The polymorphic transitions of these materials are improved based on the choice of raw material, crystallization and storage conditions of SLMs (Müller et al., 2002). In microparticles that require a fast release of their active to the environment, soy oil and canola oil can be utilized. These wall materials have structurally compact β crystals. For a controlled release, it is possible to produce SLMs with a flexible structure, with β′ crystals. For these SLMs, palm or cotton oil are good candidates as carriers (Damodaran et al., 2007).

The polymorphic transitions influence the production of SLMs and may show low values of encapsulation efficiency and release of the nucleus during storage (Ribeiro, Masuchi, et al., 2015).

## 7.4 Core materials
### 7.4.1 Oil matrices

In microencapsulation, the oil matrices are used as a core material, when presenting bioactive activity. The majority of studies found in the scientific literature talk about fish and essential oil encapsulation.

Fish oil has the most valuable omega-3 polyunsaturated fatty acids (ω-3 PUFAs), eicosapentaenoic acids (EPA), and docosahexaenoic acids (DHA). Both EPA and DHA are the most fundamental PUFAs for human health. This oil also has in its composition the ω-6 PUFAs linoleic and arachidonic acids. Essential oils, on the other hand, come from aromatic medicinal plants and have a wide variety of volatile compounds, which can be a source of other active compounds (Bakry et al., 2016).

Studies report that the role of omega-3 fatty acids in human health is related to decreased incidence of cardiovascular disease, prevention of immune response disorders, Alzheimer's disease, depression, and others. In children, DHA is fundamental in developing brain and eye functions (Bakry et al., 2016; Chang et al., 2019).

The human body does not produce the enzymes responsible for synthesizing ω-3 PUFAs, so its ingestion is necessary. Food and Agriculture Organization of the United Nations (FAO) recommends the following intake of EPA and DHA: adults at least 250 mg per day, pregnant/lactating women at least 300 mg per day, and for children, the amounts can vary according to the age from 100 to 250 mg per day (Feizollahi et al., 2018).

Essential oils are secondary metabolites of herbs and plants and can be isolated from various plant parts, like leaves, fruit, bark, root, wood, seeds, flowers, and others. They are composed of terpenes, terpenoids, and other aromatic and aliphatic compounds. The chemical composition of essential oils may vary according to the plant strain, location, and extraction method. Apart from that, essential oils have antioxidant and antimicrobial properties used as a technological alternative to increase the shelf life of food matrices. Due to its green nature and the ability to replace synthetic components, the application of essential oils in the industry has stood out (Gottschalk et al., 2018; Hosseini & Jafari, 2020).

However, fish oil and essential oils are impaired due to their bioactive components' degradation and instability. Fish oil has an unpleasant taste and odor and can undergo oxidation reactions due to its double bonds when exposed to adverse environmental conditions (Bakry et al., 2016; Chang et al., 2019).

Essential oils are quickly volatilized, high concentrations favor the appearance of unusual taste and odor, and factors such as light, water, and oxygen can lead to their oxidation. Thereby, the application of microencapsulation techniques minimizes the factors that interfere with the instability and allows the use of these compounds and their application in food matrices (Gottschalk et al., 2018; Hosseini & Jafari, 2020).

Recent research reports studies on fish oil microencapsulation. Castejón et al. (2021) performed the microencapsulation of ω-3 PUFAs extracted from oilseeds (chia, camelina, and echium oilseeds), microalgae (wet biomass of *Nannochloropsis*), and enzymatically produced fatty acid ethyl esters (FAEEs) by spray drying technique. Sodium caseinate and lactose were used as wall materials. The authors observed that microencapsulation efficiency depended on the ω-3 PUFAs. For the compound extracted from chia, the microencapsulation efficiency was the highest (76.9%), while the lowest was obtained for camelina oil (58.8%) and *Nannochloropsis gaditana* lipids (57.0%) microcapsules. All the microcapsules showed a spherical shape and smooth surface without pores, and the polyunsaturated fatty acid profile of the lipids was not affected. Linke et al. (2021) produced fish oil powders by spray drying oil/water emulsion. Maltodextrin and soy protein isolate were used as wall materials. The oil load ranged between 4.95% and 20.33% (w/w). The oxidative stability was determined at 25°C for 82 days, and for that, particle size was fixed at 50 to 80 μm, and the oxidation products of the total and encapsulated oil were determined (hydroperoxides and anisidine). The authors observed that the oxidation rate was higher for particles produced with a low oil charge. Moreover, the oxidation of the encapsulated fish oil was limited by oxygen availability and not by the amount of oil present.

This fact is explained by two effects: oxygen diffusion and an oil scavenging activity located in the outer region of the particle, consuming the penetrating oxygen and protecting the particle's oil droplets.

In the last years, several articles were published about microencapsulated essential oils using the spray drying technique. In a recent study, Hu et al. (2020) evaluated the potential of combining whey protein (WPI), maltodextrin (MD), and sodium alginate (SA) for microencapsulating cinnamon essential oil by spray drying. Results indicated that the ideal formulation consists of 70% wall materials (1:3:0.01 w/w from WPI, MD and, SA, respectively), leading to the best encapsulation efficiency (above 93%) and a retention value over 95% during storage at 50°C for 30 days. In addition, the spray drying technique also was used to microencapsulate *Mentha spicata* essential oil (Mehran et al., 2020). The authors used a blend of inulin and gum Arabic as wall material (ratio of 75:25 w/w, respectively) to optimize microencapsulation conditions. The authors concluded that optimal conditions were 35% solid wall, 4% essential oil concentration, and 110°C inlet temperature, generating particles with oil retention of 91%. They also observed that the release kinetics followed the Peppas-Sahlin mechanism, and the release phenomenon was governed by Fickian diffusion. Another methodology applied to encapsulate essential oil was spray chilling. Gottschalk et al. (2018) encapsulated an essential oil mixture using hydrogenated vegetable oil (MP: 69–73°C) as wall material. The results showed that all produced microparticle batches were free-flowing with essential oil recoveries of around 90%. The SLMs presented β-polymorphic forms leading to changes in the melting behavior and the crystalline structure.

Based on these examples, it can be said the microencapsulation technology is a method capable of minimizing the factors that interfere with the instability of oily matrices (omega-3 fatty acids and essential oils) and can facilitate its application in food matrices.

### 7.4.2 Bioactive compounds

Bioactive compounds are phytochemicals that act by modulating metabolic processes and providing health benefits. They comprise a variety of components, such as polyphenols, carotenoids, tocopherols, phytosterols, and organosulfur compounds. They have different chemical structures, distribution in nature, effectiveness against oxidative species, range of concentrations, possible site of action, specificity and biological action (Carbonell-Capella et al., 2014; Porrini & Riso, 2008).

Antioxidants act as food preservatives preventing foods from deterioration occurred through oxidation and loss of nutritional value. Moreover, they present innumerous health benefits such as antiinflammatory, antibacterial, antiallergic and antihypertensive properties apart from helping to reduce cancer, cardiovascular and neurological diseases (Ozkan et al., 2019). They have been classified into different categories based on their chemical structure and functions: hydrophilic bioactives (e.g., citrates, norbixin, betalains, phenolics, flavonoids and anthocyanins) and hydrophobic bioactives (e.g., carotenoids, tocopherols, terpenoids and vitamin E) (Carocho et al., 2018).

Carotenoids are compounds that are responsible for the yellow/red colors of many foods. Carotenoids that contain oxygen in their structure are known as xanthophylls (e.g., lutein and zeaxanthin) and those without oxygen are known as carotenes (e.g., lycopene and β-carotene) (Damodaran et al., 2007; Mcclements et al., 2009). They have different health benefits, being associated with their antioxidant activity, the decreased risk of cancer and cardiovascular diseases, strengthening of immune system and, preventing macular degeneration and risk of cataracts (Albanes, 1999; Basu & Imrhan, 2007; Gouveia & Empis, 2003; Robert et al., 2003; Stringham & Hammond, 2005). As carotenoids are susceptible to light and oxygen, they are relatively unstable when used as food additives in food systems since processing and storage conditions can result in oxidation and consequent loss of colorant properties and antioxidant activity (Nunes & Mercadante, 2007; Xianquan et al., 2005). Moreover, the high hydrophobicity of the carotenoids hinders their solubility in aqueous systems, resulting in a poor bioavailability (Ribeiro & Cruz, 2005). Encapsulation has been used to overcome these problems thus providing a physiological barrier of protection to them. Carotenoids that have been commonly subjected to microencapsulation, are astaxanthin, lycopene and β-carotene (Anandharamakrishnan & Ishwarya, 2015; Loksuwan, 2007; Rocha et al., 2012; Shen & Quek, 2014; Shu et al., 2006).

Polyphenols are a group of phytochemicals that can prevent chronic degenerative diseases and are also responsible for the color and taste characteristics of some foods (Anandharamakrishnan & Ishwarya, 2015; Scalbert et al., 2005). However, incorporating phenolic compounds into processed foods include technical challenges such as degradation of the active compounds in the processing environment and perception of bitterness at higher concentrations. Such challenges can be overcome through microencapsulation technology (Sun-Waterhouse et al., 2013).

In addition, the stability and nutritive value of vitamins have also been preserved through encapsulation techniques since they suffer from different types of instabilities. Vitamin A is a hydrophobic molecule highly prone to inactivation in aqueous medium. Vitamin E is degraded in the presence of oxygen and free radical mediated oxidative reactions. Vitamin C, in solution, is unstable in the presence of air, heat, light, moisture and oxygen (Anandharamakrishnan & Ishwarya, 2015; Rucker et al., 2007; Sauvant et al., 2012; Trindade & Grosso, 2000).

### 7.4.3 Probiotics

The consumption of probiotics presents many advantages, such as the inhibition of pathogenic microorganisms, making the intestinal tract healthy (Cavender et al., 2021; Rodrigues et al., 2020). Probiotics act on the enteric part of the gastrointestinal system, colonizing and maintaining the metabolism (Collins et al., 1998; Saarela et al., 2000). However, a great number of probiotic microorganisms does not survive in the gastric acid environment, reaching the intestine a smaller portion of what was consumed (Yeung et al., 2016). Commercial probiotics presented $10^6$-fold reduction in colony-forming units (CFU) within 5 min incubation under gastric conditions, at pH 1.2 (Dodoo et al., 2017). Aiming to avoid or minimize this problem, encapsulation of probiotic cells has been showing itself as an alternative, presenting many advantages, such as improving the resistance of microorganisms during gastric digestion, reducing cells losses that must reach the intestinal flora (Afzaal et al., 2020; Arslan-Tontul & Erbas, 2017; Puttarat et al., 2021; Silva, Tulini, et al., 2018). Encapsulation also prevents probiotic cells losses during production, transport and storage of a food in which they are applied, which can occur due to exposure to external factors, such as oxygen and changes in pH and temperature (Lopes et al., 2021; Puttarat et al., 2021; Silva et al., 2017). Some of the methods used for probiotics encapsulation are extrusion, emulsification, spray drying, spray chilling and fluidized bed. The use of extrusion associated with ionic gelation is common for encapsulation of probiotics, due to the advantages such as the maintenance of viable microorganisms' cells and the simplicity of operation and lower cost when compared to other methods (Ramos et al., 2018; Rodrigues et al., 2017, 2020). However, due to the time required for gelation of solution and formation of particles, this technique is relatively slow, making it difficult to scale up (Liu et al., 2019). Spray drying is widely used in the food industry to encapsulate probiotics due to the fast and high production rate and high reproducibility (Burgain et al., 2011; Gharsallaoui et al., 2007; Liu et al., 2015). As disadvantages, there are some process conditions, such as the use of high temperature, oxygen exposure and osmotic stress during particles production, that can reduce the viability of probiotic microorganisms (Burgain et al., 2011; Farahmandi et al., 2021; Liu et al., 2019; Ramos et al., 2018).

Analogous to spray drying, spray chilling also can be used to encapsulate probiotic cells. Lipid matrices can protect the cells under environmental conditions and control their release in the gastrointestinal tract (Đorđević et al., 2015; Silva, Tulini, et al., 2018). In addition, spray chilling presents industrial-scale applications (Rodrigues et al., 2020). This technique is less applied to microencapsulation when compared to other techniques, due to some drawbacks, such as the possibility of low encapsulation efficiency and release of the encapsulated compound from the matrix (Liu et al., 2019).

## 7.5 Food applications

### 7.5.1 Spray drying

Given the beneficial characteristics of seed and marine oils already mentioned before, some research has been done to incorporate these spray dried encapsulated oils in different types of food products (e.g., dairy and nondairy products, meat products, pastries and soups) and the authors observed greater oxidative stability and high sensory acceptability for these products (Bolger et al., 2018; Mohammed et al., 2017; Pérez-Palacios et al., 2018; Solomando et al., 2020; Stangierski et al., 2020; Vargas-Ramella et al., 2020). Solomando et al. (2020) produced cooked and dry-cured meat products added of encapsulated fish oil (with lecithin and chitosan as wall materials). In addition to obtaining a product source of EPA and DHA, the physico-chemical characteristics, oxidative stability and acceptability of the product were not affected by the encapsulated fish oil addition.

Many studies have also been done to produce food products (e.g., dairy and nondairy products) added of encapsulated antioxidants and polyphenols from plant extracts. As result, the authors concluded that the spray drying microencapsulation preserved the color, antioxidant activity and enabled sustained release of these extracts (da Silva et al., 2019; Francisco et al., 2018; Pacheco et al., 2018; Ribeiro, Ruphuy, et al., 2015; Urzúa et al., 2017). Ribeiro, Ruphuy, et al. (2015) incorporated microencapsulated mushroom extracts (with maltodextrin as wall material) in cottage cheese, resulting in products with no color modifications and with better preservation of the antioxidant activity.

In the same way, the nutritional value of vitamins added to food products (gummies and baked products) has been preserved through spray drying encapsulation (Alvim et al., 2016; Yan et al., 2021). Yan et al. (2021) produced gummies containing free or microencapsulated ascorbic acid (with casein as wall material). The authors observed that the previous encapsulation improved the stability of ascorbic acid in the gummy at accelerated storage conditions and harsher environments (92% of retention for encapsulated ascorbic acid and 79% for the unencapsulated one). During simulated gastrointestinal digestion, a slower release and more protection of the encapsulated ascorbic acid was also observed.

### 7.5.2 Spray chilling

Some requirements for encapsulating bioactives by the spray chilling technique include protecting SLMs from oxygen and moisture, leading to a controlled release of hydrophilic compounds to the external environment. Encapsulating bioactive compounds gives them better stability and bioavailability and reduces the hygroscopicity of hydrophilic materials.

Thus, the spray chilling technique can be used to obtain SLMs containing different core materials such as aromas, vitamins, minerals, probiotics, and oily matrices, favoring their application in various foodstuffs. In this section, practical examples of applying SLMs to food bases are presented.

Sartori and Menegalli (2016) produced green banana biomass biofilms with the addition of ascorbic acid (AA) microencapsulated by spray chilling to act as coatings on apple slices and delay their enzymatic browning. AA is a natural antioxidant used in food preservation, and microencapsulation can increase this compound's stability. For this, lauric acid (LA) and oleic acid (OA) mixtures were used as wall material. The authors noticed that films containing AA microparticles showed lower water vapor permeability since the particles had hydrophobic characteristics (lipidic wall material). Furthermore, after the drying process, the films containing AA microcapsules presented the highest antioxidant activity (84%), showing that the encapsulation protected the AA antioxidant activity during film production. Alvim et al. (2016) compared spray drying and spray chilling techniques to produce particles loaded with ascorbic acid to be applied in a baked product (biscuit). Thereby, the authors concluded that microencapsulation of AA using both methods was successfully performed. The AA microencapsulated inhibited dark spots on the biscuits associated with this active substance's thermal degradation during baking. Moreover, greater protection (preservation of more than 85%) of AA during the cooking step was verified by the particles obtained by spray drying followed by particles from spray chilling, when compared to the free substance.

Yin and Cadwallader (2019) studied the application of SLMs loaded with 2-acetyl-1-pyrroline zinc chloride (2AP–$ZnCl_2$). This compound is an aroma known to be unstable, and therefore its use as a flavor ingredient is limited. As a food application, the authors used the SLMs in instant rice. The authors observed the total recovery of 2AP–$ZnCl_2$ after the rice cooking stage compared to the free active compound. Controlled release of the encapsulated aroma triggered by the heat of cooked rice was also observed.

Arslan-Tontul et al. (2019) evaluated the addition of single and double-layered microcapsules of probiotics, produced by spray drying and chilling, in cake production. The microorganisms selected were *Saccharomyces boulardii*, *Lactobacillus acidophilus*, and *Bifidobacterium bifidum*. The microparticles were used in three different cake formulations: cream-filled, marmalade-filled, and chocolate-coated. Particles were added after baking and in the center of the simple cake dough (baking at 200 °C for 20 min). In plain cake, the count of *S. boulardii* and *L. acidophilus* was 2.9 log CFU/g in the double-layered microcapsules produced by spray chilling. On the other side, there were no viable *B. bifidum* detected after baking since the free forms of these probiotics did not survive in any plain cake experiments. Based on the results, the authors concluded that double-layered microcapsules produced by spray chilling in the caking process could increase the amount of probiotics in food matrices.

### 7.5.3 Ionic gelation

Encapsulated compounds produced by ionic gelation are widely used in food application due to the strong network structure of the gel, which can make the particles insoluble in water and relatively stable at different conditions of production and storage (Cardoso et al., 2020; Feltre et al., 2020; Lopes et al., 2020, 2021). Cardoso et al. (2020) added encapsulated chia oil in processed cheese and observed that its acceptability was higher than the cheese with free chia oil, indicating that the encapsulation was able to mask the unpleasant flavor of the oil.

Ionic gelation is suitable to encapsulate probiotics due to its mild process conditions, such as not using high temperatures. In a recent study, *L. acidophilus* and *B. bifidum* were encapsulated into alginate matrix by ionic gelation technique and incorporated in butter to produce a probiotic enriched product. While the butter with free microorganisms did not present any cell viability during storage, the product with 10% of encapsulated *L. acidophilus* presented a number of viable cells greater than 8 log CFU/g during 45 days. For the product with 10% of encapsulated *B. bifidum* was obtained 2 log CFU/g during 22 days of storage. Furthermore, the butter containing *L. acidophilus* was subjected to sensory analysis and did not present differences in relation to color, texture and overall appearance when compared to the control (butter without probiotic cells), indicating that the presence of capsules did not interfere with the acceptance of the product (da Silva et al., 2021).

Lopes et al. (2021) produced spreadable goat ricotta cheese added with *L. acidophilus* La-05 (free and encapsulated). The probiotics were encapsulated by external ionic gelation method, using sodium alginate as wall material and chitosan as coating. The product with encapsulated microorganisms presented higher viability of the probiotics in the product and under gastrointestinal conditions (>6 log CFU/mL) than ricotta cheese with free cells.

## 7.6 Final remarks

This chapter provided an overview of some of the most commonly used microencapsulation systems (spray drying, spray chilling and ionic gelation) for encapsulating functional compounds and highlighted some of the advantages and disadvantages of each, in addition to presenting some applications in food products. Nevertheless, as most functional compounds are sensitive to process and environmental conditions, new emerging encapsulation technologies have been of interest for the scientific community. As an example, electrohydrodynamic techniques (e.g., electrospinning and electrospraying) have stood out for being cost-effective and versatile techniques that do not require the use of high temperatures, thus contributing to avoid the degradation of the compounds.

In addition, there is a strong trend that large industries are increasingly interested in the development of functional foods for healthiness. However, their application in food systems still presents challenges that need to be overcome both in terms of the production process and the characterization of the acceptance and safety of these products.

## References

Abbas, S., da Wei, C., Hayat, K., & Xiaoming, Z. (2012). Ascorbic acid: Microencapsulation techniques and trends—A review. *Food Reviews International*, 28(4), 343–374. https://doi.org/10.1080/87559129.2011.635390.

Afzaal, M., Saeed, F., Ateeq, H., Ahmed, A., Ahmad, A., Tufail, T., Ismail, Z., & Anjum, F. M. (2020). Encapsulation of *Bifidobacterium bifidum* by internal gelation method to access the viability in cheddar cheese and under simulated gastrointestinal conditions. *Food Science & Nutrition*, 8(6), 2739–2747. https://doi.org/10.1002/fsn3.1562.

Aguilera, J. M., Lillford, P. J., & Barbosa-Cánovas, G. V. (2008). Food materials science: Principles and practice. In *Food materials science: Principles and practice* (pp. 1–616). New York: Springer. https://doi.org/10.1007/978-0-387-71947-4.

Albanes, D. (1999). β-Carotene and lung cancer: A case study. *The American Journal of Clinical Nutrition*, 69(6), 1345S–1350S. https://doi.org/10.1093/ajcn/69.6.1345S.

Albertini, B., Di Sabatino, M., Melegari, C., & Passerini, N. (2015). Formulation of spray congealed microparticles with self-emulsifying ability for enhanced glibenclamide dissolution performance. *Journal of Microencapsulation, 32*(2), 181–192.

Alvim, I. D., Stein, M. A., Koury, I. P., Dantas, F. B. H., Cruz, C. L., & d. C. V. (2016). Comparison between the spray drying and spray chilling microparticles contain ascorbic acid in a baked product application. *LWT- Food Science and Technology, 65*, 689–694. https://doi.org/10.1016/j.lwt.2015.08.049.

Anandharamakrishnan, C., & Ishwarya, S. P. (2015). *Spray drying techniques for food ingredient encapsulation.* https://doi.org/10.1002/9781118863985.

Arriola, N. D. A., Chater, P. I., Wilcox, M., Lucini, L., Rocchetti, G., Dalmina, M., Pearson, J. P., & de Mello Castanho Amboni, R. D. (2019). Encapsulation of *Stevia rebaudiana* Bertoni aqueous crude extracts by ionic gelation—Effects of alginate blends and gelling solutions on the polyphenolic profile. *Food Chemistry, 275*, 123–134. https://doi.org/10.1016/j.foodchem.2018.09.086.

Arriola, N. D. A., de Medeiros, P. M., Prudencio, E. S., Olivera Müller, C. M., & de Mello Castanho Amboni, R. D. (2016). Encapsulation of aqueous leaf extract of *Stevia rebaudiana* Bertoni with sodium alginate and its impact on phenolic content. *Food Bioscience, 13*, 32–40. https://doi.org/10.1016/j.fbio.2015.12.001.

Arslan-Tontul, S., & Erbas, M. (2017). Single and double layered microencapsulation of probiotics by spray drying and spray chilling. *LWT - Food Science and Technology, 81*, 160–169. https://doi.org/10.1016/j.lwt.2017.03.060.

Arslan-Tontul, S., Erbas, M., & Gorgulu, A. (2019). The use of probiotic-loaded single- and double-layered microcapsules in cake production. *Probiotics and Antimicrobial Proteins, 11*(3), 840–849. https://doi.org/10.1007/s12602-018-9467-y.

Assadpour, E., & Jafari, S. M. (2019). Nanoencapsulation: Techniques and developments for food applications. In A. L. Rubio, M. J. F. Rovira, M. M. Sanz, & L. G. Gómez-Mascaraque (Eds.), *Nanomaterials for food applications* Elsevier.

Augustin, M. A., & Sanguansri, L. (2008). Encapsulation of bioactives. In J. M. Aguilera, & P. J. Lillford (Eds.), *Food materials science—Principles and practice* (pp. 577–601). New York: Springer.

Bakry, A. M., Abbas, S., Ali, B., Majeed, H., Abouelwafa, M. Y., Mousa, A., & Liang, L. (2016). Microencapsulation of oils: A comprehensive review of benefits, techniques, and applications. *Comprehensive Reviews in Food Science and Food Safety, 15*(1), 143–182. https://doi.org/10.1111/1541-4337.12179.

Basu, A., & Imrhan, V. (2007). Tomatoes versus lycopene in oxidative stress and carcinogenesis: Conclusions from clinical trials. *European Journal of Clinical Nutrition, 61*(3), 295–303. https://doi.org/10.1038/sj.ejcn.1602510.

Belščak-Cvitanović, A., Bušić, A., Barišić, L., Vrsaljko, D., Karlović, S., Špoljarić, I., Vojvodić, A., Mršić, G., & Komes, D. (2016). Emulsion templated microencapsulation of dandelion (*Taraxacum officinale* L.) polyphenols and β-carotene by ionotropic gelation of alginate and pectin. *Food Hydrocolloids*, 139–152. https://doi.org/10.1016/j.foodhyd.2016.01.020.

Benavides, S., Cortés, P., Parada, J., & Franco, W. (2016). Development of alginate microspheres containing thyme essential oil using ionic gelation. *Food Chemistry, 204*, 77–83. https://doi.org/10.1016/j.foodchem.2016.02.104.

Bertoni, S., Dolci, L. S., Albertini, B., & Passerini, N. (2018). Spray congealing: A versatile technology for advanced drug-delivery systems. *Therapeutic Delivery, 9*(11), 833–845. https://doi.org/10.4155/tde-2018-0049.

Bolger, Z., Brunton, N. P., & Monahan, F. J. (2018). Impact of inclusion of flaxseed oil (pre-emulsified or encapsulated) on the physical characteristics of chicken sausages. *Journal of Food Engineering, 230*, 39–48. https://doi.org/10.1016/j.jfoodeng.2018.02.026.

Burey, P., Bhandari, B. R., Howes, T., & Gidley, M. J. (2008). Hydrocolloid gel particles: Formation, characterization, and application. *Critical Reviews in Food Science and Nutrition, 48*(5), 361–377. https://doi.org/10.1080/10408390701347801.

Burgain, J., Gaiani, C., Linder, M., & Scher, J. (2011). Encapsulation of probiotic living cells: From laboratory scale to industrial applications. *Journal of Food Engineering*, *104*(4), 467–483. https://doi.org/10.1016/j.jfoodeng.2010.12.031.

Carbonell-Capella, J. M., Buniowska, M., Barba, F. J., Esteve, M. J., & Frígola, A. (2014). Analytical methods for determining bioavailability and bioaccessibility of bioactive compounds from fruits and vegetables: A review. *Comprehensive Reviews in Food Science and Food Safety*, *13*(2), 155–171. https://doi.org/10.1111/1541-4337.12049.

Cardoso, L. G., Bordignon Junior, I. J., Vieira da Silva, R., Mossmann, J., Reinehr, C. O., Brião, V. B., & Colla, L. M. (2020). Processed cheese with inulin and microencapsulated chia oil (*Salvia hispanica*). *Food Bioscience*, *37*, 100731. https://doi.org/10.1016/j.fbio.2020.100731.

Carneiro, H. C. F., Tonon, R. V., Grosso, C. R. F., & Hubinger, M. D. (2013). Encapsulation efficiency and oxidative stability of flaxseed oil microencapsulated by spray drying using different combinations of wall materials. *Journal of Food Engineering*, *115*(4), 443–451. https://doi.org/10.1016/j.jfoodeng.2012.03.033.

Carocho, M., Morales, P., & Ferreira, I. C. F. R. (2018). Antioxidants: Reviewing the chemistry, food applications, legislation and role as preservatives. *Trends in Food Science and Technology*, *71*, 107–120. https://doi.org/10.1016/j.tifs.2017.11.008.

Carvalho, C., Pagani, A., Teles, A., Santos, J., Pacheco, T., Junior, R. C., & Pozza, M. (2020). Jamelão capsules containing bioactive compounds and its application in yoghurt. *Acta Scientiarum Polonorum. Technologia Alimentaria*, 47–56. https://doi.org/10.17306/j.afs.0744.

Carvalho, J. D. S., et al. (2019). Characterization of ascorbic acid microencapsulated by the spray chilling technique using palm oil and fully hydrogenated palm oil. *LWT*, *101*(2), 306–314.

Castejón, N., Luna, P., & Señoráns, F. J. (2021). Microencapsulation by spray drying of omega-3 lipids extracted from oilseeds and microalgae: Effect on polyunsaturated fatty acid composition. *LWT*, *148*, 111789. https://doi.org/10.1016/j.lwt.2021.111789.

Cavender, G., Jiang, N., Singh, R. K., Chen, J., & Mis Solval, K. (2021). Improving the survival of *Lactobacillus plantarum* NRRL B-1927 during microencapsulation with ultra-high-pressure-homogenized soymilk as a wall material. *Food Research International*, *139*, 109831. https://doi.org/10.1016/j.foodres.2020.109831.

Chan, L. W., Lee, H. Y., & Heng, P. W. S. (2006). Mechanisms of external and internal gelation and their impact on the functions of alginate as a coat and delivery system. *Carbohydrate Polymers*, *63*(2), 176–187. https://doi.org/10.1016/j.carbpol.2005.07.033.

Chang, C., Stone, A. K., Nickerson, M. T., Melton, L., Shahidi, F., & Varelis, P. (2019). Microencapsulated food ingredients. In L. Melton, F. Shahidi, & P. Varelis (Eds.), *Encyclopedia of food chemistry* (pp. 446–450). Academic Press. https://doi.org/10.1016/B978-0-08-100596-5.21775-7.

Chau, C. F., Wu, S. H., & Yen, G. C. (2007). The development of regulations for food nanotechnology. *Trends in Food Science and Technology*, *18*(5), 269–280. https://doi.org/10.1016/j.tifs.2007.01.007.

Chen, L. (2009). Protein micro/nanoparticles for controlled nutraceutical delivery in functional foods. In *Designing functional foods: Measuring and controlling food structure breakdown and nutrient absorption* (pp. 572–600). Elsevier Ltd. https://doi.org/10.1533/9781845696603.3.572.

Coimbra, P. P. S., de Cardoso, F. S. N., & de Gonçalves, É. C. B. A. (2020). Spray-drying wall materials: Relationship with bioactive compounds. *Critical Reviews in Food Science and Nutrition*, 1–18. https://doi.org/10.1080/10408398.2020.1786354.

Collins, J. K., Thornton, G., & Sullivan, G. O. (1998). Selection of probiotic strains for human applications. *International Dairy Journal*, *8*(5–6), 487–490. https://doi.org/10.1016/S0958-6946(98)00073-9.

Comunian, T. A., Chaves, I. E., Thomazini, M., Moraes, I. C. F., Ferro-Furtado, R., de Castro, I. A., & Favaro-Trindade, C. S. (2017). Development of functional yogurt containing free and encapsulated echium oil, phytosterol and sinapic acid. *Food Chemistry*, *237*, 948–956. https://doi.org/10.1016/j.foodchem.2017.06.071.

Comunian, T. A., & Favaro-Trindade, C. S. (2016). Microencapsulation using biopolymers as an alternative to produce food enhanced with phytosterols and omega-3 fatty acids: A review. *Food Hydrocolloids*, *61*, 442–457. https://doi.org/10.1016/j.foodhyd.2016.06.003.

Consoli, L., et al. (2016). Gallic acid microparticles produced by spray chilling technique: Production and characterization. *LWT - Food Science and Technology, 65*, 79–87.

Consoli, L., Dias, R. A. O., Rabelo, R. S., Furtado, G. F., Sussulini, A., Cunha, R. L., & Hubinger, M. D. (2018). Sodium caseinate-corn starch hydrolysates conjugates obtained through the Maillard reaction as stabilizing agents in resveratrol-loaded emulsions. *Food Hydrocolloids, 84*, 458–472. https://doi.org/10.1016/j.foodhyd.2018.06.017.

da Silva Carvalho, A. G., da Costa Machado, M. T., de Freitas Queiroz Barros, H. D., Cazarin, C. B. B., Maróstica Junior, M. R., & Hubinger, M. D. (2019). Anthocyanins from Jussara (Euterpe edulis Martius) extract carried by calcium alginate beads pre-prepared using ionic gelation. *Powder Technology, 345*, 283–291. https://doi.org/10.1016/j.powtec.2019.01.016.

da Silva, S. C., Fernandes, I. P., Barros, L., Fernandes, Â., José Alves, M., Calhelha, R. C., Pereira, C., Barreira, J. C. M., Manrique, Y., Colla, E., Ferreira, I. C. F. R., & Filomena Barreiro, M. (2019). Spray-dried Spirulina platensis as an effective ingredient to improve yogurt formulations: Testing different encapsulating solutions. *Journal of Functional Foods, 60*, 103427. https://doi.org/10.1016/j.jff.2019.103427.

da Silva, M. N., Tagliapietra, B. L., & dos Richards, N. S. P. S. (2021). Encapsulation, storage viability, and consumer acceptance of probiotic butter. *LWT, 139*, 110536. https://doi.org/10.1016/j.lwt.2020.110536.

Damodaran, S., Parkin, K. L., & Fennema, O. R. (2007). *Fennema's food chemistry* (4th ed.). Boca Raton: CRC Press.

Davarcı, F., Turan, D., Ozcelik, B., & Poncelet, D. (2017). The influence of solution viscosities and surface tension on calcium-alginate microbead formation using dripping technique. *Food Hydrocolloids, 62*, 119–127. https://doi.org/10.1016/j.foodhyd.2016.06.029.

de Moura, S. C. S. R., Berling, C. L., Germer, S. P. M., Alvim, I. D., & Hubinger, M. D. (2018). Encapsulating anthocyanins from *Hibiscus sabdariffa* L. calyces by ionic gelation: Pigment stability during storage of microparticles. *Food Chemistry, 241*, 317–327. https://doi.org/10.1016/j.foodchem.2017.08.095.

de Moura, S. C. S. R., Schettini, G. N., Garcia, A. O., Gallina, D. A., Alvim, I. D., & Hubinger, M. D. (2019). Stability of hibiscus extract encapsulated by ionic gelation incorporated in yogurt. *Food and Bioprocess Technology, 12*(9), 1500–1515. https://doi.org/10.1007/s11947-019-02308-9.

de Souza Simões, L., Madalena, D. A., Pinheiro, A. C., Teixeira, J. A., Vicente, A. A., & Ramos, Ó. L. (2017). Micro- and nano bio-based delivery systems for food applications: In vitro behavior. *Advances in Colloid and Interface Science, 243*, 23–45. https://doi.org/10.1016/j.cis.2017.02.010.

Desai, K. G. H., & Park, H. J. (2005). Recent developments in microencapsulation of food ingredients. *Drying Technology, 23*(7), 1361–1394. https://doi.org/10.1081/DRT-200063478.

Di Sabatino, M., Albertini, B., Kett, V. L., & Passerini, N. (2012). Spray congealed lipid microparticles with high protein loading: Preparation and solid state characterization. *European Journal of Pharmaceutical Sciences, 46*(5), 346–356.

Dickinson, E. (2003). Hydrocolloids at interfaces and the influence on the properties of dispersed systems. *Food Hydrocolloids, 17*(1), 25–39. https://doi.org/10.1016/S0268-005X(01)00120-5.

Ding, B., Zheng, Q., Pan, M., Chiou, Y., Yan, F., & Zhenshun, L. (2018). Microencapsulation of ammonium bicarbonate by phase separation and using palm stearin/carnauba wax as wall materials. *International Journal of Food Engineering, 14*(7–8), 1–9.

Dodoo, C. C., Wang, J., Basit, A. W., Stapleton, P., & Gaisford, S. (2017). Targeted delivery of probiotics to enhance gastrointestinal stability and intestinal colonisation. *International Journal of Pharmaceutics, 530*(1–2), 224–229. https://doi.org/10.1016/j.ijpharm.2017.07.068.

Đorđević, V., Balanč, B., Belščak-Cvitanović, A., Lević, S., Trifković, K., Kalušević, A., Kostić, I., Komes, D., Bugarski, B., & Nedović, V. (2015). Trends in encapsulation Technologies for Delivery of food bioactive compounds. *Food Engineering Reviews, 7*(4), 452–490. https://doi.org/10.1007/s12393-014-9106-7.

Estevinho, B. N., Rocha, F., Oprea, A. E., & Grumezescu, A. M. (2017). Chapter 1—A key for the future of the flavors in food industry: Nanoencapsulation and microencapsulation. In A. E. Oprea, & A. M. Grumezescu

(Eds.), *Nanotechnology applications in food: Flavor, stability, nutrition, and safety* (pp. 1–19). Academic Press. https://doi.org/10.1016/B978-0-12-811942-6.00001-7.

Farahmandi, K., Rajab, S., Tabandeh, F., Shahraky, M. K., Maghsoudi, A., & Ashengroph, M. (2021). Efficient spray-drying of *Lactobacillus rhamnosus* PTCC 1637 using Total CFU yield as the decision factor. *Food Bioscience, 40*, 100816. https://doi.org/10.1016/j.fbio.2020.100816.

Feizollahi, E., Hadian, Z., & Honarvar, Z. (2018). Food fortification with omega-3 fatty acids; microencapsulation as an addition method. *Current Nutrition & Food Science, 14*(2), 90–103. https://doi.org/10.2174/1573401313666170728151350.

Feltre, G., Sartori, T., Silva, K. F. C., Dacanal, G. C., Menegalli, F. C., & Hubinger, M. D. (2020). Encapsulation of wheat germ oil in alginate-gelatinized corn starch beads: Physicochemical properties and tocopherols' stability. *Journal of Food Science, 85*(7), 2124–2133. https://doi.org/10.1111/1750-3841.15316.

Foubert, I., Dewettinck, K., Van de Walle, D., Dijkstra, A. J., & Quinn, P. J. (2007). *Physical properties: Structural and physical characteristics. In The lipid handbook with CD-ROM* (3rd ed., pp. 471–534). Boca Raton: CRC Press.

Francisco, C. R. L., Heleno, S. A., Fernandes, I. P. M., Barreira, J. C. M., Calhelha, R. C., Barros, L., Gonçalves, O. H., Ferreira, I. C. F. R., & Barreiro, M. F. (2018). Functionalization of yogurts with Agaricus bisporus extracts encapsulated in spray-dried maltodextrin crosslinked with citric acid. *Food Chemistry, 245*, 845–853. https://doi.org/10.1016/j.foodchem.2017.11.098.

Funami, T., Fang, Y., Noda, S., Ishihara, S., Nakauma, M., Draget, K. I., Nishinari, K., & Phillips, G. O. (2009). Rheological properties of sodium alginate in an aqueous system during gelation in relation to supermolecular structures and $Ca^{2+}$ binding. *Food Hydrocolloids, 23*(7), 1746–1755. https://doi.org/10.1016/j.foodhyd.2009.02.014.

Furuta, T., & Neoh, T. L. (2021). Microencapsulation of food bioactive components by spray drying: A review. *Drying Technology*, 1–32. https://doi.org/10.1080/07373937.2020.1862181.

Gharsallaoui, A., Roudaut, G., Chambin, O., Voilley, A., & Saurel, R. (2007). Applications of spray-drying in microencapsulation of food ingredients: An overview. *Food Research International, 40*(9), 1107–1121. https://doi.org/10.1016/j.foodres.2007.07.004.

Gottschalk, P., Brodesser, B., Poncelet, D., Jaeger, H., Rennhofer, H., & Cole, S. (2018). Formation of essential oil containing microparticles comprising a hydrogenated vegetable oil matrix and characterisation thereof. *Journal of Microencapsulation, 35*(6), 513–521. https://doi.org/10.1080/02652048.2018.1515998.

Gouin, S. (2004). Microencapsulation: Industrial appraisal of existing technologies and trends. *Trends in Food Science and Technology, 15*(7–8), 330–347. https://doi.org/10.1016/j.tifs.2003.10.005.

Gouveia, L., & Empis, J. (2003). Relative stabilities of microalgal carotenoids in microalgal extracts, biomass and fish feed: Effect of storage conditions. *Innovative Food Science and Emerging Technologies, 4*(2), 227–233. https://doi.org/10.1016/S1466-8564(03)00002-X.

Grant, G. T., Morris, E. R., Rees, D. A., Smith, P. J. C., & Thom, D. (1973). Biological interactions between polysaccharides and divalent cations: The egg-box model. *FEBS Letters, 32*(1), 195–198. https://doi.org/10.1016/0014-5793(73)80770-7.

Gris, C. C. T., Frota, E. G., Guarienti, C., Vargas, B. K., Gutkoski, J. P., Biduski, B., & Bertolin, T. E. (2021). In vitro digestibility and stability of encapsulated yerba mate extract and its impact on yogurt properties. *Journal of Food Measurement and Characterization, 15*(2), 2000–2009.

Hasler, C. M. (2002). Functional foods: Benefits, concerns and challenges—A position paper from the American Council on Science and Health. *Journal of Nutrition, 132*(12), 3772–3781. https://doi.org/10.1093/jn/132.12.3772.

Helgerud, T., Olav, G., Fjæreide, T., Andersen, P. O., & Larsen, C. K. (2010). Alginates. In A. Imeson (Ed.), *Food stabilisers, thickeners and gelling agents* (pp. 50–72). Blackwell Publishing Ltd. https://doi.org/10.1002/9781444314724.ch4.

Holkem, A. T., Raddatz, G. C., Barin, J. S., Moraes Flores, É. M., Muller, E. I., Codevilla, C. F., Jacob-Lopes, E., Ferreira Grosso, C. R., & de Menezes, C. R. (2017). Production of microcapsules containing Bifidobacterium BB-12 by emulsification/internal gelation. *LWT - Food Science and Technology, 76*, 216–221. https://doi.org/10.1016/j.lwt.2016.07.013.

Hosseini, H., & Jafari, S. M. (2020). Introducing nano/microencapsulated bioactive ingredients for extending the shelf-life of food products. *Advances in Colloid and Interface Science, 282*, 102210. https://doi.org/10.1016/j.cis.2020.102210.

Hu, Q., Li, X., Chen, F., Wan, R., Yu, C.-W., Li, J., McClements, D. J., & Deng, Z. (2020). Microencapsulation of an essential oil (cinnamon oil) by spray drying: Effects of wall materials and storage conditions on microcapsule properties. *Journal of Food Processing and Preservation, 44*(11), e14805. https://doi.org/10.1111/jfpp.14805.

I Ré, M. (1998). Microencapsulation by spray drying. *Null, 16*(6), 1195–1236. https://doi.org/10.1080/07373939808917460.

Jafari, S. M., Assadpoor, E., He, Y., & Bhandari, B. (2008). Encapsulation efficiency of food flavours and oils during spray drying. *Drying Technology, 26*(7), 816–835. https://doi.org/10.1080/07373930802135972.

Jaskulski, M., Kharaghani, A., & Tsotsas, E. (2017). Encapsulation methods: Spray drying, spray chilling and spray cooling. In *Thermal and nonthermal encapsulation methods* (pp. 67–114). CRC Press. https://doi.org/10.1201/9781315267883.

Kim, J. U., Kim, B., Shahbaz, H. M., Lee, S. H., Park, D., & Park, J. (2017). Encapsulation of probiotic *Lactobacillus acidophilus* by ionic gelation with electrostatic extrusion for enhancement of survival under simulated gastric conditions and during refrigerated storage. *International Journal of Food Science and Technology, 52*(2), 519–530. https://doi.org/10.1111/ijfs.13308.

Kurozawa, L. E., & Hubinger, M. D. (2017). Hydrophilic food compounds encapsulation by ionic gelation. *Current Opinion in Food Science, 15*, 50–55. https://doi.org/10.1016/j.cofs.2017.06.004.

Kwak, H. S. (2014). Nano- and microencapsulation for foods. In *Vol. 9781118292334. Nano- and microencapsulation for foods* (pp. 1–409). Wiley Blackwell. https://doi.org/10.1002/9781118292327.

Labuschagne, P. (2018). Impact of wall material physicochemical characteristics on the stability of encapsulated phytochemicals: A review. *Food Research International, 107*, 227–247. https://doi.org/10.1016/j.foodres.2018.02.026.

Lee, Y. Y., Tang, T. K., Phuah, E. T., Alitheen, N. B. M., Tan, C. P., & Lai, O. M. (2017). New functionalities of Maillard reaction products as emulsifiers and encapsulating agents, and the processing parameters: A brief review. *Journal of the Science of Food and Agriculture, 97*(5), 1379–1385. https://doi.org/10.1002/jsfa.8124.

Linke, A., Weiss, J., & Kohlus, R. (2021). Impact of the oil load on the oxidation of microencapsulated oil powders. *Food Chemistry, 341*, 128153. https://doi.org/10.1016/j.foodchem.2020.128153.

Liu, H., Cui, S. W., Chen, M., Li, Y., Liang, R., Xu, F., & Zhong, F. (2019). Protective approaches and mechanisms of microencapsulation to the survival of probiotic bacteria during processing, storage and gastrointestinal digestion: A review. *Critical Reviews in Food Science and Nutrition, 59*(17), 2863–2878. https://doi.org/10.1080/10408398.2017.1377684.

Liu, H., Gong, J., Chabot, D., Miller, S. S., Cui, S. W., Ma, J., Zhong, F., & Wang, Q. (2015). Protection of heat-sensitive probiotic bacteria during spray-drying bysodium caseinate stabilized fat particles. *Food Hydrocolloids, 51*, 459–467. https://doi.org/10.1016/j.foodhyd.2015.05.015.

Loksuwan, J. (2007). Characteristics of microencapsulated β-carotene formed by spray drying with modified tapioca starch, native tapioca starch and maltodextrin. *Food Hydrocolloids, 21*(5–6), 928–935. https://doi.org/10.1016/j.foodhyd.2006.10.011.

Lopes, L. A. A., de Siqueira Ferraz Carvalho, R., Stela Santos Magalhães, N., Suely Madruga, M., Julia Alves Aguiar Athayde, A., Araújo Portela, I., Eduardo Barão, C., Colombo Pimentel, T., Magnani, M., & Christina Montenegro Stamford, T. (2020). Microencapsulation of *Lactobacillus acidophilus* La-05 and incorporation in vegan milks: Physicochemical characteristics and survival during storage, exposure to stress conditions, and simulated gastrointestinal digestion. *Food Research International*, *135*, 109295. https://doi.org/10.1016/j.foodres.2020.109295.

Lopes, L. A. A., Pimentel, T. C., de Carvalho, R. S. F., Madruga, M. S., de Galvão, M. S., Bezerra, T. K. A., Barão, C. E., Magnani, M., & Stamford, T. C. M. (2021). Spreadable goat Ricotta cheese added with *Lactobacillus acidophilus* La-05: Can microencapsulation improve the probiotic survival and the quality parameters? *Food Chemistry*, *346*, 128769. https://doi.org/10.1016/j.foodchem.2020.128769.

Lupo, B., Maestro, A., Gutiérrez, J. M., & González, C. (2015). Characterization of alginate beads with encapsulated cocoa extract toprepare functional food: Comparison of two gelation mechanisms. *Food Hydrocolloids*, *49*, 25–34. https://doi.org/10.1016/j.foodhyd.2015.02.023.

Marson, G. V., Saturno, R. P., Comunian, T. A., Consoli, L., da Machado, M. T. C., & Hubinger, M. D. (2020). Maillard conjugates from spent brewer's yeast by-product as an innovative encapsulating material. *Food Research International*, *136*, 109365. https://doi.org/10.1016/j.foodres.2020.109365.

Martin, M. J., Lara-Villoslada, F., Ruiz, M. A., & Morales, M. E. (2013). Effect of unmodified starch on viability of alginate-encapsulated *Lactobacillus fermentum* CECT 5716. *LWT - Food Science and Technology*, *53*(2), 480–486. https://doi.org/10.1016/j.lwt.2013.03.019.

Martirosyan, D., Kanya, H., & Nadalet, C. (2021). Can functional foods reduce the risk of disease? Advancement of functional food definition and steps to create functional food products. *Functional Foods in Health and Disease*, *11*(5), 213–221. https://doi.org/10.31989/ffhd.v11i5.788.

Maschke, A., Becker, C., Eyrich, D., Kiermaier, J., Blunk, T., & Gopferich, A. (2007). Development of a spray congealing process for the preparation of insulin-loaded lipid microparticles and characterization thereof. *European Journal of Pharmaceutics and Biopharmaceutics*, *65*(2), 175–187.

Mcclements, D. J., Decker, E. A., Park, Y., & Weiss, J. (2009). Structural design principles for delivery of bioactive components in nutraceuticals and functional foods. *Critical Reviews in Food Science and Nutrition*, *49*(6), 577–606. https://doi.org/10.1080/10408390902841529.

Medeiros, T. T. B., Aderbal, M. A. S., da Silva, A. L., Bezerra, L. R., Agostini, D. L. S., de Oliveira, D. L. V., & Mazzetto, S. E. (2019). Carnauba wax as a wall material for urea microencapsulation. *Journal of the Science of Food and Agriculture*, *99*(3), 1078–1087.

Mehran, M., Masoum, S., & Memarzadeh, M. (2020). Microencapsulation of *Mentha spicata* essential oil by spray drying: Optimization, characterization, release kinetics of essential oil from microcapsules in food models. *Industrial Crops and Products*, *154*, 112694. https://doi.org/10.1016/j.indcrop.2020.112694.

Menin, A., Zanoni, F., Vakarelova, M., Chignola, R., Donà, G., Rizzi, C., Mainente, F., & Zoccatelli, G. (2018). Effects of microencapsulation by ionic gelation on the oxidative stability of flaxseed oil. *Food Chemistry*, *269*, 293–299. https://doi.org/10.1016/j.foodchem.2018.06.144.

Mohammed, N. K., Tan, C. P., Manap, Y. A., Alhelli, A. M., & Hussin, A. S. M. (2017). Process conditions of spray drying microencapsulation of *Nigella sativa* oil. *Powder Technology*, *315*, 1–14. https://doi.org/10.1016/j.powtec.2017.03.045.

Müller, R. H., Radtke, M., & Wissing, S. A. (2002). Nanostructured lipid matrices for improved microencapsulation of drugs. *International Journal of Pharmaceutics*, *242*(1–2), 121–128. https://doi.org/10.1016/S0378-5173(02)00180-1.

Naranjo-Durán, A. M., Quintero-Quiroz, J., Rojas-Camargo, J., & Ciro-Gómez, G. L. (2021). Modified-release of encapsulated bioactive compounds from annatto seeds produced by optimized ionic gelation techniques. *Scientific Reports*, *11*(1), 1317. https://doi.org/10.1038/s41598-020-80119-1.

Nichols, D. S., & Sanderson, K. (2003). The nomenclature, structure, and properties of food lipids. In Z. Z. E. Sikorski, & A. Kolakowska (Eds.), *Chemical and functional properties of food lipids* (1st, pp. 29–59). Boca Raton: CRC Press.

Nunes, I. L., & Mercadante, A. Z. (2007). Encapsulation of lycopene using spray-drying and molecular inclusion processes. *Brazilian Archives of Biology and Technology, 50*(5), 893–900. https://doi.org/10.1590/S1516-89132007000500018.

Okuro, P. K., Eustáquio de Matos, F., & Favaro-Trindade, C. S. (2013a). Technological challenges for spray chilling encapsulation of functional food ingredients. *Food Technology and Biotechnology, 51*(2), 171–182. http://www.ftb.com.hr/51/FTB_51-2_171-182.pdf.

Okuro, P. K., et al. (2013b). Co-encapsulation of Lactobacillus acidophilus with inulin or polydextrose in solid lipid microparticles provides protection and improves stability. *Food Research International, 53*(1), 96–103.

Oriani, V. B., et al. (2016). Solid lipid microparticles produced by spray chilling technique to deliver ginger oleoresin: Structure and compound retention. *Food Research International, 80*, 41–49.

Oxley, J. D. (2012). Spray cooling and spray chilling for food ingredient and nutraceutical encapsulation. In *Encapsulation technologies and delivery systems for food ingredients and nutraceuticals* (pp. 110–130). Elsevier Ltd. https://doi.org/10.1016/B978-0-85709-124-6.50005-6.

Ozkan, G., Franco, P., De Marco, I., Xiao, J., & Capanoglu, E. (2019). A review of microencapsulation methods for food antioxidants: Principles, advantages, drawbacks and applications. *Food Chemistry, 272*, 494–506. https://doi.org/10.1016/j.foodchem.2018.07.205.

Pacheco, C., González, E., Robert, P., & Parada, J. (2018). Retention and pre-colon bioaccessibility of oleuropein in starchy food matrices, and the effect of microencapsulation by using inulin. *Journal of Functional Foods, 41*, 112–117. https://doi.org/10.1016/j.jff.2017.12.037.

Pankasemsuk, T., Apichartsrangkoon, A., Worametrachanon, S., & Techarang, J. (2016). Encapsulation of *Lactobacillus casei* 01 by alginate along with hi-maize starch for exposure to a simulated gut model. *Food Bioscience, 16*, 32–36. https://doi.org/10.1016/j.fbio.2016.07.001.

Paques, J. P., Sagis, L. M. C., van Rijn, C. J. M., & van der Linden, E. (2014). Nanospheres of alginate prepared through w/o emulsification and internal gelation with nanoparticles of CaCO3. *Food Hydrocolloids, 40*, 182–188. https://doi.org/10.1016/j.foodhyd.2014.02.024.

Passerini, N., Gavini, E., Albertini, B., et al. (2009). Evaluation of solid lipid microparticles produced by spray congealing for topical application of econazole nitrate. *Journal of Pharmacy and Pharmacology, 61*, 559–567.

Passerini, N., Perissutti, B., Albertini, B., Voinovich, D., Moneghini, M., & Rodriguez, L. (2003). Controlled release of verapamil hydrochloride from waxy microparticles prepared by spray congealing. *Journal of Controlled Release, 88*(2), 263–275.

Paucar, O. C., Tulini, F. L., Thomazini, M., Balieiro, J. C. C., Pallone, E. M. J. A., & Favaro-Trindade, C. S. (2016). Production by spray chilling and characterization of solid lipid microparticles loaded with vitamin D3. *Food and Bioproducts Processing, 100*, 344–350.

Pérez-Palacios, T., Ruiz-Carrascal, J., Jiménez-Martín, E., Solomando, J. C., & Antequera, T. (2018). Improving the lipid profile of ready-to-cook meat products by addition of omega-3 microcapsules: Effect on oxidation and sensory analysis. *Journal of the Science of Food and Agriculture, 98*(14), 5302–5312. https://doi.org/10.1002/jsfa.9069.

Poncelet, D., Lencki, R., Beaulieu, C., Halle, J. P., Neufeld, R. J., & Fournier, A. (1992). Production of alginate beads by emulsification/internal gelation. I. Methodology. *Applied Microbiology and Biotechnology, 38*(1), 39–45. https://doi.org/10.1007/BF00169416.

Porrini, M., & Riso, P. (2008). Factors influencing the bioavailability of antioxidants in foods: A critical appraisal. *Nutrition, Metabolism, and Cardiovascular Diseases, 18*(10), 647–650. https://doi.org/10.1016/j.numecd.2008.08.004.

Procopio, F. R., et al. (2018). Solid lipid microparticles loaded with cinnamon oleoresin: Characterization, stability and antimicrobial activity. *Food Research International, 113*(March), 351–361.

Puttarat, N., Thangrongthong, S., Kasemwong, K., Kerdsup, P., & Taweechotipatr, M. (2021). Spray-drying microencapsulation using whey protein isolate and nano-crystalline starch for enhancing the survivability and stability of *Lactobacillus reuteri* TF-7. *Food Science and Biotechnology*, *30*(2), 245–256. https://doi.org/10.1007/s10068-020-00870-z.

Ramos, P. E., Cerqueira, M. A., Teixeira, J. A., & Vicente, A. A. (2018). Physiological protection of probiotic microcapsules by coatings. *Critical Reviews in Food Science and Nutrition*, *58*(11), 1864–1877. https://doi.org/10.1080/10408398.2017.1289148.

Ribeiro, H. S., & Cruz, R. C. D. (2005). Highly concentrated carotenoid-containing emulsions. *Engineering in Life Sciences*, *5*(1), 84–88. https://doi.org/10.1002/elsc.200403367.

Ribeiro, A. P. B., Masuchi, M. H., Miyasaki, E. K., Domingues, M. A. F., Stroppa, V. L. Z., de Oliveira, G. M., & Kieckbusch, T. G. (2015). Crystallization modifiers in lipid systems. *Journal of Food Science and Technology*, *52*(7), 3925–3946. https://doi.org/10.1007/s13197-014-1587-0.

Ribeiro, A., Ruphuy, G., Lopes, J. C., Dias, M. M., Barros, L., Barreiro, F., & Ferreira, I. C. F. R. (2015). Spray-drying microencapsulation of synergistic antioxidant mushroom extracts and their use as functional food ingredients. *Food Chemistry*, *188*, 612–618. https://doi.org/10.1016/j.foodchem.2015.05.061.

Ribeiro, J. S., & Veloso, C. M. (2021). Microencapsulation of natural dyes with biopolymers for application in food: A review. *Food Hydrocolloids*, *112*, 106374.

Ribeiro, M. D. M. M., Arellano, D. B., & Grosso, C. R. F. (2012). The effect of adding oleic acid in the production of stearic acid lipid microparticles with a hydrophilic core by a spray-cooling process. *Food Research International*, *47*(1), 38–44.

Robert, P., Carlsson, R. M., Romero, N., & Masson, L. (2003). Stability of spray-dried encapsulated carotenoid pigments from Rosa Mosqueta (*Rosa rubiginosa*) oleoresin. *JAOCS, Journal of the American Oil Chemists' Society*, *80*(11), 1115–1120. https://doi.org/10.1007/s11746-003-0828-4.

Rocha, G. A., Fávaro-Trindade, C. S., & Grosso, C. R. F. (2012). Microencapsulation of lycopene by spray drying: Characterization, stability and application of microcapsules. *Food and Bioproducts Processing*, *90*(1), 37–42. https://doi.org/10.1016/j.fbp.2011.01.001.

Rodrigues, F. J., Cedran, M. F., Bicas, J. L., & Sato, H. H. (2020). Encapsulated probiotic cells: Relevant techniques, natural sources as encapsulating materials and food applications—A narrative review. *Food Research International*, *137*, 109682. https://doi.org/10.1016/j.foodres.2020.109682.

Rodrigues, F. J., Omura, M. H., Cedran, M. F., Dekker, R. F. H., Barbosa-Dekker, A. M., & Garcia, S. (2017). Effect of natural polymers on the survival of *Lactobacillus casei* encapsulated in alginate microspheres. *Journal of Microencapsulation*, *34*(5), 431–439. https://doi.org/10.1080/02652048.2017.1343872.

Rostamabadi, H., Falsafi, S. R., Boostani, S., Katouzian, I., Rezaei, A., Assadpour, E., & Jafari, S. M. (2021). *Design and formulation of nano/micro-encapsulated natural bioactive compounds for food applications* (pp. 1–41). Elsevier BV. https://doi.org/10.1016/b978-0-12-815726-8.00001-5.

Rucker, R., Zempleni, J., Suttie, J. W., & McCormick, D. B. (2007). *Handbook of vitamins* (4th). Boca Raton: CRC Press.

Saarela, M., Mogensen, G., Fondén, R., Mättö, J., & Mattila-Sandholm, T. (2000). Probiotic bacteria: Safety, functional and technological properties. *Journal of Biotechnology*, *84*(3), 197–215. https://doi.org/10.1016/S0168-1656(00)00375-8.

Sánchez, M. T., Ruiz, M. A., Lasserrot, A., Hormigo, M., & Morales, M. E. (2017). An improved ionic gelation method to encapsulate lactobacillus spp. bacteria: Protection, survival and stability study. *Food Hydrocolloids*, *69*, 67–75. https://doi.org/10.1016/j.foodhyd.2017.01.019.

Sartori, T., & Menegalli, F. C. (2016). Development and characterization of unripe banana starch films incorporated with solid lipid microparticles containing ascorbic acid. *Food Hydrocolloids*, *55*, 210–219. https://doi.org/10.1016/j.foodhyd.2015.11.018.

Sato, K., & Ueno, S. (2011). Crystallization, transformation and microstructures of polymorphic fats in colloidal dispersion states. *Current Opinion in Colloid and Interface Science*, *16*(5), 384–390. https://doi.org/10.1016/j.cocis.2011.06.004.

# References

Sauvant, P., Cansell, M., Hadj Sassi, A., & Atgié, C. (2012). Vitamin A enrichment: Caution with encapsulation strategies used for food applications. *Functional Foods and Nutraceuticals, 46*(2), 469–479. https://doi.org/10.1016/j.foodres.2011.09.025.

Savolainen, M., Cynthia, K., Håkan, G., Carina, D., & Anne, M. J. (2002). Evaluation of controlled-release polar lipid microparticles. *International Journal of Pharmaceutics, 244*(1–2), 151–161.

Scalbert, A., Manach, C., Morand, C., Rémésy, C., & Jiménez, L. (2005). Dietary polyphenols and the prevention of diseases. *Critical Reviews in Food Science and Nutrition, 45*(4), 287–306. https://doi.org/10.1080/1040869059096.

Schrooyen, P. M. M., van der Meer, R., & Kruif, C. G. D. (2001). Microencapsulation: Its application in nutrition. *Proceedings of the Nutrition Society, 60*(4), 475–479. https://doi.org/10.1079/pns2001112.

Selvamuthukumaran, M. (2019). *Handbook on spray drying applications for food industries*. Boca Raton: CRC Press.

Shamaei, S., Seiiedlou, S. S., Aghbashlo, M., Tsotsas, E., & Kharaghani, A. (2017). Microencapsulation of walnut oil by spray drying: Effects of wall material and drying conditions on physicochemical properties of microcapsules. *Innovative Food Science and Emerging Technologies, 39*, 101–112. https://doi.org/10.1016/j.ifset.2016.11.011.

Shen, Q., & Quek, S. Y. (2014). Microencapsulation of astaxanthin with blends of milk protein and fiber by spray drying. *Journal of Food Engineering, 123*, 165–171. https://doi.org/10.1016/j.jfoodeng.2013.09.002.

Shishir, M. R. I., Xie, L., Sun, C., Zheng, X., & Chen, W. (2018). Advances in micro and nano-encapsulation of bioactive compounds using biopolymer and lipid-based transporters. *Trends in Food Science and Technology, 78*, 34–60. https://doi.org/10.1016/j.tifs.2018.05.018.

Shu, B., Yu, W., Zhao, Y., & Liu, X. (2006). Study on microencapsulation of lycopene by spray-drying. *Journal of Food Engineering, 76*(4), 664–669. https://doi.org/10.1016/j.jfoodeng.2005.05.062.

Silva, K. C. G., Bourbon, A. I., Pastrana, L., & Sato, A. C. K. (2020). Emulsion-filled hydrogels for food applications: Influence of pH on emulsion stability and a coating on microgel protection. *Food & Function, 11*(9), 8331–8341. https://doi.org/10.1039/d0fo01198c.

Silva, K. C. G., Cezarino, E. C., Michelon, M., & Sato, A. C. K. (2018). Symbiotic microencapsulation to enhance *Lactobacillus acidophilus* survival. *LWT - Food Science and Technology, 89*, 503–509. https://doi.org/10.1016/j.lwt.2017.11.026.

Silva, M. P., Thomazini, M., Holkem, A. T., Pinho, L. S., Genovese, M. I., & Fávaro-Trindade, C. S. (2019). Production and characterization of solid lipid microparticles loaded with guaraná (*Paullinia cupana*) seed extract. *Food Research International, 123*, 144–152. https://doi.org/10.1016/j.foodres.2019.04.055.

Silva, M. P., Tulini, F. L., Marinho, J. F. U., Mazzocato, M. C., De Martinis, E. C. P., Luccas, V., & Favaro-Trindade, C. S. (2017). Semisweet chocolate as a vehicle for the probiotics *Lactobacillus acidophilus* LA3 and *Bifidobacterium animalis* subsp. lactis BLC1: Evaluation of chocolate stability and probiotic survival under in vitro simulated gastrointestinal conditions. *LWT - Food Science and Technology, 75*, 640–647. https://doi.org/10.1016/j.lwt.2016.10.025.

Silva, M. P., Tulini, F. L., Matos-Jr, F. E., Oliveira, M. G., Thomazini, M., & Fávaro-Trindade, C. S. (2018). Application of spray chilling and electrostatic interaction to produce lipid microparticles loaded with probiotics as an alternative to improve resistance under stress conditions. *Food Hydrocolloids, 83*, 109–117. https://doi.org/10.1016/j.foodhyd.2018.05.001.

Smrdel, P., Bogataj, M., Zega, A., Planinšek, O., & Mrhar, A. (2008). Shape optimization and characterization of polysaccharide beads prepared by ionotropic gelation. *Journal of Microencapsulation, 25*(2), 90–105. https://doi.org/10.1080/02652040701776109.

Sobel, R., Versic, R., Gaonkar, A. G., Gaonkar, A. G., Vasisht, N., Khare, A. R., & Sobel, R. (2014). Chapter 1—Introduction to microencapsulation and controlled delivery in foods. In A. G. Gaonkar, N. Vasisht, A. R. Khare, & R. Sobel (Eds.), *Microencapsulation in the food industry* (pp. 3–12). Academic Press. https://doi.org/10.1016/B978-0-12-404568-2.00001-7.

Solomando, J. C., Antequera, T., & Perez-Palacios, T. (2020). Evaluating the use of fish oil microcapsules as omega-3 vehicle in cooked and dry-cured sausages as affected by their processing, storage and cooking. *Meat Science, 162*, 108031. https://doi.org/10.1016/j.meatsci.2019.108031.

Stangierski, J., Rezler, R., Kawecki, K., & Peplińska, B. (2020). Effect of microencapsulated fish oil powder on selected quality characteristics of chicken sausages. *Journal of the Science of Food and Agriculture, 100*(5), 2043–2051. https://doi.org/10.1002/jsfa.10226.

Stringham, J. M., & Hammond, B. R. (2005). Dietary lutein and zeaxanthin: Possible effects on visual function. *Nutrition Reviews, 63*(2), 59–64. https://doi.org/10.1301/nr.2004.feb.59-64.

Sun-Waterhouse, D., Wadhwa, S. S., & Waterhouse, G. I. N. (2013). Spray-drying microencapsulation of polyphenol bioactives: A comparative study using different natural fibre polymers as encapsulants. *Food and Bioprocess Technology, 6*(9), 2376–2388. https://doi.org/10.1007/s11947-012-0946-y.

Trindade, M. A., & Grosso, C. R. F. (2000). The stability of ascorbic acid microencapsulated in granules of rice starch and in gum arabic. *Journal of Microencapsulation, 17*(2), 169–176. https://doi.org/10.1080/026520400288409.

Urzúa, C., González, E., Dueik, V., Bouchon, P., Giménez, B., & Robert, P. (2017). Olive leaves extract encapsulated by spray-drying in vacuum fried starch–gluten doughs. *Food and Bioproducts Processing, 106*, 171–180. https://doi.org/10.1016/j.fbp.2017.10.001.

Vargas-Ramella, M., Pateiro, M., Barba, F. J., Franco, D., Campagnol, P. C. B., Munekata, P. E. S., Tomasevic, I., Domínguez, R., & Lorenzo, J. M. (2020). Microencapsulation of healthier oils to enhance the physicochemical and nutritional properties of deer pâté. *LWT, 125*, 109223. https://doi.org/10.1016/j.lwt.2020.109223.

Vasisht, N., Gaonkar, A. G., Vasisht, N., Khare, A. R., & Sobel, R. (2014). Chapter 16—Selection of materials for microencapsulation. In A. G. Gaonkar, N. Vasisht, A. R. Khare, & R. Sobel (Eds.), *Microencapsulation in the food industry* (pp. 173–180). Academic Press. https://doi.org/10.1016/B978-0-12-404568-2.00016-9.

Vieira, F. C. R., Maria, H. A. R. A., & Lobão, P. A. L. (2013). Prolonged-release lipid microparticles prepared with decyl oleate and hydrogenated castor oil of ibuprofen. *Journal of Applied Pharmaceutical Science, 3*(10), 8–10.

Wong, P. C. H., Heng, P. W. S., & Chan, L. W. (2015a). Spray congealing as a microencapsulation technique to develop modified-release ibuprofen solid lipid microparticles: The effect of matrix type, polymeric additives and drug–matrix miscibility. *Journal of Microencapsulation, 32*(8), 725–773.

Wong, P. C. H., Heng, P. W. S., & Chan, L. W. (2015b). Determination of solid state characteristics of spray-congealed ibuprofen solid lipid microparticles and their impact on sustaining drug release. *Molecular Pharmaceutics, 12*, 1592–1604.

Xianquan, S., Shi, J., Kakuda, Y., & Yueming, J. (2005). Stability of lycopene during food processing and storage. *Journal of Medicinal Food, 8*(4), 413–422. https://doi.org/10.1089/jmf.2005.8.413.

Xu, W., Huang, L., Jin, W., Ge, P., Shah, B. R., Zhu, D., & Jing, J. (2019). Encapsulation and release behavior of curcumin based on nanoemulsions-filled alginate hydrogel beads. *International Journal of Biological Macromolecules, 134*, 210–215. https://doi.org/10.1016/j.ijbiomac.2019.04.200.

Yan, B., Davachi, S. M., Ravanfar, R., Dadmohammadi, Y., Deisenroth, T. W., Pho, T. V., Odorisio, P. A., Darji, R. H., & Abbaspourrad, A. (2021). Improvement of vitamin C stability in vitamin gummies by encapsulation in casein gel. *Food Hydrocolloids, 113*, 106414. https://doi.org/10.1016/j.foodhyd.2020.106414.

Yeung, T. W., Arroyo-Maya, I. J., McClements, D. J., & Sela, D. A. (2016). Microencapsulation of probiotics in hydrogel particles: Enhancing: *Lactococcus lactis* subsp. cremoris LM0230 viability using calcium alginate beads. *Food & Function, 7*(4), 1797–1804. https://doi.org/10.1039/c5fo00801h.

Yin, Y., & Cadwallader, K. R. (2019). Spray-chilling encapsulation of 2-acetyl-1-pyrroline zinc chloride using hydrophobic materials: Storage stability and flavor application in food. *Food Chemistry, 278*, 738–743. https://doi.org/10.1016/j.foodchem.2018.11.122.

Zhang, R., Zhou, L., Li, J., Oliveira, H., Yang, N., Jin, W., Zhu, Z., Li, S., & He, J. (2020). Microencapsulation of anthocyanins extracted from grape skin by emulsification/internal gelation followed by spray/freeze-drying techniques: Characterization, stability and bioaccessibility. *LWT, 123*, 109097. https://doi.org/10.1016/j.lwt.2020.109097.

Zuidam, N. J., & Nedović, V. A. (2010). *Encapsulation Technologies for Active Food Ingredients and Food Processing* (pp. 1–400). New York: Springer. https://doi.org/10.1007/978-1-4419-1008-0.

# CHAPTER 8

# Strategies for the reduction of salt in food products

Mirian dos Santos[a], Andrea Paola Rodriguez Triviño[a], Julliane Carvalho Barros[a], Adriano G. da Cruz[b], and Marise Aparecida Rodrigues Pollonio[a]

[a]*State University of Campinas (UNICAMP), School of Food Engineering, Cidade Universitária Zeferino Vaz, São Paulo, Brazil,* [b]*Food Department, Federal Institute of Education, Science and Technology of Rio de Janeiro (IFRJ), Rio de Janeiro, Brazil*

## 8.1 Introduction

With the growing concern for public health and well-being, sodium intake has been widely investigated due to its well-established correlation with the increased risk of some noncommunicable chronic diseases, of which hypertension is one of the most relevant. Epidemiological studies show that sodium intake is very high in most countries, far above the value recommended by the World Health Organization (WHO), particularly in industrialized countries. As the main source of sodium comes from sodium chloride, the table salt, an ingredient widely used in almost all categories of processed foods, its reduction faces many challenges that include economic, social, sensory, and technological factors (Steffensen et al., 2018). In addition, the salt added during household and restaurant meals is quite relevant, and it can be the primary source of intake in many countries.

Salt (sodium chloride) performs various physical, microbiological, and sensory properties in different processed foods. When added to food formulations, it increases the ionic strength of the system and it can thereby influence the behavior of many proteins, for instance, the solubility, impacting several of their functional properties. So, reducing the sodium content in meat, bakery, dairy products and the others, such as snacks and dehydrated products resulted from simple partial or total removal of sodium chloride can be a great problem to the food industry. In this scenario, sodium reduction in foods has been conducted in response to government campaigns to improve the nutritional profile of diets in many countries. These actions, in general, are voluntary initiatives considering the Public Health protocols with the participation of governments, industry, and consumers because salt/sodium reduction is a task force.

As the sodium content of processed foods or meals in general, is its main source in countries with many different dietary patterns, it is expected that the average sodium intake varies according to food consumption, and in fact, it happens. Bakery products, meat products, snacks, cured cheeses, snacks, and dehydrated soups are known to have higher sodium contents. The information that consumers have about the role of these foods in their diet and the relevant correlation with healthy dietary concepts is critical to make their choices and adhering to new healthier consumption patterns (Mendoza et al., 2014).

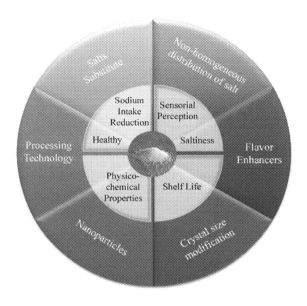

**FIG. 8.1**

Effects of sodium chloride reduction in processed foods and some strategies to reduce it.

*No permission required.*

Sodium reduction in processed products can impact flavor, aroma, texture, and safety even in well-designed strategic reformulations. Therefore, two factors become decisive for this global endeavor in the reduction of sodium in foods: a deep understanding of consumer perception considering the factors that determine its motivation to change their food consumption habits, and the technological knowledge to find an efficient substitution or reduction of sodium chloride, without loss of the quality attributes and safety. The use of innovative ingredients and new processes that allow replacing the functional and sensory properties of sodium chloride aligned to the correct communication to consumers are basic requirements for success in this challenging reformulation. Recently, nonhomogeneous distribution of salt, NaCl crystal modification and nanoparticles emerge as potential strategies to reduce NaCl. Finally, it is fundamental that governmental Public Health agencies in different countries be updated on the global panorama of sodium reduction to provide the necessary support, from a legislative point of view to proposals for new products development, always based on the principles of food safety. All these aspects are summarized in Fig. 8.1.

## 8.2 Salt, sodium, and health

Sodium is an essential nutrient for many vital metabolic functions, such as maintaining an adequate water balance in the body by continuously adjusting fluid and blood levels (Bhana et al., 2018). It acts in controlling blood pressure and in the nerve stimulation supporting muscle contraction and nutrient transport, among other functions. However, its excessive consumption is directly related to the increased risk of several chronic noncommunicable diseases, including hypertension, a major risk factor for the development of cardiovascular and stroke (Neal, 2014), cancer (D'Elia et al., 2014), bone mass loss, kidney disease, metabolic syndrome (Hoffmann & Cubeddu, 2009) and obesity (Ma et al., 2015).

Systematic reviews and meta-analyze studies have found a well-established correlation between high sodium intake and increased blood pressure, one of the main causes of these cardiovascular diseases, placing sodium reduction on the agenda of preventive measures to reduce mortality caused by this pathology (Aljuraiban et al., 2021). Data from the Global Burden of Disease study estimate that in 2017, a range between 1.4 and 5.4 million deaths was correlated to excessive dietary sodium intake (Afshin et al., 2017).

Sodium/potassium ratio is another relevant factor that affects blood pressure control. Potassium intake contributes to lower blood pressure in both hypertensive people and those who are normotensive. In general, in Western diets, potassium intake tends to be low, since its main source, fruits, and vegetables, is insufficiently consumed alongside excessive consumption of processed foods, whose sodium content is generally high. It is known that high sodium levels combined with low potassium intake can promote the contraction of smooth muscle tissue cells, increasing peripheral vascular resistance, and blood pressure, and thereby resulting in hypertension (Büssemaker et al., 2010). This information is very relevant since potassium chloride is the main salt substitute to replace sodium chloride in technological reformulation strategies resulting in increased potassium content. Despite, there is no current guideline for the sodium-to-potassium ratio, Aljuraiban et al. (2021) reported in their review, studies that show how a reduction in the sodium-to-potassium ratio from 3.09 to 1.00 lowered systolic blood pressure by 3.36 mmHg highlighting the WHO recommendations for intakes of both sodium and potassium which suggest a sodium-to-potassium ratio near 1.00, from daily intake of 2000 mg Na.

WHO (2012) recommends that sodium intake should not exceed 2 g per day, coming from salt consumption (sodium chloride), which corresponds to 5.1 g of sodium chloride. Others equivalence units regarding salt/sodium contents establish that 1 mmol of sodium equals 23 mg of sodium, and 1 g of sodium chloride contains 390 mg (17 mmol) of sodium. The Na/NaCl consumption by individuals around the world, as widely reported, and is much higher than the recommended 2 g per day. Powles et al. (2013) reported the global, regional, and national sodium intakes using 24-h urine collections and dietary sodium in adults in 1980 and 2010 by meta-analysis of 245 surveys from 66 countries (1980–2010) registering that the mean global sodium intake was estimated to be 3.95 g/d, with higher values for individuals over the age of 40. Several studies bring relevant information regarding the main factors that influence these data and they can help planning effective strategies to reduce sodium in a particular country (Kaushik et al., 2021; WHO, 2013).

In this way, sodium reduction must necessarily go through the reduction of salt (sodium chloride). Regarding this, many countries adopt different governmental programs through public health agencies, agreements with industry, and consumer education, a great challenge for all players involved. Understanding this consumption pattern is fundamental to establishing appropriate and efficient sodium reduction policies in a given country based on these variables (McKenzie et al., 2018). Thus, the focus on reducing salt consumption should occur through consumer education, cooperation between governments and industry, and regulations to prevent chronic diseases and promote good health.

## 8.3 The role of sodium in food products

Reducing dietary sodium intake is key to improving health. However, sodium reduction can be quite challenging in processed foods, since sodium chloride, the main source of sodium, is the most usual salt

Table 8.1 Main technological implications of salt reduction in processed foods.

| Processed foods | Technological implications of salt reduction | References |
|---|---|---|
| Emulsified and restructured and cured meat products | Reduced water-holding capacity, yield, texture-related properties, microbial stability, and sensory acceptance. | Arakawa and Timasheff (1991), Offer et al. (1989) |
| Fermented meat products | Starter culture overgrowth, reduced the safety throughout fermentation and drying, increased sour taste perception, modified flavor profile, and reduced acceptability of fermented meat products. | Corral et al. (2013), Hu et al. (2020) |
| Bread and related-bakery products | Weaker and faster-developing gluten network; stimulation of yeast growth, with increased fermentation rate and gas production; lighter crust color; increase off-flavor and bitterness of the products. | Antúnez et al. (2018); Avramenko et al. (2018), Belz, Ryan, et al. (2012) |
| Cheese | Overgrowth of starter cultures and secondary cultures; stimulation of spoilage and pathogenic microorganisms' growth; modification of texture-related properties and moisture in cheeses during maturation; decrease in sensory acceptance due to the appearance of bitter flavors and low saltness. | Guinee (2004), Johnson et al. (2009) |
| Fermented vegetables | Influence in the growth and activity of microorganisms responsible for fermentation process; salt contributes to preventing softening in the vegetables. | Fleming et al. (1985), Liang et al. (2020) |

*No permission required.*

applied in foods and has important functions, as highlighted in Table 8.1. Processed foods and those consumed in restaurants are considered the major responsibles for sodium intake in developed countries (Anderson et al., 2010). Therefore, this topic will address the effects of salt coming mainly from sodium chloride on the functional properties of the main groups of processed foods that contribute significantly to sodium intake, such as processed meats, bakery, and dairy products (Ni Mhurchu et al., 2011). Also, the effect of salt in the promotion of flavor and aroma in foods, and its contribution to the microbiological stability and food safety of processed food products will be discussed.

### 8.3.1 Effects on protein functional properties

Proteins play a key role in the structure, stability, and texture of foods due to their water holding capacity, emulsion, and gel formation, influencing some rheological properties. Salt at low concentration exerts a salting-in effect on polyelectrolytes, such as proteins, and stabilizes them through nonspecific electrostatic interactions. At high salt concentrations, the solubility of proteins decreases, and dehydration occurs, and this effect is known as salting-out (Arakawa & Timasheff, 1991). The influence of salt on the functional properties of proteins is quite variable depending on the protein structure and other factors that can exert a synergistic effect with salt, such as pH.

Meat contains approximately 75% water, and water-holding capacity (WHC) is a main functional property that influences technological and sensory attributes, as well as the yield of meat and meat products. The myofibrillar proteins (mainly myosin and actin) are the majority in muscle tissue, their most important functional properties are WHC, emulsification, and gelling capacity that are necessary to produce meat products with suitable texture, appearance, and adequate stability (Amiri et al., 2018). These properties are available only after their solubilization, provided by increasing the ionic strength achieved by adding sodium chloride. The chloride ion binds more efficiently than the sodium ion to the charged groups of the myofibrillar proteins, causing an increase in the negative charge and, consequently, raising the repulsive force between filaments which will result in swelling and protein solubility up to 0.5 M of ionic strength (Offer et al., 1989; Puolanne & Halonen, 2010).

In meat products, several studies have shown that NaCl reduction has a negative effect on physicochemical and sensory properties. In emulsified meat products, the reduction of NaCl decreased the emulsion stability, resulting of increasing in water loss and poor emulsification, promoting fat separation throughout the heat treatment (Horita et al., 2014). Also, the textural properties are greatly affected by the salt reduction, since, without the proper solubilization of myofibrillar proteins, there will not be the formation of a strong protein network capable of retaining water and fat, thus, the texture attributes, such as hardness, chewiness, and cohesiveness will be compromised (Rodrigues et al., 2020; Tobin et al., 2012). Restructured meat products, likewise, were negatively affected by NaCl reduction. Tamm et al. (2016) showed that reduced-salt cooked ham presents high cooking loss and the structure of the product was characterized by a high number of holes and increased cutting force which could affect the slicing.

In dry-salted meat products, salt controls the water loss and microbial growth that is important at the beginning of the process, where water activity is high. The rate of drying was investigated by Vidal et al. (2019), when the authors shown that the combination of $NaCl+CaCl_2$ (1:1 w/w) caused higher dehydration on the surface of flesh jerked beef, creating a firm and dry barrier that induced a higher moisture value in this treatment. Vidal et al. (2021) reviewed the reduction of sodium in restructured meat products, and according to their findings, the replacement of NaCl by salt substitutes, in general, showed similar effects to NaCl in controlling the growth of deteriorating microorganisms such as lactic acid bacteria and pathogens such as *Listeria* spp., *S. aureus*, and *C. perfringens*. Proteolysis can also be affected by the replacement of NaCl by other chloride salts, however, this process is quite variable and depends on the strategy addressed, and similar behavior can be extended to lipid oxidation. In fermented sausages, salt reduction influences the growth of starter bacteria, promoting a decrease in pH value and an increase of acid flavor perception after the sensory analysis, among some other effects (Hu et al., 2020).

In cheese, NaCl is an important component; it is added at a level ranging from ~0.7% to 6% (w/w), depending on the cheese type. The incorporation of salt affects the cheese quality attributes by the summarized mechanisms: (1) promoting syneresis and control final moisture of the products, (2) controlling the metabolism and survival of the starter bacteria, mainly in ripened cheeses, (3) influencing the types of secondary microorganisms, the nonstarters lactic bacteria, that may grow along the ripening period and contribute to the flavor profile, (4) regulating the enzyme activity in the final product, (5) controlling the texture of the final products as the sodium replaces calcium in the cheese microstructure, and (6) performing a component of the expected taste of the cheese (Bansal & Mishra, 2020; Guinee, 2004; Johnson et al., 2009).

Sihufe et al. (2018) evaluated the reduction of salt (0.46, 0.91, and 1.28 g/100 g) in Tybo cheese, finding a significant increase in moisture content, maturation index, and αS1-I-casein degradation, probably due to proteolysis stimulation caused by salt reduction. Rheological parameters also significantly decreased with salt reduction, which could be associated to the plasticizer effect caused by higher moisture observed in these treatments. From a sensorial view, the reduction of 30% of salt did not affect the quality of Tybo cheese. Baptista et al. (2017) studying the NaCl reduction in Prato cheese, reported that proteolysis and the intensity of known bitter-tasting peptides were not affected by salt reduction at a level of 25%, but the treatment with 50% reduction showed a decrease in hardness and it was less sensory accepted.

Hussain et al. (2012) evaluated the effect of salt on the functionality of whey proteins, at low ionic strength, the elastic modulus increased, and a strongest and stiff gel was obtained. Whereas high salt concentrations lead to a weaker gel. The increase in salt concentration also promoted a decrease in protein hydrophobicity, which may be linked to a higher thermal resistance to the gelation of whey proteins.

NaCl also plays an important role in bakery products. In bread, NaCl helps in gluten development, and thereby in texture-related properties; it controls yeast growth and the fermentation rate, improving the flavor, and reduces the spoilage. Wheat dough is a complex mixture of starch, proteins, water, fibers, fat, salt, and minor ingredients, primarily responsible for the quality of bread (Belz, Ryan, et al., 2012). The salt content can impact the level of protein–protein interactions and the strength of the gluten network, a stronger network can retain better the gas bubbles (Avramenko et al., 2018) and contribute to the bread volume. Pasqualone et al. (2019) evaluated the salt reduction in bread made with durum wheat and corroborated the strengthening of the gluten network promoted by salt addition: the saltier dough showed a more effective ability to retain the gas released by fermentation, resulting in a low $CO_2$ loss. On the other hand, the saltiest dough produced the lowest volume of gas. The salt reduction also influenced the color of the bread crust, which in general, showed lighter than the control bread. The stimulation of yeast growth promoted by the salt reduction possibly caused greater consumption of the sugars present in the formulation, so the Maillard reaction was less pronounced.

Lynch et al. (2009) evaluated the reduction of salt (1.2, 0.6, 0.3, and 0% of NaCl) in bread. Salt reduction did not have a significant impact on most rheological and textural properties, as well as the number and size of matrix cells. However, the total omission of salt in the bread formulation resulted in rheological, textural and structural changes compared to treatment without salt reduction. From a sensory point of view, the sample without added salt was correlated with negative attributes such as sourdough, yeasty, and sour/acid.

In fermented vegetables, salt has two main functions: (1) it helps to select the type and extent of microbial action, and (2) it prevents the vegetables from softening. The concentration of salt needed to avoid softening is used at the minimum that can be used and varies between products taking into account the sensory properties. The concentration for cabbage fermentation is 2–3%; cucumbers, 5–8%; and green olives, 4–7%. These concentrations of salt, in turn, at least partially guide the growth of the lactic acid bacteria, responsible for the fermentation process (Fleming et al., 1985).

Pino et al. (2018) evaluated the effect of salt concentrations (4%, 5%, 6%, and 8%) in the fermentation process of table olives. The authors found that only at the highest concentration of salt, 8%, there was inhibition in the initial development of lactic bacteria, however, at the end of the fermentation, after 120 days, no significant difference was observed in the acid lactic bacteria count between the samples. A high similarity was found between samples treated with 6% and 8% NaCl after 60 and 120 days of fermentation.

## 8.3.2 Salt as flavor enhancer: Impact on sensory properties

The flavor is one of the most relevant factors that influence food choices. It is a combination of taste, smell, and chemical sensation. Removal of a single component may lead to considerable change in the flavor profile (Israr et al., 2016). The flavor attribute defines the characteristic of the gustatory sensation, for classifying the sensations that the taste compounds elicit such as: sweet, sour, salty, bitter, and umami. Salty taste is one of the most relevant sensory attributes, as the presence of sodium improves the sensory properties of foods, increasing saltiness, decreasing bitter taste, and increasing sweetness and other congruent flavor effects (Keast & Breslin, 2003).

Sodium chloride is considered the best solution to provide saltiness and improve flavor, becoming its reduction or replacing a great challenge to industry and consumer. The perception of salt intensity is quite variable from individual to individual. The hedonic response and the consumption of salt result from a complex interplay of physiological, cultural, and environmental factors (Hayes et al., 2010). When a food product containing NaCl is ingested, sodium and chloride ions are released in the mouth, which is necessary for the activation of the physiological salt receptors. The rate of release depends on the structure and composition of the food as well as on mastication and salivation. The taste receptors are located throughout the oral cavity. Although the exact nature of the salt receptor has not yet been fully unraveled, it is believed that it consists of a sodium ion channel, which can be explained by the detection threshold or recognition threshold. The recognition threshold is the amount of sodium needed to activate the epithelial sodium channels at the taste receptors. The strength of the signals and, ultimately, the sensation of the salty taste depends mainly on the sodium concentration (Busch et al., 2013). Potassium and calcium chloride have been evaluated as sodium chloride substitutes, and despite both have some component of saltiness, they have characteristic flavors, sometimes described as "metallic" or "bitter" (Verma & Banerjee, 2012).

In addition to providing the salty taste in food, NaCl acts as a flavor enhancer in meat products, which is an important factor in its acceptability (Ruusunen & Puolanne, 2005). In general, NaCl replacers (KCl, $CaCl_2$, and $MgCl_2$) are negatively correlated with the intensity of cured flavor (Fellendorf et al., 2018), volatile profile (Corral et al., 2013); occurrence of off-flavor attributes, decrease in purchase intention (Rios-Mera et al., 2020) and bitter taste and rancid aroma (Vidal et al., 2019).

Torrico et al. (2019) evaluated the sensory acceptability of potato chips with reduced salt concentrations, and they observed that the taste preference decreased when the NaCl concentration decreased since salt plays an important role in enhancing flavors. In addition to improving the taste of food, by providing salty taste and flavor enhancer, salt can reduce the bitter taste. Antúnez et al. (2018) studied the sensory perception of toasted bread with mixtures of NaCl and KCl and verified that the partial substitution of NaCl (30%) did not result in off-flavor and bitterness of the product. However, NaCl replacement equal to or higher than 40% caused an increased bitterness, off-flavor, and metallic flavor. In bread, reducing salt tends to decrease the formation of flavor compounds resulting from the Maillard reaction and salt-potentiated lipid oxidation, which contributes to aldehyde formation, affecting the characteristic flavor profile (Pasqualone et al., 2019).

Reducing salt plays a critical role in fermented food products. In cheese, the reduction of NaCl can improve proteolysis, which results in the release of some bitter peptides, and excessive acid production during ripening. Excessive bitterness and acid profile development has been reported in salt-reduced cheeses (Guinee, 2004). The reduction of salt in Suancai, a popular fermented product of *Brassica* vegetable in China, changed the volatile profile according to the stimulation or inhibition of fermenting microorganisms, the reduction of salt was correlated to bitterness and astringency taste in this product (Liang et al., 2020).

### 8.3.3 Microbial stability

The control of microbial growth is a key factor to ensure the safety and adequate shelf life for the processed food. Adding sodium chloride in different food products acts as a relevant barrier to control microbial spoilage. The main effect of salt is to reduce the water activity ($a_w$) in the foods. The hyperosmotic shock causes shrinkage of the cytoplasmic volume, a process known as plasmolysis, which may result in a reduction in cell growth or the death of the cell (Taormina, 2010). Additionally, the effects of $Na^+$ and $Cl^-$ ions on cellular metabolism could be another reason for the inhibition of microorganisms in foods (Taormina, 2010). However, some mechanisms can be developed by microorganisms to adapt their metabolism to the high salt content in foods, for example, accumulating potassium, amino acids, or sugars in the cell, increasing the activity of sodium efflux systems, changing cell morphology and fatty acids profile in the membrane or producing specific stress response proteins (Inguglia et al., 2017). Depending on the ability of microorganisms to grow, tolerate, resist, or even require salt for their survival, the terms obligate halophile, salt-tolerant, and salt resistant have been used. Examples of bacterial pathogens that are salt-tolerant, salt resistant, and halophilic are *Listeria monocytogenes*, *S. aureus*, and *Vibrio parahaemolyticus*, respectively (Taormina, 2010).

Most bacteria, such as *Pseudomonas fluorescens*, *Escherichia coli*, *Clostridium botulinum*, *Bacillus cereus*, and *Bacillus subtilis*, are completely inhibited by an $a_w$ of 0.90 promoted by adding NaCl. However, the inhibition of *Staphylococcus aureus* needs a lower $a_w$ (0.85) (Gould & Measures, 1977). To achieve these $a_w$ values a high concentration of NaCl is required, then the taste restrictions imposed by it must be considered. Because of this, NaCl is most often used in combination with other additives and preservation techniques (Shelef & Seiter, 2005).

In meat products, the high content of nutrients, water, and a pH value, in general, close to 6.0, make these products very perishable. The main spoilage bacteria associated with meat products are *B. thermosphacta*, *Carnobacterium* spp., *Lactobacillus* spp., *Leuconostoc* spp., and *Weissella* spp., they causing defects such as sour off-flavors, discoloration, gas production, slime, and decrease in pH (Borch et al., 1991). Concerning the pathogenic microorganisms, *L. monocytogenes*, *C. botulinum*, *C. perfringens*, *E. coli O157:H7*, *Salmonella* spp., and *S. aureus* are the most common in meat products (Patarata et al., 2020) and they could have their growth stimulated for salt reduction, always associated to many other factors, such as pH, $a_w$, storage temperature, quality of raw material, and the processing conditions.

The effects of NaCl reduction and salt substitutes on the growth of *L. monocytogenes* were reported by Samapundo, Anthierens, et al. (2013) in broth and cooked ham. They found that divalent salts ($CaCl_2$ and $MgCl_2$) had higher antilisterial activities than NaCl and KCl. These results were correlated to a lower water activity of the divalent salts when compared to the same concentration in the aqueous phase of the monovalent salts.

Whiting et al. (1985) evaluated the growth of *Clostridium sporogenes* and *S. aureus* in cooked corned beef in two levels of salt under temperature abuse (5, 11, and 16°C). *C. sporogenes* (1.6% and 2.4% of NaCl) and *S. aureus* (1.24% and 2.14% of NaCl) had similar growth in all temperatures evaluated despite salt content. Otherwise, according to some studies, mesophiles, anaerobic and psychrophilic spoilage organisms are generally higher in reduced-salt samples, highlighting the refrigeration could be not enough to guarantee the meat preservation throughout the shelf life (Bower et al., 2018; Laranjo et al., 2017). One of the biggest challenges in reducing sodium in meat products is to achieve the minimum acceptable levels regarding technological, sensory, and microbiological aspects

(Delgado-Pando et al., 2018). These are the reasons why some food categories still have significant levels of sodium.

In cheese, salt has a greater influence on microbiology, impairing the growth and the control of pathogens and the population of starter lactic acid bacteria and nonstarter bacteria (Cruz et al., 2011). McCarthy et al. (2015) showed the salt reduction in cheddar cheese significantly increased the levels of lactic acid content since salt has a repressive effect on the lactic acid culture, and the reduction of salt stimulates their growth. Hystead et al. (2013) evaluated the reduction of salt (2.9%, 2.2%, and 1.3%) in cheddar using two different starter cultures: *Lactococcus lactis* ssp. *lactis* or *Lactococcus. lactis* ssp. *cremoris*. No difference was observed in total plate count (TPC) based on salt treatment in cheese stored at 4 °C monitored from 5 through 8 months of age. In the same study, *L. monocytogenes* was inoculated into cheddar cheese at a level of 4 log10 cfu/g after 2 weeks and enumerated at 4 different cheese ages to simulate postpasteurization contamination events, *L. monocytogenes* counts declined steadily, and no difference was observed in survival based on starter culture or reduced-salt treatments. Likewise, the reduction of salt did not affect the survival of *L. monocytogenes* and *E. coli* O157:H7 in nonfat cast cheese during storage (Wusimanjiang et al., 2019).

In bread, mold growth is the main deterioration factor during shelf life, and the effects of salt reduction in mold growth are relevant for producers. The study by Samapundo, Deschuyffeleer, et al. (2010) showed that the 30% reduction of NaCl in white bread did not stimulate the growth of *Penicilium roqueforti* compared to the control formulation. Belz, Mairinger, et al. (2012) studied the effect of NaCl reduction on the growth of fungi commonly found in bread, and demonstrated that reduction of NaCl (1.2%, 0.6%, 0.3%, and 0%) raised the water activity proportionally and the growth of inoculated *Fusarium culmorum* and *Aspergillus niger*, resulting in a shorter shelf-life.

In fermented vegetables, salt reduction significantly impacts the population of fermenting microorganisms (Liang et al., 2020). The effects of various concentrations of sodium chloride in fermented vegetable products on the growth of lactic bacteria (LAB) and Shiga toxin-producing *E. coli* (STEC), one of the most acid-resistant pathogens in fermented or acidified vegetables. The study showed 1% NaCl had a stimulatory effect on the growth of these bacteria compared to the unsalted control, however, at higher concentrations, the growth rates for STEC and for the LAB decreased, which demonstrates that the stimulus or inhibition of microorganisms growth is dependent of salt concentrations (Dupree et al., 2019).

## 8.4 Sodium reduction strategies for processed foods

Several strategies have been addressed to minimize the negative effects of salt reduction in foods from three main perspectives: consumer acceptance, physicochemical properties, and food stability and safety throughout the shelf life. Therefore, its reduction is a great challenge, particularly when the selected strategy is the simple salt reduction without further partial substitution by other chloride salts or combining flavor enhancers and new technological processes. To overcome the negative effects from salt reduction some strategies can be employed: selecting salt substitutes and/or flavor enhancers, optimizing the delivery of salt to taste receptors in the oral cavity, for instance, by crystals size modification, using nanotechnology science, and the nonhomogeneous spatial distribution of salt in foods. Also, ultrasound and high-pressure technologies are presented as strategies to help salt reduction in foods. Table 8.2 summarizes some of the strategies explored in the last years.

Table 8.2 Strategies to salt reduction in processed foods.

| Processed food | Strategies | Main findings | References |
|---|---|---|---|
| **Salt substitute and flavor enhancers** | | | |
| Bologna sausage | PuraQ®Arome Na4 | Reduction 43% sodium content; no compromise the microstructure; good sensory acceptance | Pires et al. (2017) |
| Frankfurter sausage | Blends of calcium, sodium, and potassium chloride | $CaCl_2$ increased the intensity of oxidative reactions | dos Santos et al. (2017) |
| Frankfurter sausage | Blends of calcium, sodium, and potassium chloride | 35% Reduction in sodium and formulations with 25% NaCl replaced by salt blends of $CaCl_2$ and KCl presented sensory performance similar to the sample with 100% NaCl | Horita et al. (2014) |
| Chicken nugget | Maintain an ionic strength equivalent to NaCl with replacement of NaCl by $CaCl_2$ | Divalent salts ($CaCl_2$) was not able to affect cooking performance and texture profile; reduce 34% sodium | Barros et al. (2019) |
| Restructured cooked ham | Ocean's Flavor—OF45, OF60 and the flavor enhancer Fonterra™ 'Savory Powder | Adding flavor enhancer did not affect negatively the flavor and residual taste | Pietrasik and Gaudette (2014) |
| Wheat bread | Blends of calcium, magnesium, and potassium chloride | The divalent salts had a greater effect on the properties of dough and bread compared to monovalent cations; the dough stiffness, bread volume, and crumb cell structure were maintained in the bread containing KCl | Reißner et al. (2019) |
| Wheat bread | Pansalt® | Noticeable increase in the bitter and residual taste in the crust at the 3% addition level; the inclusion of lysine and iodine in the formulation did not it was effective in masking the bitter taste produced by these salts | Raffo et al. (2018) |
| Cheddar cheese | Combination of salt replacer (Saloni, Saloni K, and KCl), flavor enhancer (hydrolyzed vegetable protein, yeast extract, and IMP), and bitter blocker (Lysine, AMP, and Glycine) | Flavor enhancer and bitter blocker proved to be successful in the perception of desired saltiness and without much bitterness | Khetra et al. (2016) |
| Prato cheese | Potassium chloride, Sub4salt®, and Salona™ | Reduced sodium content by up to 35% and does not interfere with the physicochemical characteristics and acceptability | Costa et al. (2018) |
| Cottage cheese | Potassium, and calcium chloride | Samples containing $CaCl_2$ had lower sensory acceptability, increasing the bitter taste; KCl had good acceptance compared to control with 100% NaCl | Fosberg (Damiano) and Joyner (Melito) (2018) |

## Table 8.2 Strategies to salt reduction in processed foods—cont'd

| Processed food | Strategies | Main findings | References |
|---|---|---|---|
| **Size crystal modification** | | | |
| Cooked ham, cooked turkey breast, and Deli type sausages | Soda-Lo® salt microspheres | Reducing salt content not affect the sensory attributes of meat products. | Raybaudi-Massilia et al. (2019) |
| Fresh pork meat | NaCl fine flakes (0.55 mm, 2.5 mm, and 5 mm) | Faster dissolution of crystals and higher degree penetration in the meat product were associated with the size of the crystals and a greater surface area. | Aheto et al. (2019) |
| Turkey ham | Micronized NaCl | NaCl reduction up to 30% was perceived by consumers regarding salty taste, designating the formulations with lower salt content as "less salty and less seasoned". Despite this, the overall acceptability was not affected in any of the samples. | Galvão et al. (2014) |
| Potato chips | Salt particles from <106, 106–425, and 425–710 μm | Increased the rate of sodium delivery rate to the tongue and sensory perception of salinity and reduce sodium in crisp snacks | Rama et al. (2013) |
| **Spray-dried salt particles** | | | |
| Bread | Spray-dried salt-yeast complex | Positive effect on saltiness intensities and reduce the amount of salt | Lee et al. (2020) |
| Cheese crackers | Nano spray drying (0.5 and 1.9 μm salt particle size) | Positive effect on the perception of salinity with microbiological control | Moncada et al. (2015) |
| Snack products | Spray drying size of chitosan/acid/NaCl ranged from 15.4–32.0 μm, raised with NaCl concentration. | Spray-dried salt has perceived as saltier than general salt and could potentially be regarded as a low-sodium salt for surface-salted foods. | Yi et al. (2017) |
| Snack products | Spray-dried salt with/without KCl | Use the spray-dried salt has perceived as saltier than general salt and could potentially be regarded as a low-sodium salt for surface-salted foods. Smaller particle size and lower bulk density allowed a significantly higher saltiness. | Chindapan et al. (2018) |

*Continued*

Table 8.2 Strategies to salt reduction in processed foods—cont'd

| Processed food | Strategies | Main findings | References |
|---|---|---|---|
| **Processing technologies** | | | |
| Chicken meat (NaCl 0–2.5%) | High hydrostatic pressure (0–300 MPa by 60, 120, and 180 s) | At 300 MPa a positive interaction of NaCl content and HHP improved the water holding capacity. HHP treatment with NaCl in low levels improved chicken meat color and texture. | Ros-Polski et al. (2015) |
| Chinese cabbage (NaCl 10%, 15%, and 20% in the brine) | Ultrasound (35 kHz, 0–180 min) | NaCl effective diffusivities increased from 147% to 812% upon ultrasound application during the brining process, also it enhanced the cabbage hardness. Ultrasound was a potential technology for accelerating the brining process. | Zhao and Eun (2018) |
| Restructured cooked ham (NaCl 0.75%, 1.12%, and 1.5%) | Ultrasound (20 kHz, 600 W cm$^{-2}$ 10 min) | Ultrasound decreased the total fluid release and increased the hardness of salt-reduced restructured cooked ham. Also, the sensory acceptance was improved. | Barretto et al. (2018) |
| Bacon (NaCl 1.5%) | Ultrasound (300/800 W for 30/60 min) | Ultrasound increased the penetration rate of the salt and was more efficient than the static dry-curing method. The hardness and the perception of salty taste were higher for the bacon submitted to the ultrasonic treatment | Pan et al. (2020) |
| **Inhomogeneous spatial salt distribution** | | | |
| Hot-served layered snacks (NaCl 0, 0.3%, and 1.0%) | Bi-layer and four-layer cream-based and cereal-based snacks | A significant enhancement of saltiness was observed in samples with a heterogeneous salt distribution for both bi-layer and four-layer products. | Emorine et al. (2013) |
| Bread (NaCl 2.0%, 1.73%, 1.50% and 1.0% on the flour) | 2, 8, or 16 dough layers were formed | The salt gradient in the bread improved the saltiness perception, some variations in hardness and volume of bread cells were still perceived through the layers with a salt gradient. | Noort et al. (2010) |
| Bread (NaCl 1.5%, and 1.05%) | Salt agglomerates with waxy starch | The salt agglomerates caused a slight increase in dough development time and a decrease in dough stability, the perception of saltiness increased, resulting in bread with 30% salt reduction but similar to the control. | Monteiro et al. (2021) |

Table 8.2 Strategies to salt reduction in processed foods—cont'd

| Processed food | Strategies | Main findings | References |
|---|---|---|---|
| Fresh sausage (NaCl 2.0%, 1.5%, and 1.0%) | Encapsulated salt with carnauba wax | 2.0% and 1.5% of encapsulated salt did not cause any difference in hardness, cohesiveness, pH, microbiological counts, TBARS index, and sensory perception of salinity compared to the control treatment (2.0% of salt without encapsulation) | Beck et al. (2021) |

*No permission required.*

### 8.4.1 Use of flavor enhancers

The flavor enhancer is a compound that modifies or increases the intensity of the perceived taste or smell of food and has no flavor of its own, contributing significantly to balance the salty taste of products with less salt, activating the taste receptors in the mouth and throat. There are several flavor enhancers and off-flavor maskers available on the market, including yeast extracts, monosodium glutamate, lactates, and flavors (Desmond, 2006; Israr et al., 2016).

PuraQ® Arome Na4, a product derived from fermentation, such as sugars, organic acids, salts, and aromas, was used by Pires et al. (2017) in the replacement of up to 60% sodium chloride in bologna sausage, resulting in a reduction in the sodium content of approximately 43%. The results also showed that replacing 40% of NaCl with PuraQ®Na4 (~34% of sodium reduction) did not compromise the microstructure and showed good sensory acceptance, even though consumers have reported that this formulation has presented "slightly less salty than ideal". Pietrasik and Gaudette (2014) reported that the 1.2% reduction in NaCl and the addition of flavor enhancer (Savory Powder) in restructured cooked hams resulted in a 30% reduction of sodium content without affecting the flavor and residual taste. Vidal et al. (2020) studied the partial reduction of NaCl and the use of lysine, yeast extract, and substitute salts KCl and $CaCl_2$ in the characteristics of salted meats and showed that lysine and yeast extract minimized the negative sensory effects due to the addition of KCl and $CaCl_2$. The addition of lysine and yeast extract increased sensory acceptance and decreased the rancid aroma, salty taste, and aftertaste of salty meats prepared with the blend of NaCl + KCl + $CaCl_2$, without changing the physicochemical quality parameters.

Monosodium glutamate (MSG) is an important contributor to enhance flavor and is widely used by the food industry. Evaluating MSG as a salty taste enhancer of the NaCl/MSG mixture [ratios 3:0, 2:1, 1:2 and 0:3 (w/w)] in pork patties, Chun et al. (2014) found that the addition of MSG was effective in reducing the level of NaCl in the formulation of the meat product enhancing the salty taste, thereby increasing overall acceptability of pork patties. Jinap et al. (2016) studied the reduction of sodium content in spicy soups using different percentages of NaCl and MSG (0–1%) and reported that it is possible to reduce the NaCl content from 0.8% to 0.3% without changing the salty taste of the soups, increasing the concentration of MSG to the optimum level of 0.7% and reducing the sodium intake by 32.5%.

Khetra et al. (2016) evaluated the sodium reduction in cheddar cheese by using potassium-based salt replacer in combination with flavor enhancers (hydrolyzed vegetable protein (soy-based), yeast extract powder, and inosine-5′ monophosphate) and bitter blockers (Lysine, AMP, and Glycine) to mask the off-flavors and bitterness. They reported that flavor-flavor interactions produced by the salt substitute, flavor enhancer, and bitter blocker proved to be successful in the perception of desired saltiness without strong bitterness in cheddar cheese.

The replacement (1.5% or 3% on flour weight) of NaCl by the mixture of inorganic salts and flavor enhancers present in Pansalt® [composed of NaCl (57%), KCl (28%), $MgSO_4$ (12%), lysine hydrochloride (2%), silica (1% added as anticaking agent) and iodine (0.0036%)] in bread was investigated. The authors found that saltiness was slightly perceived when NaCl was reduced, at the same time that it gave rise to a noticeable increase in the bitter and residual taste in the crust at the 3% addition, suggesting the need for other strategies to replace high levels of salt (Raffo et al., 2018).

### 8.4.2 Use of other salts to substitute NaCl

Different types of salts have been studied and used by the food industry, in order to reduce the sodium content by replacing sodium chloride and maintain the salty taste of products, such as potassium chloride (KCl), magnesium chloride ($MgCl_2$), calcium chloride ($CaCl_2$) and others or a blend of them (Dos Santos, Campagnol, et al., 2015; Fellendorf et al., 2018; Raffo et al., 2018). Potassium chloride has been commonly used in foods since it contains a monovalent cation and guarantees similar ionic strength to sodium, necessary for developing stable emulsions and gels in many meat products. But it results in a bitter, astringent, and metallic taste in the products (Horita et al., 2011), depending on the level used.

Fortification of minerals such as calcium is an increasingly common practice used to add nutritional value to many foods and beverages. One of the strategies addressed is replacing sodium chloride with calcium chloride (Barros et al., 2019; Lawless et al., 2004). However, the flavor properties of divalent salts such as calcium chloride, magnesium chloride are complex, as they are characterized mainly by the bitter, metallic, astringent, and irritating taste (Lawless et al., 2004). For example, in dry-cured ham salted, all the sensory attributes studied were affected by a mixture of salts $MgCl_2$ and $CaCl_2$ as a substitute for NaCl (Armenteros et al., 2012).

Aiming to evaluate the effect of 40% replacement of sodium chloride by potassium chloride on the sensory acceptability of potato chips, Torrico et al. (2019) found that this substitution level did not affect the salinity of potato chips and that the use of 40% of the KCl did not influence the bitter taste. The salinity of NaCl may have worked as a masking factor for the bitter taste of KCl, resulting in similar taste scores among potato chips treatments. Costa et al. (2018) evaluated the effect of 40% replacement of sodium chloride by KCl, and commercial salts: Salona ™ (containing $MgCl_2$, KCL, and NaCl) and Sub4salt® (containing NaCl, KCl, and sodium gluconate) in Prato cheese. They found that the sodium reduction in these conditions could be a feasible alternative for the dairy industry, since this strategy does not change the physicochemical characteristics and acceptability of Prato cheeses. Otherwise, Fosberg (Damiano) and Joyner (Melito) (2018) studied the use of KCl and $CaCl_2$ as a substitute for NaCl in a cottage cheese cream reported that samples containing $CaCl_2$ had lower sensory acceptability, which may be a result of the increased bitter taste reported by the consumers. The formulations containing KCl (particularly the 1:1 substitution ratio) had good acceptance, similar to cottage cheese cream containing 100% NaCl. The food matrix and its components could explain these conflicting results.

Reißner et al. (2019) evaluated the replacement of 50% NaCl by KCl, CaCl$_2$, and a mixture containing KCl, CaCl$_2$, and MgCl$_2$ in wheat bread, and observed that divalent cations had a greater effect on the properties of dough and bread compared to monovalent cations. The presence of CaCl$_2$ showed less dough stiffness, resulting in bread with greater volume, softer crust, crumb, and coalescent cells. CaCl$_2$ promoted weakening the protein networks after dough formation while these parameters were maintained with bread containing KCl. dos Santos et al. (2017) investigating the effect of reducing and replacing NaCl with KCl, CaCl$_2$, and a mixture of KCl and CaCl$_2$ in dry fermented sausages, found that the addition of CaCl$_2$ resulted in increased lipid oxidation during fermentation and storage. These results suggest that replacing NaCl by CaCl$_2$ increases the intensity of oxidative reactions, concluding that the addition of KCl may be a good alternative to reduce the NaCl content in fermented meat products.

One of the alternatives that can be used to reduce the undesirable effects of replacing NaCl using divalent substitute salts is to maintain the ionic strength equivalent to sodium chloride in the reformulation. Horita et al. (2014) evaluated the reduced-sodium frankfurter sausages elaborated with blends of chloride salts (NaCl, KCl, and CaCl$_2$) at ionic strength equivalent to 2% NaCl. The authors found that the replacement of 50% NaCl by mixtures of potassium (25%) and calcium (50%) chloride represented a viable strategy for reducing sodium content, showing that a 35% reduction in sodium with a sensory performance similar to that of the sample with 100% NaCl. Barros et al. (2019) reformulated chicken nuggets with the replacement of NaCl by CaCl$_2$ (25%, 50%, and 75%), considering an ionic strength equivalent to 1.5% NaCl, observed that the replacement of 75% NaCl could reduce 34% sodium in chicken nuggets. Differently, Vidal et al. (2019) found that adding 50% and 25% of CaCl$_2$, based on the ionic strength of NaCl, in jerked beef gave undesirable sensory characteristics to the product, such as bitter taste, fibrosity, rancid aroma, and aftertaste. On the other hand, the replacement of 50% NaCl by KCl provided technological and sensory characteristics very similar to the traditional jerked beef (containing 100% NaCl). These studies show the importance of food matrices to select the salt substitutes to NaCl.

### 8.4.3 Crystal size modification

The taste buds found in the specialized mucosa of the tongue, where the taste receptors are located, which act as sensors of compounds associated with the taste, recognize the perception of salty taste in food. NaCl undergoes dissociation upon contact with saliva, resulting in free ions (Na$^+$), and consequently crosses epithelial sodium channels at the apical end of taste receptor cells. This leads to the transmission of signals to the cerebral taste cortex and gives the perception of salinity. For this reason, it is possible to alter the perception of salt optimizing the stimulation of peripheral receptors/afferents. The increase of the dissolution rate could intensify the salty taste of some foods and thus reduce the required levels of salt, explained by the higher speed at which receptor stimulation occurs, increasing the perception of saltiness and, therefore, reduce the sodium intake (Hurst et al., 2021; Vinitha et al., 2021). The control of salt release in food matrices involves the enhancement of the dissolution rate of food crystals, the salt spatial distribution, and possible textural effects in the food matrices.

The dissolution rate of sodium chloride in the mouth is determined in part by the surface area exposed, which depends on the size of the crystal and the shape of the crystal. The maximum salinity of directly applied common salt is usually not reached, because the salt crystals are larger and do not dissolve fast enough to reach the sodium receptors in the mouth before a bolus is formed and swallowed

(Kilcast & Angus, 2007). The physical modification of NaCl occurs by obtaining smaller crystal sizes and a low bulk density, resulting in a large surface area that dissolves more rapidly in saliva. However, it works mostly for surface salty foods, such as French fries and potato chips (Kuo & Lee, 2014). The shape and density of crystals also influence the dissolution rate, for example, using crystal structures with a lower bulk density such as hollow pyramids or agglomerated cubes allowing them to fractionate during dissolution. Small salt crystal sizes can be obtained by mechanical crushing of large crystals (Quilaqueo & Aguilera, 2016). However, finely ground salt and with a high surface area is very hygroscopic, in that as the salt crystals are milled a hygroscopic powder is produced and therefore tends to re-agglomerates. This agglomeration can occur during milling and/or during storage and transportation, unless protected by storage systems. For this reason, simply grinding salt to a fine particle size would not be a practical option, and sometimes the particles obtained have a broad size distribution and asymmetric crystalline morphology (Shen et al., 2015).

Several commercial salts have been developed to aim this purpose such as flaked salt, which has a better solubility compared to granular salt. Because of this, the flaked salt would have more functionality in dry-cured products (Desmond, 2006). Similarly, it was reported that vacuum granulated dendritic salt (macroporous crystals) and "cubic" salt dissolve twice as fast as conventional salt, allowing water to bind better to myofibrillar proteins in meat products (CTAC - Conseil de la Transformation Agroalimentaire et des Produits de Consommation, 2009). Another ingredient generated from sea salt is SODA-LO® salt microspheres which are much smaller than normal salt crystals (average particle size of 20–30 μm) providing a greater salty taste by maximizing surface area relative to volume and reducing salt consumption (Tate & Lyel, 2014). The partial replacement of salt (NaCl) (25% (w/w), 30% (w/w), 50% (w/w) by Soda-Lo® salt microspheres in cooked ham, turkey breast, and Deli type sausages did not affect sensorial attributes. These changes in the shape of salt crystals have been recognized and widely studied as a clean label strategy to reduce salt content (Raybaudi-Massilia et al., 2019).

Studies related to salt crystal size modification in French fries show that size influenced the rate of sodium delivery rate to the tongue and sensory perception of salinity. The results evidenced that smaller crystal sizes (<106 μm) led to faster maximum intensity time, higher maximum salinity intensity, and maximum total salinity. In this way, with a controlled chewing environment, this size crystal modification could be feasible to reduce sodium in crisp snacks (Rama et al., 2013).

Hurst et al. (2021) redesigned salt crystals and applied these different salts in a model peanut system intending to increase salinity and thereby reducing sodium. They reported that dissolution and/or salinity are driven by particle size ($p < 0.05$) and hydrophobicity. In addition, it was observed that even if small particles have a high dissolution rate during oral processing, that dissolution rate is affected when the particles have a high density due to elevated levels of interaction with surface fats. On the other hand, Aheto et al. (2019) investigated the salt distribution as influenced by crystal size (fine, 0.55 mm; medium, 2.5 mm; and large, 5 mm) in cured pork meat and showed that the size of the crystals influenced the absorption and the total concentration of NaCl in salted meat. Furthermore, through hyperspectral maps of salt distribution in the samples, it was observed that the largest surface area was from the treatment with fine flake, contributing to a faster dissolution of its crystals and, therefore, to a greater penetration effect compared to the other grades of salt. However, the authors point out those factors such as the shape of the crystals, the presence of impurities, and the internal cavities of the fresh meat can also be considered in the salting process to achieve sodium reduction in salted meat.

Gaudette et al. (2019) evaluated different salt reductions levels with different crystal diameters in burgers and reported that the use of 3 mm sized salt crystals (larger diameter) was feasible to reduce sodium (0.7% salt) resulting in a salty flavor comparable to control burguer (1% table salt inclusion). Similar results were found by Rios-Mera et al. (2019), using micronized salt with 10 to 20 μm in diameter, compared to standard salt with a size of 500 μm. However, Galvão et al. (2014) reported that in turkey ham, the use of micronized salt (20 μm) as a substitute for table salt did not increase the perception of saltiness and that NaCl reduction of up to 30% resulted in products characterized by consumers as low in salt, low in seasoning. Jakopović et al. (2020) were able to reduce salt in feta cheese by up to 50% with replacement by micronized salt without presenting physicochemical differences from traditional feta cheese.

### 8.4.4 Spray-dried salt particles

The spray dryer methods are commonly applied including an atomization process that eliminates any residual moisture in the existing liquids and can control particle size distribution, dust flow, and particle morphology (Jafari et al., 2021). This technology allows combining salts with bioactive compounds bringing the multifunctional ingredients conception. Particularly, it is very interesting because reduce sodium by replacing total or partial sodium chloride can affect some attributes of food matrices stability (Kalnina & Straumite, 2019).

Yi et al. (2017) evaluated chitosan/acid/NaCl microparticles produced by spray drying in corn kernels, resulting in a saltier perception than common salt and could potentially be considered as a low-sodium salt for surface-salted foods. Similar research showed that spray-dried salt particles, in both combinations with KCl or without it, have a smaller particle size and lower bulk density, which had a significantly higher saltiness than commercial salt for industrial applications, especially in reduced-sodium dry foods such as French fries, nuts, snacks, etc. (Chindapan et al., 2018). In the same context, the spray-drying process was used to produce particles of NaCl/maltodextrin complexes, which had tasted saltier with a quick perception of salt, and it was associated with a salty intensity in bread, snacks, chips, noodles, or certain meat products (Cho et al., 2015). Likewise, Lee et al. (2020) described the influence of salt-yeast complex made by spray-dryer on enhancing the saltiness of white pan bread. The results showed a positive effect on higher saltiness intensities and reducing the amount of salt. Nevertheless, more detailed studies are required to determine the optimum concentration of salt-yeast complex in bread for the bakery industry.

Another spray-dryer technique that can be used is nano-spray drying, which produces a submicron size powder (particle sizes between 10 and 1000 nm) with high yield and reduced size distribution. By decreasing the particle size to nanometer size, the surface area/volume ratio of the particles increases inducing more intense sensory characteristics than observed in the commercial product, making it feasible reducing salt with high acceptance by consumers (Sun et al., 2021). Unlike conventional spray dryers, nano-spray drying uses a vibrating screen technology. Using a piezoelectrically driven vibrating membrane in the spray head, millions of tiny droplets are generated every second (Arpagaus, 2019; Jafari et al., 2021; Suna et al., 2014). The industrial application of this technology for reducing salt in cheese crackers shows that NaCl particles with 500 and 1900 nm, when drying a 3% ($w/v$) solution in salt concentrations of 1%, 1.5%, and 2% (w/w), resulted in positive effects on the perception of salinity, preserving the microbial stability (Moncada et al., 2015). This is a very promisor strategy to reduce sodium in food products without loss in global quality attributes.

## 8.4.5 Nonthermal processes for low/reduced sodium food products

A recent approach that has been evaluated at the development of low/reduced sodium in food products is the use of metallic salts combined with green and/or no conventional technologies. Most of the studies published relate that high pressure processing and ultrasound ate the most technologies commonly applied. They can be very useful to improve the safety and the functional quality of foods through their impact on the specific micro and macro food components (Pateiro et al., 2021).

High hydrostatic pressure (HHP) technology involves the use of pressures in the range of 100–800 MPa, combined or not with heat, for inactivating a variety of pathogenic and spoilage vegetative bacteria, yeasts, molds, viruses, and spores to ensure microbiologically safe foods (Balasubramaniam et al., 2015). The pressure applied in HHP treatments is instantaneously and uniformly transmitted throughout the food, regardless of size, shape, and composition (Martínez-Rodríguez et al., 2012). The HHP can damage microbial membranes, affecting the transport of nutrients and the disposal of cell waste, and denaturing key proteins, especially enzymes, crucial to cell metabolism, the pressure of 600 MPa can deactivate most vegetative bacteria (Wang et al., 2016).

Also, HHP induces microscopic and macroscopic physical changes in molecules, such as packaging the molecules together, inducing the unfolding of proteins and polysaccharides, which may increase the functional properties of these macro-components (Messens et al., 1997; Yamamoto, 2017), and hence, could reduce the time and exposure to the heating process while increasing the yield, texture-related parameters, safety and nutritional quality of the foods.

In cheese, particularly in ripened cheeses, the salt reduction can stimulate the growth of starter bacteria, resulting in increased proteolysis and acidity during ripening. Also, spoilage microorganisms and pathogens can find better conditions to grow from reducing salt in the original formulation. HHP can be used to decrease the count of starter microorganisms and reduce the chance of adverse effects on the texture and flavor of salt-reduced cheeses. Moreover, the application of HHP on raw milk followed by cheesemaking can reduce rennet coagulation time and increase cheese yield (Martínez-Rodríguez et al., 2012). (Ozturk et al., 2013) for instance, demonstrated that applying HHP at 450 MPa for 3 min on the milk used for cheesemaking caused a reduction in the number of starter bacteria by up to 4 logs. The over-growth of starter bacteria can cause excessive proteolysis, and thereby, raise the concentration of some bitter peptides that could influence the sensory acceptation of cheese. However, in this work, the sensory perception of acid and bitter flavors increased proportionally to the salt reduction, despite the reduction of starter bacteria with the application of HHP treatment.

In salt-reduced meat products, the HHP treatment can be a successful strategy to improve the quality properties of these products. Sikes et al. (2009), for example, studying the HHP treatments (0, 100, 200, 300, and 400 MPa) in low-salt beef sausages (0%, 0.5%, 1.0%, 1.5%, and 2.0% of NaCl), showed the 200 MPa pressure was the most effective to reduce the differences of cooking yield between low-salt sausages and the control. With increasing pressure, there was an increase in hardness and fracture forces up to 200 MPa. Similar results were found by other studies (Crehan et al., 2000; Yang et al., 2021). The effects related to pressure in meat are swelling of myofibrils, solubilization of myosin, actin, and other meat proteins, and the unfolding of protein chains. Those effects could improve the water holding capacity, emulsification, and gelation properties. Thus, HHP treatment can be a suitable strategy to reduce salt in meat products.

Ultrasound is based on mechanical waves at a frequency above the human hearing threshold and can be divided into two frequency ranges: high (100 kHz–1 MHz, power $< 1\,W\,cm^{-2}$) and low (16–100 kHz, power in the range of 10–100 $W\,cm^{-2}$). Ultrasound induces a phenomenon known as cavitation, which produces a large number of bubbles that, when collapsing, cause high local pressure and temperature that induce intense physical forces such as shear, shock waves, and turbulence (Amiri et al., 2018). These effects can cause cellular damage in microorganisms and also modify the structure and functionality of food components (Carrillo-Lopez et al., 2017). Zisu et al. (2010) demonstrated that whey protein concentrated submitted to sonication at 20 kHz combined with heat treatment improved the gelling properties and heat stability. The microstructure of sonicated whey protein gels showed a densely packed protein network.

Ultrasonic treatment at 20 kHz was evaluated on reduced-salt bacon (1.5%) by Pan et al. (2020). This process increased the penetration rate of the salt, and it was more efficient than the static dry-curing method, but it was similar to the tumbling method. However, the hardness and the perception of salty taste were higher for the bacon submitted to the ultrasonic treatment. Ultrasound operating at 25 kHz and 60% amplitude improved the protein network and their interactions with water and fat, which resulted in higher emulsion stability and texture parameters in meat emulsions (Cichoski et al., 2019), the relevant quality attributes to be preserved in low sodium meat products.

### 8.4.6 Heterogeneous distribution of salt

The heterogeneous (or inhomogeneous) spatial distribution of salt in food products has been studied as a possible solution for a global reduction of salt in processed foods. Studies have shown that the rapid perception of salt can lead to the sensation that the product has higher salt content than in reality, and the samples with heterogeneous distribution usually had seemed as saltier due to the strong release of salt at the beginning of mastication, which promotes a short and intense stimulus that can increase salt perception, possibly through a reduction of adaptation (Busch et al., 2009; Emorine et al., 2013; Noort et al., 2010).

The heterogeneous spatial distribution can be achieved by two main strategies: (1) the structuring of food in layers with different salt contents, which was evaluated, for example, in snacks (Emorine et al., 2013, 2015), and bread (Noort et al., 2010; Sinesio et al., 2019), and (2) the encapsulation, coating or agglomerated salt to create an effect known as salt spots, which was evaluated, in different food categories such as in fresh sausage (Beck et al., 2021) and bread (Li et al., 2021; Monteiro et al., 2021).

Emorine et al. (2013) showed a significant saltiness enhancement in hot-served layered snacks with a heterogeneous salt distribution. As pointed out in this study, the process of cooking, storage, and preparation by the consumer, for reheated products, can be a limiting factor in the heterogeneous distribution of salt because of the modifications that occur in the food matrix, such as diffusion of water with protein denaturation and the salt itself. These modifications tend to minimize the differences in salt concentrations through the layers, and the sensory stimulus could be different. The authors showed that salt migration from higher concentration layers to lower ones happened after a reheated process for snacks formed by two layers (cereal and cream-based layers). These conditions reduced the contrast stimulus and, for products containing 1% of salt in the cereal-based layer, nearly complete homogeneity between the final concentrations of the two layers was observed after reheating.

Salt migration between layers occurred in the study of Noort et al. (2010) throughout the preparation (formation, fermentation, baking, and cooling), but not during the freezing of bread. Concerning the perception of saltiness in bread, the layers with salt gradient content improved it significantly, and the authors also demonstrated that the magnitude of the saltiness enhancement was proportional to the sensory contrast. However, salt reduction affected the gluten network development and yeast activity, resulting in some variations in hardness and volume of bread cells perceived through the layers.

Inhomogeneous salt agglomerates with waxy starch were evaluated as a strategy for salt reduction in bread by Monteiro et al. (2021); according to the authors, salt agglomerate did not modify the physicochemical characteristics of bread and also improved its color parameters. The strategy employed allowed 30% sodium reduction with no saltiness differences. Li et al. (2021) evaluated the arabinoxylans impregnated with salt in bread to create an inhomogeneous spatial distribution of salt. The results showed that this process allowed 20% salt reduction without impacting the saltiness. According to the authors, the moisture content, crumb structure, water distribution, dough rheology, and bread texture suggest the strategy proposed mitigated the detrimental effect of dietary fiber on the dough and bread quality.

Another important issue is that salt in many foods products provides functional effects, for instance, in meat products, where the homogeneous distribution of salt is a crucial factor to ensure adequate protein solubilization and consequently adequate yield and texture. Thus, heterogeneous salt distribution should be a challenge on these products. However, Beck et al. (2021) evaluated the encapsulated salt with carnauba wax in fresh sausages to create a taste contrast of salt perception. The authors showed that the incorporation of 2.0% and 1.5% of encapsulated salt did not cause any difference in hardness, cohesiveness, pH, $a_w$, microbiological counts, TBARS index, and sensory perception of salinity compared to the control treatment (2.0% of salt). However, as expected, the use of 1% encapsulated salt caused a decrease in textural parameters and a lower perception of the salty taste in fresh sausage samples.

## 8.5 Challenges to reduce sodium in food products

Reducing sodium intake worldwide is a task with great challenges for health agencies, food industries, and consumers in general, as they need to change some consumption patterns. Thus, several areas of scientific, social, and economic development need to be stimulated to work together with feasible proposals to overcome this challenge.

The perception of sensory attributes of foods and consumer acceptance are important factors for the successful reformulation of processed foods with sodium reduction (Nguyen & Wismer, 2019). Using salt substitutes, mainly KCl, seems the most studied strategy for reducing sodium in food products. However, depending on food composition, these salts reduce global acceptance and they promote the decreasing of physicochemical properties depending on the levels used. So, to compensate this sensorial or stability loss, the industry must find other solutions such as adding flavor enhancers, extenders, or bioactive compounds, which, in general, increase the final production costs. In this way, it is necessary to bring a deep view of the strategies that were studied up to now.

Understanding consumer perceptions can lead to more effective responses from relevant sensory descriptors to optimize low-sodium meat product formulations. Dos Santos, Bastianello Campagnol, et al. (2015) studying the reduction in NaCl, replaced by KCl, $CaCl_2$, and a blend of KCl and $CaCl_2$ in dry fermented sausages, emphasized the importance of the flavor attribute in the consumer's perception, showing the negative characteristic were attributed to substitute salts, such as bitter taste, rancid flavor, acid, and rancid aroma and without salt. Similar results were found by Vidal et al. (2019), with the addition of $CaCl_2$ based on the ionic strength of NaCl in jerked beef.

Sensory evaluation tools since the classical quantitative descriptive analysis (QDA) until the sensory methods based in consumer perception as free listing, projective mapping, ultra-flash profile, preference attribute elicitation, among others besides temporal methods as temporal check-all-that-apply questionnaire (TCATA) and temporal dominance of sensations (TDS), and can be used more effectively to characterize the significance of NaCl reduction on the sensory attributes of food products since found limited studies at the literature using these innovative sensory methodologies (Teixeira & Rodrigues, 2021). A recent study demonstrates the simultaneous use of QDA and TDS presented as an effective approach to elucidate the sensory profiling of low sodium probiotic Prato cheeses with fractional substitution of NaCl with KCl and addition of flavor enhancers (Silva et al., 2018).

Assessing the impact of reducing or replacing the content of sodium chloride with potassium chloride in soft cheeses (type "Camembert") and semihard (type "Reblochon"), Dugat-Bony et al. (2019) observed that reduced sodium (20%) cheeses were perceived to be significantly more acidic and bitter, with more intense color, a stronger ammonia flavor, animal/cowshed odor and global intensity of odor than control cheeses. Antúnez et al. (2016) verified that consumers' perceptions were affected by more than a 10% reduction of NaCl in bread. Above this level, the salt reduction resulted in bread being less salty, tasty, and smoother, however, the small salt reduction can be a viable strategy to reduce the sodium content in bread, without affecting the consumer's perception. In addition, the authors were able to verify that although the hedonic perception was not influenced by the information regarding the % salt reduction, consumers were more discriminatory in the informed condition.

The intrinsic properties that correlate food products with health benefits can be a key factor in increasing purchase intent. Torrico et al. (2019), studied the reduction and replacement of NaCl by KCl in potato chips and observed that the purchase intention of NaCl-reduced potato chips has been significantly improved after the sodium information was shown to consumers. Kongstad and Giacalone (2020) evaluated the effect of simple reduction NaCl (10–30%) and/or its substitution (KCl and MSG) on the salt perception of young consumers (18–30 years), also in potato chips, using a reference product as a basis for systematic reformulation. They verified a strong labeling effect when consumers were informed about the product's characteristics. The reference sample was perceived to be significantly less healthy when compared to all samples with reduced salt, which means that information on overall sodium reduction is effective in increasing the healthiness perception.

Although studies have shown that consumers have related food information about reducing salt content with the term "healthier", Rodrigues et al. (2017) reported that the consumers having little knowledge about the risk of sodium intake do not use salt content information on food labels to make food product choices.

Alongside the highly relevant sensory impact, the importance of salt in processed foods is due to the positive effects on technological processes considering safety, texture, and shelf life. On the other hand, the attempt to reduce salt content requires a modification in the ingredients list or using new technologies during production, aiming to replace the technological sodium chloride functions (Fraqueza et al., 2021). When sodium chloride is reduced without other ingredient replacement, the water activity increased, and the ionic strength reduced making the food product more susceptible to loss of stability and spoilage. This salt reduction can be counteracted by using several alternative methods, such as: active packaging, modified atmosphere packaging, high isostatic pressures, irradiation, and the addition of chemical additives, among others (Fraqueza et al., 2021), as previously commented.

Hurdle technology is an intelligent approach to improve the microbial stability of reformulated foods as well as their nutritional (Leistner, 2000). The adoption of hurdle technology has become more prevalent since the food industries have focused on healthier foods with minimal processing which are often kept under chilled storage to achieve their shelf-lives (Singh & Shalini, 2016).

Khanipour et al. (2016) investigated the effects of NaCl (0–3% w/w), potassium sorbate (0–0.2% w/w), and biopreservative nisin (0–240 ppm) on inhibiting the growth of *C. sporogenes* in low-salt cheese analogue. The results demonstrated that it is feasible to produce high moisture and low salt product, but only if sufficient levels of potassium sorbate and nisin are added to the product. On the other hand, Mehyar et al. (2018) studied the application of edible coatings and films incorporated antimicrobial agents in low sodium Halloumi cheese with chitosan coatings with or without lysozyme or natamycin on the shelf-life, microbial quality, and sensory properties. Coatings increased cheese shelf-life by $\geq 5$ days compared to the control treatment when they were brined in 5% and 10% NaCl at 3 °C but did not affect the shelf-life of cheese in 15% NaCl. Tan et al. (2020) studied the structural integrity, texture, sensory properties, and shelf life in the fresh low salt wheat noodles adding mixtures of hydrocolloids, and salt substitutes. The authors concluded that mixtures of locust bean gum successfully reduced sodium content and retard the growth of microorganisms and shelf life was extended at 4 °C. Also, at this temperature, there was possible to slow down the pH and color changes as compared to storage at room temperature, which was 25 °C.

The hurdles such as HHP have been investigated for extending the shelf life of the low salt processed meat products. In this regard, O'Neill et al. (2018) showed the combination of HHP with organic acids or salt replacers in frankfurters and cooked ham was a very interesting strategy of such barriers compensated for the significant salt reduction and extended the shelf life. Fulladosa et al. (2012) indicated that using K-lactate combined with high-pressure (600 MPa) treatments provided an additional reduction in microbiological counts in restructured dry-cured ham. Similarly, Ferrini et al. (2012) showed that NaCl reduction or the substitution of salt using KCl or K-lactate mixed with a quick-dry-slice process and HHP allowed the production of dry-cured meat with reduced Na content, without lower microbiological risks. Also, the addition of K-lactate (25 g/kg) increased microbiological safety.

In sodium-reduced meat products, HHP has been applied at different stages of the process. Orel et al. (2020) showed with HHP at 600 MPa for 3 min applied in the ready-to-eat chicken breast with 50% NaCl replacement extended its microbiological shelf life. Pietrasik et al. (2017) applied HHP in reduced-sodium and nitrite-free naturally cured wieners and reported that HHP at 600 MPa was able to increase the shelf life and did not affect the processing characteristics of sausages. In addition, HHP showed a positive effect on the sensory quality of products, including sodium-reduced formulations containing natural forms of nitrite.

## 8.6 Final considerations

Salt reduction figures as one of the most necessary strategic reformulations to make food products healthier. However, its implementation is a great challenge for the food industry and health agencies, since replacing sodium chloride, the main source of sodium, requires a lot of effort to accomplish since it performs many functions that influence the technological, sensory, and microbiological safety properties of foods.

The success of approaches to reduce sodium intake must consider consumers' sensory perception accompanied by clear communication about the correlation between sodium intake and health. From a food industry perspective, the development of strategic solutions to compensate the different functions of sodium chloride in food processing is required. The use of extender ingredients, flavor enhancers, bioactive compounds, and the application of new process technologies have been widely reported in the literature, but they are still far from an ideal solution. Strategies for total or partial replacement of sodium chloride, finally, need to take into account the final cost of reformulated food products and regulatory aspects in order not to compromise the safety and consumer access to healthier foods.

## References

Afshin, A., Sur, P. J., Fay, K. A., Cornaby, L., Ferrara, G., Salama, J. S., Mullany, E. C., Abate, K. H., Abbafati, C., Abebe, Z., Afarideh, M., Aggarwal, A., Agrawal, S., Akinyemiju, T., Alahdab, F., Bacha, U., Bachman, V. F., Badali, H., Badawi, A., & Murray, C. J. L. (2017). Health effects of dietary risks in 195 countries, 1990–2017: A systematic analysis for the global burden of disease study. *The Lancet*, *393*, 30041–30048. https://doi.org/10.1016/S0140-6736(19.

Aheto, J. H., Huang, X., Xiaoyu, T., Bonah, E., Ren, Y., Alenyorege, E. A., & Chunxia, D. (2019). Investigation into crystal size effect on sodium chloride uptake and water activity of pork meat using hyperspectral imaging. *Journal of Food Processing and Preservation*, *43*(11). https://doi.org/10.1111/jfpp.14197.

Aljuraiban, G. S., Jose, A. P., Gupta, P., Shridhar, K., & Prabhakaran, D. (2021). Sodium intake, health implications, and the role of population-level strategies. *Nutrition Reviews*, *79*(3), 351–359. https://doi.org/10.1093/nutrit/nuaa042.

Amiri, A., Sharifian, P., & Soltanizadeh, N. (2018). Application of ultrasound treatment for improving the physicochemical, functional and rheological properties of myofibrillar proteins. *International Journal of Biological Macromolecules*, *111*, 139–147. https://doi.org/10.1016/j.ijbiomac.2017.12.167.

Anderson, C. A. M., Appel, L. J., Okuda, N., Brown, I. J., Chan, Q., Zhao, L., Ueshima, H., Kesteloot, H., Miura, K., Curb, J. D., Yoshita, K., Elliott, P., Yamamoto, M. E., & Stamler, J. (2010). Dietary sources of sodium in China, Japan, the United Kingdom, and the United States, women and men aged 40 to 59 years: The INTERMAP study. *Journal of the American Dietetic Association*, *110*(5), 736–745. https://doi.org/10.1016/j.jada.2010.02.007.

Antúnez, L., Giménez, A., & Ares, G. (2016). A consumer-based approach to salt reduction: Case study with bread. *Food Research International*, *90*, 66–72. https://doi.org/10.1016/j.foodres.2016.10.015.

Antúnez, L., Giménez, A., Vidal, L., & Ares, G. (2018). Partial replacement of NaCl with KCl in bread: Effect on sensory characteristics and consumer perception. *Journal of Sensory Studies*, *33*(5). https://doi.org/10.1111/joss.12441.

Arakawa, T., & Timasheff, S. N. (1991). *The interactions of proteins with salts, amino acids, and sugars at high concentration* (pp. 226–245). Springer Science and Business Media LLC. https://doi.org/10.1007/978-3-642-76226-0_8.

Armenteros, M., Aristoy, M. C., Barat, J. M., & Toldrá, F. (2012). Biochemical and sensory changes in dry-cured ham salted with partial replacements of NaCl by other chloride salts. *Meat Science*, *90*(2), 361–367. https://doi.org/10.1016/j.meatsci.2011.07.023.

Arpagaus, C. (2019). Nano spray drying of bioactive food ingredients. In *Proceedings of Eurodrying*.

Avramenko, N. A., Tyler, R. T., Scanlon, M. G., Hucl, P., & Nickerson, M. T. (2018). The chemistry of bread making: The role of salt to ensure optimal functionality of its constituents. *Food Reviews International*, *34*(3), 204–225. https://doi.org/10.1080/87559129.2016.1261296.

Balasubramaniam, V. M. B., Martínez-Monteagudo, S. I., & Gupta, R. (2015). Principles and application of high pressure-based technologies in the food industry. *Annual Review of Food Science and Technology*, *6*, 435–462. https://doi.org/10.1146/annurev-food-022814-015539.

Bansal, V., & Mishra, S. K. (2020). Reduced-sodium cheeses: Implications of reducing sodium chloride on cheese quality and safety. *Comprehensive Reviews in Food Science and Food Safety*, *19*(2), 733–758. https://doi.org/10.1111/1541-4337.12524.

Baptista, D. P., Araújo, F. D.d. S., Eberlin, M. N., & Gigante, M. L. (2017). Reduction of 25% salt in Prato cheese does not affect proteolysis and sensory acceptance. *International Dairy Journal*, *75*, 101–110. https://doi.org/10.1016/j.idairyj.2017.08.001.

Barretto, T. L., Pollonio, M. A. R., Telis-Romero, J., & da Silva Barretto, A. C. (2018). Improving sensory acceptance and physicochemical properties by ultrasound application to restructured cooked ham with salt (NaCl) reduction. *Meat Science*, *141*, 55–62. https://doi.org/10.1016/j.meatsci.2018.05.023.

Barros, J. C., Gois, T. S., Pires, M. A., Rodrigues, I., & Trindade, M. A. (2019). Sodium reduction in enrobed restructured chicken nuggets through replacement of NaCl with CaCl2. *Journal of Food Science and Technology*, *56*(8), 3587–3596. https://doi.org/10.1007/s13197-019-03777-8.

Beck, P. H. B., Matiucci, M. A., Neto, A. A. M., & Feihrmann, A. C. (2021). Sodium chloride reduction in fresh sausages using salt encapsulated in carnauba wax. *Meat Science*, *175*. https://doi.org/10.1016/j.meatsci.2021.108462.

Belz, M. C. E., Mairinger, R., Zannini, E., Ryan, L. A. M., Cashman, K. D., & Arendt, E. K. (2012). The effect of sourdough and calcium propionate on the microbial shelf-life of salt reduced bread. *Applied Microbiology and Biotechnology*, *96*(2), 493–501. https://doi.org/10.1007/s00253-012-4052-x.

Belz, M. C. E., Ryan, L. A. M., & Arendt, E. K. (2012). The impact of salt reduction in bread: A review. *Critical Reviews in Food Science and Nutrition*, *52*(6), 514–524. https://doi.org/10.1080/10408398.2010.502265.

Bhana, N., Utter, J., & Eyles, H. (2018). Knowledge, attitudes and behaviours related to dietary salt intake in high-income countries: A systematic review. *Current Nutrition Reports*, *7*(4), 183–197. https://doi.org/10.1007/s13668-018-0239-9.

Borch, E., Berg, H., & Holst, O. (1991). Heterolactic fermentation by a homofermentative Lactobacillus sp. during glucose limitation in anaerobic continuous culture with complete cell recycle. *Journal of Applied Bacteriology*, *71*(3), 265–269. https://doi.org/10.1111/j.1365-2672.1991.tb04457.x.

Bower, C. G., Stanley, R. E., Fernando, S. C., & Sullivan, G. A. (2018). The effect of salt reduction on the microbial community structure and quality characteristics of sliced roast beef and Turkey breast. *LWT*, *90*, 583–591. https://doi.org/10.1016/j.lwt.2017.12.067.

Busch, J. L., Tournier, C., Knoop, J. E., Kooyman, G., & Smit, G. (2009). Temporal contrast of salt delivery in mouth increases salt perception. *Chemical Senses*, *34*(4), 341–348. https://doi.org/10.1093/chemse/bjp007.

Busch, J. L. H. C, Yong, F. Y. S., & Goh, S. M. (2013). Sodium reduction: Optimizing product composition and structure towards increasing saltiness perception. *Trends in Food Science and Technology*, *29*(1), 21–34. https://doi.org/10.1016/j.tifs.2012.08.005.

Büssemaker, E., Hillebrand, U., Hausberg, M., Pavenstädt, H., & Oberleithner, H. (2010). Pathogenesis of hypertension: interactions among sodium, potassium, and aldosterone. *American Journal of Kidney Diseases*, *55*(6), 1111–1120. https://doi.org/10.1053/j.ajkd.2009.12.022.

Carrillo-Lopez, L. M., Alarcon-Rojo, A. D., Luna-Rodriguez, L., & Reyes-Villagrana, R. (2017). Modification of food systems by ultrasound. *Journal of Food Quality*, *2017*. https://doi.org/10.1155/2017/5794931.

Chindapan, N., Niamnuy, C., & Devahastin, S. (2018). Physical properties, morphology and saltiness of salt particles as affected by spray drying conditions and potassium chloride substitution. *Powder Technology*, *326*, 265–271. https://doi.org/10.1016/j.powtec.2017.12.014.

Cho, H. Y., Kim, B., Chun, J. Y., & Choi, M. J. (2015). Effect of spray-drying process on physical properties of sodium chloride/maltodextrin complexes. *Powder Technology*, *277*, 141–146. https://doi.org/10.1016/j.powtec.2015.02.027.

Chun, J. Y., Kim, B., Lee, J. G., Cho, H. Y., Min, S. G., & Choi, M. J. (2014). Effects of NaCl replacement with gamma-aminobutyric acid (GABA) on the quality characteristics and sensory properties of model meat products. *Korean Journal for Food Science of Animal Resources, 34*(4), 552–557. https://doi.org/10.5851/kosfa.2014.34.4.552.

Cichoski, A. J., Silva, M. S., Leães, Y. S. V., Brasil, C. C. B., de Menezes, C. R., Barin, J. S., Wagner, R., & Campagnol, P. C. B. (2019). Ultrasound: A promising technology to improve the technological quality of meat emulsions. *Meat Science, 148,* 150–155. https://doi.org/10.1016/j.meatsci.2018.10.009.

Corral, S., Salvador, A., & Flores, M. (2013). Salt reduction in slow fermented sausages affects the generation of aroma active compounds. *Meat Science, 93*(3), 776–785. https://doi.org/10.1016/j.meatsci.2012.11.040.

Costa, R. G. B., Sobral, D., Teodoro, V. A. M., Costa, L. C. G., de Paula, J. C. J., Landin, T. B., & de Oliveira, M. B. (2018). Sodium substitutes in Prato cheese: Impact on the physicochemical parameters, rheology aspects and sensory acceptance. *LWT- Food Science and Technology, 90,* 643–649. https://doi.org/10.1016/j.lwt.2017.12.051.

Crehan, C. M., Troy, D. J., & Buckley, D. J. (2000). Effects of salt level and high hydrostatic pressure processing on frankfurters formulated with 1.5 and 2.5% salt. *Meat Science, 55*(1), 123–130. https://doi.org/10.1016/S0309-1740(99)00134-5.

Cruz, A. G., Faria, J. A. F., Pollonio, M. A. R., Bolini, H. M. A., Celeghini, R. M. S., Granato, D., & Shah, N. P. (2011). Cheeses with reduced sodium content: Effects on functionality, public health benefits and sensory properties. *Trends in Food Science and Technology, 22*(6), 276–291. https://doi.org/10.1016/j.tifs.2011.02.003.

CTAC - Conseil de la Transformation Agroalimentaire et des Produits de Consommation. (2009). *Reformulation of products to reduce sodium: salt reduction guide for the food industry* (pp. 1–69). Canada: Édikom. https://www.foodtechcanada.ca/wp-content/uploads/2018/05/Salt-reduction-guide-for-the-food-industry.pdf.

D'Elia, L., Galletti, F., & Strazzullo, P. (2014). Dietary salt intake and risk of gastric cancer. *Cancer Treatment and Research, 159,* 83–95. https://doi.org/10.1007/978-3-642-38007-5_6.

Delgado-Pando, G., Fischer, E., Allen, P., Kerry, J. P., O'Sullivan, M. G., & Hamill, R. M. (2018). Salt content and minimum acceptable levels in whole-muscle cured meat products. *Meat Science, 139,* 179–186. https://doi.org/10.1016/j.meatsci.2018.01.025.

Desmond, E. (2006). Reducing salt: A challenge for the meat industry. *Meat Science, 74*(1), 188–196. https://doi.org/10.1016/j.meatsci.2006.04.014.

Dos Santos, B. A., Bastianello Campagnol, P. C., da Cruz, A. G., Galvão, M. T. E. L., Monteiro, R. A., Wagner, R., & Pollonio, M. A. R. (2015). Check all that apply and free listing to describe the sensory characteristics of low sodium dry fermented sausages: Comparison with trained panel. *Food Research International, 76,* 725–734. https://doi.org/10.1016/j.foodres.2015.06.035.

Dos Santos, B. A., Campagnol, P. C. B., Cavalcanti, R. N., Pacheco, M. T. B., Netto, F. M., Motta, E. M. P., Celeguini, R. M. S., Wagner, R., & Pollonio, M. A. R. (2015). Impact of sodium chloride replacement by salt substitutes on the proteolysis and rheological properties of dry fermented sausages. *Journal of Food Engineering, 151,* 16–24. https://doi.org/10.1016/j.jfoodeng.2014.11.015.

dos Santos, B. A., Campagnol, P. C. B., Fagundes, M. B., Wagner, R., & Pollonio, M. A. R. (2017). Adding blends of NaCl, KCl, and CaCl2 to low-sodium dry fermented sausages: Effects on lipid oxidation on curing process and shelf life. *Journal of Food Quality, 2017.* https://doi.org/10.1155/2017/7085798.

Dugat-Bony, E., Bonnarme, P., Fraud, S., Catellote, J., Sarthou, A. S., Loux, V., Rué, O., Bel, N., Chuzeville, S., & Helinck, S. (2019). Effect of sodium chloride reduction or partial substitution with potassium chloride on the microbiological, biochemical and sensory characteristics of semi-hard and soft cheeses. *Food Research International, 125.* https://doi.org/10.1016/j.foodres.2019.108643.

Dupree, D. E., Price, R. E., Burgess, B. A., Andress, E. L., & Breidt, F. (2019). Effects of sodium chloride or calcium chloride concentration on the growth and survival of *Escherichia coli* o157:H7 in model vegetable fermentations. *Journal of Food Protection, 82*(4), 570–578. https://doi.org/10.4315/0362-028X.JFP-18-468.

Emorine, M., Septier, C., Andriot, I., Martin, C., Salles, C., & Thomas-Danguin, T. (2015). Combined heterogeneous distribution of salt and aroma in food enhances salt perception. *Food & Function, 6*(5), 1449–1459. https://doi.org/10.1039/c4fo01067a.

Emorine, M., Septier, C., Thomas-Danguin, T., & Salles, C. (2013). Heterogeneous salt distribution in hot snacks enhances saltiness without loss of acceptability. *Food Research International, 51*(2), 641–647. https://doi.org/10.1016/j.foodres.2013.01.006.

Fellendorf, S., Kerry, J. P., Hamill, R. M., & O'Sullivan, M. G. (2018). Impact on the physicochemical and sensory properties of salt reduced corned beef formulated with and without the use of salt replacers. *LWT, 92*, 584–592. https://doi.org/10.1016/j.lwt.2018.03.001.

Ferrini, G., Comaposada, J., Arnau, J., & Gou, P. (2012). Colour modification in a cured meat model dried by quick-dry-slice process® and high pressure processed as a function of NaCl, KCl, K-lactate and water contents. *Innovative Food Science & Emerging Technologies, 13*, 69–74. https://doi.org/10.1016/j.ifset.2011.09.005.

Fleming, H. P., McFeeters, R. F., & Daeschel, M. A. (1985). The lactobacilli, pediococci, and leuconostocs: Vegetable products. In *Bacterial starter cultures for foods* (1st ed.). CRC Press. https://doi.org/10.1201/9781351070065-8.

Fosberg (Damiano), H., & Joyner (Melito), H. S. (2018). The impact of NaCl replacement with KCl and CaCl2 on cottage cheese cream dressing rheological behavior and consumer acceptance. *International Dairy Journal, 78*, 73–84. https://doi.org/10.1016/j.idairyj.2017.10.003.

Fraqueza, M. J., Laranjo, M., Elias, M., & Patarata, L. (2021). Microbiological hazards associated with salt and nitrite reduction in cured meat products: Control strategies based on antimicrobial effect of natural ingredients and protective microbiota. *Current Opinion in Food Science, 38*, 32–39. https://doi.org/10.1016/j.cofs.2020.10.027.

Fulladosa, E., Sala, X., Gou, P., Garriga, M., & Arnau, J. (2012). K-lactate and high pressure effects on the safety and quality of restructured hams. *Meat Science, 91*(1), 56–61. https://doi.org/10.1016/j.meatsci.2011.12.006.

Galvão, M. T. E. L., Moura, D. B., Barretto, A. C. S., & Pollonio, M. A. R. (2014). Effects of micronized sodium chloride on the sensory profile and consumer acceptance of Turkey ham with reduced sodium content. *Food Science and Technology, 34*(1), 189–194. https://doi.org/10.1590/S0101-20612014005000009.

Gaudette, N. J., Pietrasik, Z., & Johnston, S. P. (2019). Application of taste contrast to enhance the saltiness of reduced sodium beef patties. *LWT, 116*. https://doi.org/10.1016/j.lwt.2019.108585.

Gould, G. W., & Measures, J. C. (1977). Water relations in single cells. *Philosophical Transactions of the Royal Society of London. Series B, Biological Sciences, 278*(959), 151–166. https://doi.org/10.1098/rstb.1977.0035.

Guinee, T. P. (2004). Salting and the role of salt in cheese. *International Journal of Dairy Technology, 57*(2–3), 99–109. https://doi.org/10.1111/j.1471-0307.2004.00145.x.

Hayes, J. E., Sullivan, B. S., & Duffy, V. B. (2010). Explaining variability in sodium intake through oral sensory phenotype, salt sensation and liking. *Physiology & Behavior, 100*(4), 369–380. https://doi.org/10.1016/j.physbeh.2010.03.017.

Hoffmann, I. S., & Cubeddu, L. X. (2009). Salt and the metabolic syndrome. *Nutrition, Metabolism and Cardiovascular Diseases, 19*(2), 123–128. https://doi.org/10.1016/j.numecd.2008.02.011.

Horita, C. N., Messias, V. C., Morgano, M. A., Hayakawa, F. M., & Pollonio, M. A. R. (2014). Textural, microstructural and sensory properties of reduced sodium frankfurter sausages containing mechanically deboned poultry meat and blends of chloride salts. *Food Research International, 66*, 29–35. https://doi.org/10.1016/j.foodres.2014.09.002.

Horita, C. N., Morgano, M. A., Celeghini, R. M. S., & Pollonio, M. A. R. (2011). Physico-chemical and sensory properties of reduced-fat mortadella prepared with blends of calcium, magnesium and potassium chloride as partial substitutes for sodium chloride. *Meat Science, 89*(4), 426–433. https://doi.org/10.1016/j.meatsci.2011.05.010.

Hu, Y., Zhang, L., Zhang, H., Wang, Y., Chen, Q., & Kong, B. (2020). Physicochemical properties and flavour profile of fermented dry sausages with a reduction of sodium chloride. *LWT, 124*, 109061. https://doi.org/10.1016/j.lwt.2020.109061.

Hurst, K. E., Ayed, C., Derbenev, I. N., Hewson, L., & Fisk, I. D. (2021). Physicochemical design rules for the formulation of novel salt particles with optimised saltiness. *Food Chemistry, 360*. https://doi.org/10.1016/j.foodchem.2021.129990.

Hussain, R., Gaiani, C., Jeandel, C., Ghanbaja, J., & Scher, J. (2012). Combined effect of heat treatment and ionic strength on the functionality of whey proteins. *Journal of Dairy Science, 95*(11), 6260–6273. https://doi.org/10.3168/jds.2012-5416.

Hystead, E., Diez-Gonzalez, F., & Schoenfuss, T. C. (2013). The effect of sodium reduction with and without potassium chloride on the survival of *Listeria monocytogenes* in Cheddar cheese. *Journal of Dairy Science, 96*(10), 6172–6185. https://doi.org/10.3168/jds.2013-6675.

Inguglia, E. S., Zhang, Z., Tiwari, B. K., Kerry, J. P., & Burgess, C. M. (2017). Salt reduction strategies in processed meat products—A review. *Trends in Food Science and Technology, 59*, 70–78. https://doi.org/10.1016/j.tifs.2016.10.016.

Israr, T., Rakha, A., Sohail, M., Rashid, S., & Shehzad, A. (2016). Salt reduction in baked products: Strategies and constraints. *Trends in Food Science and Technology, 51*, 98–105. https://doi.org/10.1016/j.tifs.2016.03.002.

Jafari, S. M., Arpagaus, C., Cerqueira, M. A., & Samborska, K. (2021). Nano spray drying of food ingredients; materials, processing and applications. *Trends in Food Science and Technology, 109*, 632–646. https://doi.org/10.1016/j.tifs.2021.01.061.

Jakopović, K. L., Barukčić, I., Božić, A., & Božanić, R. (2020). Production of feta cheese with a reduced salt content. In *Vol. 9. Hrana u Zdravlju i Bolesti: Znanstveno-Stručni Časopis Za Nutricionizam i Dijetetiku* (pp. 9–15).

Jinap, S., Hajeb, P., Karim, R., Norliana, S., & Abdul-Kadir, S. Y. R. (2016). Reduction of sodium content in spicy soups using monosodium glutamate. *Food & Nutrition Research, 60*. https://doi.org/10.3402/fnr.v60.30463.

Johnson, M. E., Kapoor, R., McMahon, D. J., McCoy, D. R., & Narasimmon, R. G. (2009). Reduction of sodium and fat levels in natural and processed cheeses: Scientific and technological aspects. *Comprehensive Reviews in Food Science and Food Safety, 8*(3), 252–268. https://doi.org/10.1111/j.1541-4337.2009.00080.x.

Kalnina, I., & Straumite, E. (2019). A review–effect of salt on the sensory perception of snacks. In *Baltic conference on food science and technology: Conference proceedings*.

Kaushik, A, Peralta-Alvarez, F., Gupta, P., Bazo-Alvarez, J. C., Ofori, S., Bobrow, K., … Mohan, S. (2021). Assessing the policy landscape for salt reduction in South-East Asian and Latin American countries – An initiative towards developing an easily accessible, integrated, searchable online repository. *Global Heart, 16*(1), 1–12. https://doi.org/10.5334/gh.929. 49.

Keast, R. S. J., & Breslin, P. A. S. (2003). An overview of binary taste-taste interactions. *Food Quality and Preference, 14*(2), 111–124. https://doi.org/10.1016/S0950-3293(02)00110-6.

Khanipour, E., Flint, S. H., McCarthy, O. J., Palmer, J., Golding, M., Ratkowsky, D. A., Ross, T., & Tamplin, M. (2016). Modelling the combined effect of salt, sorbic acid and nisin on the probability of growth of *Clostridium sporogenes* in high moisture processed cheese analogue. *International Dairy Journal, 57*, 62–71. https://doi.org/10.1016/j.idairyj.2016.02.039.

Khetra, Y., Kanawjia, S. K., & Puri, R. (2016). Selection and optimization of salt replacer, flavour enhancer and bitter blocker for manufacturing low sodium Cheddar cheese using response surface methodology. *LWT- Food Science and Technology, 72*, 99–106. https://doi.org/10.1016/j.lwt.2016.04.035.

Kilcast, D., & Angus, F. (2007). Reducing salt in foods: Practical strategies. In *Reducing salt in foods: Practical strategies* (pp. 1–383). Elsevier Ltd. https://doi.org/10.1533/9781845693046.

Kongstad, S., & Giacalone, D. (2020). Consumer perception of salt-reduced potato chips: Sensory strategies, effect of labeling and individual health orientation. *Food Quality and Preference, 81*, 103856. https://doi.org/10.1016/j.foodqual.2019.103856.

Kuo, W. Y., & Lee, Y. (2014). Effect of food matrix on saltiness perception-implications for sodium reduction. *Comprehensive Reviews in Food Science and Food Safety*, *13*(5), 906–923. https://doi.org/10.1111/1541-4337.12094.

Laranjo, M., Gomes, A., Agulheiro-Santos, A. C., Potes, M. E., Cabrita, M. J., Garcia, R., Rocha, J. M., Roseiro, L. C., Fernandes, M. J., Fraqueza, M. J., & Elias, M. (2017). Impact of salt reduction on biogenic amines, fatty acids, microbiota, texture and sensory profile in traditional blood dry-cured sausages. *Food Chemistry*, *218*, 129–136. https://doi.org/10.1016/j.foodchem.2016.09.056.

Lawless, H. T., Rapacki, F., Horne, J., Hayes, A., & Wang, G. (2004). The taste of calcium chloride in mixtures with NaCl, sucrose and citric acid. *Food Quality and Preference*, *15*(1), 83–89. https://doi.org/10.1016/S0950-3293(03)00099-5.

Lee, J., Bae, J., Jeong, H., Cho, Y., & Choi, M. J. (2020). Saltiness enhancement in white pan bread supplemented with spray-dried salt-yeast complex. *Powder Technology*, *367*, 115–121. https://doi.org/10.1016/j.powtec.2020.03.024.

Leistner, L. (2000). Basic aspects of food preservation by hurdle technology. *International Journal of Food Microbiology*, *55*(1–3), 181–186. https://doi.org/10.1016/S0168-1605(00)00161-6.

Li, Y. L., Han, K. N., Feng, G. X., Wan, Z. L., Wang, G. S., & Yang, X. Q. (2021). Salt reduction in bread via enrichment of dietary fiber containing sodium and calcium. *Food & Function*, *12*(6), 2660–2671. https://doi.org/10.1039/d0fo03126g.

Liang, H., He, Z., Wang, X., Song, G., Chen, H., Lin, X., Ji, C., & Li, S. (2020). Effects of salt concentration on microbial diversity and volatile compounds during suancai fermentation. *Food Microbiology*, *91*, 103537. https://doi.org/10.1016/j.fm.2020.103537.

Lynch, E. J., Dal Bello, F., Sheehan, E. M., Cashman, K. D., & Arendt, E. K. (2009). Fundamental studies on the reduction of salt on dough and bread characteristics. *Food Research International*, *42*(7), 885–891. https://doi.org/10.1016/j.foodres.2009.03.014.

Ma, Y., He, F. J., & MacGregor, G. A. (2015). High salt intake: Independent risk factor for obesity? *Hypertension*, *66*(4), 843–849. https://doi.org/10.1161/HYPERTENSIONAHA.115.05948.

Martínez-Rodríguez, Y., Acosta-Muñiz, C., Olivas, G. I., Guerrero-Beltrán, J., Rodrigo-Aliaga, D., & Sepúlveda, D. R. (2012). High hydrostatic pressure processing of cheese. *Comprehensive Reviews in Food Science and Food Safety*, *11*(4), 399–416. https://doi.org/10.1111/j.1541-4337.2012.00192.x.

McCarthy, C. M., Wilkinson, M. G., Kelly, P. M., & Guinee, T. P. (2015). Effect of salt and fat reduction on the composition, lactose metabolism, water activity and microbiology of Cheddar cheese. *Dairy Science and Technology*, *95*(5), 587–611. https://doi.org/10.1007/s13594-015-0245-2.

McKenzie, B., Santos, J. A., Trieu, K., Thout, S. R., Johnson, C., Arcand, J. A., Webster, J., & McLean, R. (2018). The science of salt: A focused review on salt-related knowledge, attitudes and behaviors, and gender differences. *Journal of Clinical Hypertension*, *20*(5), 850–866. https://doi.org/10.1111/jch.13289.

Mehyar, G. F., Al Nabulsi, A. A., Saleh, M., Olaimat, A. N., & Holley, R. A. (2018). Effects of chitosan coating containing lysozyme or natamycin on shelf-life, microbial quality, and sensory properties of halloumi cheese brined in normal and reduced salt solutions. *Journal of Food Processing and Preservation*, *42*(1). https://doi.org/10.1111/jfpp.13324.

Mendoza, J. E., Schram, G. A., Arcand, J. A., Henson, S., & L'Abbe, M. (2014). Assessment of consumers' level of engagement in following recommendations for lowering sodium intake. *Appetite*, *73*, 51–57. https://doi.org/10.1016/j.appet.2013.10.007.

Messens, W., Van Camp, J., & Huyghebaert, A. (1997). The use of high pressure to modify the functionality of food proteins. *Trends in Food Science and Technology*, *8*(4), 107–112. https://doi.org/10.1016/S0924-2244(97)01015-7.

Moncada, M., Astete, C., Sabliov, C., Olson, D., Boeneke, C., & Aryana, K. J. (2015). Nano spray-dried sodium chloride and its effects on the microbiological and sensory characteristics of surface-salted cheese crackers. *Journal of Dairy Science*, *98*(9), 5946–5954. https://doi.org/10.3168/jds.2015-9658.

Monteiro, A. R. G., Nakagawa, A., Pimentel, T. C., & Sousa, I. (2021). Increasing saltiness perception and keeping quality properties of low salt bread using inhomogeneous salt distribution achieved with salt agglomerated by waxy starch. *LWT, 146*. https://doi.org/10.1016/j.lwt.2021.111451.

Neal, B. (2014). Dietary salt is a public health hazard that requires vigorous attack. *Canadian Journal of Cardiology, 30*(5), 502–506. https://doi.org/10.1016/j.cjca.2014.02.005.

Nguyen, H., & Wismer, W. V. (2019). A comparison of sensory attribute profiles and liking between regular and sodium-reduced food products. *Food Research International, 123*, 631–641. https://doi.org/10.1016/j.foodres.2019.05.037.

Ni Mhurchu, C., Capelin, C., Dunford, E. K., Webster, J. L., Neal, B. C., & Jebb, S. A. (2011). Sodium content of processed foods in the United Kingdom: Analysis of 44,000 foods purchased by 21,000 households1-3. *American Journal of Clinical Nutrition, 93*(3), 594–600. https://doi.org/10.3945/ajcn.110.004481.

Noort, M. W. J., Bult, J. H. F., Stieger, M., & Hamer, R. J. (2010). Saltiness enhancement in bread by inhomogeneous spatial distribution of sodium chloride. *Journal of Cereal Science, 52*(3), 378–386. https://doi.org/10.1016/j.jcs.2010.06.018.

Offer, G., Knight, P., Jeacocke, R., Almond, R., Cousins, T., Elsey, J., … Purslow, P. (1989). The structural basis of the water-holding, appearance and toughness of meat and meat products. *Food Structure, 8*(1), 151–170. 17.

O'Neill, C. M., Cruz-Romero, M. C., Duffy, G., & Kerry, J. P. (2018). Shelf life extension of vacuum-packed salt reduced frankfurters and cooked ham through the combined application of high pressure processing and organic acids. *Food Packaging and Shelf Life, 17*, 120–128. https://doi.org/10.1016/j.fpsl.2018.06.008.

Orel, R., Tabilo-Munizaga, G., Cepero-Betancourt, Y., Reyes-Parra, J. E., Badillo-Ortiz, A., & Pérez-Won, M. (2020). Effects of high hydrostatic pressure processing and sodium reduction on physicochemical properties, sensory quality, and microbiological shelf life of ready-to-eat chicken breasts. *LWT, 127*, 109352. https://doi.org/10.1016/j.lwt.2020.109352.

Ozturk, M., Govindasamy-Lucey, S., Jaeggi, J. J., Johnson, M. E., & Lucey, J. A. (2013). The influence of high hydrostatic pressure on regular, reduced, low and no salt added Cheddar cheese. *International Dairy Journal, 33*(2), 175–183. https://doi.org/10.1016/j.idairyj.2013.01.008.

Pan, Q., Yang, G.h., Wang, Y., Wang, X.x., Zhou, Y., Li, P.j., & Chen, C.g. (2020). Application of ultrasound-assisted and tumbling dry-curing techniques for reduced-sodium bacon. *Journal of Food Processing and Preservation, 44*(8). https://doi.org/10.1111/jfpp.14607.

Pasqualone, A., Caponio, F., Pagani, M. A., Summo, C., & Paradiso, V. M. (2019). Effect of salt reduction on quality and acceptability of durum wheat bread. *Food Chemistry, 289*, 575–581. https://doi.org/10.1016/j.foodchem.2019.03.098.

Patarata, L., Novais, M., Fraqueza, M. J., & Silva, J. A. (2020). Influence of meat spoilage microbiota initial load on the growth and survival of three pathogens on a naturally fermented sausage. *Food, 9*(5). https://doi.org/10.3390/foods9050676.

Pateiro, M., Munekata, P. E., Cittadini, A., Domínguez, R., & Lorenzo, J. M. (2021). Metallic-based salt substitutes to reduce sodium content in meat products. *Current Opinion in Food Science, 38*, 21–31. https://doi.org/10.1016/j.cofs.2020.10.029.

Pietrasik, Z., & Gaudette, N. J. (2014). The impact of salt replacers and flavor enhancer on the processing characteristics and consumer acceptance of restructured cooked hams. *Meat Science, 96*(3), 1165–1170. https://doi.org/10.1016/j.meatsci.2013.11.005.

Pietrasik, Z., Gaudette, N. J., & Johnston, S. P. (2017). The impact of high hydrostatic pressure on the functionality and consumer acceptability of reduced sodium naturally cured wieners. *Meat Science, 129*, 127–134. https://doi.org/10.1016/j.meatsci.2017.02.020.

Pino, A., De Angelis, M. D., Todaro, A., Van Hoorde, K. V., Randazzo, C. L., & Caggia, C. (2018). Fermentation of Nocellara Etnea table olives by functional starter cultures at different low salt concentrations. *Frontiers in Microbiology, 9*. https://doi.org/10.3389/fmicb.2018.01125.

Pires, M. A., Munekata, P. E. S., Baldin, J. C., Rocha, Y. J. P., Carvalho, L. T., dos Santos, I. R., Barros, J. C., & Trindade, M. A. (2017). The effect of sodium reduction on the microstructure, texture and sensory acceptance of Bologna sausage. *Food Structure, 14*, 1–7. https://doi.org/10.1016/j.foostr.2017.05.002.

Powles, J., Fahimi, S., Micha, R., Khatibzadeh, S., Shi, P., Ezzati, M., Engell, R. E., Lim, S. S., Danaei, G., & Mozaffarian, D. (2013). Global, regional and national sodium intakes in 1990 and 2010: A systematic analysis of 24 h urinary sodium excretion and dietary surveys worldwide. *BMJ Open, 3*(12). https://doi.org/10.1136/bmjopen-2013-003733.

Puolanne, E., & Halonen, M. (2010). Theoretical aspects of water-holding in meat. *Meat Science, 86*(1), 151–165. https://doi.org/10.1016/j.meatsci.2010.04.038.

Quilaqueo, M., & Aguilera, J. M. (2016). Crystallization of NaCl by fast evaporation of water in droplets of NaCl solutions. *Food Research International, 84*, 143–149. https://doi.org/10.1016/j.foodres.2016.03.030.

Raffo, A., Carcea, M., Moneta, E., Narducci, V., Nicoli, S., Peparaio, M., Sinesio, F., & Turfani, V. (2018). Influence of different levels of sodium chloride and of a reduced-sodium salt substitute on volatiles formation and sensory quality of wheat bread. *Journal of Cereal Science, 79*, 518–526. https://doi.org/10.1016/j.jcs.2017.12.013.

Rama, R., Chiu, N., Carvalho Da Silva, M., Hewson, L., Hort, J., & Fisk, I. D. (2013). Impact of salt crystal size on in-mouth delivery of sodium and saltiness perception from snack foods. *Journal of Texture Studies, 44*(5), 338–345. https://doi.org/10.1111/jtxs.12017.

Raybaudi-Massilia, R., Mosqueda-Melgar, J., Rosales-Oballos, Y., Citti de Petricone, R., Frágenas, N. N., Zambrano-Durán, A., Sayago, K., Lara, M., & Urbina, G. (2019). New alternative to reduce sodium chloride in meat products: Sensory and microbiological evaluation. *LWT, 108*, 253–260. https://doi.org/10.1016/j.lwt.2019.03.057.

Reißner, A. M., Wendt, J., Zahn, S., & Rohm, H. (2019). Sodium-chloride reduction by substitution with potassium, calcium and magnesium salts in wheat bread. *LWT, 108*, 153–159. https://doi.org/10.1016/j.lwt.2019.03.069.

Rios-Mera, J. D., Saldaña, E., Cruzado-Bravo, M. L. M., Martins, M. M., Patinho, I., Selani, M. M., Valentin, D., & Contreras-Castillo, C. J. (2020). Impact of the content and size of NaCl on dynamic sensory profile and instrumental texture of beef burgers. *Meat Science, 161*. https://doi.org/10.1016/j.meatsci.2019.107992.

Rios-Mera, J. D., Saldaña, E., Cruzado-Bravo, M. L. M., Patinho, I., Selani, M. M., Valentin, D., & Contreras-Castillo, C. J. (2019). Reducing the sodium content without modifying the quality of beef burgers by adding micronized salt. *Food Research International, 121*, 288–295. https://doi.org/10.1016/j.foodres.2019.03.044.

Rodrigues, I., Gonçalves, L. A., Carvalho, F. A. L., Pires, M., JP Rocha, Y., Barros, J. C., Carvalho, L. T., & Trindade, M. A. (2020). Understanding salt reduction in fat-reduced hot dog sausages: Network structure, emulsion stability and consumer acceptance. *Food Science and Technology International, 26*(2), 123–131. https://doi.org/10.1177/1082013219872677.

Rodrigues, J. F., Pereira, R. C., Silva, A. A., Mendes, A. O., & Carneiro, J.d. D. S. (2017). Sodium content in foods: Brazilian consumers' opinions, subjective knowledge and purchase intent. *International Journal of Consumer Studies, 41*(6), 735–744. https://doi.org/10.1111/ijcs.12386.

Ros-Polski, V, Koutchma, T., Xue, J., Defelice, C., & Balamurugan, S. (2015). Effects of high hydrostatic pressure processing parameters and NaCl concentration on the physical properties, texture and quality of white chicken meat. *Innovative Food Science & Emerging Technologies, 30*, 31–42. https://doi.org/10.1016/j.ifset.2015.04.003.

Ruusunen, M., & Puolanne, E. (2005). Reducing sodium intake from meat products. In *Vol. 70(3). Meat Science* (pp. 531–541). Elsevier Ltd. https://doi.org/10.1016/j.meatsci.2004.07.016.

Samapundo, S., Anthierens, T., Ampofo-Asiama, J., Xhaferi, R., Van Bree, I., Szczepaniak, S., Goemaere, O., Steen, L., Dhooge, M., Paelinck, H., & Devlieghere, F. (2013). The effec. of NaCl reduction and replacement on the growth of *Listeria monocytogenes* in broth, cooked ham and white sauce. *Journal of Food Safety, 33*(1), 59–70. https://doi.org/10.1111/jfs.12023.

Samapundo, S., Deschuyffeleer, N., Van Laere, D., De Leyn, I., & Devlieghere, F. (2010). Effect of NaCl reduction and replacement on the growth of fungi important to the spoilage of bread. *Food Microbiology*, *27*(6), 749–756. https://doi.org/10.1016/j.fm.2010.03.009.

Shelef, L., & Seiter, J. (2005). Indirect and miscellaneous antimicrobials. In P. M. Davidson, J. N. Sofos, & A. L. Branen (Eds.), *Antimicrobials in food* (3rd ed.). CRC Press. https://doi.org/10.1201/9781420028737.ch17.

Shen, S., Hoffman, A. J., & Butler, S. E. (2015). *Method of producing salt composition*. WO Patent WO2015015151.

Sihufe, G. A., De Piante Vicín, D. A., Marino, F., Ramos, E. L., Nieto, I. G., Karlen, J. G., & Zorrilla, S. E. (2018). Effect of sodium chloride reduction on physicochemical, biochemical, rheological, structural and sensory characteristics of Tybo cheese. *International Dairy Journal*, *82*, 11–18. https://doi.org/10.1016/j.idairyj.2018.02.006.

Sikes, A. L., Tobin, A. B., & Tume, R. K. (2009). Use of high pressure to reduce cook loss and improve texture of low-salt beef sausage batters. *Innovative Food Science and Emerging Technologies*, *10*(4), 405–412. https://doi.org/10.1016/j.ifset.2009.02.007.

Silva, H. L. A., Balthazar, C. F., Silva, R., Vieira, A. H., Costa, R. G. B., Esmerino, E. A., Freitas, M. Q., & Cruz, A. G. (2018). Sodium reduction and flavor enhancer addition in probiotic Prato cheese: Contributions of quantitative descriptive analysis and temporal dominance of sensations for sensory profiling. *Journal of Dairy Science*, *101*(10), 8837–8846. https://doi.org/10.3168/jds.2018-14819.

Sinesio, F., Raffo, A., Peparaio, M., Moneta, E., Saggia Civitelli, E., Narducci, V., Turfani, V., Ferrari Nicoli, S., & Carcea, M. (2019). Impact of sodium reduction strategies on volatile compounds, sensory properties and consumer perception in commercial wheat bread. *Food Chemistry*, *301*, 125252. https://doi.org/10.1016/j.foodchem.2019.125252.

Singh, S., & Shalini, R. (2016). Effect of hurdle technology in food preservation: A review. *Critical Reviews in Food Science and Nutrition*, *56*(4), 641–649. https://doi.org/10.1080/10408398.2012.761594.

Steffensen, I. L., Frølich, W., Dahl, K. H., Iversen, P. O., Lyche, J. L., Lillegaard, I. T. L., & Alexander, J. (2018). Benefit and risk assessment of increasing potassium intake by replacement of sodium chloride with potassium chloride in industrial food products in Norway. *Food and Chemical Toxicology*, *111*, 329–340. https://doi.org/10.1016/j.fct.2017.11.044.

Sun, R., Lu, J., & Nolden, A. (2021). Nanostructured foods for improved sensory attributes. *Trends in Food Science & Technology*, *108*, 281–286. https://doi.org/10.1016/j.tifs.2021.01.011.

Suna, S., Sinir, G.Ö., & Copur, Ö. (2014). Nano-spray drying applications in food industry. In *Proceedings of the 11th international conference of food physicists food physics and innovative technologies* (pp. 10–12).

Tamm, A., Bolumar, T., Bajovic, B., & Toepfl, S. (2016). Salt (NaCl) reduction in cooked ham by a combined approach of high pressure treatment and the salt replacer KCl. *Innovative Food Science and Emerging Technologies*, *36*, 294–302. https://doi.org/10.1016/j.ifset.2016.07.010.

Tan, H. L., Tan, T. C., & Easa, A. M. (2020). The use of selected hydrocolloids and salt substitutes on structural integrity, texture, sensory properties, and shelf life of fresh no salt wheat noodles. *Food Hydrocolloids*, *108*. https://doi.org/10.1016/j.foodhyd.2020.105996.

Taormina, P. J. (2010). Implications of salt and sodium reduction on microbial food safety. *Critical Reviews in Food Science and Nutrition*, *50*(3), 209–227. https://doi.org/10.1080/10408391003626207.

Tate & Lyel. (2014). *Soda-lo® salt microspheres*. https://www.tateandlyle.com/ingredient/soda-lo-salt-microspheres.

Teixeira, A., & Rodrigues, S. (2021). Consumer perceptions towards healthier meat products. *Current Opinion in Food Science*, *38*, 147–154. https://doi.org/10.1016/j.cofs.2020.12.004.

Tobin, B. D., O'Sullivan, M. G., Hamill, R. M., & Kerry, J. P. (2012). Effect of varying salt and fat levels on the sensory quality of beef patties. *Meat Science*, *91*(4), 460–465. https://doi.org/10.1016/j.meatsci.2012.02.032.

Torrico, D. D., Nguyen, P. T., Li, T., Mena, B., Gonzalez Viejo, C., Fuentes, S., & Dunshea, F. R. (2019). Sensory acceptability, quality and purchase intent of potato chips with reduced salt (NaCl) concentrations. *LWT*, *102*, 347–355. https://doi.org/10.1016/j.lwt.2018.12.050.

Verma, A. K., & Banerjee, R. (2012). Low-sodium meat products: Retaining salty taste for sweet health. *Null, 52*(1), 72–84. https://doi.org/10.1080/10408398.2010.498064.

Vidal, V. A. S., Biachi, J. P., Paglarini, C. S., Pinton, M. B., Campagnol, P. C. B., Esmerino, E. A., da Cruz, A. G., Morgano, M. A., & Pollonio, M. A. R. (2019). Reducing 50% sodium chloride in healthier jerked beef: An efficient design to ensure suitable stability, technological and sensory properties. *Meat Science, 152*, 49–57. https://doi.org/10.1016/j.meatsci.2019.02.005.

Vidal, V. A. S., Lorenzo, J. M., Munekata, P. E. S., & Pollonio, M. A. R. (2021). Challenges to reduce or replace NaCl by chloride salts in meat products made from whole pieces—A review. *Critical Reviews in Food Science and Nutrition, 61*(13), 2194–2206. https://doi.org/10.1080/10408398.2020.1774495.

Vidal, V. A. S., Paglarini, C. S., Freitas, M. Q., Coimbra, L. O., Esmerino, E. A., Pollonio, M. A. R., & Cruz, A. G. (2020). Q methodology: An interesting strategy for concept profile and sensory description of low sodium salted meat. *Meat Science, 161*. https://doi.org/10.1016/j.meatsci.2019.108000.

Vinitha, K., Leena, M. M., Moses, J. A., & Anandharamakrishnan, C. (2021). Size-dependent enhancement in salt perception: Spraying approaches to reduce sodium content in foods. *Powder Technology, 378*, 237–245. https://doi.org/10.1016/j.powtec.2020.09.079.

Wang, C. Y., Huang, H. W., Hsu, C. P., & Yang, B. B. (2016). Recent advances in food processing using high hydrostatic pressure technology. *Critical Reviews in Food Science and Nutrition, 56*(4), 527–540. https://doi.org/10.1080/10408398.2012.745479.

Whiting, R. C., Benedict, R. C., Kunsch, C. A., & Blalock, D. (1985). Growth of *Clostridium sporogenes* and *Staphylococcus aureus* at different temperatures in cooked corned beef made with reduced levels of sodium chloride. *Journal of Food Science, 50*(2), 304–307. https://doi.org/10.1111/j.1365-2621.1985.tb13387.x.

WHO. (2012). *Guideline: Potassium intake for adults and children*. Geneva, Switzerland: World Health Organization. WHO Document Production Services http://www.who.int/nutrition/publications/guidelines/potassium_intake/en/.

WHO. (2013). *Mapping salt reduction initiatives in the WHO European region* (pp. 1–50). World Health Organ. Tech. Rep. https://www.euro.who.int/__data/assets/pdf_file/0009/186462/Mapping-salt-reduction-initiatives-in-the-WHO-European-Region.pdf.

Wusimanjiang, P., Ozturk, M., Ayhan, Z., & Çagri Mehmetoglu, A. (2019). Effect of salt concentration on acid- and salt-adapted Escherichia coli O157:H7 and listeria monocytogenes in recombined nonfat cast cheese. *Journal of Food Processing and Preservation, 43*(11). https://doi.org/10.1111/jfpp.14208.

Yamamoto, K. (2017). Food processing by high hydrostatic pressure. *Bioscience, Biotechnology, and Biochemistry, 81*(4), 672–679. https://doi.org/10.1080/09168451.2017.1281723.

Yang, H. J., Han, M. Y., Wang, H.f., Cao, G.t., Tao, F., Xu, X. L., Zhou, G. H., & Shen, Q. (2021). HPP improves the emulsion properties of reduced fat and salt meat batters by promoting the adsorption of proteins at fat droplets/water interface. *LWT, 137*. https://doi.org/10.1016/j.lwt.2020.110394.

Yi, C., Tsai, M. L., & Liu, T. (2017). Spray-dried chitosan/acid/NaCl microparticles enhance saltiness perception. *Carbohydrate Polymers, 172*, 246–254. https://doi.org/10.1016/j.carbpol.2017.05.066.

Zhao, C. C., & Eun, J. B. (2018). Influence of ultrasound application and NaCl concentrations on brining kinetics and textural properties of Chinese cabbage. *Ultrasonics Sonochemistry, 49*, 137–144. https://doi.org/10.1016/j.ultsonch.2018.07.039.

Zisu, B., Bhaskaracharya, R., Kentish, S., & Ashokkumar, M. (2010). Ultrasonic processing of dairy systems in large scale reactors. *Ultrasonics Sonochemistry, 17*(6), 1075–1081. https://doi.org/10.1016/j.ultsonch.2009.10.014.

# CHAPTER 9

# Strategies for the reduction of sugar in food products

Ana Gomes[a], Ana I. Bourbon[b], Ana Rita Peixoto[c], Ana Sanches Silva[d,e], Ana Tasso[f], Carina Almeida[d], Clarisse Nobre[g], Cláudia Nunes[h,i], Claudia Sánchez[j], Daniela A. Gonçalves[g], Diogo Castelo-Branco[f], Diogo Figueira[f], Elisabete Coelho[h], Joana Gonçalves[a], José A. Teixeira[g], Lorenzo Miguel Pastrana Castro[b], Manuel A. Coimbra[h], Manuela Pintado[a], Miguel Ângelo Parente Ribeiro Cerqueira[b], Pablo Fuciños[b], Paula Teixeira[a], Pedro A.R. Fernandes[h,i], and Vitor D. Alves[k]

[a]*CBQF-Centro de Biotecnologia e Química Fina—Laboratório Associado, Escola Superior de Biotecnologia, Universidade Católica Portuguesa, Porto, Portugal,* [b]*International Iberian Nanotechnology Laboratory, Braga, Portugal,* [c]*Vieira de Castro Produtos Alimentares, S.A, V. N. Famalicão, Portugal,* [d]*National Institute for Agricultural and Veterinary Research (INIAV), I.P., Vila do Conde, Portugal,* [e]*Center for Study in Animal Science (CECA), University of Oporto, Oporto, Portugal,* [f]*Mendes Gonçalves SA, Golegã, Portugal,* [g]*CEB—Centre of Biological Engineering, University of Minho, Braga, Portugal,* [h]*LAQV-REQUIMTE, Department of Chemistry, University of Aveiro, Campus Universitário de Santiago, Aveiro, Portugal,* [i]*CICECO, Department of Chemistry, University of Aveiro, Aveiro, Portugal,* [j]*National Institute for Agricultural and Veterinary Research (INIAV), I.P., Alcobaça, Portugal,* [k]*Frulact SA, Maia, Portugal*

## 9.1 Introduction

Nowadays consumers tend to prefer "natural" food ingredients rather than "synthetic" counterparts (Maruyama et al., 2021). In fact, there is a dichotomy associating natural compounds (frequently designated with the prefix bio-) to "healthy" ingredients and synthetic ones to "unhealthy" (Bast & Haenen, 2013). Understandably, this is not always accurate, for instance many toxins (e.g., mycotoxins, plant toxins) are widespread in natural sources, and associated with innumerous severe disorders/diseases, including cancer. The general population commonly relates the word "chemical" with danger, therefore "chemofobia" is established in our society. Due to these conceptions, organic farming, known for not using pesticides such as fungicides, insecticides, and herbicides, has gained considerable attention and many followers in the last years developing from a niche to become a true alternative to cope with societal and environmental challenges.

Some consumers are very demanding with food quality standards. Some of these have also economic capacity to buy more expensive and higher quality food products. However, many times, they are also influenced by the information released by food industry, many times disseminated through mass media, which can be misinterpreted. A good example is the case of food antioxidants; although there are a considerable number of scientific studies associating their intake to positive health effects, some studies report a lack of activity or even toxic/harmful effects at high doses (Boaz et al., 2000; Lloret et al., 2019; Middha et al., 2019). However, some actors within the food industry advertise food products based on their content on antioxidants and potential benefits for human health, even though in

Europe there is no approved health claim concerning antioxidant activity except those associated with antioxidant vitamins and olive oil polyphenols (European Parliament, 2012).

In this context, clean label is a recent concept used within the food industry, which results from the reduction or total removal of artificial ingredients or synthetic chemicals (e.g., preservatives, colorants, flavors). However, there is still no common and objective official definition of clean label. Asioli et al. (2017) defined clean label in a broad (front-of-pack) and strict (back-of-pack) sense. The main idea of this concept is to reduce the list of ingredients in order to have a short list of known (familiar) ingredients instead of additives codes (E-numbers). Sometimes this concept can be very narrow, therefore, there are those who defend the use of the term "clean(er) label" in order to demonstrate constant progress (Puratos, 2021).

Non communicable diseases (NCDs) are the leading causes of death and were responsible for 41 million (71%) of the world's 55 million deaths in 2019 (WHO, 2021). Modifiable risk factors such as inadequate dietary habits and physical inactivity are some of the most common causes of NCDs; they are also risk factors for obesity (WHO, 2018). This constitutes a serious public health concern and is currently considered a global epidemic, affecting all age groups. In addition, obesity is associated with an increased risk of death (Prada et al., 2020). Excessive sugar intake is of concern because of its association with poor dietary quality, obesity and risk of NCDs (WHO, 2015), as it contributes to an increase in total energy consumption, usually without other nutritional benefits (e.g., vitamins or minerals) (Prada et al., 2020). Consequently, sugars are among the ingredients that the food industry tries most to reduce (Struck et al., 2014). In fact, to prevent weight disorders and chronic diseases, the World Health Organization (WHO) strongly recommends that the energy intake from free sugars per day in both adults and children represents less than 10% of the total energy intake (WHO, 2015). Moreover, a further reduction up to 5% is also recommended.

Dietary recommendations include an appropriate diet rich in fruits and vegetables and the reduction of the intake of added sugar foods and drinks. However, the compliance with these recommendations often compromises the products' sensory properties whereas poor compliance with these recommendations makes food less healthy. So, the effective way to attain sugar reduction targets lies in decreasing the added sugar content of the processed products to levels that do not compromise the product properties.

The main role of sugar in foods is to provide sweetness, but it also contributes to the flavor, due to its involvement in Maillard reactions, texture and preservation. In addition, sugar also helps to balance other tastes such as salty, sour, and spicy in less sweet products (Clemens et al., 2016). In beverages, sugar is also utilized to provide the desired mouthfeel (Gwinn, 2013). Therefore, sugar reduction represents an important challenge for the food industry, in different technological aspects, in order to maintain the sensory quality of the product and ensure that it fulfills the consumer's expectations.

Among the various strategies used nowadays for achieving sugar reduction, product reformulation by partial or total replacement with sweeteners is the most common approach in both solid and liquid foods (Di Monaco et al., 2018). Although, this approach allows to substantially reduce the amount of sugar in the products, the sensory profile and satiety value of sucrose cannot be fully replaced. Thus, a combination of different sweeteners is commonly used to attain a taste and flavor similar to that of sucrose.

Several nutritive sweeteners (e.g., fructose, glucose, lactose, high-fructose corn syrup, trehalose, and polyols such as erythritol, sorbitol, mannitol, xylitol, etc.) and non-nutritive sweeteners (e.g., natural compounds such as stevia, thaumatin and monk fruit, and artificial compounds like saccharin, aspartame, sucralose, etc.) are currently available for sugar substitution.

An alternative strategy is the gradual reduction of sugar (MacGregor & Hashem, 2014). This is a relatively easy approach, applied both in solid and liquid foods, where sugar is reduced slowly and progressively. Although consumers may notice a difference in the sweetness of the products, they gradually adapt to the changes and there is no negative effect on product acceptance.

In dairy products, lactose hydrolysis can also be used as a method for sugar reduction (Harju et al., 2012). Enzymatic hydrolysis of lactose (lactase/β-galactosidase) in milk can cause up to 70% of the lactose breakdown, increasing the sweetness equivalent and, consequently, the sweetening in comparison to regular milk. However, the single-use of enzymes is an expensive strategy. Ultra- and nanofiltration are membrane filtration methodologies that, in addition to allowing enzymes to be recovered and reused, also allow the removal of lactose (Jelen & Tossavainen, 2003). Then, used in conjunction with lactose hydrolysis, filtration methodologies represent a relatively quick and cost-effective process to reduce sugar in milk and milk-based products.

The use of multisensory interactions is another useful method for sugar reduction. Although a high level of sugar reduction cannot be achieved, the formulation of the product is easy and achievable without sweeteners. The method is based on the enhanced perception of sweetness intensity produced by aroma, color and other stimuli (Hutchings et al., 2019).

One more innovative alternative for sugar reduction is the heterogeneous distribution method, which uses stimulation of taste receptors to enhance sweetness perception. This method can make use of serum release from solid foods, as well as the particle size and viscosity in solid and liquid foods (Hutchings et al., 2019) for an optimized stimulation of taste receptors.

This chapter aims to address in further detail the main strategies to reduce sugar content in food products. In this frame, the main food grade alternatives to sugar will be discussed. Moreover, the methods to improve sweetening will be discussed and the products able to reduce sugar content already in the market, or with potential to be commercialized in a near future, will be reviewed, considering their characteristics, main advantages and drawbacks.

## 9.2 Functional and technological role of sugar in food products

Sugars, defined as monosaccharides and disaccharides, are naturally present in several foods namely fruit, vegetables, cereals, milk and honey. For thousands of years, human beings have consumed sugars as part of the intact tissue matrix of fruits and vegetables. The availability and sources of sugars in the diet markedly increased throughout history, as humans started to extract sugars, mainly sucrose, from natural sources to sweeten beverages and culinary preparations (Deliza et al., 2021).

The most notable function of sugar is its sweet taste. The perception of sugars sweetness is attributed to their interaction with the taste receptors along the surface of the tongue (Hutchings et al., 2019). Besides, it contributes to the flavor profile and affects mouthfeel and texture properties as well. It is involved in the Maillard reaction, affects freezing point, acts as bulking and preserving agent and promotes lightness among others (Di Monaco et al., 2018).

Additionally, in confectionery products sucrose, as hygroscopic, binds to the water present in the cake batter, which results in an increase in the viscosity of the batter helping to retain gas bubbles and consequently increasing the final volume of the cake. As sugar binds with water, this prevents the full hydration of the gluten proteins (found in the flour), preventing the formation of a gluten network, thus acting as a tenderizer of baked goods (Richardson et al., 2021). Sucrose, also increases the temperature of starch gelatinization and egg protein denaturation, allowing gas bubbles to expand before the gel formation. Sucrose is also important for the browning of the crust, since when exposed to heat, it degrades to fructose and glucose. These monosaccharides are reducing sugars and participate in Maillard browning. Sucrose also undergoes caramelization at high temperatures and produces a brown color (Milner et al., 2020).

In beverages, sucrose performs different functions besides sweetness. It also helps to balance other tastes such as sour, salty, and spicy in less sweet products. It is cheap and is an efficient way to increase the liking and acceptance of foods (Mahato et al., 2020).

## 9.3 Food reformulation to reduce free sugar intake

The term "sugars" includes those that are consumed as a naturally occurring component of many foods namely intrinsic sugars incorporated within the structure of intact fruit and vegetables or sugars from milk (lactose and galactose); and free sugars, which are monosaccharides and disaccharides added to foods and beverages during processing, preparation, or at the table, and sugars naturally present in honey, syrups, fruit juices and fruit juice concentrates (Faruque et al., 2019; R. K. Johnson et al., 2009; World Health Organization, 2018).

Total sugar intake is mostly determined by the consumption of sweet products such as chocolates, candies, cakes, biscuits, soft beverages among many others. These products are typically high in free sugars that have particularly adverse effects on health (Prada et al., 2020). A healthy, well-balanced diet contains naturally occurring sugars as previously mentioned. In addition, sugars add desirable sensory effects to many foods and beverages, and a sweet taste promotes enjoyment of meals and snacks (R. K. Johnson et al., 2009). Thus, understanding the process by which ingested sugars are broken down and then converted to fat and stored in the human body is of utmost importance to understand the harm that excess sugar consumption may cause (Faruque et al., 2019).

The negative health implications of excessive sugar consumption led the WHO, in 2015, to issue sugar intake guidelines to address this problem. Other evidence-based reports by the European Food Safety Authority, Public Health England, and the World Cancer Research Fund International support the WHO recommendation to limit the consumption of free sugars (Luo et al., 2019). The American Heart Association (AHA) recommends limiting daily sugar consumption to 25 g/day for an adult woman and 37.5 g/day for an adult man (R. K. Johnson et al., 2009).

Since the beginning of the current century (2000) the reduction in sugar consumption has slowed down the annual increase of obesity in both adults and children. While sugar is necessary for a healthy life and the consumption trend is going in the right direction, we still consume more than 300% of the daily recommended amount of added sugar (Faruque et al., 2019). A general strategy that has been used to reduce sugar intake is to increase consumers' knowledge about sugar, based on the assumption that knowledge can change awareness, attitudes and consequently behaviors (Prada et al., 2020).

While most processed foods include added sugars, the actual added or free sugar content is rarely presented in nutritional labels and could be a challenging task for consumers to find due to the multiple types of sugar that exist in the product (Prada et al., 2020).

In the EU, nutrition and health claims are used by the food and beverage industry to inform consumers about the beneficial health attributes of food products, with the premise that food legislation is to protect consumers. The two key objectives are to ensure that consumers are not misled with regard to claims made on, or about, food and to facilitate cross border trade within the EU (Gilsenan, 2011).

A food systems approach tackling the incentives and creating disincentives to the production and commercialization of high-sugar products is necessary to accomplish healthier and more sustainable diets (Deliza et al., 2021). So, some governments have implemented taxes and recommendations for food manufacturers and consumers for high sucrose products (Milner et al., 2020).

For instance, in Portugal, health authorities, taking into account the WHO guidelines, have been implementing strategies to reduce sugar consumption. A recent law that introduced restrictions on advertising of foodstuffs containing high energy value, sugar, salt, and processed and saturated fatty acids (Law No. 30/2019, Diário da República, 23 April 2019), is a good example. Other measures included limiting unhealthy products in vending machines available in institutions under the auspices of the Ministry of Health (Order no. 7516-A/2016, Diário da República), and increasing taxes on beverages with high sugar content (Order no. 42/2016, Diário da República) (Prada et al., 2020).

These strategies aim to encourage the food industry to reduce the sucrose content of a variety of products. Those kinds of initiatives are pressuring the food industry to produce reduced sucrose-containing products, while consumers continue to demand high quality products. Sugar reduction or removal in confectionary products is an important research objective for the food industry, considering negative press, consumer awareness around civilization diseases and government strategies for sugar reduction in high sugar products (Milner et al., 2020). However, relying exclusively on the commitment of the food industry to reduce sugar consumption will likely not be enough to achieve the required reduction in sugar intake.

Although the need to reduce the sugar content of processed products has been widely acknowledged, progress has been slow (Deliza et al., 2021). Product reformulation may be an acceptable way of reducing sugar intake by some consumers, even though significant improvements in the sensory quality of sugar reduced products are required (Di Monaco et al., 2018). So far, sugar reduction strategies pursued by food chemists and industrial researchers (see Table 9.1) have been targeted at minimizing changes in their sensory characteristics, that is, reducing sugar while maintaining sweetness

Table 9.1 Sugar reduction strategies used in the food industry.

| Strategy | Objective | Molecules in use | Disadvantages | Reference |
|---|---|---|---|---|
| Non-nutritive sweeteners (NNS) | Most NNS possess high-potency sweetness (except sugar alcohols). The use of a reduced amount of these molecules leads to a negligible impact on nutrition and energy. | Advantame, aspartame, acesulfame-K, neotame, saccharin, and sucralose, cyclamates, stevia (active compounds: stevioside and rebaudioside) and monk fruit extracts (active compounds: mogrosides); thaumatin, erythritol; neohesperidin dihydrochalcone. | Still exist common issues of differences in the temporal sensory profile and bitter aftertastes. | Mahato et al. (2020), Tucker and Tan (2017) |

*Continued*

**Table 9.1** Sugar reduction strategies used in the food industry—cont'd

| Strategy | Objective | Molecules in use | Disadvantages | Reference |
|---|---|---|---|---|
| Bulking agents | To provide energy although influenced by the food recipe, the legislative restrictions and consumer preference. Compensates the loss in texture and volume when sugar is reduced. | Inulin, maltodextrin polydextrose, oligofructose | Do not contribute to the sweetness and have a negative effect on the sensorial characteristics. | Di Monaco et al. (2018) |
| Flavors with modifying properties (FMPs) | Help modify or enhance the flavor profile from taste and odor perspectives | Homoeriodictyol and (R)-citronellal (bitter masking effect), annurcoic acid (sweetness enhancer); cinacalcet (positive allosteric modulators) | Many flavoring substances are self-limiting in their use in food. Nearly all flavoring substances are at least some-what volatile. PAMs of natural origin, with broader spectrum and higher intensity are still required. | Arthur et al. (2015), Beltrami et al. (2018), Guentert (2018), Hallagan and Hall (2009), Servant et al. (2011) |
| Gradual reduction | Gradually reduce the sugar quantity from products | | Consumers should accept the changed sensory profile | Mahato et al. (2020) |
| Multisensory interactions | A technique where sugar reduction is achieved without the use of NNSs or any other sweeteners | It uses aroma, color and other stimuli to perceive sweetness intensity | High level of sugar reduction cannot be achieved | Mahato et al. (2020) |
| Heterogenous distribution | Stimulation of taste receptors, serum release, as well as particle size and viscosity of foods to enhance the sweetness in foods | Particle-size and viscosity play an important role in sweetness perception | Achievable only in a small scale | Mahato et al. (2020) |

(compounds that could trigger sweetness in the same way as sugars), color and texture. This has been achieved mainly through total or partial replacement of sugar by non-nutritive sweeteners (NNS) (low-calorie sweeteners, artificial sweeteners, and noncaloric sweeteners with high sweetness potency that offer no nutritional benefits as vitamins and minerals do) and bulking agents (usually provide energy and their use depends on the food recipe, on the legislative restrictions and on consumer preference), or by the use of cross-modal interactions (flavor-sweet or texture-sweet interactions) and non-homogeneous distribution of sugar in the food matrix (Deliza et al., 2021; Mahato et al., 2020). Another approach for maintaining perceived sweetness while reducing the total amount of carbohydrates is to decrease the polymerization degree through enzymatic hydrolysis, which allows reducing the total saccharide content and leads to an increased glycemic index (GI) of the concerned food (Palzer, 2017).

The use of flavors with modifying properties (FMPs) have recently emerged as an alternative to the non-nutritive sweeteners that are widely used by the flavor industry to modify or enhance the flavor profile of foods and beverages. These FMPs may not necessarily have or impart a specific characteristic flavor of their own but can modify the flavor profile by altering flavor attributes such as intensifying specific flavor characteristics (e.g., perceived fruitiness), reducing specific flavor characteristics, masking of off-notes or bitterness, or changing the time onset and duration of the perception of specific aspects of the flavor profile. One type of FMPs are positive allosteric modulators, which are small molecules such as 3-((4-amino-2,2-dioxido-1*H*-benzo[*c*][1,2,6]thiadiazin-5-yl)oxy)-2,2-dimethyl-*N*-propylpropanamide (S6973; CAS 1093200-92-0) and (*S*)-1-(3-(((4-amino-2,2-dioxido-1*H*-benzo[*c*][1,2,6]thiadiazin-5-yl)oxy)methyl)piperidin-1-yl)-3-methylbutan-1-one (S617; CAS 1469426-64-9), that interact with human sweet taste receptors, enhancing their activity and consequently sweetness perception. Although FMPs have a different mechanism of action from non-nutritive sweeteners, they still seek to reduce sugar content (and consequently caloric content) while maintaining sweetness perception (Arthur et al., 2015; Deliza et al., 2021; Harman & Hallagan, 2013; Mahato et al., 2020). Notably, all such sweeteners need also to have thermal and pH stability, water solubility, low production cost, and with minimal impact on other sensorial properties of foods (DuBois & Prakash, 2012). Although, more than 50 classes of organic compounds classified as "sweet" or sweeteners have emerged (DuBois & Prakash, 2012; Hutchings et al., 2019), most of them are not commercialized as they need to address the safety requirements to be generally recognized as safe (GRAS), an approval established by the authorities as the European Food Safety Authority (EFSA) and Food and Drug Administration (FDA). The food safety authorities also regulate the sweeteners occurrence in food by establishing the acceptable daily intake (ADI) values, expressed as mg/(kg body weight)/day, and the foods they can be added. According to European Parliament (2008) most sweeteners are classified as food additives and thus they are found in food labels at the ingredient lists by referring to the name or the appropriate E number, i.e., code of the food additive.

The abovementioned strategies, which will be further detailed in Section 9.5 of this chapter, are related to the innate preference for sweet taste. In addition to this, repeated exposure to foods with high sweetness intensity may contribute to the formation of preferences, and the choice for products with lower sweetness intensity is not developed (Deliza et al., 2021; Di Monaco et al., 2018). Due to this sweet preference from consumers that leads to a critical acceptability of sugar reduction products, another interesting approach can be the utilization and impact of repeated exposure to sweet odors as added ingredients to enhance the perceptible sweetness in low sugar foods (Di Monaco et al., 2018).

## 9.4 Sugar structure modification and encapsulation for enhanced sweet perception

The successful replacement of sugar presents a great challenge for food researchers and the industry because of the multiple functions performed by this ingredient (Kistler et al., 2021).

As previously mentioned, different strategies are being explored to reduce the amount of sugar in food products without affecting the consumer "sweet perception." One of these approaches is the reduction of sugar particles size, in which the sub-micro scale restructuring of sucrose aims to increase the surface/volume ratio per amount of sugar consumed, thus increasing the perception of sweet for the same amount of sugar when compared to larger sizes (20–100 µm). There are currently two companies (https://www.douxmatok.com/ and https://www.nestle.com/) that try to address this issue through the production of sugar particles with high porosity or hollowness, with sizes above the 0.8 µm, or through the use of an "inert" material as the core of the particles, but in this case with sizes between 3 and 100 nm (Acutis et al., 2017; Baniel, 2007). However, these patents reported the use of a sweetener composition comprising a core particle in association with a sweetener carbohydrate coating. Spray dryer and nano spray dryer are techniques used to obtain these micro- and nanoparticles. Spray drying is a well-established versatile method used in the pharmaceutical and food industries to generate dry powder from a liquid state where drug solution is sprayed into air by atomization to evaporate the solvent (Gharsallaoui et al., 2007). By utilizing the nano spray drying process, it is possible to produce spray-dried particles in the submicron scale down to nanoscale (350–500 nm). In addition, it can produce remarkably high yields of up to 94% for powder amounts down to the milligram scale (e.g., 3.0–500 mg) (X. Li et al., 2010). It is possible by applying the nano spray dryer to produce nanoparticles in a single step, within a continuous and scalable process. Hence, spray drying techniques using nano spray dryer have been investigated by several authors as an optimum method for the preparation of thermosensitive compounds such as proteins and enzymes in a stable dry form, in the presence of different types of additives such as surfactants and sugars (Whelehan et al., 2014). This technique is a potential process to obtain crystalline powders in a single step, rather than multi-stage conventional crystallization processes where crystallization is normally followed by a further drying step.

Spray drying of sucrose is known to be difficult because its glass-transition temperature is low (62 °C), making this material sticky at low drying temperatures. However, few works reported that combining the drying and crystallization processes allows that drying sucrose at temperatures above the glass-transition temperature hastens the subsequent transition from a sticky amorphous material to a less sticky crystalline one (Chiou & Langrish, 2008). The transformation from a sticky amorphous material right through to the less sticky crystalline state by operating the dryer at a sufficiently high temperature may assist in the spray drying of sucrose by creating more crystalline sucrose at these high temperatures (Imtiaz-Ul-Islam & Langrish, 2009). Another strategy is the addition of carriers to the spray drying feed solution to increase Tg and avoid the stickiness problem, such as inulin, modified starch, pectin, sodium alginate, maltodextrin, isolated whey protein and lecithin (Kanojia et al., 2016; Kumar et al., 2014; Langrish & Wang, 2009; Shakiba et al., 2016). However, there are a limited number of publications that focus on the use of nano spray drying that uses sucrose to obtain particles. Abdel-Mageed et al. (2019) obtained α-amylase nanoparticles, in combination with sucrose by the nano spray drying technique. These particles have a spherical morphology with a size of around 600 nm (Abdel-Mageed et al., 2019).

Another strategy to enhance the sweetener perception by sugar modification is the use of alternative sugar compounds, using encapsulation as methodology. The encapsulation process allows the creation of structures with a controlled release of sweetness in food products and masks the intense flavor of these alternative compounds. In the literature, it is possible to identify different works, mostly patents, carried out to encapsulate sweeteners and evaluate their effect to enhance their application (Favaro-Trindade et al., 2015). An example of this is the case of acesulfame-K, a dietary sweetener, with a flavor intensity about 200 times higher than sucrose but with a strong aftertaste flavor, which stability depends on the pH and temperature. Yatka et al. (1989) developed a chewing gum with high consumer acceptance, using acesulfame-K encapsulated in microcapsules, allowing a controlled release of the sweet flavor over time.

Aspartame, a low-calorie sweetener, is also commonly used to replace conventional sugar. However, its stability in aqueous solutions is a limiting factor, with its loss of flavor being reported over. One of the strategies found to overcome such limitations is the use of microcapsules produced by spray drying with the application of a matrix with low solubility, allowing the use of high concentrations of this sweetener and its controlled release for longer periods. Rocha-Selmi et al. (2013) observed that encapsulation of these two sweeteners (acesulfame-K and aspartame) by double emulsion followed by a complex coacervation reduced water solubility and hygroscopicity. Another type of sweetener, used commercially, is thaumatin. This sweetener is a protein that has a high sensitivity under different conditions leading to its denaturation. One of the strategies found to reduce these disadvantages was the use of different encapsulation techniques to promote its application in new products (even when subjected to higher temperatures). Cherukuri and Mansukhani (1989) found that there was a need to create a matrix to encapsulate this type of sweeteners, allowing them to reduce their intense flavor and be able to be used in food products. The use of a coating with a biopolymer using the spray drying technique allowed to protect and extend the stability of this sweetener over time in a controlled manner. Numerous examples are found reporting the need for encapsulation of these sweeteners (acesulfame-k, aspartame, thaumatin, xylitol) which are pointed out as an alternative to conventional sugar and the need for a solution to improve their stability and sensory properties in food products is unanimous to obtain new healthier solutions for the consumer.

## 9.5 Food grade alternatives to sugar

As mentioned in Section 9.3 sweetening agents can be of natural or synthetic origin, being usually classified as nutritive and high-intensity sweeteners. Nutritive sweeteners encompass molecules that can be absorbed and metabolized, acting as a source of energy, while high-intensity sweeteners account for compounds which sweetening power surpasses hundreds of folds that of sucrose and which caloric contribution is negligible (Miele et al., 2017).

### 9.5.1 Nutritive sweeteners

Nutritive sweeteners comprehend carbohydrate related compounds characterized by comparable sweetening power to sucrose and with a contribution to the bulk of foods. Honey, molasses, and syrups comprehend alternative bulk sweeteners to sucrose, usually aiming to provide an additional nutritional value to foods. However, their high caloric contribution led to a widespread preference for low caloric nutritive sweeteners such as polyols, tagatose, and hydrogenated starch hydrolysates (Hutchings et al., 2019; Miele et al., 2017).

Polyols are food additives obtained by the reduction of the carbonyl functional group of sugars to an alcohol. These include sorbitol, mannitol, maltitols, lactitol, xylitol and polyglycitol, obtained by catalytic hydrogenation of glucose, mannose, maltose, lactose, xylose and starch hydrolysates, respectively. The erythritol is an exception, since it is produced from glucose fermentation by yeasts (Dasgupta et al., 2017). The sweetening power of polyols is generally lower than sucrose (30–90%), a feature compensated by their lower caloric contribution, 0.2–3 kcal/g (Table 9.2). Polyols

**Table 9.2 Currently approved sweeteners by the European Food Safety Authority (EFSA) for food applications.**

| Compound (E-number) | Sweetening power | Caloric value (kcal/g) | Solubility[a] (g/100 mL) | ADI (mg/kg) | Year of approval | Applications |
|---|---|---|---|---|---|---|
| Sucrose | 1 | 4 | 210 | – | – | Transversal applications to food where it acts as bulking, preservative and sweetening agent. |
| *Nutritive sweeteners* | | | | | | |
| Sorbitol (E420) | 0.6 | 2.6 | 235 | – | 1984 | Drinks, gums, baked foods, sweets, sausages, oral hygiene products, drugs |
| Mannitol (E421) | 0.5 | 1.6 | 22 | – | 1984 | |
| Isomaltitol (E953) | 0.4–0.5 | 2.0 | 39 | – | 1984 | Chocolate, gums, pastry, yogurts |
| Maltitol (E965) | 0.9 | 3.0 | 175 | – | 1984 | Chocolates, gum, dairy products, jam, jellies. |
| Lactitol (E966) | 0.4 | 2.0 | 140 | – | 1984 | Gums, ice creams, chocolate, bakes goods |
| Xylitol (E967) | 1 | 2.4 | 200 | – | 1984 | Refreshments, gums |
| Erythritol (E968) | 0.6–0.7 | 0.2 | 61 | – | 2003 | Dairy products, candies, chocolate, baked goods |
| Polyglycitol (964) | 0.3–0.8 | 3 | n.d. | – | 2012 | Chocolates, jams, jellies and other food/ vegetable spreads |

Table 9.2 Currently approved sweeteners by the European Food Safety Authority (EFSA) for food applications—cont'd

| Compound (E-number) | Sweetening power | Caloric value (kcal/g) | Solubility (g/100 mL) | ADI (mg/kg) | Year of approval | Applications |
|---|---|---|---|---|---|---|
| Tagatose (E963) | 0.9 | 1.5 | 160 | – | 2005 | Cereals, beverages, confectionery, and dairy products. |

*High intensity sweeteners*

*Synthetic*

| Compound (E-number) | Sweetening power | Caloric value (kcal/g) | Solubility (g/100 mL) | ADI (mg/kg) | Year of approval | Applications |
|---|---|---|---|---|---|---|
| Acesulfame K (E950) | 150–200 | – | 27 | 15 | 1984 | Sweets, baked goods, marmalade, jams, soft drinks, dairy products, tabletop sweetener |
| Aspartame (E951) | 200 | – | 1 | 40 | 1984 | Dairy products, soft drinks |
| Cyclamates (E952) | 30–80 | – | 13 | 11 | 1984 | Tabletop sweetener, jellies, canned fruits, baked goods, soft drinks |
| Saccharin (E954) | 240–300 | – | 0.2 | 5 | 1977 | Jellies, marmalade, jams, tabletop sweetener, desserts, soft drinks |
| Sucralose (E955) | 750 | | 28 | 5 | 2000 | Dairy products, ice cream, biscuits, canned fruits, soft drinks, baked goods, jellies, marmalades, gums |
| Neotame (E961) | 8000 | – | 1 | 0.3 | 2009 | Soft drinks, dairy products, tea, tabletop sweetener |
| Advantame (E969) | 20,000 | – | 0.1 | 5 | 2013 | coffee, tea, gum, dairy products |

*Continued*

Table 9.2 Currently approved sweeteners by the European Food Safety Authority (EFSA) for food applications—cont'd

| Compound (E-number) | Sweetening power | Caloric value (kcal/g) | Solubility (g/100 mL) | ADI (mg/kg) | Year of approval | Applications |
|---|---|---|---|---|---|---|
| *Natural* | | | | | | |
| Steviol glycosides (E960) | 300 | – | 0.4 | 4 | 2011 | Tabletop sweetener, ice cream, cakes, yogurts, drinks, pastry, flavored milk. |
| Thaumatin (E957) | 2000–3000 | – | 60 | 50 | 1984 | Sauces, processed vegetables, soups, egg derived products |
| Neohesperidin dihydrochalcone (E959) | 1500 | – | 0.05 | 35 | 1994 | Ice creams, pastry, bubble gum, soft drinks, dairy products, snacks, tabletop sweetener |

[a]*Water solubility at 20°C. n.d. – not defined as is dependent on the proportion of sorbitol, mannitol and higher molecular weight polyols. ADI – acceptable daily intake.*
*Information adapted from Hartel, R. W., von Elbe, J. H., & Hofberger, R. (2018). Chemistry of bulk sweeteners, in:* Confectionery science and technology. *Springer International Publishing, Springer, Cham, pp. 3–37. https://doi.org/10.1007/978-3-319-61742-8_1. Hutchings, S. C., Low, J. Y. Q., & Keast, R. S. J. (2019). Sugar reduction without compromising sensory perception. An impossible dream?* Critical Reviews in Food Science and Nutrition, 59, *2287–2307. https://doi.org/10.1080/10408398.2018.1450214.*

are also thermostable compounds, not participating in browning reactions due to the absence of reducing capacity provided by the carbonyl group of sugars (Aidoo et al., 2013; Hartel et al., 2018). Polyols also act as bulking agents for flavor stabilization, texture modulation, water activity control, and crystal formation provided by their capability to retain water (Rice et al., 2020). These features, together with polyols non-interference on insulin levels and their non-cariogenic properties, make these additives relevant in dietetic and diabetic food markets. Nevertheless, if ingested in excess, polyols have laxative effects. Besides, due to their lower heat of solution than sugars, polyols provide a mouthfeel cooling sensation, only appreciated by the consumers in a restricted group of foods as occurs with chewing gum. To overcome this, polyols are often combined with other sweeteners (Hartel et al., 2018; Rice et al., 2020).

Tagatose is a ketohexose industrially produced from lactose by using enzymes or through galactose isomerization under alkali conditions. In the food industry, tagatose has been applied to foods as an effective sucrose replacer due to their comparable sweetening power, solubility, and bulking properties (Table 9.2). Tagatose also has a lower caloric contribution (3 kcal/g vs 4 kcal/g for sucrose) and does not interfere with blood glucose levels and caries formation. Tagatose is currently approved by EFSA as an ingredient and by FDA as a food additive for food products where sucrose is relevant for the bulk of foods (Jayamuthunagai et al., 2017). Arabinitol is another alternative food grade sweetening agent allowed to be used as sugar replacer by FDA but not by EFSA.

### 9.5.2 High intensity sweeteners

High intensity sweeteners comprehend groups of substances that might be obtained by chemical synthesis, as the approved acesulfame K, aspartame, cyclamates, saccharin, sucralose, neotame, aspartame-acesulfame salts and advantame, or from natural origin following extraction and purification processes, as steviol glycosides, thaumatin, and neohesperidin dihydrochalcone (Hutchings et al., 2019; Miele et al., 2017). Owing to their high sweetness, which might surpass thousands of folds that of sucrose (Table 9.2), high intensity sweeteners are used in minute amounts when compared to sugars and thus, have no contribution to the caloric intake. This, together with the fact that they are poorly fermentable compounds, has allowed health claims regarding the prevention of tooth demineralization and control of blood glucose levels in products where these compounds replace sugars (Miele et al., 2017). Natural sweeteners are generally more attractive, mostly because consumers mistrust regarding the effects of synthetic sweeteners on Human health. Even at an industrial perspective, natural sweeteners are advantageous due to their higher stability than most of the synthetic ones at the pH ranges commonly found in foods and high temperatures applied during thermal processing (Hutchings et al., 2019). However, similarly to synthetic sweeteners, natural sweeteners fail to perfectly mimic sugars by providing off-tastes and presenting a slow-onset sweet taste that lingers (Komes et al., 2016). For this reason, high intensity sweeteners are combined with bulking agents of lower sweetness to minimize the impact on food taste (Antenucci & Hayes, 2015; Hartel et al., 2018; Hutchings et al., 2019). Other natural sweeteners are gaining renewed interest by industries, boosted by the demands of the consumers for more natural and synthetic free ingredients. These include monellin (serendipity berry), pentadin (oubli plant), brazzein (oubli climbing plant), glycyrrhizin (licorice root plant) and luo han guo mogrosides (monk fruit). All these natural sweeteners, except for glycyrrhizinic acid, are still unable to be used as sugar replacers in the European Union countries, and only the monk fruit was considered as safe by FDA. However, due to their advantages over sweeteners already implemented in the market, especially the synthetic ones, they will likely take part of a food grade sweeteners list, allowing a more widespread production of sweet foods minimizing caloric intake.

## 9.6 Enzymatic and innovative methods to improve sweetening

An innovative strategy being used to reduce sugar content in food is the application of enzymes into the product formulation to directly convert its intrinsic sugar into other valuable compounds. Under optimal conditions, enzymes may transform sugar-rich food products into new healthier food versions with additional functionalities, even during the manufacture process of the product itself

(Sanz-Valero & Cheng, 2012). The *in-situ* enzymatic sugar conversion into non-digestible saccharides with prebiotic features will be further discussed.

### 9.6.1 Sugar reduction in milk and dairy products

Lactose is commonly found in dairy products, such as milk, yogurt, cream, butter, ice cream, and cheese (Facioni et al., 2020). Besides the common hydrolysis toward glucose and galactose, through the use of β-galactosidases, the lactose content of these dairy products may also be converted *in-situ* into prebiotics. Apart from catalyzing lactose hydrolysis, β-galactosidases can produce galactosylated products through their inherent transgalactosylation activity. The transfer of a galactosyl moiety from lactose into acceptor molecules containing hydroxyl groups yield prebiotic galacto-oligosaccharides (GOS), with varied degree of polymerization (DP) and different glycosidic linkages (Xavier et al., 2018). GOS are non-digestible carbohydrates, with reduced caloric value (2 kcal/g), pleasant taste and with bulk properties similar to sucrose (Torres et al., 2010).

The direct transformation of lactose into GOS for the manufacture of low-lactose milk has been successfully accomplished. Using 0.1% (w/w) of a commercial β-galactosidase enzyme (Maxilact® LGI 5000), 98.16% of the skim milk lactose content was reduced after 1 h reaction, at 37°C, resulting in $0.71 \pm 0.01$ g of GOS per 100 g of product were produced, accounting for 14.7% of the final carbohydrates content. The prebiotic-enriched skim milk was mainly composed of 6-galactosyllactose (0.23 g), 6-galactobiose (0.18 g) and allolactose (0.17 g) per 100 g of product (Oh et al., 2019). In another study, milk lactose was reduced by 50%. The *in-situ* reaction was carried at 40°C, for 2.5 h using 0.1% (v/v) of a β-galactosidase from *Bacillus circulans* (Biolactase NTL-CONC). An amount of 8.1 g/L of GOS was synthesized, resulting in 17.6% of the total carbohydrates of the mixture (Rodriguez-Colinas et al., 2014).

A yogurt was prepared using hydrolyzed milk with reduced lactose content. After 2 h reaction, at 42°C, 63% of the milk lactose was reduced by a commercial β-galactosidase from *Kluyveromyces lactis* (Lactozyme 2600L), applied at 0.1% (v/v). From the reaction, 9.7 g/L of GOS were produced alongside with 12 g/L of glucose and 8.9 g/L of galactose. Lactose conversion did not induce any change in color, consistency, and overall acceptability of the yogurt. However, taste and texture were significantly improved ($p < 0.05$) and remained stable over 28 days (Raza et al., 2018). In drinkable yogurts, when 0.16 g/L of β-galactosidase enzyme were applied, lactose content (4.08 g per 100 g) was reduced to 1.26 g (per 100 g), enriching the yogurt with 0.62 g (per 100 g) of GOS. Also, 0.36 g (per 100 g) of GOS were synthetized in stirred yogurts using 0.25 g/L of the same β-galactosidase. In this case, lactose was reduced from 5.55 g to 1.52 (per 100 g). No changes were found in the GOS content throughout the storage period at 5°C for 21 days (Vénica et al., 2015).

A low-lactose buttermilk was produced from a trans-glycosylated milk. GOS were produced by addition of enzyme to pasteurized milk. Lactose was reduced from $3.906 \pm 0.057$ to $2.763 \pm 0.057$ g in the prebiotic buttermilk. No significant differences were found after a lactose reduction in comparison with the control buttermilk. An improvement in color, flavor, taste and consistency was verified, raising overall acceptability score from 7.2 to 10.4 (in a scale of 15). The produced buttermilk stored in pouch and glass packaging maintained its stability for 6 days (Tahir et al., 2019).

An innovative strategy to reduce lactose amount of food products is the application of cellobiose 2-epimerase enzymes, which catalyze the epimerization of lactose to epilactose and the isomerization of lactose to lactulose (Kim et al., 2012). In contrast to its substrate lactose, both lactulose and epilactose cannot be hydrolyzed by human small intestinal enzymes, thus have a lower caloric value (Mu et al., 2013; Olano & Corzo, 2009). The conversion of lactose milk by application of a thermophilic cellobiose 2-epimerase from *Caldicellulosiruptor saccharolyticus* was carried out at 50 °C for 24 h. The reaction yielded 57.7% (28.0 g/L) of lactulose and 15.5% (7.49 g/L) of epilactose, from 48.5 ± 2.1 g/L of initial lactose content. Similar results were obtained at 5 °C, where lactose conversion resulted in 56.7% (27.5 g/L) lactulose and 13.6% (6.57 g/L) epilactose, after 72 h (Rentschler et al., 2015). In another study, two different cellobiose 2-epimerases from *Flavobacterium johnsoniae* DSM 2064 (*FjCE*) and *Pedobacter heparinus* DSM 2366 (*PhCE*) were used for cow milk lactose conversion into epilactose. Both conversions were performed at 8 °C for 24 h. Lactose conversion by *FjCE* yielded 33.6% epilactose (14.3 g/L) whilst a yield of 30.5% (13.9 g/L) was obtained with *PhCE* enzyme (Krewinkel et al., 2014).

### 9.6.2 Sugar reduction in juices and beverages

A high amount of sugar is present in sugar-sweetened beverages, which includes soda; fruit, sports, and energy drinks; and sweetened teas (Krieger et al., 2020). To reduce their sugar content, the application of dextransucrase enzymes into beverages seems to be a feasible strategy. The enzyme catalyzes the synthesis of D-glucose polymers, or the transfer of D-glucosyl units from sucrose to an acceptor molecule (Q. P. Li et al., 2017). Hence, dextransucrases can be used to convert caloric sugars into prebiotic oligosaccharides, such as isomalto-oligosaccharides (IMO) and malto-oligosaccharides (MOS).

The sugar content of tangerine and orange juices was significantly decreased by the addition of 3 U/mL of dextransucrase from *Leuconostoc mesenteroides*, at 16 °C. Sucrose was almost depleted (99%) within the final tangerine juice formulation and sugars, including fructose and glucose, were converted into IMO, MOS and leucrose. Oligosaccharides were produced in a yield of 47% (315.7 g/L), which resulted in a reduction of 30.5% of the caloric value of the juice. No significant differences in overall acceptability were detected, apart from a slight reduction of sweetness (Nguyen, Cho, et al., 2015). In a similar study, using the same enzyme and operational conditions, over 97% of the sucrose of an orange juice was converted into oligosaccharides (157.4 g/L) yielding 31.7%. A caloric reduction of 24.2% was obtained in the modified juice (Nguyen, Seo, et al., 2015).

The intrinsic sucrose from orange and apple juice concentrates was significantly reduced when treated with 14.47 U/g$_{sucrose}$ dextransucrase from *Lactobacillus reuteri* 180, for 90 min, at 50 °C and pH 4.5. In the apple juice formula, sucrose level was reduced by 96.8%, with 8% less total sugars. For the orange juice, a reduction of 99.5% of sucrose was verified, with a decrease of total sugars by 23.2%. The main identified products in the juice were leucrose, isomaltose, and isomaltotriose, being maltose, maltotriose and panose also formed but at lower concentrations (Johansson et al., 2016). A complete depletion of sucrose in both mango and pineapple juices was possible with the application of a dextransucrase from *Weissella cibaria* RBA12. Several IMO (DP 3–5), isomaltose, and leucrose were synthesized, which means that some fructose and glucose also acted as acceptor molecules for the formation of the oligosaccharides. In mango juice, 14.6 g/L oligosaccharide were produced resulting in less 21.03 g/L of sugars in the final product. Similarly, 11.12 g/L of oligosaccharides were synthetized in a pineapple juice, with a sugar reduction of 18.3 g/L. The removal of the native sucrose, alongside with a total sugar content reduction, decreased the juices caloric value (Baruah et al., 2017).

Fructosyltransferases can also be applied for intrinsic sucrose conversion in beverages. These enzymes, with transfructosylating activity, catalyze the synthesis of fructo-oligosaccharides (FOS) from sucrose. FOS have a sweet taste similar to that of sucrose, and it has been estimated that the caloric value of FOS (1.5–2.0 kcal/g) represents 40–50% of that of sucrose (Sabater-Molina et al., 2009).

An orange juice concentrate was treated with 0.036% (w/w) dextransucrase from *Leuconostoc mesenteroides* B-5 12F at 30°C and pH 5.2 for 4 h, resulting in the reduction of total sugar content from 512 to 426 g/L and conversion to gluco-oligosaccharides (GuOS). Using the same juice, a two-step strategy by a bi-enzymatic system was also evaluated. First, the juice concentrate was treated with 8% (w/w) fructosyltransferase enzyme, at 50°C, to convert part of the sucrose into FOS. This step resulted in a juice with 22.6% FOS, and sucrose content was reduced from 47.9% to 15.5%, increasing the monosaccharides amount from 52.1% to 61.8%. After, 0.014% (w/w) dextransucrase from *L. mesenteroides* was added, and the reaction was carried at 30°C for 24 h. This second enzyme produced GuOS, reducing the sucrose content. Using this strategy, the obtained sucrose and monosaccharides content in the juice was 5% and 65%, respectively, and the juice was enriched with 30% of oligosaccharides (Sanz-Valero & Cheng, 2012). Similarly, orange and apple juice concentrates were incubated with a fructosyltransferase (14 U/Gds) from *Aspergillus japonicas* EB001. Initial sucrose content was reduced from 15.99% to 0.59% (w/w), in the orange juice, with an increase in glucose from 4.46% to 12.94% (w/w). In apple juice, sucrose was reduced from 4.68% to 0.50% (w/w) and glucose levels increased from 6.85% to 15.3% (w/w) (Henderson et al., 2007).

The use of commercial enzyme preparations in juice manufacturing process was firstly reported in 1930. Most of those preparations contain cellulase and pectic enzymes (Sharma et al., 2017). Apart from those, other preparations such as Viscozyme L®, also contain fructosyltransferases enzymes showing remarkable activity toward fructo-oligosaccharides (FOS) synthesis from sucrose (Virgen-Ortíz et al., 2016). Viscozyme L® was added to longan pulp, during the pressing stage, to produce a juice with reduced sugar content and rich in FOS. After 5 h treatment, sucrose was reduced from 110.98 to 17.11 mg/g, while glucose increased from 36.87 to 69.05 mg/L, and fructose remained unchanged (∼28 mg/L). The final juice had an oligosaccharide concentration of 61.59 mg/g, representing ∼35% of the total carbohydrates (Cheng et al., 2018).

Less reported, is the conversion of fructose content of food products into sorbitol. Usually regarded as a nutritive sweetener, sorbitol has approximately 60% the sweetness of sucrose and provides a caloric value of 2.6 kcal/g, which is 30% fewer when compared with that of sucrose (Zhang et al., 2020). Using a by-enzymatic system of a commercial invertase and an oxidoreductase from *Zymomonas mobilis*, the biotransformation of pineapple juice sugars was accomplished. First, the juice was treated with 3.5 U/mL invertase for 0.5 h at 25°C. About 70% of sucrose was hydrolyzed into glucose and fructose, increasing its contents in the pineapple juice to 39 g and 41 g/L, respectively. In a second step, glucose–fructose oxidoreductase (GFOR) was used to transform glucose into gluconolactone, and fructose into sorbitol. After 24 h reaction, at pH 6.2, only 7.8 g/L glucose and 9.0 g/L fructose remained in the juice mixture (Aziz et al., 2011).

## 9.7 Products and market

As a consequence of the growing trends in the reduction of sugar consumption, the global market size of sugar substitutes is estimated to increase at a compound annual growth rate (CAGR) of 4.5%, from a value of USD 16.5 billion in 2020 to an estimated USD 20.6 billion by 2025 (Sugar Substitutes Market, 2021). This growth is driven by some regulatory limitations and also by the health consciousness among consumers to encourage the search for healthier natural food choices. The relation between sugar consumption and health problems has increased in the last years (R. J. Johnson et al., 2017; Lustig et al., 2012) but it was at the beginning of the 20th century that one of the first studies related sugar consumption with health problems (Charles, 1907). Since then, the food industry has presented several synthetic and natural alternatives, some with no calories associated and others with less calories than sucrose. Currently, there is a huge number of available commercial solutions that can be used as sugar substitutes, and while the trend of using these compounds is increasing it is also true that some of the solutions presented in the market have faced several constraints related to regulatory approval and technological challenges (as explained before in this chapter). All these alternatives have advantages and drawbacks that can be related to sensorial perception, toxicity and price.

Without any doubt, the market demand for more natural and sustainable compounds have led companies to look for new solutions that could help reduce the sucrose amount in foods. Also, consumer perception is a key factor, therefore products based on stevia, fibers, or monk fruit are welcomed and understood by the consumers as good alternatives when compared with others such as aspartame and erythritol. Table 9.3 presents some of the commercially available natural alternatives to sucrose.

Nowadays, stevia is perhaps the most popular natural sweetener with a higher margin to growth in the next years. For example, in 2019 the global stevia market was evaluated at 539 million USD and is expected to grow to 900 million USD by 2025 (CBI, 2021). One interesting strategy is presented by the DouxMatok company (https://www.douxmatok.com/). The use of sucrose to build new micro and submicro structures allows having the same sweet sensation with a low amount of sucrose.

Table 9.3 Commercially available natural alternatives to sucrose.

| Source | Trade mark | Company website |
| --- | --- | --- |
| Xylitol | Xylisorb | https://www.roquette.com/ |
| Stevia rebaudiana | Nutrasweetnatural | https://nutrasweetnatural.com/ |
| Stevia rebaudiana | – | https://sweegen.com/ |
| Stevia rebaudiana | Eversweet | https://www.avansya.com/ |
| Stevia rebaudiana | Purecircle | https://www.purecircle.com/ |
| Stevia rebaudiana | PureVia | https://purevia.com/ |
| Stevia rebaudiana | Enliten | https://www.ingredion.com/ |
| Isomaltooligosaccharide | Vitafiber | https://www.bioneutra.ca |
| Inulin | Frutafit | http://www.inspiredbyinulin.com/frutafit-/-frutalose.html |
| Monk fruit (mogroside V) | Sugarlike | http://sugarlike.org/ |
| Fiber from sugar cane | Fosvitae | https://zukan.es |
| Ammoniated glycyrrhizin and Monoammonium glycyrrhizinate | Magnasweet | https://magnasweet.com |
| Thaumatin | Thaumatin | https://www.silverfernchemical.com/ |
| Thaumatin | Talin | https://www.naturex.com/ |
| Tagatose | Bonumose | http://bonumose.com/ |
| Neohesperidin dihydrochalcone | – | https://www.bordas-sa.com/ |

## 9.8 Conclusions and future outlook

Designing foods by changing their composition and structure can be part of the solution for the development of foods with improved nutrition. They can be used to improve health and well-being and be sustainable at the same time. However, understanding and directing consumer behavior will also be crucial to overcoming these challenges. Any newly designed foods that are healthier and more sustainable must also be affordable, delicious, and convenient, otherwise consumers will not adopt them. Therefore, creating new solutions is an important step to designing a new generation of more healthy food products. This chapter reports different effective innovative strategies for substantially reducing sugar content such as sugar substitutes or in the field of sugar structure modification and the advances in legislation. Gradual sugar reduction has been demonstrated to have potential in a lab setting however the long-term effectiveness of this approach is yet unknown.

The food industry is acknowledging the role it has to play in sugar reduction, with many companies setting targets for a reduction in some of their product categories. Some sugar reduction strategies will suit certain products and consumer segments and not others. However, a substantial reduction in sugar content in any one food without changing sensory perception is an incredibly challenging technical task.

## Acknowledgments

This work supported by the project cLabel+ (POCI-01-0247-FEDER-046080) cofinanced by Compete 2020, Lisbon 2020, Portugal 2020 and the European Union, through the European Regional Development Fund (ERDF). We would also like to thank the scientific collaboration under the FCT project UID/Multi/50016/2019.

## References

Abdel-Mageed, H. M., Fouad, S. A., Teaima, M. H., Abdel-Aty, A. M., Fahmy, A. S., Shaker, D. S., & Mohamed, S. A. (2019). Optimization of nano spray drying parameters for production of α-amylase nanopowder for biotheraputic applications using factorial design. *Drying Technology, 37*, 2152–2160. https://doi.org/10.1080/07373937.2019.1565576.

Acutis, R.D., Whitehouse, A.S., Forny, L., Meunier, V.D.M., Dupas-Langlet, M., & Mahieux, J.P. (2017). Amorphous porous particles for reducing sugar in food. WO 2017/093309 A1, issued 2017.

Aidoo, R. P., Depypere, F., Afoakwa, E. O., & Dewettinck, K. (2013). Industrial manufacture of sugar-free chocolates—Applicability of alternative sweeteners and carbohydrate polymers as raw materials in product development. *Trends in Food Science and Technology, 32*, 84–96. https://doi.org/10.1016/j.tifs.2013.05.008.

Antenucci, R. G., & Hayes, J. E. (2015). Nonnutritive sweeteners are not supernormal stimuli. *International Journal of Obesity, 39*, 254–259. https://doi.org/10.1038/ijo.2014.109.

Arthur, A. J., Karanewsky, D. S., Luksic, M., Goodfellow, G., & Daniels, J. (2015). Toxicological evaluation of two flavors with modifying properties: 3-((4-Amino-2,2-dioxido-1H-benzo[c][1,2,6]thiadiazin-5-yl)oxy)-2,2-dimethyl-N-propylpropanamide and (S)-1-(3-(((4-amino-2,2-dioxido-1H-benzo[c][1,2,6]thiadiazin-5-yl)oxy)methyl)piperidin-1-yl)-3-methylbutan-1-one. *Food and Chemical Toxicology, 76*, 33–45. https://doi.org/10.1016/j.fct.2014.11.018.

Asioli, D., Aschemann-Witzel, J., Caputo, V., Vecchio, R., Annunziata, A., Næs, T., & Varela, P. (2017). Making sense of the "clean label" trends: A review of consumer food choice behavior and discussion of industry implications. *Food Research International, 99*, 58–71. https://doi.org/10.1016/j.foodres.2017.07.022.

Aziz, M. G., Michlmayr, H., Kulbe, K. D., & del Hierro, A. M. (2011). Biotransformation of pineapple juice sugars into dietetic derivatives by using a cell free oxidoreductase from *Zymomonas mobilis* together with commercial invertase. *Enzyme and Microbial Technology, 48*, 85–91. https://doi.org/10.1016/j.enzmictec.2010.09.012.

Baniel, A. (2007). Sweetener compositions. WO2007007310A1.

Baruah, R., Deka, B., & Goyal, A. (2017). Purification and characterization of dextransucrase from *Weissella cibaria* RBA12 and its application in in vitro synthesis of prebiotic oligosaccharides in mango and pineapple juices. *LWT, 84*, 449–456. https://doi.org/10.1016/j.lwt.2017.06.012.

Bast, A., & Haenen, G. R. M. M. (2013). Ten misconceptions about antioxidants. *Trends in Pharmacological Sciences, 34*, 430–436. https://doi.org/10.1016/j.tips.2013.05.010.

Beltrami, M. C., Döring, T., & de Dea Lindner, J. (2018). Sweeteners and sweet taste enhancers in the food industry. *Food Science and Technology, 38*, 181–187. https://doi.org/10.1590/fst.31117.

Boaz, M., Smetana, S., Weinstein, T., Matas, Z., Gafter, U., Iaina, A., Knecht, A., Weissgarten, Y., Brunner, D., Fainaru, M., & Green, M. (2000). Secondary prevention with antioxidants of cardiovascular disease in end-stage renal disease (SPACE): Randomised placebo-controlled trial. *Lancet, 356*, 1213–1218. https://doi.org/10.1016/S0140-6736(00)02783-5.

CBI. (2021). *The European market potential for soap* (WWW Document) https://www.cbi.eu/market-information/natural-food-additives/stevia/market-potential (Accessed 21 October 2021).

Charles, R. (1907). Diabetes in the tropics. *BMJ, 2*, 1051–1064.

Cheng, Y., Lan, H., Zhao, L., Wang, K., & Hu, Z. (2018). Characterization and prebiotic potential of longan juice obtained by enzymatic conversion of constituent sucrose into fructo-oligosaccharides. *Molecules, 23*. https://doi.org/10.3390/molecules23102596.

Cherukuri, S., & Mansukhani, G. (1989). Sweetener delivery systems containing polyvinyl acetate. US Patent 4816265.

Chiou, D., & Langrish, T. A. G. (2008). A comparison of crystallisation approaches in spray drying. *Journal of Food Engineering, 88*, 177–185. https://doi.org/10.1016/j.jfoodeng.2008.02.004.

Clemens, R. A., Jones, J. M., Kern, M., Lee, S. Y., Mayhew, E. J., Slavin, J. L., & Zivanovic, S. (2016). Functionality of sugars in foods and health. *Comprehensive Reviews in Food Science and Food Safety, 15*, 433–470. https://doi.org/10.1111/1541-4337.12194.

Dasgupta, D., Bandhu, S., Adhikari, D. K., & Ghosh, D. (2017). Challenges and prospects of xylitol production with whole cell bio-catalysis: A review. *Microbiological Research, 197*, 9–21. https://doi.org/10.1016/j.micres.2016.12.012.

Deliza, R., Lima, M. F., & Ares, G. (2021). Rethinking sugar reduction in processed foods. *Current Opinion in Food Science, 40*, 58–66. https://doi.org/10.1016/j.cofs.2021.01.010.

Di Monaco, R., Miele, N. A., Cabisidan, E. K., & Cavella, S. (2018). Strategies to reduce sugars in food. *Current Opinion in Food Science, 19*, 92–97. https://doi.org/10.1016/j.cofs.2018.03.008.

DuBois, G. E., & Prakash, I. (2012). Non-caloric sweeteners, sweetness modulators, and sweetener enhancers. *Annual Review of Food Science and Technology, 3*, 353–380. https://doi.org/10.1146/annurev-food-022811-101236.

European Parliament, C.o.t. E. U. (2008). Regulation (EU) No 1333/2008. *Official Journal of the European Union, L 304*(16), 18–33.

European Parliament, C.o.t. E. U. (2012). Commission Regulation (EU) No 432/2012 establishing a list of permitted health claims made on foods, other than those referring to the reduction of disease risk and to children's development and health. *Official Journal of the European Union, L 136*, 1–40.

Facioni, M. S., Raspini, B., Pivari, F., Dogliotti, E., Cena, H., & Cena, H. (2020). Nutritional management of lactose intolerance: The importance of diet and food labelling. *Journal of Translational Medicine, 18*. https://doi.org/10.1186/s12967-020-02429-2.

Faruque, S., Tong, J., Lacmanovic, V., Agbonghae, C., Minaya, D., & Czaja, K. (2019). The dose makes the poison: Sugar and obesity in the United States—A review. *Polish Journal of Food and Nutrition Sciences, 69*, 219–233. https://doi.org/10.31883/pjfns/110735.

Favaro-Trindade, C. S., Rocha-Selmi, G. A., & dos Santos, M. G. (2015). Microencapsulation of sweeteners. In *Microencapsulation and microspheres for food applications* (pp. 333–349). Elsevier. https://doi.org/10.1016/B978-0-12-800350-3.00022-4.

Gharsallaoui, A., Roudaut, G., Chambin, O., Voilley, A., & Saurel, R. (2007). Applications of spray-drying in microencapsulation of food ingredients: An overview. *Food Research International, 40*, 1107–1121. https://doi.org/10.1016/j.foodres.2007.07.004.

Gilsenan, M. B. (2011). Nutrition and health claims in the European Union: A regulatory overview. *Trends in Food Science and Technology, 22*, 536–542. https://doi.org/10.1016/j.tifs.2011.03.004.

Guentert, M. A. (2018). Flavorings with modifying properties. *Journal of Agricultural and Food Chemistry, 66*, 3735–3736. https://doi.org/10.1021/acs.jafc.8b00909.

Gwinn, R. (2013). *Technology and ingredients to assist with the reduction of sugar in food and drink*. https://pt.scribd.com/document/395342789/Technology-and-Ingredients-to-Assist-With-the-Reduction-of-Sugar (Accessed 25 November 2021).

Hallagan, J. B., & Hall, R. L. (2009). Under the conditions of intended use—New developments in the FEMA GRAS program and the safety assessment of flavor ingredients. *Food and Chemical Toxicology, 47*, 267–278. https://doi.org/10.1016/j.fct.2008.11.011.

Harju, M., Kallioinen, H., & Tossavainen, O. (2012). Lactose hydrolysis and other conversions in dairy products: Technological aspects. *International Dairy Journal, 22*, 104–109. https://doi.org/10.1016/j.idairyj.2011.09.011.

Harman, C. L., & Hallagan, J. B. (2013). Sensory testing for flavorings with modifying properties. *Food Technology, 67*, 44–47.

Hartel, R. W., von Elbe, J. H., & Hofberger, R. (2018). *Chemistry of bulk sweeteners, in: Confectionery science and technology* (pp. 3–37). Cham: Springer International Publishing, Springer. https://doi.org/10.1007/978-3-319-61742-8_1.

Henderson, W.E., King, W., & Shetty, J.K. (2007). In situ Fructooligosaccharide Production and Sucrose Reductio. Patent No. WO 2007/061918 A2.

Hutchings, S. C., Low, J. Y. Q., & Keast, R. S. J. (2019). Sugar reduction without compromising sensory perception. An impossible dream? *Critical Reviews in Food Science and Nutrition, 59*, 2287–2307. https://doi.org/10.1080/10408398.2018.1450214.

Imtiaz-Ul-Islam, M., & Langrish, T. A. G. (2009). Comparing the crystallization of sucrose and lactose in spray dryers. *Food and Bioproducts Processing, 87*, 87–95. https://doi.org/10.1016/j.fbp.2008.09.003.

Jayamuthunagai, J., Gautam, P., Srisowmeya, G., & Chakravarthy, M. (2017). Biocatalytic production of D-tagatose: A potential rare sugar with versatile applications. *Critical Reviews in Food Science and Nutrition, 57*, 3430–3437. https://doi.org/10.1080/10408398.2015.1126550.

Jelen, P., & Tossavainen, O. (2003). Low lactose and lactose-free milk and dairy products—Prospects, technologies and applications. *Australian Journal of Dairy Technology, 58*, 161–165.

Johansson, S., Diehl, B., Christakopoulos, P., Austin, S., & Vafiadi, C. (2016). Oligosaccharide synthesis in fruit juice concentrates using a glucansucrase from *Lactobacillus reuteri* 180. *Food and Bioproducts Processing, 98*, 201–209. https://doi.org/10.1016/j.fbp.2016.01.013.

Johnson, R. K., Appel, L. J., Brands, M., Howard, B. V., Lefevre, M., Lustig, R. H., Sacks, F., Steffen, L. M., & Wylie-Rosett, J. (2009). Dietary sugars intake and cardiovascular health a scientific statement from the American Heart Association. *Circulation, 120*, 1011–1020. https://doi.org/10.1161/CIRCULATIONAHA.109.192627.

Johnson, R. J., Sánchez-Lozada, L. G., Andrews, P., & Lanaspa, M. A. (2017). Perspective: A historical and scientific perspective of sugar and its relation with obesity and diabetes. *Advances in Nutrition, 8*, 412–422. https://doi.org/10.3945/an.116.014654.

Kanojia, G., Have, R.t., Bakker, A., Wagner, K., Frijlink, H. W., Kersten, G. F. A., & Amorij, J.-P. (2016). The production of a stable infliximab powder: The evaluation of spray and freeze-drying for production. *PLoS One, 11*, e0163109. https://doi.org/10.1371/journal.pone.0163109.

Kim, J. E., Kim, Y. S., Kang, L. W., & Oh, D. K. (2012). Characterization of a recombinant cellobiose 2-epimerase from *Dictyoglomus turgidum* that epimerizes and isomerizes β-1,4- and α-1,4-gluco-oligosaccharides. *Biotechnology Letters*, *34*, 2061–2068. https://doi.org/10.1007/s10529-012-0999-z.

Kistler, T., Pridal, A., Bourcet, C., & Denkel, C. (2021). Modulation of sweetness perception in confectionary applications. *Food Quality and Preference*, *88*, 104087. https://doi.org/10.1016/j.foodqual.2020.104087.

Komes, D., Belščak-Cvitanović, A., Jurić, S., Bušić, A., Vojvodić, A., & Durgo, K. (2016). Consumer acceptability of liquorice root (*Glycyrrhiza glabra* L.) as an alternative sweetener and correlation with its bioactive content and biological activity. *International Journal of Food Sciences and Nutrition*, *67*, 53–66. https://doi.org/10.3109/09637486.2015.1126563.

Krewinkel, M., Gosch, M., Rentschler, E., & Fischer, L. (2014). Epilactose production by 2 cellobiose 2-epimerases in natural milk. *Journal of Dairy Science*, *97*, 155–161. https://doi.org/10.3168/jds.2013-7389.

Krieger, J., Bleich, S. N., Scarmo, S., & Ng, S. W. (2020). Sugar-sweetened beverage reduction policies: Progress and promise. *Annual Review of Public Health*, *42*, 439–461. https://doi.org/10.1146/annurev-publhealth-090419-103005.

Kumar, S., Gokhale, R., & Burgess, D. J. (2014). Sugars as bulking agents to prevent nano-crystal aggregation during spray or freeze-drying. *International Journal of Pharmaceutics*, *471*, 303–311. https://doi.org/10.1016/j.ijpharm.2014.05.060.

Langrish, T. A. G., & Wang, S. (2009). Crystallization rates for amorphous sucrose and lactose powders from spray drying: A comparison. *Drying Technology*, *27*, 606–614. https://doi.org/10.1080/07373930802716391.

Li, X., Anton, N., Arpagaus, C., Belleteix, F., & Vandamme, T. F. (2010). Nanoparticles by spray drying using innovative new technology: The Büchi Nano Spray Dryer B-90. *Journal of Controlled Release*, *147*, 304–310. https://doi.org/10.1016/j.jconrel.2010.07.113.

Li, Q. P., Wang, C., Zhang, H. B., Hu, X. Q., Li, R. H., & Hua, J. H. (2017). Designing of a novel dextransucrase efficient in synthesizing oligosaccharides. *International Journal of Biological Macromolecules*, *95*, 696–703. https://doi.org/10.1016/j.ijbiomac.2016.11.114.

Lloret, A., Esteve, D., Monllor, P., Cervera-Ferri, A., & Lloret, A. (2019). The effectiveness of vitamin E treatment in Alzheimer's disease. *International Journal of Molecular Sciences*, *20*. https://doi.org/10.3390/ijms20040879.

Luo, X., Arcot, J., Gill, T., Louie, J. C. Y., & Rangan, A. (2019). A review of food reformulation of baked products to reduce added sugar intake. *Trends in Food Science and Technology*, *86*, 412–425. https://doi.org/10.1016/j.tifs.2019.02.051.

Lustig, R. H., Schmidt, L. A., & Brindis, C. D. (2012). The toxic truth about sugar. *Nature*, *482*, 27–29. https://doi.org/10.1038/482027a.

MacGregor, G. A., & Hashem, K. M. (2014). Action on sugar-lessons from UK salt reduction programme. *Lancet*, *383*, 929–931. https://doi.org/10.1016/S0140-6736(14)60200-2.

Mahato, D. K., Keast, R., Liem, D. G., Russell, C. G., Cicerale, S., & Gamlath, S. (2020). Sugar reduction in dairy food: An overview with flavoured milk as an example. *Foods*, *9*, 1400. https://doi.org/10.3390/foods9101400.

Maruyama, S., Streletskaya, N. A., & Lim, J. (2021). Clean label: Why this ingredient but not that one? *Food Quality and Preference*, *87*, 104062. https://doi.org/10.1016/j.foodqual.2020.104062.

Middha, P., Weinstein, S. J., Männistö, S., Albanes, D., & Mondul, A. M. (2019). β-Carotene supplementation and lung cancer incidence in the alpha-tocopherol, beta-carotene cancer prevention study: The role of tar and nicotine. *Nicotine & Tobacco Research*, *21*, 1045–1050. https://doi.org/10.1093/ntr/nty115.

Miele, N. A., Cabisidan, E. K., Galiñanes Plaza, A., Masi, P., Cavella, S., & Di Monaco, R. (2017). Carbohydrate sweetener reduction in beverages through the use of high potency sweeteners: Trends and new perspectives from a sensory point of view. *Trends in Food Science and Technology*, *64*, 87–93. https://doi.org/10.1016/j.tifs.2017.04.010.

Milner, L., Kerry, J. P., O'Sullivan, M. G., & Gallagher, E. (2020). Physical, textural and sensory characteristics of reduced sucrose cakes, incorporated with clean-label sugar-replacing alternative ingredients. *Innovative Food Science and Emerging Technologies*, *59*, 102235. https://doi.org/10.1016/j.ifset.2019.102235.

Mu, W., Li, Q., Fan, C., Zhou, C., & Jiang, B. (2013). Recent advances on physiological functions and biotechnological production of epilactose. *Applied Microbiology and Biotechnology*, *97*, 1821–1827. https://doi.org/10.1007/s00253-013-4687-2.

Nguyen, T. T. H., Cho, J. Y., Seo, Y. S., Woo, H. J., Kim, H. K., Kim, G. J., Jhon, D. Y., & Kim, D. (2015). Production of a low calorie mandarin juice by enzymatic conversion of constituent sugars to oligosaccharides and prevention of insoluble glucan formation. *Biotechnology Letters*, *37*, 711–716. https://doi.org/10.1007/s10529-014-1723-y.

Nguyen, T. T. H., Seo, Y. S., Cho, J. Y., Lee, S., Kim, G. J., Yoon, J. W., Ahn, S. H., Hwang, K. H., Park, J. S., Jang, T. S., & Kim, D. (2015). Synthesis of oligosaccharide-containing orange juice using glucansucrase. *Biotechnology and Bioprocess Engineering*, *20*, 447–452. https://doi.org/10.1007/s12257-014-0741-x.

Oh, N. S., Kim, K., Oh, S., & Kim, Y. (2019). Enhanced production of galactooligosaccharides enriched skim milk and applied to potentially synbiotic fermented Milk with *Lactobacillus rhamnosus* 4B15. *Food Science of Animal Resources*, *39*, 725–741. https://doi.org/10.5851/kosfa.2019.e55.

Olano, A., & Corzo, N. (2009). Lactulose as a food ingredient. *Journal of the Science of Food and Agriculture*, *89*, 1987–1990. https://doi.org/10.1002/jsfa.3694.

Palzer, S. (2017). Technological solutions for reducing impact and content of health sensitive nutrients in food. *Trends in Food Science and Technology*, *62*, 170–176. https://doi.org/10.1016/j.tifs.2016.11.022.

Prada, M., Saraiva, M., Garrido, M. V., Rodrigues, D. L., & Lopes, D. (2020). Knowledge about sugar sources and sugar intake guidelines in Portuguese consumers. *Nutrients*, *12*, 1–17. https://doi.org/10.3390/nu12123888.

Puratos. (2021). *Reliable partners in innovation*. Clean(er) Label [WWW Document].

Raza, A., Iqbal, S., Ullah, A., Khan, M. I., & Imran, M. (2018). Enzymatic conversion of milk lactose to prebiotic galacto-oligosaccharides to produce low lactose yogurt. *Journal of Food Processing & Preservation*, *42*. https://doi.org/10.1111/jfpp.13586.

Rentschler, E., Schuh, K., Krewinkel, M., Baur, C., Claaßen, W., Meyer, S., Kuschel, B., Stressler, T., & Fischer, L. (2015). Enzymatic production of lactulose and epilactose in milk. *Journal of Dairy Science*, *98*, 6767–6775. https://doi.org/10.3168/jds.2015-9900.

Rice, T., Zannini, E., Arendt, E. K., & Coffey, A. (2020). A review of polyols—Biotechnological production, food applications, regulation, labeling and health effects. *Critical Reviews in Food Science and Nutrition*, *60*, 2034–2051. https://doi.org/10.1080/10408398.2019.1625859.

Richardson, A. M., Tyuftin, A. A., O' Sullivan, M. G., Kilcawley, K. N., Gallagher, E., & Kerry, J. P. (2021). The impact of sugar particles size and natural substitutes for the replacement of sucrose and fat in chocolate brownies: Sensory and physicochemical analysis. *European Journal of Engineering and Technology Research*, *6*, 104–113. https://doi.org/10.24018/ejers.2021.6.1.2294.

Rocha-Selmi, G. A., Bozza, F. T., Thomazini, M., Bolini, H. M. A., & Fávaro-Trindade, C. S. (2013). Microencapsulation of aspartame by double emulsion followed by complex coacervation to provide protection and prolong sweetness. *Food Chemistry*, *139*, 72–78. https://doi.org/10.1016/j.foodchem.2013.01.114.

Rodriguez-Colinas, B., Fernandez-Arrojo, L., Ballesteros, A. O., & Plou, F. J. (2014). Galactooligosaccharides formation during enzymatic hydrolysis of lactose: Towards a prebiotic-enriched milk. *Food Chemistry*, *145*, 388–394. https://doi.org/10.1016/j.foodchem.2013.08.060.

Sabater-Molina, M., Larqué, E., Torrella, F., & Zamora, S. (2009). Dietary fructooligosaccharides and potential benefits on health. *Journal of Physiology and Biochemistry*, *65*, 315–328. https://doi.org/10.1007/BF03180584.

Sanz-Valero, J., & Cheng, P.-S. (2012). Intrinsic sugar reduction of juices and ready to drink products. Patent No. WO 2012/059554 A1.

Servant, G., Tachdjian, C., Li, X., & Karanewsky, D. S. (2011). The sweet taste of true synergy: Positive allosteric modulation of the human sweet taste receptor. *Trends in Pharmacological Sciences*, *32*, 631–636. https://doi.org/10.1016/j.tips.2011.06.007.

Shakiba, S., Mansouri, S., Selomulya, C., & Woo, M. W. (2016). In-situ crystallization of particles in a countercurrent spray dryer. *Advanced Powder Technology*, *27*, 2299–2307. https://doi.org/10.1016/j.apt.2016.07.001.

Sharma, H. P., Patel, H., & Sugandha. (2017). Enzymatic added extraction and clarification of fruit juices—A review. *Critical Reviews in Food Science and Nutrition, 57*, 1215–1227. https://doi.org/10.1080/10408398.2014.977434.

Struck, S., Jaros, D., Brennan, C. S., & Rohm, H. (2014). Sugar replacement in sweetened bakery goods. *International Journal of Food Science and Technology, 49*, 1963–1976. https://doi.org/10.1111/ijfs.12617.

Sugar Substitutes Market. (2021). (Accessed 20 October 2021) https://www.marketsandmarkets.com/Market-Reports/sugar-substitute-market-1134.html.

Tahir, H., Bacha, U., Iqbal, S., Iqbal, S. S., & Tanweer, A. (2019). Development of prebiotic galactooligosaccharide enriched buttermilk and evaluation of its storage stability. *Progress in Nutrition, 21*, 935–942. https://doi.org/10.23751/pn.v21i4.7579.

Torres, D. P. M., Gonçalves, M.d. P. F., Teixeira, J. A., & Rodrigues, L. R. (2010). Galacto-oligosaccharides: Production, properties, applications, and significance as prebiotics. *Comprehensive Reviews in Food Science and Food Safety, 9*, 438–454. https://doi.org/10.1111/j.1541-4337.2010.00119.x.

Tucker, R. M., & Tan, S. Y. (2017). Do non-nutritive sweeteners influence acute glucose homeostasis in humans? A systematic review. *Physiology & Behavior, 182*, 17–26. https://doi.org/10.1016/j.physbeh.2017.09.016.

Vénica, C. I., Bergamini, C. V., Rebechi, S. R., & Perotti, M. C. (2015). Galacto-oligosaccharides formation during manufacture of different varieties of yogurt. Stability through storage. *LWT - Food Science and Technology, 63*, 198–205. https://doi.org/10.1016/j.lwt.2015.02.032.

Virgen-Ortíz, J. J., Ibarra-Junquera, V., Escalante-Minakata, P., Centeno-Leija, S., Serrano-Posada, H., de Jesús Ornelas-Paz, J., Pérez-Martínez, J. D., & Osuna-Castro, J. A. (2016). Identification and functional characterization of a fructooligosaccharides-forming enzyme from *Aspergillus aculeatus*. *Applied Biochemistry and Biotechnology, 179*, 497–513. https://doi.org/10.1007/s12010-016-2009-8.

Whelehan, M., Plüss, R., & John, P. (2014). *Microencapsulation: Prilling by vibration* (pp. 1–8).

WHO. (2021). *The Global Health Observatory: Noncommunicable diseases*. WHO Library.

World Health Organization. (2015). *Guideline: Sugars intake for adults and children*. Geneva, Switzerland.

World Health Organization. (2018). *Guideline: Sugars intake for adults and children* (pp. 1716–1722). World Health Organization 57.

Xavier, J. R., Ramana, K. V., & Sharma, R. K. (2018). β-Galactosidase: Biotechnological applications in food processing. *Journal of Food Biochemistry, 42*. https://doi.org/10.1111/jfbc.12564.

Yatka, R.J., Foster, B.E., & Orland, P. (1989). Alitame stability using hydrogenated starch hydrolysate syrups. US5034231A.

Zhang, W., Chen, J., Chen, Q., Wu, H., & Mu, W. (2020). Sugar alcohols derived from lactose: Lactitol, galactitol, and sorbitol. *Applied Microbiology and Biotechnology, 104*, 9487–9495. https://doi.org/10.1007/s00253-020-10929-w.

CHAPTER

# New technological strategies for improving the lipid content in food products

## 10

**S. Cofrades[a] and M.D. Alvarez[b]**

[a]*Department of Products, Institute of Food Science, Technology and Nutrition (ICTAN-CSIC), Madrid, Spain,*
[b]*Department of Characterisation, Quality and Safety, Institute of Food Science, Technology and Nutrition (ICTAN-CSIC), Madrid, Spain*

## 10.1 Introduction

In recent decades, there has been an important advance in the knowledge about the role of dietary lipids in human health, particularly in relation to some of the most prevalent non-communicable diseases in our society. It is known that diets that are rich in fat (particularly saturated fat) are related to an increase in the risk of developing coronary heart diseases (Barros et al., 2020), which in turn are the major cause of mortality in type II diabetes mellitus, as there is an excess of saturated fatty acids (SFAs), cholesterol and trans-fatty acids (TFAs) in the diet. The quality of dietary fatty acids (FAs) can differentially modulate circulating cholesterol levels and the risk of cardiovascular diseases (CVDs) and other metabolic diseases. For instance, long-chain SFAs such as myristic acid (C14:0) and palmitic acid (C16:0) are believed to increase circulating cholesterol levels, increasing the risk of heart disease (Siri-Tarino et al., 2010). Also, trans fats are known to significantly increase the risk of heart disease by increasing circulating cholesterol levels in humans (Hammad et al., 2016). Both Saturated and trans fat is widely used in food processing, partially due to their unique physicochemical properties, such as their solid-like appearance, desirable flavor, and texture (American Heart Association, 2017). The main source of saturated and trans fats in the diet comes from processed foods, including meat, dairy and bakery products.

In this sense, health authorities worldwide are encouraging people to decrease their consumption of SFAs, TFAs, and cholesterol. In fact, the addition of TFAs is already limited in EU and USA markets. Consequently, several current nutritional recommendations suggest that the dietary fat intake should ideally account for between 15% and 30% of the total diet energy; no more than 10% of the total energy intake should come from SFAs; polyunsaturated fatty acids (PUFAs) should represent 6–10% of the caloric intake (*n*-6, 5–8%, *n*-3, 1–2%); and approximately 10–15% should be from monounsaturated fatty acids (MUFAs) (World Health Organization, 2003). Additionally, a PUFA *n*-6/*n*-3 ratio of less than four is recommended. In summary, the recommendations are aimed at reducing the consumption of foods with a high content of SFAs, such as meat, dairy and bakery products, since they are rich sources of SFAs, TFAs and cholesterol, and their consumption has been associated with a negative impact on human health; therefore, the modification of their lipid fraction is a major goal for the food industry.

Although reducing the consumption of saturated fats is a mainstay of international dietary recommendations, growing evidence indicates that lowering saturated fats provides convincing cardiovascular benefits only when they are replaced by PUFAs (Simopoulos, 1999). This evidence indicates the need to increase the consumption of healthy PUFA rich vegetable/marine oils to reduce coronary heart disease mortality (Connor, 2000). In this sense, the increasing consumers' knowledge of the effect of lipids on their health explains the growing tendency to look at the nutritional information provided in a product, information that in many cases determines their decision on whether to buy it or not.

Despite the open debate on the dietary recommendation of fats, reformulating common foods rich in fat, mainly SFAs with a reduced content of total and saturated fats, represents a useful tool to control the daily energy intake in the general population. Indeed, in this type of foods, solid fats, which are rich in saturated fats, are responsible for a large number of high-quality attributes, mainly in relation to texture, aroma, flavor and stability. This positive and fundamental contribution of SFAs to the sensory properties and technological manufacturing of food is the main reason why, today, the substitution of certain saturated fats (e.g., palm fat and pork fat), or the general reduction of these, is such a complex matter. In this sense, one of the most important challenges for the processed-food industry is finding solutions to produce high nutritional and healthier products while maintaining traditional flavors, consumer expectations and acceptability. Due to the current trend to reduce fat consumption, the food industry has been increasing strategies to produce and commercialize products where the reduction/elimination of saturated fat and, if possible, the increase of unsaturated fat, are important goals.

## 10.2 Modification of the lipid fraction in food products

There are different strategies associated with animal production systems (genetic and nutritional), manufacturing processes (technological), conservation and consumption, which are applied to qualitative and/or quantitatively modify the lipid fraction of foods. Of all strategies, the most common method to design new healthier foods is using technological reformulation strategies, as they are faster than genetic and nutritional strategies as they have a direct effect during the development of the final product (Jiménez-Colmenero, 2007). Among the technological strategies to enhance food composition, the one that has been most studied for its health implications is the improvement of fat content, and the main objectives that have been identified are the following: "total fat reduction", "cholesterol reduction" and "FA profile modification". In this sense, products of animal origin, such as meat derivatives and some dairy products, as well as pastry products, are good candidates to undergo changes in their composition. Thus, they can become healthy products through reformulation processes aimed at reducing/eliminating harmful components, such as some fats, as previously mentioned (Patel et al., 2020).

As regards the modification of the lipid fraction in foods, it has been extensively studied over the last decades and can be achieved by using different fat replacers to ensure the functionality of the replaced fat. Fat replacers are classified as *fat substitutes* and as *fat mimetics* depending on their chemical conformation and functionalities (Peng & Yao, 2017). *Fat substitutes* are macromolecules which physically and chemically resemble triglycerides and, therefore, could be used as fat replacers in a 1:1

ratio. These fat replacers can be classified into two groups: liquids and plastics. *Fat mimetics* are normally protein macromolecules or carbohydrate-based compounds (modified starches and hydrocolloids) that are considered to imitate the organoleptic and physical properties of fats mainly by water binding. Both types of substances exhibit polar and water-soluble properties, which create a state of creaminess and lubricity similar to that of full-fat products, but which cannot replace fat gram-for-gram. Their main advantage is that *fat mimetics* reduce the caloric content of the food. The selection of one fat substitute or another is based on its properties and its function within the food matrix (improving technological and/or sensory properties), as well as on different consumer safety aspects. Thus, fat replacers should be recognized as safe and healthy, and possessing certain sensory and functional properties.

Cholesterol is an essential component of animal cell membranes and a precursor of biologically active compounds, such as important hormones, bile salts and vitamin D3. Although cholesterol is frequently related to risk factors associated with the development of cardiovascular diseases, it would be more correct to refer the association between cholesterol and high-density lipoproteins (HDL) and low-density lipoproteins (LDL), considering that a direct relationship between high plasmatic cholesterol levels, especially LDL types, and the risk of CVDs has been found (Calpe-Berdiel et al., 2009). Animal-origin foods like meat and dairy products are important cholesterol sources and this cholesterol content may be reduced in humans by the inclusion of plant sterols in this type of products. Plant sterols could be used for "cholesterol reduction" since they are substances similar to cholesterol that can be divided into sterols and stanols, stanols being the saturated form of sterols. As stated by the European Food Safety Authority (EFSA) (2008), scientific studies indicate that the consumption of 1.5–3 g of plant sterols per day could significantly reduce the level of LDL-C in individuals if consumed as part of a healthy diet. Additionally, in meat derivates, the cholesterol content can be reduced by substituting meat raw materials with cholesterol-free compounds, such as plant-derived proteins, carbohydrates and a mixture of vegetable oils (Fernández-Gines et al., 2005). For instance, intact soy (with isoflavones) has a greater effect on reducing LDL and total cholesterol concentrations than extracted soy.

It should be noted that trans fats contribute to approximately 1.2% of the total energy intake in Western diets (Kris-Etherton et al., 2012) and are produced as a by-product industrially through the partial hydrogenation of vegetable oils in the presence of a metal catalyst, at a high temperature or under vacuum. They are also produced naturally in meat and dairy products through the biohydrogenation of unsaturated fatty acids by bacterial enzymes in ruminant animals. As mentioned above, TFA consumption is associated with a higher risk of coronary heart disease via blood lipids and pro-inflammatory processes (Risérus et al., 2009) and the development of obesity, diabetes, neurodegenerative diseases (Alzheimer's), some types of cancer (mainly breast cancer), impaired fertility, endometriosis and cholelithiasis (Mozaffarian et al., 2009). Consequently, in processed foods, TFAs should be kept as low as possible and should always constitute <1% of energy. One of the major improvements achieved to date is the elimination of trans-fats in margarine, pastries, cakes and biscuits in Europe (Rippin et al., 2017). Manufacturers have begun to replace saturated fats with MUFAs and PUFAs. The substitution of hydrogenated vegetable oils for other vegetable oils, such as olive oil, which are characterized by having no trans-fats or low a content of trans-fats and a high content of cis-unsaturated FAs, would improve the nutritional quality of food and add the health benefits associated with lower intakes of trans-fats. In this sense, oilseed producers and agriculturists must increase the supply of substitute oils.

## 10.2.1 Decreasing fat and cholesterol contents by using ingredients that can serve as fat replacers

Currently, the greatest effort in reformulation is focused on the reduction of saturated fat in processed meats, milk and milk products, bakery, pastry and ready-made foods (FEEDcities Project, 2017), since they are the main source of dietary fat (mainly saturated fat), and they are foods frequently consumed by many sectors of the population. In this context, numerous reformulated low/reduced-fat products have been obtained by replacing saturated fat (animal and vegetable origin) with different fat replacers.

In this sense, reformulated meat products with just "total fat reduction" tend to be firmer, rubberized and less succulent, with an unpleasant dark color and usually presenting a rubbery skin formation as well as fat and moisture loss, all of which represent important qualitative characteristics for consumers. Also, as it is in meat products, the production of low-fat dairy products is not an easy task. For example, acidified milk products with a low-fat content are more difficult to accept by consumers due to undesirable rheological properties and high syneresis, whereas the problems with low-fat cheese are the flavor and the rubbery, hard texture. In turn, the reduction of fat content in bakery formulations results in changes in the textural and physical properties in terms of volume, air content, density, hardness, etc., which are highly dependent on the type of bakery product that is being reformulated, and may have an important impact on product processing, shelf-life and consumer acceptance (Gutierrez-Luna et al., 2021).

In order to overcome the negative impact of the reduction of saturated fat and to produce quality reduced-fat products, different *fat mimetics* are used, such as protein-based and carbohydrate-based compounds, which provide a lower caloric value, and intend to ensure the same technological and sensorial characteristics in the final product (Bom Frost et al., 2001; Jonnalagadda & Jones, 2005). Among the protein-based fat replacers, proteins from milk, soy, wheat, whey or egg are usually used. Carbohydrate-based fat replacers are categorized into fiber-based, starch-derived, cellulose-based, gum-based and others (Chavan et al., 2016).

Plant proteins are commonly used in processed products for their functional properties, as they act as good emulsifiers by improving binding properties (water and fat) and stabilizing the oil-water interfaces of the emulsions (Rodriguez Furlán et al., 2014), as well as for their relative low cost. In addition to their technological properties, plant proteins lower blood lipid levels compared with animal proteins, therefore presenting beneficial effects in the prevention and treatment of CVDs, cancer and osteoporosis, as well as in alleviating menopausal symptoms. This is the case of soy protein, for which in 1999 the U.S. Food and Drug Administration approved a health claim according to which diets that are low in saturated fat and cholesterol and that include 25 g soy protein per day, may reduce the risk of heart disease (Sadler, 2004). In this context, different meat and dairy products with a low-fat content have been developed by using different plant proteins, such as soy proteins in the form of soy flour, concentrates and isolates, resulting in products with good final characteristics due to the formation of a strong gel as a consequence of their water binding capacity and gelling properties (Montowska & Fornal, 2018). Additionally, whey proteins that can form thermally induced gels, when heated above 70 °C, provide excellent functional properties and have been used in low-fat animal-origin products (Youssef & Barbut, 2011).

Among the most frequently used carbohydrate-based *fat mimetics* there is dietary fiber from cereals, legumes, fruits, roots, etc., as it improves the technological properties of the products in which it is used (water and oil retention, emulsion stability and texture) (Hygreeva et al., 2014) and because of its health benefits. Fiber-rich foods have a lower energy density and high satiety effects, contributing to reduce the energy intake or appetite (Clark & Slavin, 2013). Also, it is known that dietary fiber intake (soluble and insoluble) plays a key role in chronic disease prevention, helping to prevent CVDs, diabetes mellitus and even some types of cancer, such as colorectal cancer (Brownlee et al., 2017). The daily intake of fiber recommended in Europe for adults is 30 g (Macdiarmid et al., 2018), nevertheless, the intake of fiber in adults between the ages of 19–64 years is considerably low (6 g/day), according to a survey conducted in the UK (Scientific Advisory Committee on Nutrition (SACN), 2015). Therefore, a way to increase its consumption without changing eating habits would be by introducing it in different foods that are frequently consumed. In this sense, various types of dietary fibers at different concentrations have been used as partial *fat mimetics* alone or combined with other ingredients for formulations of reduced-fat meat products, particularly in ground (fresh sausages, burgers) and restructured products (Desmond et al., 1998; Mansour & Khalil, 1999), and cooked meat emulsions (Cofrades et al., 2000). Fibers used in the reformulation of low-fat meat products include inulin from cereals (oats, wheat, rice), fruits (apple, lemon, orange), legumes (soybeans, peas), roots (carrot, konjac), tubers (potato) and algae (red, green and brown) (Chavan et al., 2016; Sayago-Ayerdi et al., 2009), which could improve the binding properties, textural characteristics, cooking yield and also enhance the nutritional value of the products. For example, inulin at levels of 1.0–6.0% has been used as a fat replacer in many meat products due to its ability to increase viscosity and water-holding capacity, to form gels and provide optimal texture (Felisberto et al., 2015; Rodriguez Furlán et al., 2014). Moreover, the addition of inulin in food is known to reduce the risk of colon cancer, diabetes, obesity and cardiovascular diseases (Y. Zhang et al., 2010).

Fat reduction in dairy products can also be achieved by replacing it with carbohydrate-based fat replacers to ensure the functionality of the replaced fat. Brennan and Tudorica (2008) used dietary fibers as thickeners and stabilizers to produce sauces and dressing assortments. So far, a series of studies have been conducted on the use of cereal, vegetable and fruit fibers in yogurt (Garcia-Perez et al., 2006; Hashim et al., 2009; Sendra et al., 2010). In acidified milk products, fibers have benefits such as improving texture and firmness, presenting low syneresis and providing sensory characteristics that are accepted by consumers (McCann et al., 2011). Concerning the development of healthier bakery products with a reduced-fat content, Espert (2019) investigated the partial substitution of cocoa butter with different hydrocolloids (xanthan gum, methylcellulose and hydroxypropylmethylcellulose) in the production of reduced-fat creams and cocoa fillings with suitable technological characteristics and reduced lipid digestibility.

Other carbohydrate-based fat replacers are starch derivates, being notable the role of starch-based fat replacers that can be produced either chemically or enzymatically, as modified starch and maltodextrins, respectively, which have been demonstrated to have a promising ability to improve the overall quality of reduced-fat foods. Modified starch granules act directly as fat globules modulating the structure and sensory characteristics of foods, whereas maltodextrins can form thermoreversible gels. Both modified-starch granules and maltodextrins can create a fat-like mouthfeel, and therefore are potential fat replacers. Starch-based fat replacers have been widely used in various food products, such as dairy products, baked goods, salad dressings, and mayonnaise (Puligundla et al., 2015). In addition, the use of native or chemically modified tapioca and maize starch has the potential to improve the stability of milk gelled systems in acidified yogurt, improving the overall quality of reduced-fat yogurt (Lobato-Calleros et al., 2014). Also, Rodriguez-Sandoval et al. (2017) tested the efficacy of cross-linked cassava starch as a fat replacer in muffins, indicating that up to 8% fat could be replaced without compromising the overall quality or stability of the muffin, yielding a texture similar to the control. Extensive research on the use of starch-based fat replacers in the food industry is still ongoing. For their part, maltodextrins are capable of improving the viscosity of low-fat yogurts (Costa et al., 2016). In addition, Fuangpaiboon and Kijroongrojana (2017) compared the efficacy of various commercial protein-based and carbohydrate-based fat replacers in substituting fat from ice cream. Maltodextrin and inulin have also been tested as replacers for shortening and margarine in cakes and puff pastries, respectively. In addition to reducing the overall energy density of food products by using maltodextrins as fat replacers, other physiological benefits such as promoting oral health, reducing appetite, and enhancing brain function have also been linked to the consumption of maltodextrins (Costa et al., 2016).

In general, these works concluded that although the microstructural and rheological properties of *fat mimetics* are quite different from fat, when used to replace fat in food products, the final products show qualities that are comparable to the standard formulations, and in some cases, the properties of the reformulated products are even better than those of the standards.

Some examples of the application of different fat replacers used to reduce the fat content in the reformulation of different processed foods are presented in Table 10.1.

**Table 10.1 Fat replacers used in processed products in order to develop foods with reduced fat content.**

| Food product | Fat replacer | Consequences in product characteristics | Reference |
|---|---|---|---|
| **Meat products** | | | |
| Turkish meatballs | Textured soy protein | Up to 10% improvement in meatball quality with no adverse effects on the final product | Kilic et al. (2010) |
| Meat patties | Milk proteins | Acceptable textural properties | Andiç et al. (2010) |
| Chicken breast | Whey proteins | Increase in quality characteristics, increased water-holding capacity, decreased cooking loss and hardness | Ha et al. (2019) |
| Dry fermented sausages | Konjac gel | Reduction of weight loss and hardness and improved juiciness. Acceptable sensory characteristics | Ruiz-Capillas et al. (2012) |
| Low-fat pork sausages | Oat β-glucan/marine collagen peptide | Increased springiness, chewiness and juiciness. Improved safety and nutritional value | Fan et al. (2020) |
| Chicken patties | Mango peel powder | Suitable fat substitute at 2% to replace 50% added vegetable fat without affecting quality parameters | Chappalwar et al. (2020) |
| Low-fat bolognas | Lemon albedo fiber | Good quality characteristics and improved nutritional value | Fernández-Gines et al. (2005) |
| Reduced-fat sausages | Oatmeal (previously hydrated) | Increased water holding capacity, cooking yield; product with softer texture | Yang et al. (2007) |

**Table 10.1** Fat replacers used in processed products in order to develop foods with reduced fat content—cont'd

| Food product | Fat replacer | Consequences in product characteristics | Reference |
|---|---|---|---|
| Beef gels | κ-Carrageenan | Increased hardness and fracturability and decreased cooking and purge loss | Pietrasik (2003) |
| Chicken patties | Cashew apple fiber | Decreased lipid content | Guedes-Oliveira et al. (2016) |
| Cooked meat model system | Cellulose fiber (Z-trim) | Technological characteristics similar to the standard sample | Schmiele et al. (2015) |
| Brazilian cooked sausages | Inulin with oat fiber | Addition of up to 6% of inulin with up to 0.85% oat fiber do not compromise technological parameters and sensory acceptably | Souza et al. (2019) |
| Dry fermented sausage | Inulin gel with water | Decreased springiness, chewiness and hardness and increased adhesiveness. Acceptability in terms of all sensory attributes | Glisic et al. (2019) |
| Chicken sausage | Inulin | Best physicochemical, textural, colorimetric, and sensory results with 100% substitution | Alaei et al. (2018) |
| Beef burger | Fiber pectin from tomato pomace | Decreased cooking loss; reduction in burger diameter | Namir et al. (2015) |
| **Dairy products** | | | |
| Yogurt | Starches | Lowest syneresis sensitivity and most desirable sensory properties in yogurts with 20% acetylated distarch adipate | Cais-Sokolinska et al. (2006) |
| Yogurt | Barley β-glucan, guar gum, inulin | Improved texture and rheological properties; reduction of syneresis | Brennan and Tudorica (2008) |
| Reduced-fat yogurt | Native or chemically modified tapioca and maize starch | Improved stability and overall quality | Lobato-Calleros et al. (2014) |
| Reduced-fat yogurt | Maltodextrins | Improved viscosity | Costa et al. (2016) |
| Low-fat cheese | Various types of starches | Increased viscosity and firmness; acceptable sensory properties | Gampala and Brennan (2008), Cais-Sokolinska et al. (2006) |
| Yogurt | Fibers from cereal, vegetables and fruits | Improved texture and firmness, low syneresis and acceptable sensory properties | Dello Staffolo et al. (2004), Aportela-Palacios et al. (2005), Garcia-Perez et al. (2006), Hashim et al. (2009), Sendra et al. (2010) |
| Cheese | Commercial hydrocolloid fat Simplesse® D-100 and Novagel™ NC-200 | Increased moisture content and yield leading to an improvement of texture | Romeih et al. (2002) |

*Continued*

**Table 10.1** Fat replacers used in processed products in order to develop foods with reduced fat content—cont'd

| Food product | Fat replacer | Consequences in product characteristics | Reference |
|---|---|---|---|
| Iranian white brined cheese | Gum tragacanth | Increasing gum tragacanth concentration reduced hardness and increased whiteness | Rahimi et al. (2007) |
| Low-fat paneer | Soy protein isolate | Increased yield, protein, ash, moisture contents and decreased fat, moisture protein ratio, lactose and calorie contents | Kumar et al. (2011) |
| Low-fat Oaxaca cheese | κ- and λ-Carrageenan | Acceptable overall quality with low quantities of carrageenan | Totosaus and Guemes-Vera (2008) |
| Non-fat kefir | Inulin | Higher viscosity and no negative effects on quality | Ertekin and Guzel-Seydim (2010) |
| Fat-free plain yogurt | Inulin | Inulins of various chain lengths did not affect viscosity, color, and product appearance | Aryana et al. (2007) |
| Non-fat goat milk yogurt | Heat-treated whey protein concentrate and pectin | Higher viscosity and lower syneresis compared with other yogurts | T. H. Zhang et al. (2015) |
| Low-fat whipped cream | Sodium caseinate–carboxymethyl cellulose/locust bean gum | Improved physical and sensory characteristics | Rezvani et al. (2020) |
| **Other products** | | | |
| Baked goods and bread | Inulin | Smaller loaves with harder crumb and darker color | Morris and Morris (2012) |
| Confectionery, fruit preparations and desserts | Inulin | Improved consistency without rough sensation | Arcia et al. (2011) |
| Reduced-fat puff pastry | Maltodextrin gel | High score in sensory attributes | Pimdit et al. (2008) |
| Reduced-fat filling creams | Xanthan gum and cellulose ethers cellulose | Improved quality characteristics and reduced lipid digestibility | Espert (2019) |
| Muffin | Cross-linked cassava starch | Texture similar to that of the full-fat control | Rodriguez-Sandoval et al. (2017) |

## 10.2.2 Improving the lipid profile in food products by using lipids with a healthy fatty acid profile

The most appropriate strategy to improve the lipid composition of most processed products is, in addition to reducing the total fat content, to improve the lipid profile of these products. This can be done by substituting all or part of the saturated fat (vegetable or animal) commonly added during the preparation of processed foods such as meat, bakery, dairy, etc. for healthier lipids whose characteristics are more in line with the nutritional recommendations, that is, a lower proportion of SFAs and a higher proportion of MUFAs and PUFAs (especially long-chain $n$-3 PUFAs). Therefore, according to the aforementioned evidence, MUFAs and PUFAs are two healthy lipids that may be used as substitutes for saturated fat in the diet, and also in the reformulation of processed meats, dairy and bakery products. For that purpose, oils of vegetable and marine origin (rich in MUFAs and PUFAs) (Table 10.2) have been used in the reformulation of numerous healthy products.

The most important MUFA found in the diet is oleic acid (OA; C18:1n-9c) and it is present in olive oil (approx. 70%), which has been shown to increase the HDL-/LDL-cholesterol ratio, decrease aggregation of thrombocytes and have anti-atherogenic and anti-thrombotic properties (Tvrzicka et al., 2011). Other oils rich in MUFAs are canola oil, peanut oil, high oleic sunflower

Table 10.2 Main MUFA and PUFA (including LC PUFA) content of some vegetable and marine (fish and microalgae) oils.

| Vegetables | Oleic | LA | n-6 | ALA | n-3 | | | Reference |
| --- | --- | --- | --- | --- | --- | --- | --- | --- |
| | (g/100 g) | | | | | | | |
| Canola | 59.5 | 18.8 | – | 11.9 | – | | | Kostik et al. (2013) |
| Chia | 10.5 | 20.4 | – | 60 | – | | | Ciftci et al. (2012) |
| Flax | 18.1 | 15.3 | – | 58.2 | – | | | Ciftci et al. (2012) |
| Hemp | 11.5 | 59.4 | 62.4 | 0.36 | 0.4 | | | Orsavova et al. (2015) |
| Olive | 66.4 | 16.4 | 16.4 | 1.6 | 1.6 | | | Orsavova et al. (2015) |
| Peanut | 71.1 | 18.2 | 18.2 | – | 0 | | | Orsavova et al. (2015) |
| Perilla | 16.2 | 17.9 | – | 60.4 | – | | | Osakabe et al. (2002) |
| Sesame | 41.5 | 40.9 | 40.9 | 0.21 | 0.2 | | | Orsavova et al. (2015) |
| Sunflower | 28 | 62.2 | 62.2 | 0.16 | 0.4 | | | Orsavova et al. (2015) |
| Walnut | 27.3 | 50 | – | 15 | – | | | Dogan and Akgul (2005) |
| **Fish** | | | | | | EPA | DHA | |
| | | | | | | (mg/g oil) | | |
| Mackerel | – | – | – | – | – | 140 | 100 | Valenzuela et al. (2012) |
| Anchovy | – | – | – | – | – | 110 | 100 | Valenzuela et al. (2012) |
| Trout | – | – | – | – | – | 45 | 86 | Valenzuela et al. (2012) |
| Sardine | – | – | – | – | – | 160 | 100 | Valenzuela et al. (2012) |
| Jumbo squid liver | – | – | – | – | – | 127 | 130 | Rubio-Rodríguez et al. (2012) |

*Continued*

## Table 10.2 Main MUFA and PUFA (including LC PUFA) content of some vegetable and marine (fish and microalgae) oils—cont'd

| Vegetables | Oleic | LA | n-6 | ALA | n-3 | | | Reference |
|---|---|---|---|---|---|---|---|---|
| | (g/100 g) | | | | | | | |
| Salmon offcuts | – | – | – | – | – | 79 | 63 | Rubio-Rodríguez et al. (2012) |
| Hake offcuts | – | – | – | – | – | 36 | 82 | Rubio-Rodríguez et al. (2012) |
| **Microalgae** | | | | | | | | |
| *Crypthecodinium cohnii* | | | | | | | 400–450 | Alves Martins et al. (2013) |
| *Ulkenia* sp. Gaertner | | | | | | | 350–400 | Alves Martins et al. (2013) |
| *Schizochytrium* sp. | | | | | | | 350–400 | Yin et al. (2020) |

oil and sesame oil (see Table 10.2). Besides their nutritional benefits, the use of MUFA enriched oils has technological advantages since they present a neutral aroma and flavor that makes them a great choice to be included in a wide variety of food products. On the other hand, PUFAs present in vegetables (seeds and oils) are short-chain PUFAs such as α-linolenic acid (ALA, C18:3n-3) and linoleic acid (LA, C18:2n-6). Regarding the PUFAs present in fish and seaweeds (meal and oils), these are long-chain *n*-3 PUFAs (LC *n*-3 PUFAs) such as eicosapentaenoic acid (EPA; C20:5n-3), docosapentaenoic acid (DPA; C22:5n-3) and docosahexaenoic acid (DHA; C22:6n-3). The biological effects of EPA and DHA are broad, comprising cardiac and endothelial function, blood pressure, cardiac electrophysiology and vascular reactivity, in addition to potent antiplatelet and anti-inflammatory effects (Hooper et al., 2004). ALA and LA are known as true essential FAs (Timilsena et al., 2017) since the human body cannot synthesize LC *n*-3 PUFAs, and they are used as precursors. However, the degree of conversion of ALA to EPA and DHA is small in the human body (Burdge & Wootton, 2002), and hence, EPA and DHA should also be included in the diet to maintain health and well-being.

Healthier oils rich in MUFAs and PUFAs can be incorporated into foods in different ways, generally depending on the product to which they are to be added. In this sense, the simplest way to incorporate them in the reformulation of products is by direct addition, however, this presents limitations in terms of the characteristics of the final product, mainly due to the fact that most oils of vegetable and marine origin are in a liquid state at room temperature, and therefore do not have the physicochemical, technological, or oxidative stability that is characteristic of solid fats with high contents of SFAs and TFAs. Some studies have conducted the simple incorporation of vegetable and marine oils into meat products using different methods, such as microinjection (Domazakies, 2005) or the direct incorporation of the oil into the matrix of different products such as pork frankfurters, burger patties and baked products (Álvarez et al., 2012; Manaf et al., 2019). Most of the studies showed complications when incorporating these oils in their liquid form, as changes in the texture of the product occurred as a consequence of problems in the integration of the ingredients due to low viscosity. As well, increased susceptibility to lipid oxidation due to the high susceptibility of MUFAs and PUFAs against oxidation was observed, causing rancidity (off-flavors) and consumer rejection, together with the production of toxic compounds.

Aimed at improving consistency and reducing oxidative processes, there are various techniques that can be used, such as encapsulation, emulsification and structuring liquid oils into solid fats.

### 10.2.2.1 Incorporation of encapsulated oils

Regarding oil and/or, more specifically, PUFA $n$-3 FA microencapsulation and nanoencapsulation, it is a procedure that enables its addition in small amounts to different products, providing protection from lipid oxidation in a very effective manner and masking undesirable odors and flavors, also enhancing the protective efficacy and bioavailability of PUFA $n$-3 in the gastrointestinal tract (Kaushik et al., 2015). The most frequent process is freeze-drying and spray-drying, which generally comprises forming an emulsion by using proteins, polysaccharides, as well as other emulsifying elements and then eliminating the aqueous phase by atomization in a dry air flow at high temperature, transforming the dispersed phase into powder. Also, ultrasonication and high-pressure homogenization can adequately form physicochemically stable micro/nanoemulsions containing PUFA $n$-3 (Gumus et al., 2021). Other technologies have also been employed to encapsulate oils, such as complex coacervation, extrusion, etc., which have been widely described (Timilsena et al., 2017). Fish, flaxseed and microalgae oils have been the most commonly used oils in encapsulation processes. In the last two decades, dairy products such as cheese, milk and butter, meat products, ice cream and milkshake have shown their potential to be fortified with PUFA $n$-3 in order to produce healthier products (Barrow et al., 2009; Bermúdez-Aguirre & Barbosa-Cánovas, 2011; Kolanowski & Weiβbrodt, 2007; Ullah et al., 2017). In general, encapsulation of healthier lipids increases the bioavailability of functional FA rich oils but also reduces the serum lipidemic profile and obesity-related risk factors. However, the sensory quality of fortified products should be improved.

### 10.2.2.2 Incorporation of emulsified oils

Using oil emulsions to incorporate vegetable and marine oils as ingredients to replace saturated fat in the development of healthier products, is one of the most convenient and versatile procedures on account of the multiple advantages it offers (Jiménez-Colmenero, 2007). Conventional emulsions are composed of two immiscible phases (usually oil and water), where one phase is dispersed within the other in the form of small particles, and are generally classified according to the arrangement of the two immiscible phases, as either oil-in-water O/W or water-in-oil W/O systems. Among others, oil emulsions limit matrix separation, they are easily dispersed and favor product stability from processing to consumption (Djordjevic et al., 2004). Several procedures have been tested to elaborate emulsions with different types of oils of vegetable and/or marine origin, or a combination of both, which later will be used as animal/vegetable fat substitutes in different products. This has allowed obtaining foods with an improved lipid profile while maintaining acceptable technological properties (Freire et al., 2016). Although the incorporation of O/W emulsions into different products is a very common strategy, such systems are generally prone to suffer from physical instability phenomena such as coalescence and phase separation, not being able to provide a solid texture unless the concentration of droplets in the emulsion results in a tightly packed emulsion (McClements, 2010). This limitation has led to the design of new strategies based on the structuring of liquid oils for the substitution of significant amounts of saturated fats in different food products (meat, dairy and bakery derivatives) without changing the chemical properties of the vegetable/marine oil.

### 10.2.2.3 Incorporation of healthy oils based on new emerging structuring methods

Liquid oil stabilization and structuring strategies include those based on the application of *interesterification* processes, *creation of structured emulsions*, *organogelation* and the formation of *oil bulking agents*. Using materials with a solid structure (similar to that of saturated fat) and an improved fat content (amount and FA profile, with a reduction of saturated fat and an increase of unsaturated fatty acids (UFAs)) as compared with solid fats (i.e., pork lard, butter, shortenings and margarines) that are commonly used in food applications, will lead to the production of healthier products with adequate technological and sensory properties.

As the properties of the lipid material determine the properties of the product in which is to be used, these must be considered when designing alternatives to saturated fats. Color, consistency, stability to lipid oxidation, taste, etc. are among the main quality attributes associated with this type of fats. Therefore, quality criteria, in terms of fatty acid composition and their properties, must be taken into account in the design of new lipid materials. The choice of suitable lipid modification technologies will depend on the target lipid structure, production costs, and consumer demand. A combination of some or all of the present lipid modification techniques may be required.

Oil structuring can be achieved by different chemical and physical means, as reported in various reviews (Jiménez-Colmenero et al., 2015; Martins et al., 2018). Briefly, chemical means include hydrogenation and interesterification. Besides the fact that TFAs can be generated during hydrogenation, this chemical process modifies the nature of the native oil and the high unsaturation, and correlated health benefits are lost. Physical means consist either of oil entrapment into a gelled continuous phase through polymer gelation or in the formation of structured emulsions with the support of an appropriate emulsifying agent. Structured oils considered as fat substitutes are therefore addressed as organogels or hydrogels depending on whether the continuous phase is lipophilic or hydrophilic, respectively.

Although most studies on fat replacers have mainly focused on either understanding structuring principles or on their rheological characterization, in recent years, reports which explore the functionality of structured lipids in potential applications in real food formulations have been published (Chaves et al., 2018; Patel et al., 2020).

Due to the relevance of structured lipids in the development of healthier products based on their improved lipid content, their main structuring mechanisms as well as some of their applications in the development of healthier foods are described below.

#### Enzymatic modification of triacylglycerols

Enzymatic interesterification is a process used to modify the physical properties of oils and fats, and it is widely used in the synthesis of triacylglycerols (TAGs) in applications such as shortenings in baked products, fat replacers in meat and confectionery products, $n$-3 FA-enriched TAGs and modified digestibility fats. The enzymatic interesterification reaction generally takes place between a fat with a high melting point and a liquid oil, leading to an exchange of fatty acids within and between TAGs, which results in the formation of new TAG molecules with unique properties (desirable plasticity, texture and mouthfeel). This process does not have the adverse effects found in hydrogenation and that are related to the formation of SFAs and, particularly, TFAs.

The enzymatic interesterification process presents certain advantages over the chemical process, as it is less aggressive, can be conducted at relatively low temperatures and it is more specific than chemical reactions, allowing the incorporation of a particular fatty acid at a specific position on the glycerol backbone, depending on the selected enzymes (lipases and esterases). In this way, it is possible to obtain specific lipids with defined structure and chemical composition, as opposed to chemical methods which do not allow this possibility due to the random nature of the reaction. Some commercially available lipase sources used for food processing include *Candida cylindracea*, *Rhizomucor miehei*, *Mucor miehei*, and *Penicillium roqueforti*. However, the interesterification process in most cases presents the disadvantage of the deteriorating oxidative quality, especially when the oil sources used are susceptible to oxidation.

In addition to creating plastic fats with a healthy lipid profile, this process can be used to valorize by-products of the food industry, i.e., beef tallow in the meat industry because of the high melting temperature and low level of UFAs; tallow is considered as a less valuable fat not suitable for direct human consumption. Combinations of enzymatically interesterified vegetable oils (sunflower, soybean, palm, canola, cotton, safflower, olive and maize oils in different proportions) have been tailored to simulate the characteristics of pork backfat (hard fat) and meet the technological requirements of the meat processing industry. These materials have been used as an alternative to animal fat and have lowered SFA content resulting in safe, tasty and healthier meat products (Javidipour et al., 2005; Ozvural & Vural, 2008; Vural et al., 2004). For instance, Cheong et al. (2010) formulated frankfurters using enzymatically interesterified blends of pork fat and rapeseed oil, and the products presented adequate physicochemical characteristics and sensory attributes as compared with the reference frankfurters elaborated with pork fat. Kowalska et al. (2015) created new fats by chemical and enzymatic interesterification of fats such as mutton tallow and vegetable oils, which contained certain amounts of monoacylglycerols and diacylglycerols (MAGs and DAGs, respectively) with potential applications in the food industry.

Similarly, this enzymatic lipid modification process has been widely used in bakery products. For instance, palm stearin is a hard stock that is usually blended with soft oils such as soybean, canola, sunflower, and cottonseed oils through an enzymatic interesterification process to produce trans-free plastic fats, as well as to improve the solid fat content (SFC) of the product. The presence of palm stearin helps to enhance plasticity and maintain the shape and structure of the product to withstand temperature fluctuations (Nor Aini & Miskandar, 2007). Medium and long chain triglycerides (MLCT) produced from the esterification of lipase, stearic acid, capric acid and glycerol with a 3:1 fatty acid/glycerol molar ratio, can be used as functional hard stock in shortenings and margarines, which may be used for obesity management purposes, as they have a high SFC at 25°C. Besides, MLCT produced from interesterification between a hard stock (fully hydrogenated soybean oil) and a soft oil (rice bran oil, coconut oil) give the newly formed oil plasticity properties suitable to be converted into margarine and shortening (Adhikari et al., 2012). Similarly, Y. Zhang et al. (2010) studied the production of margarine using Lipozyme IM lipase-catalyzed interesterification of palm stearin and coconut oil (75/25, w/w) in a 1 kg batch stirred tank reactor. Also, recently, sunflower oil-based cocoa butter equivalents (CBE) have been produced enzymatically by Kadivar et al. (2016). In their study, sunflower oil-based CBE were blended with cocoa butter (CB) in various proportions to be used in confectionery products.

The authors claimed that the melting heat (72.3–77.5 J/g) and onset temperatures (12.8–15.5 °C) of certain blends were comparable to those of CB (76.7 J/g and 14 °C). Moreover, the isosolid phase diagram showed satisfactory compatibility.

### Gelled or structured emulsions

The technological limitations of conventional emulsions have led to the development of more complex structured or gelled emulsions with novel functional properties and broad industrial applications (Jiménez-Colmenero et al., 2015; McClements, 2012). These new systems can form from a stable liquid-like emulsion through a process of gelation of the continuous phase (hydrogelation) and/or the aggregation of the emulsion droplets (organogelation). These can be stabilized by polymer molecules such as polysaccharides and proteins (Chen et al., 2016; Lin et al., 2020). The rheological properties of this material (solid or viscous) are mainly determined by the properties of the hydrogelified or organogelified phase. With the appropriate formulation and processing conditions, organogelified emulsions can be used as low-fat spreadable products or to control the release of hydrophilic and hydrophobic bioactive compounds. For instance, Chen et al. (2016) formulated a zein stabilized high oil-in-glycerol (O/G) emulgel enriched with β-carotene and observed that the formation of the emulsion significantly improved the photo-stability of β-carotene, which was retained in a significant quantity under light exposure for a period of time, and thus its oxidation was delayed. Furthermore, zein-based O/G emulgels, as an alternative to margarine, showed comparable functionalities (texture and sensory attributes) to those of a standard cake. Moschakis et al. (2016) formulated an organogelated emulsion in which sterols were incorporated into sunflower oil as the oil phase while the aqueous phase was made with Tween 20 and xanthan gum. This system could be utilized in foods containing relatively high amounts of solid fat such as margarine, spreadable and processed cheese, mayonnaise and dressings. Some commercial fibers have been used to develop structured emulsions as reported by Curti et al. (2018). These authors used a vegetable fiber HI-FIBER WF (HI-FOO, Parma, Italy), which is a natural emulsifier and stabilizer formed by various polysaccharide fractions (including cellulose, pectins, mucilages, etc.). When this HI-FIBER is mixed with water and oil it is able to bind water in a very stable way and could be used as a fat replacer. Blake and Marangoni (2015) successfully demonstrated the use of structured emulsions for replacing roll-in shortenings or laminate fats.

In relation to the gelation of O/W emulsions, one of the main applications is the obtaining of structured double emulsions. To understand this type of application, it is first necessary to know that double emulsions are multicompartmentalized systems characterized by the coexistence of O/W and W/O emulsions, in which the globules of the dispersed phase contain smaller droplets inside them (Jiménez-Colmenero, 2013). The most common types are W/O/W (water droplets dispersed in oil and surrounded by an aqueous phase) and O/W/O (oil droplets dispersed in an aqueous phase and surrounded by an oil phase) (Jiménez-Colmenero, 2013). In this type of double emulsions, there are two water-oil interfaces and therefore two types of emulsifying/stabilizing agents are required. Briefly, the formation of multiple W/O/W emulsions starts with the preparation of a W/O emulsion, for which a lipophilic emulsifying agent (e.g., polyglycerol polyricinoleate, PGPR) is required. This system is then emulsified with an aqueous solution in which hydrophilic emulsifying agents are dissolved (being sodium caseinate the most common) (Cofrades et al., 2013; Serdaroğlu et al., 2016). Due to their properties, including the ability to trap and protect various substances and to produce their controlled release from inside one phase into the other, these systems can be potentially useful in the production of healthy foods. In the food industry, including the meat industry, they have been employed for fat reduction, flavor masking and improvement of the sensory properties of products, as well as for the controlled release and protection of labile ingredients during food processing and preservation, or even for the action of certain enzymatic activity during gastrointestinal digestion (Jiménez-Colmenero, 2013; Jiménez-Colmenero et al., 2015; McClements, 2012). For example, in cooked emulsion-type meat products, they have been used as substitutes of animal fat to improve their lipid content, obtaining encouraging results (Cofrades et al., 2013; Freire et al., 2016; Serdaroğlu et al., 2016).

Despite the advantages of using these new lipid materials in the formulation of healthy products in relation to their lipid profile, the applications tested in real products are scarce, and some examples of meat products are shown in Table 10.3.

**Table 10.3 Examples of oil-structuring technology applications to improve the fat content of food products.**

| Oil-structuring strategy | Oil | Food product | Reference |
|---|---|---|---|
| Emulsion gel: carrageenan | Microalgal oil emulsified | Beef patties | Alejandre et al. (2019) |
| Emulsion gel: microbial transglutaminase (MTG) | Combination of olive, flaxseed and fish oils | Frankfurter | Delgado-Pando et al. (2010) |
| Emulsion gel: κ-carrageenan | Linseed oil | Bologna-type sausage | Poyato et al. (2014) |
| Emulsion gel: MTG, alginate or gelatin | Virgin olive oil | Frankfurter | Pintado et al. (2015) |
| Gelled double emulsions: MTG and gelatin | Perilla oil | Pork patties | Freire et al. (2017) |
| Emulsion gel and oleogel | Olive/Chia oils | Dry fermented sausage | Pintado and Cofrades (2020) |
| Konjac-based oil bulking system | Combination of olive, flaxseed and fish oils | Dry fermented sausages | Jiménez-Colmenero et al. (2013) |
| Konjac-based oil bulking system | Combination of olive, flaxseed and fish oils | Pork patties | Salcedo-Sandoval et al. (2015) |
| Alginate-based oil bulking system | Virgin olive oils | Frankfurter | Herrero et al. (2014) |
| Oleogel: ethyl cellulose | Soy, flaxseed and canola oils | Frankfurter | Zetzl et al. (2012) |
| Oleogel: MAG or soy lecithin | Combination of virgin olive oil and sunflower oils | Meat suspension | Lupi et al. (2012) |
| Oleogel: ethyl cellulose or ethyl cellulose/sorbitan monostearate | Canola oil | Frankfurter (beef) and fresh pork sausage | Wood (2013) |
| Oleogel: ethyl cellulose or beeswax | Combination of olive, linseed and fish oils | Pork burgers | Gómez-Estaca, Pintado, et al. (2019) |

*Continued*

**Table 10.3** Examples of oil-structuring technology applications to improve the fat content of food products—cont'd

| Oil-structuring strategy | Oil | Food product | Reference |
|---|---|---|---|
| Oleogel: ethyl cellulose or beeswax | Combination of olive, linseed and fish oils | Pâté | Gómez-Estaca, Herrero, et al. (2019) |
| Oleogel: beeswax | Linseed oil | Frankfurter | Franco et al. (2019) |
| Oleogel: beeswax | Linseed oil | Pâté | Martins et al. (2020) |
| Oleogel: ethyl cellulose, sorbitan monostearate | Canola and high oleic sunflower oils | Creams for various fillings | Stortz et al. (2012) |
| Olegoel: ethyl cellulose in ethanol/ethanol mix | Hydrogenated palm kernel oil | Heat resistant chocolate | Stortz and Marangoni (2013) |
| Oleogel: rice bran wax, candelilla wax or carnauba wax | High oleic sunflower oil | Ice cream | Zulim Botega et al. (2013) |
| Oleogel: carnauba wax and MAGs | Virgin olive oil | Spreadable products | Öğütcü and Yılmaz (2014) |
| Oleogel: shellac wax | Canola oil | Chocolate spread and cakes | Patel, Rajarethinem, et al. (2014) |
| Oleogel: sunflower wax; candelilla wax | Soybean, almond, sorn, flaxseed, peanut, sunflower, walnut oils | Margarines | Hwang et al. (2014) Da Silva et al. (2018) |
| Oleogel: rice bran wax or ethyl cellulose | High oleic sunflower, soy oils | Cheese spread | Bemer et al. (2016) |
| Oleogel: sunflower, rice bran, bee and candelilla waxes | Olive, soybean and flaxseed oils | Cookies | Hwang et al. (2016) |
| Oleogel: candelilla wax | Grapeseed oil | Muffins | Lim et al. (2017) |
| Oleogel: ethyl cellulose | Soybean oil | Shortening for bread | Ye et al. (2019) |

### Oil bulking agents

An alternative way to stabilize/structure oils is the formation of oil bulking agents, which consist of a dispersion of a large number of oil droplets in an aqueous gel matrix. In this way, the oil is physically trapped within a hydrogel network structure that provides strength to the system, making it suitable for its use as a fat analogue in various products (Jiménez-Colmenero et al., 2015).

In general, the formation of these systems is relatively simple as it is based on dispersing and homogenizing the oil in the aqueous phase, and subsequently inducing gelation by the action of a gelling agent. Such structures help to immobilize the oil particles, thus acting as bulking agents. There is a significant number of hydrocolloids that can be used as gelling agents, either individually or in combination, to create a variety of gel structures. These include konjac glucomannan, alginate or alginate-inulin and alginate-dextrin mixtures (Herrero et al., 2014; Salcedo-Sandoval et al., 2013; Stajic et al., 2014; Triki et al., 2013a, 2013b). Among them, it is to be noted that konjac gels can simulate the organoleptic properties of fat (mouthfeel) and connective tissue in meat systems, so they have been used as animal fat replacers in various meat derivatives, such as boiled (Salcedo-Sandoval et al., 2015), fermented ( Jiménez-Colmenero et al., 2013; Triki et al., 2013a) and fresh (Salcedo-Sandoval et al., 2015; Triki et al., 2013b) products; some of them are shown in Table 10.3.

### Organogels/oleogels

An organogel (named oleogel when it is used a food-grade oil) can be defined as an organic liquid (oil) trapped within a three-dimensional gel structure that is induced by a gelling agent (organogelator/oleogelator); it is thermoreversible and presents viscoelastic properties (Stortz et al., 2012). These materials can contain more than 97% oil by weight and a high consistency structure (Patel, Cludts, et al., 2014). Organogels require specific thermal (high temperature) and/or shear conditions to form, and these conditions will vary depending on the type of organogelator. The formation of oleogels occurs mainly by direct methods such as self-assembly (formed by self-organization at the molecular level in the oil phase), and crystallization (crystal particles are produced through nucleation and subsequent crystal growth in the oil phase). In addition, it is possible to form oleogels by indirect methods from stable O/W emulsions from which the water is removed (by applying thermal treatments) to obtain systems consisting of more than 95% stabilized oil in a network (Patel, Rajarethinem, et al., 2014; Tavernier et al., 2018). Another classification of organogels is based on whether the organogelators used are low molecular weight compounds or polymers (Pehlivanoğlu et al., 2018). The former include TAGs, DAGs, etc. However, polymers such as ethyl cellulose offer great possibilities for food applications considering their potential as gelling agents (Patel, Cludts, et al., 2014).

The use of oleogels in food presents different possibilities, including a greater capacity to bind the oil in the food matrix, to reduce the presence of SFAs and TFAs or to improve the stability of the food emulsions in which they are incorporated. Furthermore, the development of such structures can favor the formation of micelles during food digestion, improving the bioavailability of fat-soluble bioactive compounds and, in addition, regulating postprandial fat levels in the blood ( Jiménez-Colmenero et al., 2015). However, despite their promising applications in food, including molecular gastronomy, there are still relatively few foods in which they have been employed. These include some sauces, biscuits, spreads, margarines as well as certain meat products, some of which are presented in Table 10.3. For example, in relation to the application of oleogels in meat products, a partial replacement of beef fat in frankfurters was carried out using an ethyl cellulose oleogel with canola oil as the fat mimetic, observing a similar texture as that of the frankfurters elaborated with beef fat, as well as acceptable sensory properties and an improved lipid profile (Zetzl et al., 2012). Recently, Oh et al. (2019) made oleogels from hydrophilic polymers (using the foam-templated approach) that were used to replace animal fat in cooked products such as patties. The highest overall acceptability was obtained at a 50% replacement level, which led to an important reduction in the ratio of saturated to unsaturated fats and therefore, producing nutritionally superior patties. In addition, Gómez-Estaca, Pintado, et al. (2019) and Gómez-Estaca, Herrero, et al. (2019) successfully structured a healthy oil mixture into solid-like structures using ethyl cellulose or beeswax as oleogelators, which were used to elaborate pork liver pâtés and pork burgers by partial or total substitution of back fat. The technological behavior of the reformulated products was not significantly affected as compared to the control sample, and the same was also true for product characteristics such as color and texture. From the sensory acceptability point of view, fat substitution by the beeswax oleogel had no significant effects on any of the parameters evaluated, whereas the substitution with ethyl cellulose had a negative effect that was directly related to the level of substitution (see Table 10.3).

Regarding the application in other food products, several authors have studied the use of oleogels employing different types of liquid oils and organogelators in the preparation of dairy products such as margarine, ice creams and cheese, as well as some bakery products like cakes, muffins, chocolate spreads, with reduced SFA and TFA contents, as it has been reported recently by Gutierrez-Luna et al. (2021). For instance, Zulim Botega et al. (2013) reported that wax-based gels had strong potential to replace milk fat in ice-cream formulations and in a recent study, Moriano and Alamprese (2017) developed ice cream formulated with sterol-based oleogels as milk fat substitutes. For cream cheese products, rice bran wax and ethyl cellulose oleogels were used to formulate reduced-fat cream cheese spreads aimed at nutritionally enhancing the FA profile (Bemer et al., 2016) (see Table 10.3). Textural properties of the reformulated samples showed comparable hardness, spreadability, and stickiness to the full-fat cream cheese control. The samples were also subjected to sensory evaluation to compare the organoleptic properties to those of the full-fat control and the results suggested that the palatability of the tested samples could be improved to reduce their strong flavor and bitterness. In bakery products, for instance, Hwang et al. (2016) studied the replacement of bakery fat with oleogels made with four different natural waxes (sunflower wax, beeswax, rice bran wax, and candelilla wax) and three different vegetable oils (olive, soybean, and flaxseed oils) in cookies. They found that while oleogel hardness and melting behavior were significantly affected by the type of wax and oil, the properties of the cookies (hardness, spread factor, and fracturability) were not significantly affected by the different waxes and oils. Several of these cookies showed similar properties to those made with commercial margarine. Recently, Ye et al. (2019) formulated shortenings using ethyl cellulose 100 as an organogelator and triglyceryl monostearate at 1% (w/w) as an emulsifier, which yielded a shortening with excellent air-incorporation ability. When the formulated shortening was incorporated into the matrix, breads with excellent specific volume and a stable soft texture were produced.

Other applications of oleogels in the development of different dairy and bakery products with improved lipid profiles are presented in Table 10.3.

Although wax-based oleogels showed great promise in replicating most of the functions of solid fats in food, there is still a need to solve some critical issues when reformulating margarines with wax-based mimetics, such as residual waxy taste, low firmness due to the absence of strain-hardening properties in wax gels and low long-term emulsion stability (Patel et al., 2020).

In general, the different strategies based on oil structuring have been studied according to their design, formation, structure and final properties. However, regardless of the numerous potential applications, there are currently few examples of structured lipids being incorporated into real foods. Table 10.3 and Fig. 10.1 show some of the applications of oil structuring technology to improve the fat content (quantitative and qualitatively) of some foods.

Based on everything described above, it should be noted that obtaining lipid materials by employing healthy oils with technological properties that are similar to those of animal fat, constitutes one of the main challenges for the food industry. They could allow obtaining products with an improved lipid profile that could include nutrition and health claims on the labelling without reducing technological, sensory and safety properties.

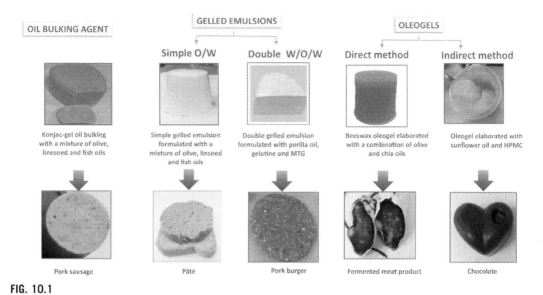

**FIG. 10.1**
Different options to structure healthy oils and their application into different food products developed in our laboratory.

## 10.3 Decreasing fat digestibility in food products

Obtaining healthier foods can also be achieved by decreasing the digestibility of the fat in food. In this context, it is currently known that the structure of food can facilitate or delay the release of fat during digestion and its absorption by the human body. Therefore, understanding how food structure modulates the kinetics of lipid digestion and/or digestibility has become an important focus of research, as it may lead to more effective ways of maintaining a healthy diet (Mat et al., 2016). The main form in which lipids are found in food is as emulsions, and after ingestion, emulsified lipids undergo a complex series of physical and chemical changes while they pass through the mouth, stomach, small and large intestine, affecting their ability to be digested and/or absorbed. The composition, structure and properties of emulsion droplets can be modified during chewing, digestion and absorption due to mechanical forces, the presence of enzymes and bile salts and changes in pH and/or temperature (McClements, 2015). Thus, the structure of emulsions and their stability play an important role in the digestion and absorption of lipids. It is in the intestinal stage in which most lipid digestion occurs (70–90%) (Armand et al., 1996) due to the action of pancreatic lipase. In order for this enzyme to exert its action, it must first be adsorbed on the surface of the fat droplets; for this reason, the characteristics of the interface between fat and the continuous aqueous phase are decisive when it comes to lipid digestion. In recent years, different studies have demonstrated that the rate of lipid digestion, and thus, the bioaccessibility of FAs depend on the delivery matrix (Singh et al., 2009). The term bioaccessibility has been defined as the fraction of the components (nutrients and other compounds) that are released from the food matrix into the juices of the gastrointestinal tract, which directly influence the fraction of these components that end up in the systemic circulation. Consequently, it is important to measure the concentration of a lipid component after digestion. One of the most important parameters when measuring bioaccessibility in an *in vitro* digestion model is the rate and extent of lipid digestion by the action of pancreatic lipase, i.e., the conversion of TAGs and DAGs into MAGs and free fatty acids (FFAs). *In vitro* digestion assays simulate the physiological conditions of *in vivo* digestion and are useful tools to study and understand the changes and interactions of nutrients, drugs and other compounds, as well as their bioaccessibility. One of the most commonly used methods to measure lipid digestion is the determination of the amount of free fatty acids produced during *in vitro* intestinal digestion by titration with an alkaline solution (Mun et al., 2007).

When fat is in the form of an emulsion, the initial properties of the emulsion directly influence the extent of digestion of the lipids present (Mun et al., 2006) and one of the main factors controlling the rate of lipolysis is the surface area of the droplet available for the adsorption of pancreatic lipase (Golding et al., 2011). The type of fat present also determines the extent of digestibility. Bonnaire et al. (2008) showed that the conversion rate of TAGs into FFAs was higher in an emulsion that contained liquid oil than in an emulsion with solid droplets. Another variable that influences lipolysis is the initial particle size of the matrix; small fat globules, but coated with surfactants in a viscous medium, undergo a slower hydrolysis than large fat globules which are coated with surfactants that are more weakly bound to the globule and in a less viscous medium (McClements & Decker, 2009). Seimon et al. (2009) concluded that the effect of fat emulsions on gastrointestinal motility, hormone release, appetite and energy ingestion is related to the size of the fat droplets. If the interfacial layer that surrounds the droplets is not strong enough, the droplets may coalesce in the gastrointestinal tract, which would lead to an increase in particle size and loss of emulsion stability. These destabilization processes are modulated by adding substances that modify the interfacial properties, such as emulsifiers. The incorporation of emulsifiers with high surfactant capacity hinders other substances with surface properties present in digestion (bile, pancreatic lipase) from adhering to the fat droplets, thus reducing their digestion. In this sense, the incorporation of hydrocolloids in the continuous phase of emulsions has been shown to be an effective tool to reduce fat lipolysis (Beysseriat et al., 2006; Gidley, 2013; Qin et al., 2017). In this way, Espinal-Ruiz et al. (2014) investigated the influence of different dietary fibers on lipid digestion in corn oil/water emulsions by using an *in vitro* digestion model and found that the rate and extent of lipid digestion decreased with increasing concentrations of methylcellulose, pectin, and chitosan. In this context, different studies have also shown that the incorporation of cellulose ethers or xanthan gum as stabilizers in O/W emulsions is an effective manner to reduce fat digestibility in *in vitro* digestion systems (Espert et al., 2016; Espert et al., 2017; Espert, Salvador, & Sanz, 2019). Additionally, Malinauskyte et al. (2014) studied the effect of carboxymethyl cellulose on the lipid digestion and physicochemical properties of whey protein-stabilized emulsions during digestion. These authors concluded that the thick network formed by carboxymethyl cellulose in the continuous phase limits the interaction of fat droplets with intestinal fluids, hence slowing down the rate of lipid digestion. The ability of cellulose ethers to compete with bile salts for the oil/water interface has also been studied, concluding that cellulose ethers bind bile salts, thereby reducing digestion and the subsequent absorption of lipids (Torcello-Gómez & Foster, 2014). In this regard, the main factor associated with reduced lipid digestibility is a physical impediment caused by the hydrocolloid network present in the continuous phase of emulsions, which limits the proper function of the digestive juices (Espert et al., 2016). The authors cited above, also observed that higher structural resistance during the *in vitro* digestion is associated with greater viscoelasticity and lower fat bioavailability and digestibility. Additionally, the chemical substitution of cellulose ethers affects the structure of the emulsion, the structural changes that occur during digestion and fat bioavailability (Espert et al., 2017). A higher content of methoxyl was associated with a lower decrease in viscoelasticity during digestion, indicating greater structural strength and lower fat digestibility, while cellulose with a lower content of methoxyl showed a higher structure loss and greater digestibility. In turn, it has also been observed that the digestibility-reducing effect of xanthan gum is enhanced when Tween 80 polysorbate is present as an emulsifier (Espert, Salvador, & Sanz, 2019). In the absence of this emulsifier, the xanthan gum emulsion also undergoes an increase in consistency at the gastric level, which is considered an interesting property for alternative applications such as increased satiety.

Regarding the oleogels, these could represent a strategy to modulate lipid digestion and deliver health benefits. In this context, different authors have studied how the different structuring agents modulate the oleogel structure which affect the rate and extend of lipid digestibility (Ashkar et al., 2020; Calligaris et al., 2020).

Nonetheless, while many studies have produced emulsions with hydrocolloids and oleogels capable of influencing fat digestion, these investigations have focused on model emulsions. Hence, the study of the digestibility of emulsions in more complex matrices, such as food, is a topic that has been more rarely addressed. Structural changes in these more complex matrices in which interactions occur between all components are determinant for the degree of lipid digestibility and physical properties. In this regard, Espert, Constantinescu, et al. (2019) developed an emulsion stabilized with xanthan gum for the reformulation of a conventional low-fat and low-saturated fat cream filler with adequate consistency and creaminess, which also exhibited improved structural resistance to *in vitro* digestion and, consequently, a significant decrease in fat digestibility (Espert et al., 2017; Espert, Bresciani, et al., 2019). A panna cotta prepared with a hydroxypropyl methylcellulose-based emulsion was also well accepted by consumers and presented a lower initial rate and extent of digestion (Borreani et al., 2020).

Considering that meat products are the main source of saturated fats in the diet and given their important consumption in industrialized countries, one of the main challenges would be to reduce the digestibility of the animal fat commonly added during the production of this type of products. Nevertheless, there are hardly any bioavailability studies to understand the release and digestion of animal fat lipids from the ingestion of reformulated meat products. Based on the knowledge regarding the fact that hydrocolloids are capable of reducing the digestibility of emulsified oils, Santiaguín-Padilla et al. (2019) suggested encapsulating pork fat with a non-digestible carbohydrate, such as pectin, for incorporation into the formulation of a Frankfurter-type sausage. In an *in vitro* digestion study, these authors observed that during gastrointestinal digestion, pectin-encapsulated pork fat offered greater protection against the hydrolytic action of lipases on TAGs as compared with the direct addition of pectin, and without affecting its sensory acceptability. Therefore, this may be a potential strategy to reduce fat intake from meat products.

## 10.4 Future perspectives

The reduction and/or improvement of the lipid content of foods rich in saturated fat is a challenge for both research and the food industry, as they would help to reduce the incidence of the main chronic degenerative diseases associated with the diet. To this end, within the technological strategies, structured lipids are systems with great potential to be used as fat replacers, as well as carriers of functional ingredients. When used as fat replacers, it is essential to understand their role as part of the food product. The replacement of saturated fats by structured oils such as gelled emulsions or oleogels is an interesting strategy to make healthier products with an improved fatty acid profile (low in saturated fatty acids and TFAs and high in UFAs). Despite the several studies published in which these structured lipids were used in meat and bakery products as fat substitutes, research is still needed to explore the best suitable emulsifiers and oleogelators that are food-grade, economical and efficient at low concentrations. On the other hand, it is also necessary to expand the number of applications in processed foods rich in saturated fat. In turn, bioaccessibility and bioavailability studies are required to understand the release and digestion of nutrients, especially lipids, from the intake of these reformulated food products. Finally, more studies are needed to confirm the claimed health effects, so they can be incorporated into foods' labelling and communicate the health benefits of these reformulated foods to consumers.

## Acknowledgments

Thanks to Projects PID2019-103872RB-I00/AEI/10.13039/501100011033 and RTI-2018-099738-B-C21 of the Plan Nacional de Investigación Cientifica, Desarrollo e Innovación Tecnológica (I+D+i), Ministerio de Ciencia e Innovación, and Intramural project CSIC: 202070E177.

## References

Adhikari, P., Shin, J. A., Lee, J. H., Kim, H. R., Kim, I. H., Hong, S. T., & Lee, K. T. (2012). Crystallization, physicochemical properties, and oxidative stability of the interesterified hard fat from rice bran oil, fully hydrogenated soybean oil, and coconut oil through lipase-catalyzed reaction. *Food and Bioprocess Technology*, *5*, 1–14.

Alaei, F., Hojjatoleslamy, M., & Dehkordi, S. M. H. (2018). The effect of inulin as a fat substitute on the physicochemical and sensory properties of chicken sausages. *Food Science & Nutrition*, *6*, 512–519.

Alejandre, M., Ansorena, D., Calvo, M. I., Cavero, R. Y., & Astiasarán, I. (2019). Influence of a gel emulsion containing microalgal oil and a blackthorn (*Prunus spinosa* L.) branch extract on the antioxidant capacity and acceptability of reduced-fat beef patties. *Meat Science*, *148*, 219–222.

Álvarez, D., Xiong, Y. L., Castillo, M., Payne, F. A., & Garrido, M. D. (2012). Textural and viscoelastic properties of pork frankfurters containing canola-olive oils, rice bran, and walnut. *Meat Science*, *92*, 8–15.

Alves Martins, D., Custódio, L., Barreira, L., Pereira, H., Ben-Hamadou, R., Varela, J., & Abu-Salah, K. M. (2013). Alternative sources of n-3 long-chain polyunsaturated fatty acids in marine microalgae. *Marine Drugs*, *11*(7), 2259–2281.

American Heart Association. (2017). *Advisory: Replacing saturated fat with healthier fat could lower cardiovascular risk*. AHA News. Available from: https://www.heart.org/en/news/2018/07/17/advisory-replacing-saturated-fat-with healthier-fat-could-lower-cardiovascular-risks (17 July 2018).

Andiç, S., Zorba, Ö., & Tunçtürk, Y. (2010). Effect of whey powder, skim milk powder and their combination on yield and textural properties of meat patties. *International Journal of Agriculture and Biology*, *12*(6), 871–876.

Aportela-Palacios, A., Sosa-Morales, M. E., & Velez-Ruiz, J. F. (2005). Rheological and physicochemical behavior of fortified yogurt, with fiber and calcium. *Journal of Texture Studies*, *36*, 333–349.

Arcia, P., Costell, E., & Tárrega, A. (2011). Inulin blend as prebiotic and fat replacer in dairy desserts: Optimization by response surface methodology. *Journal of Dairy Science*, *94*(5), 2192–2200.

Armand, M., Borel, P., Pasquier, B., Dubois, C., Senft, M., Andre, M., Peyrot, J., Salducci, J., & Lairon, D. (1996). Physicochemical characteristics of emulsions during fat digestion in human stomach and duodenum. *American Journal of Physiology-Gastrointestinal and Liver Physiology*, *271*(1), G172–G183.

Aryana, K. Y., Plauche, S., Rao, R. M., McGrew, P., & Shah, N. P. (2007). Fat-free plain yogurt manufactured with inulins of various chain lengths and *Lactobacillus acidophilus*. *Journal of Food Science, 72*, M79–M84.

Ashkar, A., Rosen-Kligvasser, J., Lesmes, U., & Davidovich-Pinhas, M. (2020). Controlling lipid intestinal digestibility using various oil structuring mechanisms. *Food & Function, 11*, 7495–7508.

Barros, J. C., Munekata, P. E. S., Carvalho, F. A. L., Pateiro, M., Barba, F. J., Domínguez, R., Trindade, M. A., & Lorenzo, J. M. (2020). Use of tiger nut (*Cyperus esculentus* L.) oil emulsion as animal fat replacement in beef burgers. *Foods, 44*, 1–15.

Barrow, C. J., Nolan, C., & Holub, B. J. (2009). Bioequivalence of encapsulated and microencapsulated fish-oil supplementation. *Journal of Functional Foods, 1*, 38–43.

Bemer, H. L., Limbaugh, M., Cramer, E. D., Harper, W. J., & Maleky, F. (2016). Vegetable organogels incorporation in cream cheese products. *Food Research International, 85*, 67–75.

Bermúdez-Aguirre, D., & Barbosa-Cánovas, G. V. (2011). Quality of selected cheeses fortified with vegetable and animal sources of omega-3. *LWT - Food Science and Technology, 44*, 1577–1584.

Beysseriat, M., Decker, E. A., & McClements, D. J. (2006). Preliminary study of the influence of dietary fiber on the properties of oil-in-water emulsions passing through an in vitro human digestion model. *Food Hydrocolloids, 20*(6), 800–809.

Blake, A. I., & Marangoni, A. G. (2015). Factors affecting the rheological properties of a structured cellular solid used as a fat mimetic. *Food Research International, 74*, 284–293.

Bom Frost, M., Dijksterhuis, G., & Martens, M. (2001). Sensory perception of fat in milk. *Food Quality and Preference, 12*(5–7), 327–336.

Bonnaire, L., Sandra, S., Helgason, T., Decker, E. A., Weiss, J., & McClements, D. J. (2008). Influence of lipid physical state on the in vitro digestibility of emulsified lipids. *Journal of Agricultural and Food Chemistry, 56* (10), 3791–3797.

Borreani, J., Hernando, I., & Quiles, A. (2020). Cream replacement by hydrocolloid-stabilized emulsions to reduce fat digestion in panna cottas. *LWT - Food Science and Technology, 119*, 108896.

Brennan, C. S., & Tudorica, C. M. (2008). Carbohydrate-based fat replacers in the modification of the rheological, textural and sensory quality of yoghurt: Comparative study of the utilisation of barley beta-glucan, guar gum and inulin. *International Journal of Food Science and Technology, 43*(5), 824–833.

Brownlee, I., Chater, P. I., Pearson, J. P., & Wilcox, M. D. (2017). Dietary fibre and weight loss: Where are we now? *Food Hydrocolloids, 68*, 186–191.

Burdge, G. C., & Wootton, S. A. (2002). Conversion of α-linolenic acid to eicosapentaenoic, docosapentaenoic and docosahexaenoic acids in young women. *British Journal of Nutrition, 88*(4), 411–420.

Cais-Sokolinska, D., Pikul, J., Wojtowski, J., & Dankow, R. (2006). The effect of structure forming starch additives on the physicochemical and sensory properties of yoghurts. *Milchwissenschaft - Milk Science International, 61*, 173–176.

Calligaris, S., Alongi, M., Lucci, P., & Anese, M. (2020). Effect of different oleogelators on lipolysis and curcuminoid bioaccessibility upon in vitro digestion of sunflower oil oleogels. *Food Chemistry, 314*, 126146.

Calpe-Berdiel, L., Escolà-Gil, J. C., & Blanco-Vaca, F. (2009). New insights into the molecular actions of plant sterols and stanols in cholesterol metabolism. *Atherosclerosis, 203*(1), 18–31.

Chappalwar, A. M., Pathak, V., Goswami, M., & Verma, A. K. (2020). Development of functional chicken patties with incorporation of mango peel powder as fat replacer. *Nutrition & Food Science, 50*(6), 1063–1073.

Chavan, R. S., Khedkar, C. D., & Bhatt, S. (2016). Fat replacer. In B. Caballero, P. Finglas, & F. Toldrá (Eds.), *Vol. 2. The encyclopedia of food and health* (pp. 589–595). Oxford: Academic Press.

Chaves, K. F., Barrera-Arellano, D., & Ribeiro, A. P. B. (2018). Potential application of lipid organogels for food industry. *Food Research International, 105*, 863–872.

Chen, X. W., Fu, S. Y., Hou, J. J., Guo, J., Wang, J. M., & Yang, X. Q. (2016). Zein based oil-in-glycerol emulgels enriched with β-carotene as margarine alternatives. *Food Chemistry, 211*, 836–844.

Cheong, L. Z., Zhang, H., Nersting, L., Jensen, K., Haagensen, J. A. J., & Xu, X. (2010). Physical and sensory characteristics of pork sausages from enzymatically modified blends of lard and rapeseed oil during storage. *Meat Science, 85*, 691–699.

Ciftci, O. N., Przybylski, R., & Rudzińska, M. (2012). Lipid components of flax, perilla, and chia seeds. *European Journal of Lipid Science and Technology, 114*(7), 794–800.

Clark, M. J., & Slavin, J. L. (2013). The effect of fiber on satiety and food intake: A systematic review. *The Journal of the American College of Nutrition, 32*, 200–211.

Cofrades, S., Antoniou, I., Solas, M. T., Herrero, A. M., & Jiménez-Colmenero, F. (2013). Preparation and impact of multiple (water-in-oil-in-water) emulsions in meat systems. *Food Chemistry, 141*, 338–346.

Cofrades, S., Hughes, E., & Troy, D. J. (2000). Effects of oat fibre and carrageenan on texture of frankfurtersformulated with low and high fat. *European Food Research and Technology, 211*, 19–26.

Connor, W. E. (2000). Importance of n−3 fatty acids in health and disease. *The American Journal of Clinical Nutrition, 71*(1), 171S–175S.

Costa, M., Frasao, B., Rodrigues, B., Silva, A., & Conte-Junior, C. (2016). Effect of different fat replacers on the physicochemical and instrumental analysis of low-fat cupuassu goat milk yogurts. *Journal of Dairy Research, 83*, 493–496.

Curti, E., Federici, E., Diantom, A., Carini, E., Pizzigalli, E., Wu Symon, V., Pellegrini, N., & Vittadini, E. (2018). Structured emulsions as butter substitutes: Effects on physicochemical and sensory attributes of shortbread cookies. *Journal of the Science of Food and Agriculture, 98*, 3836–3842.

Da Silva, T. L. T., Chaves, K. F., Fernandes, G. D., Rodrigues, J. B., Bolini, H. M. A., & Arellano, D. B. (2018). Sensory and technological evaluation of margarines with reduced saturated fatty acid contents using oleogel technology. *Journal of the American Oil Chemists' Society, 95*, 673–685.

Delgado-Pando, G., Cofrades, S., Ruiz-Capillas, C., & Jiménez-Colmenero, F. (2010). Healthier lipid combination as functional ingredient influencing sensory and technological properties of low-fat frankfurters. *European Journal of Lipid Science and Technology, 112*(8), 859–870.

Dello Staffolo, M., Bertola, N., Martino, M., & Bevilacqua, A. (2004). Influence of dietary fiber addition on sensory and rheological properties of yogurt. *International Dairy Journal, 14*, 263–268.

Desmond, E. M., Troy, D. J., & Buckley, D. J. (1998). Comparative studies of nonmeat adjuncts used in the manufacture of low-fat beef burgers. *Journal of Muscle Foods, 9*, 221–241.

Djordjevic, D., Kim, H. J., McClements, D. J., & Decker, E. A. (2004). Physical stability of whey protein-stabilized oil-in-water emulsions at pH 3: Potential ω-3 fatty acid delivery systems (Part A). *Journal of Food Science, 69*(5), C351–C355.

Dogan, M., & Akgul, A. (2005). Fatty acid composition of some walnut (*Juglans regia* L.) cultivars from east Anatolia. *Grasas y Aceites, 56*(4), 328–331.

Domazakies, E. (2005). *Method for the production of meat products from entire muscular tissue and direct integration of olive oil*, GR Patent GR1004991B.

Ertekin, B., & Guzel-Seydim, Z. B. (2010). Effect of fat replacers on kefir quality. *Journal of the Science and Food Agriculture, 90*, 543–548.

Espert, M. (2019). *Funcionalidad de hidrocoloides en la digestibilidad de emulsiones aceite/agua* (Ph.D. thesis). Universidad de Valencia España.

Espert, M., Borreani, J., Hernando, I., Quiles, A., Salvador, A., & Sanz, T. (2017). Relationship between cellulose chemical substitution, structure and fat digestion in o/w emulsions. *Food Hydrocolloids, 69*, 76–85.

Espert, M., Bresciani, A., Sanz, T., & Salvador, A. (2019). Functionality of low digestibility emulsions in cocoa creams. Structural changes during in vitro digestion and sensory perception. *Journal of Functional Foods, 54*, 146–153.

Espert, M., Constantinescu, L., Sanz, T., & Salvador, A. (2019). Effect of xanthan gum on palm oil in vitro digestion. Application in starch-based filling creams. *Food Hydrocolloids, 86*, 87–94.

Espert, M., Salvador, A., & Sanz, T. (2016). In vitro digestibility of highly concentrated methylcellulose O/W emulsions: Rheological and structural changes. *Food & Function*, *7*(9), 3933–3942.

Espert, M., Salvador, A., & Sanz, T. (2019). Rheological and microstructural behavior of xanthan gum and xanthan gum-Tween 80 emulsions during in vitro digestion. *Food Hydrocolloids*, *95*, 454–461.

Espinal-Ruiz, M., Parada-Alfonso, F., Restrepo-Sánchez, L. P., Narváez-Cuenca, C. E., & McClements, D. J. (2014). Impact of dietary fibers [methyl cellulose, chitosan, and pectin] on digestion of lipids under simulated gastrointestinal conditions. *Food & Function*, *5*(12), 3083–3095.

European Food Safety Authority (EFSA). (2008). Consumption of food and beverages with added plant sterols in the European Union. *The EFSA Journal*, *133*, 1–21.

Fan, R., Zhou, D., & Cao, X. (2020). Evaluation of oat beta-glucan-marine collagen peptide mixed gel and its application as the fat replacer in the sausage products. *PLoS One*, *15*(5), e0233447.

FEEDcities Project. (2017). *The food environment description in cities in Eastern Europe and Central Asia-Kyrgyzstan*. Geneva: WHO Health Organization.

Felisberto, M. H. F., Wahanik, A. L., Gomes-Ruffi, C. R., Clerici, M. T. P. S., Chang, Y. K., & Steel, C. J. (2015). Use of chia (*Salvia hispanica* L.) mucilage gel to reduce fat in pound cakes. *LWT – Food Science and Technology*, *63*(2), 1049–1055.

Fernández-Gines, J. M., Fernandez-Lopez, J., Sayas-Barbera, E., & Perez-Alvarez, J. A. (2005). Meat products as functional foods: A review. *Journal of Food Science*, *70*(2), R37–R43.

Franco, D., Martins, A. J., López-Pedrouso, M., Purriños, L., Cerqueira, M. A., Vicente, A. A., … Lorenzo, J. M. (2019). Strategy towards replacing pork backfat with a linseed oleogel in frankfurter sausages and its evaluation on physicochemical, nutritional, and sensory characteristics. *Foods*, *8*, 366.

Freire, M., Bou, R., Cofrades, S., Solas, M. T., & Jiménez-Colmenero, F. (2016). Double emulsions to improve frankfurter lipid content: Impact of perilla oil and pork backfat. *Journal of the Science of Food and Agriculture*, *96*, 900–908.

Freire, M., Cofrades, S., Serrano-Casas, V., Pintado, T., Jimenez, M. J., & Jimenez-Colmenero, F. (2017). Gelled double emulsions as delivery systems for hydroxytyrosol and n-3 fatty acids in healthy pork patties. *Journal of Food Science and Technology*, *54*, 3959–3968.

Fuangpaiboon, N., & Kijroongrojana, K. (2017). Sensorial and physical properties of coconut-milk ice cream modified with fat replacers. *International Journal of Science and Technology*, *11*(2), 133–147.

Gampala, P., & Brennan, C. S. (2008). Potential starch utilisation in a model processed cheese system. *Starch-Starke*, *60*, 685–689.

Garcia-Perez, F. J., Sendra, E., Lario, Y., Fernandez-Lopez, J., Sayas-Barbera, E., & Perez-Alvarez, J. A. (2006). Rheology of orange fiber enriched yogurt. *Milchwissenschaft - Milk, Science International*, *61*(1), 55–59.

Gidley, M. J. (2013). Hydrocolloids in the digestive tract and related health implications. *Current Opinion in Colloid & Interface Science*, *18*(4), 371–378.

Glisic, M., Baltic, M., Glisic, M., Trbovic, D., Jokanovic, M., Parunovic, N., Dimitrijevic, M., Suvajdzic, B., Boskovic, M., & Vasilev, D. (2019). Inulin-based emulsion-filled gel as a fat replacer in prebiotic- and PUFA-enriched dry fermented sausages. *International Journal of Food Science and Technology*, *54*, 787–797.

Golding, M., Wooster, T. J., Day, L., Xu, M., Lundin, L., Keogh, J., & Clifton, P. (2011). Impact of gastric structuring on the lipolysis of emulsified lipids. *Soft Matter*, *7*(7), 3513–3523.

Gómez-Estaca, J., Herrero, A. M., Herranz, B., Álvarez, M. D., Jiménez-Colmenero, F., & Cofrades, S. (2019). Characterization of ethyl cellulose and beeswax oleogels and their suitability as fat replacers in healthier lipid pates development. *Food Hydrocolloids*, *87*, 960–969.

Gómez-Estaca, J., Pintado, T., Jiménez-Colmenero, F., & Cofrades, S. (2019). Assessment of a healthy oil combination structured in ethyl cellulose and beeswax oleogels as animal fat replacers in low-fat, PUFA-enriched pork burgers. *Food and Bioprocess Technology*, *12*, 1068–1081.

Guedes-Oliveira, J. M., Salgado, R. L., Costa-Lima, B. R. C., Guedes-Oliveira, J., & Conte-Junior, C. A. (2016). Washed cashew apple fiber (*Anacardium occidentale* L.) as fat replacer in chicken patties. *LWT - Food Science and Technology, 71*, 268–273.

Gumus, C. E., Mohammad, S., & Gharibzahedi, T. (2021). Yogurts supplemented with lipid emulsions rich in omega-3 fatty acids: New insights into the fortification, microencapsulation, quality properties, and health-promoting effects. *Trends in Food Science & Technology, 110*, 267–279.

Gutierrez-Luna, K., Astiasarán, I., & Ansorena, D. (2021). Gels as fat replacers in bakery products: A review. *Critical Reviews in Food Science and Nutrition, 7*, 1–14.

Ha, J. H., Lee, J., Lee, J. J., Choi, Y. I., & Lee, H. J. (2019). Effects of whey protein injection as a curing solution on chicken breast meat. *Food Science of Animal Resources, 39*, 494–502.

Hammad, S., Pu, S., & Jones, P. (2016). Current evidence supporting the link between dietary fatty acids and cardiovascular disease. *Journal of Lipids, 51*(5), 507–517.

Hashim, I. B., Khalil, A. H., & Afifi, H. S. (2009). Quality characteristics and consumer acceptance of yogurt fortified with date fiber. *Journal of Dairy Science, 92*(11), 5403–5407.

Herrero, A. M., Carmona, P., Jiménez-Colmenero, F., & Ruiz-Capillas, C. (2014). Polysaccharide gels as oil bulking agents: Technological and structural properties. *Food Hydrocolloids, 36*, 374–381.

Hooper, L., Thompson, R. L., Harrison, R. A., Summerbell, C. D., Moore, H., Worthington, H. V., Ness, A., Capps, N., Davey Smith, G., Riemersma, R., & Ebrahim, S. (2004). Omega 3 fatty acids for prevention and treatment of cardiovascular disease. *Cochrane Database of Systematic Reviews*, (4), CD003177.

Hwang, H.-S., Singh, M., & Lee, S. (2016). Properties of cookies made with natural wax–vegetable oil organogels. *Journal of Food Science, 81*, C1045–C1054.

Hwang, H.-S., Singh, M., Winkler-Moser, J. K., Bakota, E. L., & Liu, S. X. (2014). Preparation of margarines from organogels of sunflower wax and vegetable oils. *Journal of Food Science, 79*, C1926–C1932.

Hygreeva, D., Pandey, M. C., & Radhakrishna, K. (2014). Potential applications of plant based derivatives as fat replacers, antioxidants and antimicrobials in fresh and processed meat products. *Meat Science, 98*(1), 47–57.

Javidipour, I., Vural, H., Ozbas, O. O., & Tekin, A. (2005). Effects of interesterified vegetable oils and sugar beet fibre on the quality of Turkish-type salami. *International Journal of Food Science and Technology, 40*(2), 177–185.

Jiménez-Colmenero, F. (2007). Healthier lipid formulation approaches in meat-based functional foods technological options for replacement of meat fats by non-meat fats. *Trends in Food Science & Technology, 18*(11), 567–578.

Jiménez-Colmenero, F. (2013). Potential applications of multiple emulsions in the development of healthy and functional foods. *Food Research International, 52*, 64–74.

Jiménez-Colmenero, F., Cofrades, S., Herrero, A. M., Solas, M. T., & Ruiz-Capillas, C. (2013). Konjac gel for use as potential fat analogue for healthier meat product development: Effect of chilled and frozen storage. *Food Hydrocolloids, 30*(1), 351–357.

Jiménez-Colmenero, F., Salcedo-Sandoval, L., Bou, R., Cofrades, S., Herrero, A. M., & Ruiz-Capillas, C. (2015). Novel applications of oil-structuring methods as a strategy to improve the fat content of meat products'. *Trends in Food Science & Technology, 44*(2), 177–188.

Jonnalagadda, S. S., & Jones, J. M. (2005). ADA reports, position of the American dietetic association: Fat replacers. *Journal of the American Dietetic Association, 105*, 266–275.

Kadivar, S., De Clercq, N., Mokbul, M., & Dewettinck, K. (2016). Influence of enzymatically produced sunflower oil based cocoa butter equivalents on the phase behavior of cocoa butter and quality of dark chocolate. *LWT - Food Science and Technology, 66*, 48–55.

Kaushik, P., Dowling, K., Barrow, C. J., & Adhikari, B. (2015). Microencapsulation of omega-3 fatty acids: A review of microencapsulation and characterization methods. *Journal of Functional Foods, 19*, 868–881.

Kilic, B., Kankaya, T., Ekici, Y. K., & Orhan, H. (2010). Effect of textured soy protein on quality characteristics of low fat cooked kofte (Turkish meatball). *Journal of Animal and Veterinary Advances, 9*(24), 3048–3054.

Kolanowski, W., & Weiβbrodt, J. (2007). Sensory quality of dairy products fortified with fish oil. *International Dairy Journal, 17*, 1248–1253.

Kostik, V., Memeti, S., & Bauer, B. (2013). Fatty acid composition of edible oils and fats. *Journal of Hygienic Engineering and Design, 4*, 112–116.

Kowalska, D., Gruczynska, E., & Kowalska, M. (2015). The effect of enzymatic interesterification on the physicochemical properties and thermo-oxidative stabilities of beef tallow stearin and rapeseed oil blends. *Journal of Thermal Analysis and Calorimetry, 120*, 507–517.

Kris-Etherton, P. M., Lefevre, M., Mensink, R. P., Petersen, B. J., Fleming, J. A., & Flickinger, B. D. (2012). Trans fatty acid intakes and food sources in the U.S. population: Nhanes 1999–2002. *Lipids, 47*, 931–940.

Kumar, S. S., Balasubramanian, S., Biswas, A. K., Chatli, M. K., Devatkal, S. K., & Sahoo, J. (2011). Efficacy of soy protein isolate as a fat replacer on physico-chemical and sensory characteristics of low-fat paneer. *Journal of Food Science and Technology-Mysore, 48*, 498–501.

Lim, J., Jeong, S., Lee, J. H., Park, S., Lee, J., & Lee, S. (2017). Effect of shortening replacement with oleogels on the rheological and tomographic characteristics of aerated baked goods. *Journal of the Science of Food and Agriculture, 97*, 3727–3732.

Lin, D., Kelly, A. L., & Miao, S. (2020). Preparation, structure-property relationships and applications of different emulsion gels: Bulk emulsion gels, emulsion gel particles, and fluid emulsion gels. *Trends in Food Science & Technology, 102*, 123–137.

Lobato-Calleros, C., Ramirez-Santiago, C., Vernon-Carter, E., & Alvarez- Ramirez, J. (2014). Impact of native and chemically modified starches addition as fat replacers in the viscoelasticity of reduced-fat stirred yogurt. *Journal of Food Engineering, 131*, 110–115.

Lupi, F. R., Gabriele, D., Baldino, N., Seta, L., de Cindio, B., & De Rose, C. (2012). Stabilization of meat suspensions by organogelation: A rheological approach. *European Journal of Lipid Science and Technology, 114*, 1381–1389.

Macdiarmid, J., Clark, H., Whybrow, S., De Ruiter, H., & McNeill, G. (2018). Assessing national nutrition security: The UK reliance on imports to meet population energy and nutrient recommendations. *PLoS One, 13*, e0192649.

Malinauskyte, E., Ramanauskaite, J., Leskauskaite, D., Devold, T. H., Schüller, R. B., & Vegarud, G. E. (2014). Effect of human and simulated gastric juices on the digestion of wheyproteins and carboxymethylcellulose-stabilised O/W emulsions. *Food Chemistry, 165*, 104–112.

Manaf, Y. N., Marikkar, J. M. N., Mustafa, S., Van Bockstaele, F., & Nusantoro, B. P. (2019). Effect of three plant-based shortenings and lard on cookie dough properties and cookies quality. *International Food Research Journal, 26*(6), 1795–1802.

Mansour, E. H., & Khalil, A. H. (1999). Characteristics of low-fat beef burgers as influenced by various types of wheat fibres. *Journal of the Science of Food and Agriculture, 79*(4), 493–498.

Martins, A. J., Lorenzo, J. M., Franco, D., Pateiro, M., Dominguez, R., Munekata, P. E. S., Pastrana, L. M., Vicente, A. A., Cunha, R. L., & Cerqueira, M. A. (2020). Characterization of enriched meat-based pâté manufactured with oleogels as fat substitutes. *Gels, 6*(2), 17.

Martins, A. J., Vicente, A. A., Cunha, R. L., & Cerqueira, M. A. (2018). Edible oleogels: An opportunity for fat replacement in foods. *Food & Function, 9*(2), 758–773.

Mat, D. J. L., Le Feunteuna, S., Michon, C., & Souchon, I. (2016). In vitro digestion of foods using pH-stat and the INFOGEST protocol: Impact of matrix structure on digestion kinetics of macronutrients, proteins and lipids. *Food Research International, 88*, 226–233.

McCann, T. H., Fabre, F., & Day, L. (2011). Microstructure, rheology and storage stability of low-fat yoghurt structured by carrot cell wall particles. *Food Research International, 44*(4), 884–892.

McClements, D. J. (2010). Emulsion design to improve the delivery of functional lipophilic components. *Annual Review of Food Science and Technology, 1*(1), 241–269.

McClements, D. J. (2012). Advances in fabrication of emulsions with enhanced functionality using structural design principles. *Current Opinion in Colloid & Interface Science, 17*(5), 235–245.

McClements, D. J. (2015). Reduced-fat foods: The complex science of developing diet-based strategies for tackling overweight and obesity. *Advances in Nutrition, 6*(3), 338S–352S.

McClements, D. J., & Decker, E. A. (Eds.). (2009). *Designing functional foods: Measuring and controlling food structure breakdown and nutrient absorption*. Cambridge: Woodhead Publishing.

Montowska, M., & Fornal, E. (2018). Detection of peptide markers of soy, milk and egg white allergenica proteins in poultry products by LC-Q-TOF-MS/MS. *LWT - Food Science and Technology, 87*, 310–317.

Moriano, M. E., & Alamprese, C. (2017). Organogels as novel ingredients for low saturated fat ice creams. *LWT - Food Science and Technology, 86*, 371–376.

Morris, C., & Morris, G. A. (2012). The effect of inulin and fructo-oligosaccharide supplementation on the textural, rheological and sensory properties of bread and their role in weight management: A review. *Food Chemistry, 133*(2), 237–248.

Moschakis, T., Panagiotopoulou, E., & Katsanidis, E. (2016). Sunflower oil organogels and organogel-in-water emulsions (part I): Microstructure and mechanical properties. *LWT - Food Science and Technology, 73*, 153–161.

Mozaffarian, D., Aro, A., & Willett, W. C. (2009). Health effects of trans-fatty acids: Experimental and observational evidence. *European Journal of Clinical Nutrition, 63*, 5–21.

Mun, S., Decker, E. A., & McClements, D. J. (2007). Influence of emulsifier type on in vitro digestibility of lipid droplets by pancreatic lipase. *Food Research International, 40*(6), 770–781.

Mun, S., Decker, E. A., Park, Y., Weiss, J., & McClements, D. J. (2006). Influence of interfacial composition on in vitro digestibility of emulsified lipids: Potential mechanism for chitosan's ability to inhibit fat digestion. *Food Biophysics, 1*(1), 21–29.

Namir, M., Siliha, H., & Ramadan, M. F. (2015). Fiber pectin from tomato pomace: Characteristics, functional properties and application in low-fat beef burger. *Journal of Food Measurement and Characterization, 9*, 305–312.

Nor Aini, I., & Miskandar, M. S. (2007). Utilization of palm oil and palm products in shortenings and margarines. *European Journal of Lipid Science and Technology, 109*(4), 422–432.

Öğütcü, M., & Yılmaz, E. (2014). Oleogels of virgin olive oil with carnauba wax and monoglyceride as spreadable products. *Grasas y Aceites, 65*(3), e040.

Oh, I., Lee, J., Lee, H. G., & Lee, S. (2019). Feasibility of hydroxypropyl methylcellulose oleogel as an animal fat replacer for meat patties. *Food Research International, 122*, 566–572.

Orsavova, J., Misurcova, L., Ambrozova, J. V., Vicha, R., & Mlcek, J. (2015). Fatty acids composition of vegetable oils and its contribution to dietary energy intake and dependence of cardiovascular mortality on dietary intake of fatty acids. *International Journal of Molecular Sciences, 16*, 12871–12890.

Osakabe, N., Yasuda, A., Natsume, M., Sanbongi, C., Kato, Y., Osawa, T., & Yoshikawa, T. (2002). Rosmarinic acid, a major polyphenolic component of *Perilla frutescens*, reduces lipopolysaccharide (LPS)-induced liver injury in D-galactosamine (D-GalN)-sensitized mice. *Free Radical Biology and Medicine, 33*(6), 798–806.

Ozvural, E. B., & Vural, H. (2008). Utilization of interesterified oil blends in the production of frankfurters. *Meat Science, 78*(3), 211–216.

Patel, A. R., Cludts, N., Bin Sintang, M. D., Lesaffer, A., & Dewettinck, K. (2014). Edible oleogels based on water soluble food polymers: Preparation, characterization and potential application. *Food & Function, 5*, 2833–2841.

Patel, A. R., Nicholson, R. A., & Marangoni, A. G. (2020). Applications of fat mimetics for the replacement of saturated and hydrogenated fat in food products. *Current Opinion in Food Science, 33*, 61–68.

Patel, A. R., Rajarethinem, P. S., Gredowska, A., Turhan, O., Lesaffer, A., De Vos, W. H., Van de Walle, D., & Dewettinck, K. (2014). Edible applications of shellac oleogels: Spreads, chocolate paste and cakes. *Food & Function, 5*, 645–652.

Pehlivanoğlu, H., Demirci, M., Toker, S., Konar, N., Karasu, S., & Sagdic, O. (2018). Oleogels, a promising structured oil for decreasing saturated fatty acid concentrations: Production and food-based applications. *Critical Reviews in Food Science and Nutrition, 58*, 1330–1341.

Peng, X., & Yao, Y. (2017). Carbohydrates as fat replacers. *The Annual Review of Food Science and Technology, 8*, 331–351.

Pietrasik, Z. (2003). Binding and textural properties of beef gels processed with kappa-carrageenan, egg albumin and microbial transglutaminase. *Meat Science, 63*, 317–324.

Pimdit, K., Therdathai, N., & Jangchud, K. (2008). Effects of fat replacers on the physical, chemical and sensory characteristics of puff pastry. *Kasetsart Journal-Natural Science, 42*, 739–746.

Pintado, T., & Cofrades, S. (2020). Quality characteristics of healthy dry fermented sausages formulated with a mixture of olive and chia oil structured in oleogel or emulsion gel as animal fat replacer. *Foods, 9*, 830.

Pintado, T., Herrero, A. M., Ruiz-Capillas, C., Triki, M., Carmona, P., & Jiménez-Colmenero, F. (2015). Effects of emulsion gels containing bioactive compounds on sensorial, technological, and structural properties of frankfurters. *Food Science and Technology International, 22*, 132–145.

Poyato, C., Ansorena, D., Berasategi, I., Navarro-Blasco, I., & Astiasarán, I. (2014). Optimization of a gelled emulsion intended to supply ω-3 fatty acids into meat products by means of response surface methodology. *Meat Science, 98*, 615–621.

Puligundla, P., Cho, Y., & Lee, Y. (2015). Physicochemical and sensory properties of reduced-fat mayonnaise formulations prepared with rice starch and starch-gum mixtures. *European Journal of Food and Agriculture, 27*(8), 463–468.

Qin, D., Yang, X., Gao, S., Yao, J., & McClements, D. J. (2017). Influence of dietary fibers on lipid digestion: Comparison of single-stage and multiple-stage gastrointestinal models. *Food Hydrocolloids, 69*, 382–392.

Rahimi, J., Khosrowshahi, A., Madadlou, A., & Aziznia, S. (2007). Texture of low-fat Iranian White cheese as influenced by gum tragacanth as a fat replacer. *Journal of Dairy Science, 90*, 4058–4070.

Rezvani, F., Abbasi, H., & Nourani, M. (2020). Effects of protein-polysaccharide interactions on the physical and textural characteristics of low-fat whipped cream. *Journal of Food Processing and Preservation, 44*(10), e14743.

Rippin, H., Jyh, J. H., Ocke, M., Jewell, J., Breda, J., & Cade, J. E. (2017). An exploration of socio-economic and food characteristics of high trans fatty acid consumers in the Dutch and UK national surveys after voluntary product reformulation. *Food & Nutrition Research, 61*, 1412793.

Risérus, U., Willett, W. C., & Hu, F. B. (2009). Dietary fats and prevention of type 2 diabetes. *Progress in Lipid Research, 48*, 44–51.

Rodriguez Furlán, L. T., Pérez Padilla, A., & Campderrós, M. E. (2014). Development of reduced fat minced meats using inulin and bovine plasma proteins as fat replacers. *Meat Science, 96*, 762–768.

Rodriguez-Sandoval, E., Prasca-Sierra, I., & Hernandez, V. (2017). Effect of modified cassava starch as a fat replacer on the texture and quality characteristics of muffins. *Food Measure, 11*, 1630–1639.

Romeih, E. A., Michaelidou, A., Biliaderis, C. G., & Zerfiridis, G. K. (2002). Low fat white-brined cheese made from bovine milk and two commercial fat mimetics: Chemical, physical and sensory attributes. *International Dairy Journal, 12*(6), 525–540.

Rubio-Rodríguez, N., de Diego, S. M., Beltrán, S., Jaime, I., Sanz, M. T., & Rovira, J. (2012). Supercritical fluid extraction of fish oil from fish by-products: A comparison with other extraction methods. *Journal of Food Engineering, 109*(2), 238–248.

Ruiz-Capillas, C., Triki, M., Herrero, A. M., Rodriguez-Salas, L., & Jimenez-Colmenero, F. (2012). Konjac gel as pork backfat replacer in dry fermented sausages: Processing and quality characteristics. *Meat Science, 92*(2), 144–150.

Sadler, M. J. (2004). Meat alternatives—Market developments and health benefits. *Trends in Food Science & Technology, 15*, 250–260.

Salcedo-Sandoval, L., Cofrades, S., Ruiz-Capillas Perez, C., Carballo, J., & Jimenez-Colmenero, F. (2015). Konjac-based oil bulking system for development of improved-lipid pork patties: Technological, microbiological and sensory assessment. *Meat Science, 101*, 95–102.

Salcedo-Sandoval, L., Cofrades, S., Ruiz-Capillas Perez, C., Solas, M. T., & Jimenez-Colmenero, F. (2013). Healthier oils stabilized in konjac matrix as fat replacers in n-3 PUFA enriched frankfurters. *Meat Science, 93*(3), 757–766.

Santiaguín-Padilla, A. J., Peña-Ramos, E. A., Pérez-Gallardo, A., Rascñib-Chu, A., González-Ávila, M., González-Rios, H., González-Noriega, J. A., & Islava-Lagarda, T. (2019). In vitro digestibility and quality of an emulsified meat product formulated with animal fat encapsulated with pectin. *Journal of Food Science, 84*(6), 1331–1339.

Sayago-Ayerdi, S. G., Brenes, A., & Goñi, I. (2009). Effect of grape antioxidant dietary fiber on the lipid oxidation of raw and cooked chicken hamburgers. *LWT - Food Science and Technology, 42*(5), 971–976.

Schmiele, M., Mascarenhas, M. C. C. N., Barretto, A. C. D., & Pollonio, M. A. R. (2015). Dietary fiber as fat substitute in emulsified and cooked meat model system. *LWT - Food Science and Technology, 61*, 105–111.

Scientific Advisory Committee on Nutrition (SACN). (2015). *Carbohydrates and health*. Available from: https://assets.publishing.service.gov.uk/government/uploads/system/uploads/attachment_data/file/445503/SACN_Carbohydrates_and_Health.pdf (15 July 2021).

Seimon, R. V., Feltrin, K. L., Meyer, J. H., Brennan, I. M., Wishart, J. M., Horowitz, M., & Feinle-Bisset, C. (2009). Effects of varying combinations of intraduodenal lipid and carbohydrate on antropyloroduodenal motility, hormone release, and appetite in healthy males. *American Journal of Physiology-Regulatory, Integrative and Comparative Physiology, 296*(4), 912–920.

Sendra, E., Kuri, V., Fernandez-Lopez, J., Sayas-Barbera, E., Navarro, C., & Perez-Alvarez, J. A. (2010). Viscoelastic properties of orange fiber enriched yogurt as a function of fiber dose, size and thermal treatment. *LWT - Food Science and Technology, 43*(4), 708–714.

Serdaroğlu, M., Ozturk, B., & Urgu, M. (2016). Emulsion characteristics, chemical and textural properties of meat systems produced with double emulsions as beef fat replacers. *Meat Science, 117*, 187–195.

Simopoulos, A. P. (1999). Essential fatty acids in health and chronic disease. *The American Journal of Clinical Nutrition, 70*(3), 560s–569s.

Singh, H., Ye, A., & Horne, D. (2009). Structuring food emulsions in the gastrointestinal tract to modify lipid digestion. *Progress in Lipid Research, 48*(2), 92–100.

Siri-Tarino, P., Sun, Q., Hu, F., & Krauss, R. (2010). Saturated fatty acids and risk of coronary heart disease: Modulation by replacement nutrients. *Current Atherosclerosis Reports, 12*(6), 384–390.

Souza, C. V. B., Bellucci, E. R. B., Lorenzo, J. M., & Barretto, A. C. D. (2019). Low-fat Brazilian cooked sausage-Paio—With added oat fiber and inulin as a fat substitute: Effect on the technological properties and sensory acceptance. *Food Science and Technology, 39*, 295–303.

Stajic, S., Zivkovic, D., Tomovic, V., Nedovic, V., Perunovic, M., Kovjanic, N., Levic, S., & Stanisic, N. (2014). The utilisation of grapeseed oil in improving the quality of dry fermented sausages. *International Journal of Food Science and Technology, 49*(11), 2356–2363.

Stortz, T. A., & Marangoni, A. G. (2013). Ethylcellulose solvent substitution method of preparing heat resistant chocolate. *Food Research International, 51*(2), 797–803.

Stortz, T. A., Zetzl, A. K., Barbut, S., Cattaruzza, A., & Marangoni, A. G. (2012). Edible oleogels in food products to help maximize health benefits and improve nutritional profiles. *Lipid Technology, 24*, 151–154.

Tavernier, I., Doan, C. D., Van der Meeren, P., Heyman, B., & Dewettinck, K. (2018). The potential of waxes to alter the microstructural properties of emulsion-templated oleogels. *European Journal of Lipid Science and Technology, 120*(3), 170039.

Timilsena, Y. P., Wang, B., Adhikari, R., & Adhikari, B. (2017). Advances in microencapsulation of polyunsaturated fatty acids (PUFAs)-rich plant oils using complex coacervation: A review. *Food Hydrocolloids, 69*, 369–381.

Torcello-Gómez, A., & Foster, T. J. (2014). Interactions between cellulose ethers and a bile salt in the control of lipid digestion of lipid-based systems. *Carbohydrate Polymers, 113*, 53–61.

Totosaus, A., & Guemes-Vera, N. (2008). Effect of kappa- and lambda-carrageenans as fat-replacers in low-fat Oaxaca cheese. *International Journal of Food Properties, 11*, 656–668.

Triki, M., Herrero, A. M., Rodríguez-Salas, L., Jiménez-Colmenero, F., & Ruiz-Capillas, C. (2013a). Chilled storage characteristics of low-fat, n-3 PUFA-enriched dry fermented sausage reformulated with a healthy oil combination stabilized in a konjac matrix. *Food Control, 31*, 158–165.

Triki, M., Herrero, A. M., Rodríguez-Salas, L., Jiménez-Colmenero, F., & Ruiz-Capillas, C. (2013b). Effect of preformed konjac gels, with and without olive oil, on the technological attributes and storage stability of merguez sausage. *Meat Science, 93*(3), 351–360.

Tvrzicka, E., Kremmuda, L. S., Stankova, B., & Zak, A. (2011). Fatty acids as biocompounds: Their role in human metabolism, health and disease—A review. Part 1: Classification, dietary sources and biological functions. *Biomedical papers of the Medical Faculty of the University Palacky, Olomouc, Czechoslovakia, 155*(2), 117–130.

Ullah, R., Nadeem, M., & Imran, M. (2017). Omega-3 fatty acids and oxidative stability of ice cream supplemented with olein fraction of chia (*Salvia hispanica* L.) oil. *Lipids in Health and Disease, 16*, 34.

Valenzuela, B., Sanhueza, J., & de la Barra, F. (2012). El aceite de pescado: ayer un desecho industrial, hoy un producto de alto valor nutricional. *Revista Chilena de Nutrición, 39*(2), 201–209.

Vural, H., Javidipour, I., & Ozbas, O. O. (2004). Effects of interesterified vegetable oils and sugarbeet fiber on the quality of frankfurters. *Meat Science, 67*(1), 65–72.

Wood, J. (2013). *Reduction of saturated fat in finely comminuted and ground meat products by use of canola oil organogels and the effect on organoleptic qualities, texture and microstructure*. (Ph.D. thesis), Guelph, ON.

World Health Organization. (2003). *Diet, nutrition and the prevention of chronic diseases: Report of a Joint WHO/FAO Expert Consultation*. Geneva. WHO Technical Report Series, No. 916.

Yang, H. S., Choi, S. G., Jeon, J. T., Park, G. B., & Joo, S. T. (2007). Textural and sensory properties of low fat pork sausages with added hydrated oatmeal and tofu as texture-modifying agents. *Meat Science, 75*, 283–289.

Ye, X., Li, P. X., Lo, Y. M., Fu, H., & Cao, Y. P. (2019). Development of novel shortenings structured by ethylcellulose oleogels. *Journal of Food Science, 84*, 1456–1464.

Yin, F., Sun, X., Zheng, W., Luo, X., Peng, C., Jia, Q., & Fu, Y. (2020). Improving the quality of microalgae DHA-rich oil in the deodorization process using deoxygenated steam. *Journal of Food Processing and Preservation, 44*, e14602.

Youssef, M. K., & Barbut, S. (2011). Effects of two types of soy protein isolates, native and preheated whey protein isolates on emulsified meat batters prepared at different protein levels. *Meat Science, 87*, 54–60.

Zetzl, A. K., Marangoni, A. G., & Barbut, S. (2012). Mechanical properties of ethylcellulose oleogels and their potential for saturated fat reduction in frankfurters. *Food & Function, 3*(3), 327–337.

Zhang, Y., Liu, Y., Wang, J., Zhang, R., Jing, H., Yu, X., Zhang, Y., Xu, Q., Zhang, J., Zheng, Z., Nosaka, N., Arai, C., Kasai, M., Aoyama, T., Wu, J., & Xue, C. (2010). Medium- and long-chain triacylglycerols reduce body fat and blood triacylglycerols in hypertriacylglycerolemic, overweight but not obese, Chinese individuals. *Lipids, 45*(6), 501–510.

Zhang, T. H., McCarthy, J., Wang, G. R., Liu, Y. Y., & Guo, M. R. (2015). Physiochemical properties, microstructure, and probiotic survivability of nonfat goats' milk yogurt using heat-treated whey protein concentrate as fat replacer. *Journal of Food Science, 80*, M788–M794.

Zulim Botega, D. C., Marangoni, A. G., Smith, A. K., & Douglas Goff, H. (2013). Development of formulations and processes to incorporate wax oleogels in ice cream. *Journal of Food Science, 78*(12), C1845–C1851.

# PART IV

# Health in vitro and in vivo studies

# CHAPTER 11

# Understanding food structure modifications during digestion and their implications in nutrient release

Alejandra Acevedo-Fani, Debashree Roy, Duc Toan Do, and Harjinder Singh

*Riddet Institute, Massey University, Palmerston North, New Zealand*

## 11.1 Introduction

The increase in obesity and noncommunicable chronic diseases (type-2 diabetes, metabolic syndrome and cardiovascular diseases) over the past half century has given rise to strong consumer awareness of the role of nutrition and high-quality foods in maintaining good health and wellness. Public health nutrition has identified potential dietary contributors to diseases through epidemiological research and human intervention studies and this knowledge has been the basis of food-based dietary guidelines that shape public health nutrition policies. However, food manufacturers are challenged to develop more nutritious and healthier food products while retaining satisfactory sensory and textural characteristics.

Foods are made up of several macro-components (water, proteins, lipids, carbohydrates) and micro-components (minerals, vitamins, enzymes, nutraceuticals). The assembly and the interactions of these components define the overall structure of food materials. There is also growing evidence that postprandial behaviour is not governed solely by the nutrient composition of food; additional factors, such as food structure and its modification, also play an important role in the digestive process and subsequent metabolic responses.

The term "food structure" is used to describe the organization of several similar or dissimilar structural elements, their binding into a unit and the interrelationship between the individual elements and their groups (Aguilera, 2005, 2006). The different structural elements consist of biomaterials (proteins, polysaccharides and fats/oils) with different sizes and shapes over a range of different length scales, ranging from nanometres to centimetres. The concept of food structure is applicable to the inherent biological structures present in natural foods, such as fruits, vegetables and muscle foods. These structures are often classified as fibrous structures (e.g., meat), fleshy materials (e.g., fruits and vegetables), encapsulated embryos (e.g., grains and pulses) and complex fluids (e.g., milk) (Aguilera, 2005) (Fig. 11.1). The principal structural elements present in natural foods are cell walls, starch granules,

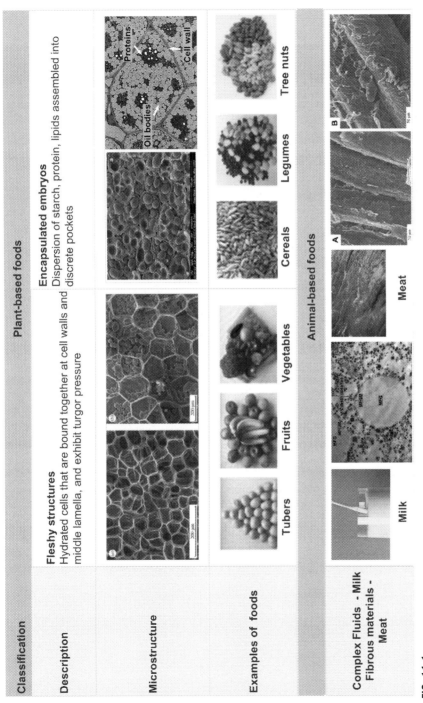

**FIG. 11.1**

Structures in natural plant-based and animal-based foods.

*Adapted from Do, D. T., Singh, J., Oey, I. & Singh, H. (2018), Biomimetic plant foods: Structural design and functionality. Trends in Food Science & Technology, 82, 46–59. Available from: https://doi.org/10.1016/j.tifs.2018.09.010; Sharma, P., Oey, I. & Everett, D. W. 2015, Interfacial properties and transmission electron microscopy revealing damage to the milk fat globule system after pulsed electric field treatment. Food Hydrocolloids, 47, 99–107. Available from: https://doi.org/10.1016/j.foodhyd.2015.01.023 and Ozuna, C., Puig, A., García-Pérez, J. V., Mulet, A. & Cárcel, J. A. (2013). Influence of high intensity ultrasound application on mass transport, microstructure and textural properties of pork meat (Longissimus dorsi) brined at different NaCl concentrations. Journal of Food Engineering, 119, 1, 84–93. Available from: https://doi.org/10.1016/j.jfoodeng.2013.05.016, with permission from Elsevier Inc.*

protein bodies, lipid bodies, cells and tissues. These structures are designed by nature for the benefit of the plant or the animal. For instance, fibrous muscle structures are involved in animal movement, dense fat structures are involved in energy storage in animals, fleshy starch-rich structures are involved in energy storage in plants and encapsulated embryos in grains are involved in providing reproductive functions for plants. Processing and cooking methods used in food preparation convert these structures into palatable material for human consumption.

In the production of manufactured or processed foods, a variety of ingredients, including flours, starches, oils and protein isolates, are assembled using different processing operations, such as thermal treatment, homogenization and shear, as well as microbiological or biochemical transformations. In this case, the food structure is defined by the nature and proportions of the different structural elements and their interactions within the desired formulation. Processing can be used to create new structures that are completely different from those encountered in any biological systems, e.g., complex structures in foods such as ice cream, which is a solid milk fat emulsion containing air bubbles, ice, sugar and fat crystals, or the aerated structures of cakes and bread.

The design and the creation of structures in processed foods have been of great scientific interest in food science and product development during the last decade, not only in developing greater understanding as to the relationship between food structure and sensory properties but also in relation to food structure and digestion (Norton et al., 2014; Singh et al., 2015; Singh & Gallier, 2014). This is due to substantial growth in multidisciplinary research directed towards understanding the physical and biochemical processes that occur during the passage of food through the various stages of the human gastrointestinal tract (GIT). The development of *in vitro* and *ex vivo* digestion models has facilitated mechanistic understanding of the changes in food structures within the GIT and understanding of the connection between the food structure/matrix and the release and digestion of nutrients (Bornhorst & Singh, 2014). There is now increasing evidence that the physiological and metabolic responses to foods are influenced in part by the food structure. As mentioned earlier, the main building blocks of the food structure are macronutrients (e.g., proteins and lipids) and micronutrients (e.g., minerals and vitamins), and other bioactive compounds are often contained within the food structure. Upon ingestion, the original structures in natural or processed foods are processed in the GIT through a series of mechanical, chemical and enzymatic processes, which allow the release of nutrients into a form that can be easily absorbed by the human body (Singh et al., 2015). The released nutrients are absorbed from the gut lumen to the enteric circulation, through the intestinal epithelium and then dispatched to the targeted tissues. Efforts have been made to structure foods to control the kinetics of the delivery of nutrients and to achieve targeted physiological responses associated with health benefits, with a major focus on modulating the release of macronutrients. Different food structures and matrices generally lead to different digesta compositions, leading to variable delivery outcomes in the GIT, which may consequently trigger different physiological responses (Moughan, 2020). However, despite ongoing rapid advances in knowledge, there remain large gaps in our understanding of how various foods with different compositions, physical properties and structures affect the rates of nutrient digestion and bioavailability.

This chapter provides an overview of current research into how proteins, lipids and carbohydrates in different structural formats respond to gastrointestinal conditions and how this may affect the kinetics of digestion of macronutrients. Several case studies on natural plant-based foods and animal-based foods are presented to illustrate the impact of the food matrix on digestion.

## 11.2 Overview of the digestion process

The human digestive tract consists of a number of hollow organs that are connected from the mouth to the anus; several accessory glands and organs that secrete fluids to these hollow organs to digest foods are attached. The mouth, stomach and small and large intestines have evolved to elicit very specialized functions in the processing of food structures, mostly involving mechanical and chemical processes (Fig. 11.2).

The mechanical disruption of foods starts in the mouth with chewing and mixing with salivary fluids. Solid and semisolid foods are broken into small particles, forming a bolus with saliva that is safe for swallowing, whereas liquid foods have very little residence time in the oral cavity. In the mouth, food undergoes temperature changes, mechanical shearing because of mastication and enzymatic hydrolysis induced by salivary α-amylase, overall facilitating a further breakdown of the food structure. During oral processing, sensory stimulation also occurs, which may trigger some metabolic responses such as satiety; this is because the sensory stimuli inform the brain and gut about the uptake of nutrients (Ahima & Antwi, 2008; Miquel-Kergoat et al., 2015).

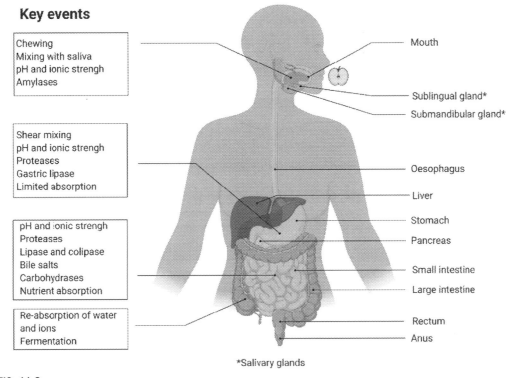

**FIG. 11.2**

Schematic representation of the digestive system and the key events intervening in food digestion.

*Adapted from "Digestive system", by BioRender.com (2021). Retrieved from https://app.biorender.com/biorender-templates*

Once the bolus reaches the stomach it interacts with the gastric juices, which are composed mainly of water, hydrochloric acid, digestive enzymes and mucus (Feher, 2017; Soybel, 2005). Parietal cells secrete hydrochloric acid, which lowers the pH of the stomach, whereas chief cells secrete pepsinogen, which is a precursor of pepsin that is activated when the stomach reaches a low pH (Waldum et al., 2014). The pH of the stomach in the fasted state varies from 1.5 to 2.0 but increases to 2.5–6.7 in the fed state and decreases again to < 3.1 after eating (Simonian et al., 2005). The rate of the pH decrease after the consumption of a meal depends on the food composition, its buffering capacity and the efficiency of mixing in the stomach. The pH decrease not only increases the activities of pepsin (pH 2–5) and gastric lipase (pH 4–6), which hydrolyse the food components (Bornhorst & Singh, 2014; Minekus et al., 2014), but also causes considerable changes in the food structure, such as aggregation of proteins or destabilization of oil droplets (Gallier et al., 2013). The stomach is divided into two functional regions: proximal and distal. The proximal region serves as a reservoir for the swallowed boluses, whereas the distal region is where the mechanical breakdown of foods, which is caused by antral contraction waves or peristaltic contractions of the stomach wall, occurs. These movements facilitate the grinding, mixing and re-dispersion of food particles to create a mixture called "chyme". Gastric motility also enables the emptying of chyme to the duodenum (Goyal et al., 2019). The pylorus controls the rate of delivery of chyme from the stomach to the duodenum and avoids backflow (Tortora & Derrickson, 2008). Only food particles less than 1–2 mm in size pass through the pyloric sphincter into the duodenum; those of bigger size are pushed back to the stomach by retropulsion for further grinding (Kong & Singh, 2009b).

Once the chyme enters the duodenum, the gastric enzymes in the chyme continue the digestion for a short period of time until the pancreas secretes bicarbonate, which neutralizes the pH of the intestinal contents (Sullivan, 2009). Other secretions from the pancreas, the gallbladder and glands located in the intestinal wall are released into the lumen; they include proteases and peptidases (trypsin, chymotrypsin, carboxypeptidases, etc.), lipases and esterases (pancreatic lipase, cholesterol esterase, phospholipase A2, etc.) and pancreatic amylases (Singh et al., 2009). In the process, the liver produces bile, which consists of bile salts, phospholipids, cholesterol, bilirubin, electrolytes and water. The pH of the pancreatic secretions ranges from 5 to 7, which provides an optimal environment for the enzymatic hydrolysis of the food components into small molecules. The length of the small intestine is 6–7 m and it is joined to the stomach via the pylorus and to the colon via the ileocaecal valve. There are three main segments of the small intestine: the duodenum, the jejunum and the ileum. The duodenum is 23–28 cm in length, whereas the jejunum and ileum are around 2 and 3 m in length, respectively (Sensoy, 2021). Most of the pancreatic secretions occur in the duodenum; they are very limited in the other sections of the small intestine. During intestinal digestion, the segmenting and peristaltic movements facilitate mixing of the chyme with intestinal juices and moves it to the large intestine. Nutrients are then absorbed through the brush border of the epithelium of the villi in the intestinal wall.

Any food material that was not digested and absorbed in the small intestine will move to the large intestine for further processing. The large intestine is 150 cm in length and extends from the ileocaecal valve to the anus. The cells of the large intestinal wall are protected with a thick mucus layer (approximately 120 μm thick) and a loosely adherent mucus layer on top (approximately 700 μm thick) (Bergstrom et al., 2017). The colon is characterized by an anaerobic environment that is colonized by more than $10^{10}$ microorganisms per gram of intestinal content, and has three main functions: (1) the absorption of large quantities of water and electrolytes, which leads to conversion of the liquid intestinal contents into a semisolid stool; (2) the absorption of the free fatty acids that are produced by microbial fermentation of the undigested carbohydrates and proteins; the microbiota also synthesize essential vitamins such as niacin, vitamin $B_1$ and vitamin K (Rogers, 2010; Sensoy, 2021); (3) the elimination of faecal material in a regulated and controlled manner, largely under voluntary control. The large intestine has two distinct segments: the proximal colon and the distal colon. Most of the water and electrolytes are absorbed in the proximal colon and microbial catabolism also occurs. The distal colon provides final desiccation of the colonic material.

## 11.3 Digestion of macronutrients

### 11.3.1 Proteins

Dietary proteins are found in animal and plant sources, forming part of a food matrix. These foods often undergo processing or cooking treatments that cause denaturation and aggregation of the proteins or their interactions with other food components (e.g., sugars, lipids, polyphenols, vitamins). These modifications of the food structure may influence the accessibility of proteins to proteolytic enzymes in the GIT. Moreover, the native molecular structures of proteins significantly affect their susceptibility to hydrolysis because of their different molecular conformations (i.e., secondary, ternary and quaternary structures) and regions of affinity to hydrophobic and hydrophilic environments. For instance, caseins have a disordered conformation that contributes to a rapid digestion, but other highly folded proteins such as β-lactoglobulin are more resistant to the action of proteases in their native form. Therefore, the rate and the extent of protein digestion depend on both the molecular structure and shape of proteins and the food structure modifications that are induced by processing.

Proteins are initially digested in the stomach, by interacting with pepsins. These enzymes preferably cleave peptide bonds close to phenylalanine, tyrosine or leucine residues but they cannot hydrolyse the peptide bonds of the side chains of amino acids (the γ- and β-carboxyl groups of glutamate and aspartate and the ε-amino group of lysine) (Emery, 2015). At least three distinct pepsins are secreted in the stomach with pHs of activity varying from neutral to extremely acidic. In the stomach, proteins are mostly cleaved into polypeptides and oligopeptides and very limited amounts of free amino acids. Therefore, it is generally recognized that there is no absorption of amino acids in the stomach (Moughan, 2009; Zebrowska et al., 1983). The proteins and peptides move into the small intestine for a thorough digestion that is facilitated by pancreatic proteases secreted into the gut lumen. These proteases are located at the brush border membrane of the intestinal wall or are active within the mucosal cells. Pancreatic proteases are composed of endopeptidases, such as chymotrypsin, trypsin and elastase, and exopeptidases including carboxypeptidases A and B (Moughan, 2009). The proteases produced in the small intestinal wall are aminopeptidases and di- and tri-peptidases, and are classified as exopeptidases (Emery, 2015). In general, endopeptidases produce a large number of small fragments of peptides that are further hydrolysed by exopeptidases of the brush border and the mucosal cells. At this stage, small peptides are hydrolysed into amino acids, which can pass to the bloodstream.

### 11.3.2 Lipids

The digestion and the absorption of dietary lipids are complex processes involving enzymatic activity and physicochemical events occurring along the length of the GIT. Dietary lipids are ingested in the form of mixed triacylglycerols (TAGs) from different natural foods including meat, nuts, milk, etc., where they are contained in different matrices and structural formats. Lipolysis is an interfacial process; for efficient lipolysis to take place, the lipid must be in an emulsified state. The first step in lipid digestion is the dispersion of the lipid into finely divided emulsion droplets, which occurs in the stomach because of the shear and mechanical action generated by peristaltic movements. Gastric lipase is then able to bind to the emulsion–water interface, resulting in hydrolysis of 5–40% of the ingested TAGs in healthy adults (Armand, 2007). The pH optimum of gastric lipase is between 3.0 and 6.0 and it acts at the sn-3 position of the fatty acids of TAGs, generating mainly free fatty acids and 1,2-diacylglycerols (Mu & Hoy, 2004).

Partially hydrolysed emulsified lipid then enters the duodenum lumen, where the majority of fat digestion occurs by colipase-dependent pancreatic lipase, in conjunction with colipase and bile salts (Armand et al., 1996). Colipase forms a stoichiometric complex with lipase, allowing the pancreatic lipase to anchor firmly to the substrate (hydrophobic lipid core) at the oil–water interface. Pancreatic lipase specifically hydrolyses the sn-1 and sn-3 position fatty acids of TAGs, leading to the formation of free fatty acids and 2-monoacylglycerols (2-MAGs). In healthy adults, the combined lipolytic ability is in slight excess relative to the amount of lipid ingested; hence fat digestion is highly efficient ($>97\%$) (Armand, 2007; Carrière et al., 2005).

The next step involves the solubilization and uptake of the lipolytic products by enterocytes. The resulting lipolytic products, including 2-MAGs and free fatty acids, are then incorporated into self-assembled structures, such as bile salt micelles and phospholipid vesicles, and are transited to intestinal epithelial cells for absorption (Mu & Hoy, 2004). These self-assembled structures increase the aqueous concentration of lipolytic products, which enhances the rate of absorption of lipids. Long-chain free fatty acids and 2-MAGs are then transported into the endoplasmic reticulum inside the enterocyte, where they are synthesized into TAGs. Finally, the TAGs are packaged with cholesterol, lipoproteins and other lipids into particles called chylomicrons, which are distributed to different cells via the lymphatic system (Iqbal & Hussain, 2009). In contrast, short-chain and medium-chain fatty acids are absorbed directly through portal veins (Bach & Babayan, 1982).

### 11.3.3 Starch

Starch is the main carbohydrate that is found in natural raw foods such as grains, cereals, fruits and vegetables. It consists of amylopectin (70–80%) and amylose (20–30%), both of which are glucose polymers. The relative amounts of amylose and amylopeptin depends on the source of starch. The rate and the extent of the hydrolysis of starches depend on the botanical origin, the physicochemical properties and the type of processing. For example, starch granules absorb large amounts of water during cooking because of gelatinization. Swollen granules burst, releasing amylose and amylopectin into the aqueous phase. The heating and subsequent cooling of starchy foods result in retrogradation. After cooling, amylopectin adopts a highly amorphous structure and amylose recrystallises, which may limit the access of enzymes and reduces starch hydrolysis (Brownlee et al., 2018; Patel et al., 2017).

The digestion of starch starts in the mouth, with mastication and hydrolysis by salivary α-amylase, which randomly cleaves α-1,4-glycosidic bonds (Holmes, 1971). The partially digested bolus moves to the stomach for mechanical shearing and acidification, and digestion may or may not continue. If the food enters the stomach as a compact bolus, the salivary amylase may remain active for longer, but, if the food is swallowed as particulates, α-amylase will be inhibited by the acidic environment of the stomach. Starch digestion continues in the small intestine. The α-amylase secreted by pancreatic cells and enzymes of the brush border cleaves the mixture of polysaccharides, dextrins and monosaccharides coming from the stomach into maltose, maltotriose, isomaltose, α-limit dextrins and several linear α-(1→4) linked polyglycan chains (Sim et al., 2008). Pancreatic α-amylase secreted into the intestinal lumen converts starch into maltose, maltotriose and α-limit dextrins. Then, these disaccharides are further hydrolysed to monosaccharides by maltase-glucoamylase and sucrase-isomaltase; these exohydrolase double-headed enzymes are present in the brush border of the mucosal cells (Nichols et al., 2003). The result of this process is monosaccharides that consist of 80% glucose, 15% fructose and 5% galactose (Holmes, 1971). Absorption of glucose into the bloodstream occurs in the duodenum and the proximal jejunum; this requires membrane-associated transporters located in the brush border and basolateral membranes of the enterocyte cells (Chen et al., 2016; Smith & Morton, 2010).

## 11.4 Modification of plant-based food structures in the GIT

During processing and passage through the GIT, plant-based food organs are mechanically reduced into smaller particles of tissues and cell clusters that are further disintegrated into individual cells. Intracellular nutrients are released from ruptured cells but remain encapsulated within intact cells (Do et al., 2018). Modifications of these plant structures in the GIT have a great impact on the digestion and absorption of nutrients.

### 11.4.1 Starchy legumes

Starchy legumes are made up of two cotyledons. The cotyledon tissue comprises numerous cells. The middle lamella, rich in pectin, glues together primary walls of adjacent cells. Each cell consists of multiple starch granules that are tightly embedded in a protein matrix, all encapsulated by a thick robust outer cell wall (Berg et al., 2012; Do et al., 2019). Starchy legumes are a classical case of the processing and digestive dynamics of the plant tissue structure affecting the digestion of starch (Berg et al., 2012; Do et al., 2019).

Typically, legumes are hydrothermally processed (e.g., boiling, steaming) prior to consumption. Cooking promotes dissolution of pectic polymers in the middle lamella and subsequent separation of intact cells upon application of mechanical forces (e.g., mastication, peristaltic contractions). In contrast, high-pressure treatment causes cell rupture, rendering the intracellular contents readily available for enzymatic attack. During cooking, starch granules and proteins undergo partial gelatinization and thermal denaturation, respectively (Do et al., 2018).

Oral processing, the first unit operation in the digestion process, modifies legume structures by mechanically converting them into ready-to-swallow boluses of certain particle size distribution (PSD). Pallares et al. (2019) investigated the human mastication of cooked legumes with different hardness levels, which were induced by varying the thermal treatment time. The boluses generated contained two distinct fractions. Specifically, the first fraction (>2000 μm) contained mostly seed coat material, whereas the second fraction consisted of mostly individual cotyledon cells (40–125 μm). With respect to the latter fraction, the boluses of harder legumes contained some large cell clusters because of the strong cell-cell adhesion. In contrast, softer legumes had a higher degree of pectin solubilization in the middle lamella, which was induced by a longer cooking time, resulting in weakening of the adhesion forces and efficient separation of individual cells during chewing. The presence of a larger fraction of individual cells in the boluses of the softer legumes could lead to a larger surface area for the binding of amylase, thus favouring a higher rate of starch digestion *in vitro*. Moreover, it was shown that the PSD of boluses generated by individual human subjects with different masticatory parameters (i.e., number of chewing cycles, chewing duration and chewing frequency) was relatively similar within a specific hardness level. This implies that the starch digestion of cooked legumes is primarily controlled by process-induced structural properties/modifications rather than by mechanical disintegration in the mouth (e.g., the inter-individual variability in mastication behaviour) (Pallares et al., 2019).

The boluses of cooked legumes are broken up into a large number of individual cotyledon cells during their transit through the GIT. Previous studies have revealed that the intactness of the cell walls is maintained to a great extent throughout cooking and in the following simulated gastric and small intestinal phases. This is perhaps due to the high resistance of plant cell walls against thermal degradation and enzymatic hydrolysis. The resilient cell walls could act as efficient physical barriers, restricting both intracellular starch gelatinization during cooking and access of amylases to starch substrates. The beneficial consequences of this include a reduction in the rate and extent of starch hydrolysis and a slow steady release of glucose (Berg et al., 2012; Brummer et al., 2015).

Noah et al. (1998) examined human digestive effluents at the end of the small intestine following the consumption of cooked white bean meals. These effluents contained some intact-starch-encapsulating cells with no detection of free starch outside the cells at 3h after ingestion, suggesting that the cell wall barriers could obstruct the access of digestive enzymes and consequently could protect starch from digestion. After 11.5h, only cell wall fragments were detected in the ileal digesta. This implied that the prolonged mechanical action and the acidity of the stomach could cause cell wall breakdown, rendering starch and protein available for digestion (Noah et al., 1998).

Using in situ optical microscopy, Do, Singh, Oey, Singh, Yada, and Frostad (2020) were the first to monitor the structural modifications of raw, intact navy bean cotyledon cells (Fig. 11.3A) through each stage of hydrothermal cooking, followed by simulated digestion, in one continuous sequence. As shown in Fig. 11.3B, the cells maintained their structural integrity with little or no swelling after cooking. Intracellular starch granules swelled, but only to a limited extent, because of their confinement within the cell walls. In contrast, free starch granules (released from broken cells) swelled and gelatinized to their full extent. After 30min of gastric digestion, the cells and starch exhibited no apparent changes in their structures because of the absence of starch-degrading enzymes in the gastric fluid (Fig. 11.3C). As soon as the free starch encountered pancreatic amylase in the small intestine, it rapidly contracted (Fig. 11.3D) and virtually disappeared before the onset of the digestion of most cells, in which the encapsulated starch shrank radially inward towards the centre of each cell. As the shrinkage slowly progressed, the cellular contents became gradually emptier over time (Fig. 11.3E). At the end of 2h of small intestinal digestion, some residual starch remained encapsulated within the intact cells, whereas free starch outside the cells was no longer detected (Fig. 11.3F) (Do, Singh, Oey, Singh, Yada, & Frostad, 2020). This is consistent with the aforementioned *in vivo* observations by Noah et al. (1998) of the contents of the human ileal digesta at 3h after the ingestion of cooked white beans. In addition, the navy bean cell walls remained predominantly intact throughout the cooking and digestive processes. Therefore, it was postulated that the mechanism of amylolysis may entail diffusion of α-amylase into the cells through the cell walls. The enzyme then catalyses the hydrolysis of starch from the periphery towards the centre of the cells into smaller molecules, which, in turn, diffuse out of the cells for intestinal absorption, albeit at a slow rate (Do, Singh, Oey, Singh, Yada, & Frostad, 2020).

In short, the modifications of the unique cotyledon cell structure in the upper GIT have received increasing scientific attention in the last few years because of their important implications for slow glucose absorption and attenuated postprandial glycaemia (i.e., a slow and steady rise in blood sugar

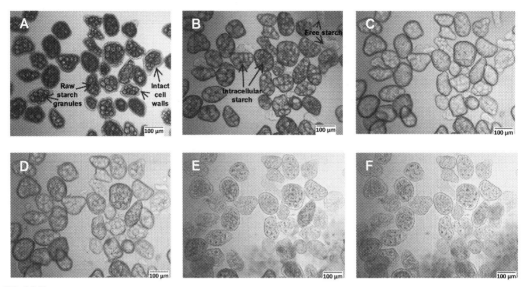

**FIG. 11.3**

Representative light photomicrographs of navy bean cotyledon cells taken from a time-lapse imaging sequence of a simulated cooking and digestion experiment: (A) raw intact cells at room temperature (24 °C); (B) after heating from 24 to 90 °C (heating rate, 6.2 °C/min) and holding for an additional 10 min at 90 °C; (C) after 30 min of gastric digestion by pepsin at 37 °C and pH 1.2; after (D) 20 s, (E) 30 min and (F) 120 min of small intestinal digestion by pancreatin at 37 °C and pH 6.8. Scale bar = 100 μm.

*Images are reprinted from Do, D. T., Singh, J., Oey, I., Singh, H., Yada, R. Y. & Frostad, J. M. (2020). A novel apparatus for time-lapse optical microscopy of gelatinisation and digestion of starch inside plant cells.* Food Hydrocolloids, 104, *105551., with permission from Elsevier Inc.*

levels) (Dhital et al., 2016; Do, 2020; Do et al., 2019; Rovalino-Córdova et al., 2019). The combined physical barriers of the intact cell walls and the compact protein matrix are considered to be a predominant factor for limiting amylolysis through hindering the access of amylase to entrapped starch (Dhital et al., 2016; Do, 2020; Do et al., 2019). However, food processing, gastric conditions and mechanical actions in the gut could cause dissolution of cell wall polymers, especially pectic polysaccharides. This could subsequently result in enlargement of the pore size on the cell walls for easier amylase diffusion and thus could improve the efficiency of starch digestion (Li et al., 2021; Pallares et al., 2018).

The undigested starch in the upper gut, also referred to as "resistant starch", escapes to the large intestine, where it is extensively fermented by the intestinal microbiota. Human *in vivo* evidence indicated that almost all of the encapsulated white bean resistant starch was digested after transit through the colon. Interestingly, a large amount of intact cotyledon cells was recovered in the stool of one particular human subject, possibly because of inefficient chewing causing incomplete tissue breakdown during passage through the GIT. Because of the natural encapsulation of macronutrients by the cell walls, legumes contain an abundance of resistant starch, which provides many health benefits in humans (Noah et al., 1998).

### 11.4.2 Cereals

Cereals, including rice, wheat, barley and sorghum, are important staple foods for human consumption worldwide. Structurally, the majority of the nutrients are naturally stored in starchy endosperm cells. Each cell contains starch granules that are embedded in a protein matrix and enclosed by a cell wall (Zheng & Wang, 2015).

Cereal grains are generally milled into flour prior to further processing and consumption. Coarse flours prepared from sorghum, barley or wheat grains possess a substantially lower surface area/volume ratio and a higher degree of starch encapsulation by the intact cell wall/protein matrix than fine flours. Hence, coarse flours limit amylase diffusion through the grain fragments and slow down starch hydrolysis more effectively than fine flours (Al-Rabadi et al., 2009; Farooq et al., 2018; Guo et al., 2018; Korompokis et al., 2019).

Oral processing causes physical breakdown of cereal grains into smaller particles for downstream gastrointestinal digestion. Several studies demonstrated that a higher chewing efficiency (duration/cycle) of cooked rice correlated with a higher degree of particle fragmentation and a higher percentage of smaller particles in the bolus. Decreasing the particle size of the rice bolus increased both the *in vitro* starch digestion rate and the *in vivo* glycaemic responses in human subjects (Ranawana et al., 2010, 2014; Sivakamasundari et al., 2021). Hence, preserving the intactness of cereal structures through either minimal processing or lowering the mastication intensity provides opportunities for reducing the glycaemic index of cereal-based foods.

Masticated grain particles are softened by the gastric juices and continue to be broken down in the stomach. Using dynamic human and animal gastric simulators, multiple studies investigated the relationship between the physical structure of cooked grains (brown rice versus white rice) and their disintegration behaviour. An interesting structural feature of rice grains is the bran layer on brown rice, which is removed during polishing to produce white rice. The rigid bran layer was found to inhibit the diffusion of moisture and gastric acids into the rice kernels, limiting both textural softening and rice particle disintegration. Consequently, brown rice digesta contained larger-sized particulates with bran dietary fibre attached, which may contribute to its higher viscosity compared with white rice digesta. These gastric modifications of brown rice could lead to slower gastric emptying and a reduced rate of starch digestion (Kong et al., 2011; Wang et al., 2015; Wu et al., 2017). In agreement with the *in vitro* data, there was a slower gastric emptying rate of protein *in vivo* in brown rice than in white rice. This was attributable to an accumulation of bran layer fragments in the stomach of growing pigs (Bornhorst, Chang, et al., 2013). In humans, the presence of the bran layer in brown rice led to delayed gastric emptying. The physiological effects pertaining to this may include attenuated postprandial glycaemic responses and enhanced satiety (Pletsch & Hamaker, 2018).

Apart from the physical form, the varieties of grains also affect their gastric breakdown behaviours. For instance, indica rice has higher levels of endosperm hardness, grain thickness and protein content than japonica rice. Using a dynamic human gastric simulator, Goebel et al. (2019) showed that indica rice developed larger-sized pores on the grain surface by enzymatic action. In contrast, japonica rice developed a larger number of tiny surface pores, allowing easier diffusion of the gastric juices into the starchy endosperm. As a result, japonica rice was found to disintegrate more extensively into smaller-sized particulates in the gastric digesta, which could explain its higher rate of starch hydrolysis in the small intestine compared with indica rice.

During small intestinal digestion, compelling *in vitro* evidence suggests that the structural integrity of the cereal endosperm plays an important role in modulating the rate and the extent of starch digestion. The intact cell walls and the protein matrix, such as those present in isolated sorghum cells (Bhattarai et al., 2018) and flour particles of wheat and buckwheat (Chen et al., 2020; Korompokis et al., 2019; Roman et al., 2017), could restrict intracellular starch gelatinization during cooking as well as the diffusion of digestive enzymes into endosperm cells. Therefore, these cell structural barriers hinder the accessibility of entrapped starch by α-amylase and efficiently slow down starch digestion. When processed into porridge, coarsely milled wheat endosperm (containing intact cells) exhibited lower *in vitro* bioaccessibility for both starch and protein than finely milled endosperm (lacking intact cells) (Mandalari, Merali, et al., 2018). *In vivo*, the coarse porridge retained endosperm integrity during gastro-ileal transit and elicited significantly lower human glycaemic responses than the smooth porridge containing fine particles (Edwards et al., 2015).

### 11.4.3 Vegetables and fruits

Fleshy parenchyma tissue is a distinctive structural feature that is typically found in many vegetables and fruits. The tissue matrix comprises hydrated cells (spherical or polyhedral) that are bound together at the cell wall/middle lamella and exhibit turgor pressure (Do et al., 2018). Nutrients are located in thin-walled cells, e.g., potato cells encapsulate numerous starch granules (Do, Singh, Oey, & Singh, 2020) and carrot and mango cells encapsulate carotenoid crystals (Low et al., 2015; Tydeman, Parker, Wickham, et al., 2010).

To understand the structural modifications of fruits/vegetables in the GIT, their process-induced destructuring must be understood first. The processing of vegetables (e.g., boiling, roasting, mashing) and fruits (e.g., cutting, chopping, grinding) induces changes in tissue microstructures, especially disruption of plant cell walls and cell membranes. During the cutting of fresh raw fruits and vegetables, the cells tend to rupture and release their contents because of the breakage of the cell walls. In contrast, as often found in cooked vegetables and mealy fruits, the cells tend to separate and remain intact, thus effectively encapsulating nutrients (Brett & Waldron, 1996). The degree of cellular separation/rupture can be modulated through processing methods/intensity, which significantly affects the bioaccessibility of nutrients (Do et al., 2018; Lemmens et al., 2010). Furthermore, hydrothermal processing involving the application of high water contents and elevated temperatures could cause loss of turgor pressure, pectin solubilization in the cell wall/middle lamella (Ormerod et al., 2002), starch gelatinization (Bordoloi, Kaur, & Singh, 2012), coalescence of lipophilic carotenoids (Tydeman, Parker, Wickham, et al., 2010) and relocation of polyphenols from vacuoles to around cell walls (Liu, Lopez-Sanchez, & Gidley, 2019).

In the first step of the digestion process, mastication causes the breakdown of intact cell walls and the release of encapsulated nutrients. This step is crucial for extracting bioactive compounds from fruits, with potential implications for their subsequent absorption and health benefits. Low et al. (2015) reported that human *in vivo* mastication of mango tissues yielded boluses consisting of different particle size fractions of fragmented cells, cell clusters and vascular fibres. Smaller particles sustained greater structural damage and released more carotenoids following *in vitro* digestion. However, neither the bolus particle size nor the degree of chewing (coarse versus fine) appeared to have a significant impact on the bioaccessibility of carotenoids from the mango tissues. This highlights that the effect of the cell wall barriers restricting carotenoid release is much weaker from soft tissues (e.g., mango) than from firm tissues (e.g., carrot), in which more robust cell walls efficiently entrap carotenoids and prevent them from being incorporated into micelles for absorption (Low et al., 2015; Tydeman, Parker, Wickham, et al., 2010).

During static *in vitro* gastric digestion, the cell walls, the cell plasma membranes and the vacuolar membranes of mango and apple tissues remain mostly intact under acidic conditions, thus effectively impeding the release of carotenoids and polyphenols, respectively, from the cell vacuoles. In contrast, using a dynamic stomach model incorporating additional peristalsis movement, Liu, Dhital, et al. (2019) demonstrated that the two main physical forces (crushing and shearing) applied to apple tissues disrupted surface cells and compacted the centre of the tissues. These mechanical forces caused the apple particles to break up and release encapsulated polyphenols. Nevertheless, the release was influenced by the binding of polyphenols to broken cell wall fragments and mucin. In line with these findings, gastric juice alone was found to have no significant impact on the breakdown of the cell walls of cooked sweet potatoes during static *in vitro* digestion. However, when combined with mechanical forces in a human gastric simulator, the cell wall breakdown was much more extensive, resulting in a greater release of β-carotene into the gastric digesta (Somaratne et al., 2020). Therefore, mechanical forces (e.g., peristaltic contractions), rather than biochemical digestion (e.g., acidic hydrolysis), are the primary determinant influencing the disintegration of fruits/vegetables and the bioaccessibility of nutrients in the stomach (Liu, Dhital, et al., 2019; Low et al., 2015; Somaratne et al., 2020).

Taking carrots as an example, the disintegration of particulates during dynamic gastric digestion is governed by two competitive processes, namely surface erosion and texture softening (tenderization). The former is caused by the friction and shear forces exerted on the carrot surfaces by the gastric fluid containing particulates. Peristaltic forces generated in the stomach also contribute to solid disintegration by causing surface erosion. In contrast, the latter involves the diffusion of gastric acids and moisture into the carrot tissues, which is followed by the hydrolysis of pectic materials and the mass transfer of leached solids into the gastric medium. Cooking reduces the internal cohesive force in the carrot matrix and softens the texture. Faster disintegration of the cooked tissues ensues, which is predominantly driven by the surface erosion effect rather than by the tenderization effect (Kong & Singh, 2008, 2011).

Following small intestinal digestion, *in vitro* and *in vivo* evidence showed that the gross structures of raw and cooked carrot tissues remained relatively unchanged. However, swelling and pectin solubilization of the cell walls occurred extensively close to the tissue surfaces, probably because of the acid hydrolysis of glycosidic linkages in the gastric environment and resulting changes in the cell wall viscoelasticity. Despite this, the robust cell walls resisted digestion in the upper gut to a large extent. Accordingly, some carotene-containing cells passed into human ileal effluents in their intact form. These findings emphasize the importance of the cell wall barriers in restricting the release and absorption of carotenes (Tydeman, Parker, Faulks, et al., 2010; Tydeman, Parker, Wickham, et al., 2010).

Lastly, the structural modifications of starchy vegetables play a critical role in the digestion of carbohydrates and the absorption of glucose in the small intestine. As shown in Fig. 11.4, the parenchyma cell walls of Agria potatoes retain their structural integrity to a large extent not only after cooking (Fig. 11.4B) but also throughout simulated digestion (Fig. 11.4C–E). As such, pancreatic amylases are most likely to diffuse into the intact potato cells through the cell walls and catalyse starch hydrolysis. This could explain the formation of empty cavities that were formerly occupied by the hydrolysed starch in most of the cells at the end of the small intestinal digestion (Fig. 11.4E) (Bordoloi, Singh, & Kaur, 2012; Do, Singh, Oey, & Singh, 2020). Despite the retention of intact cellular structures, the

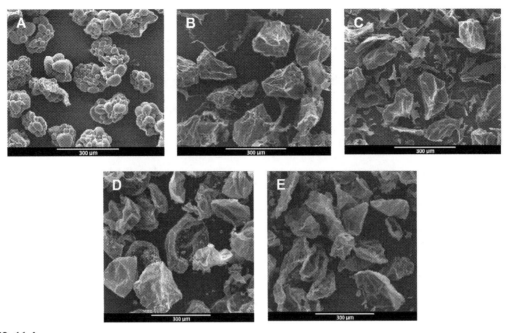

**FIG. 11.4**

Representative scanning electron micrographs of isolated Agria potato parenchyma cells sampled at different time points: (A) raw; (B) after cooking in a boiling water bath (approximately 95 °C) for 20 min; (C) after 30 min of simulated gastric digestion; after (D) 10 min and (E) 120 min of simulated small intestinal digestion. Images were acquired with an accelerating voltage of 20 kV. Scale bar = 300 μm.

*Images are reprinted from Do, D. T., Singh, J., Oey, I. & Singh, H. 2020, Isolated potato parenchyma cells: Physico-chemical characteristics and gastro-small intestinal digestion in vitro.* Food Hydrocolloids, 108, *105972, with permission from Elsevier Inc.*

permeable, thin cell walls and the lack of an intracellular protein matrix could combine to contribute to the fast amylolysis rate and the rapid release of glucose from potato parenchyma cells (Do, Singh, Oey, & Singh, 2020), as opposed to legume cotyledon cells, in which the robust structures are effective in curbing starch digestion (Do et al., 2019).

## 11.4.4 Tree nuts

Shelled tree nuts are made up of two edible cotyledons. Each cotyledon consists primarily of numerous thin-walled parenchyma cells. The cells of raw nuts contain numerous globoid starch grains, protein bodies and oil bodies (oleosomes) (OBs) of various sizes that are densely packed within a cytoplasmic network (Muniz et al., 2013; Young et al., 2004). The OBs, which are stabilized by oleosins, are the most representative storage components of the cells (McArthur & Mattes, 2020).

Processing techniques (e.g., grinding, roasting, blanching) can cause cell wall rupture, alterations in the cytoplasmic network, disruption of OB membranes, coalescence of oil droplets, lipid oxidation and distortion/aggregation of protein bodies. These process-induced microstructural changes influence the stability, release and bioavailability of intracellular nutrients (Altan et al., 2011; Grundy, Lapsley, & Ellis, 2016; Mandalari et al., 2014). There has been immense interest in understanding the structural modifications of edible nuts that are caused by processing and occur in the GIT and that modulate the bioaccessibility and absorption of nutrients. In particular, almonds have been the subject of extensive research in the last decade, both *in vitro* and *in vivo*, because of their considerable health benefits (Grundy, Lapsley, & Ellis, 2016).

Both randomized human trials and *in vitro* studies concur that mastication plays a key role in the physical disintegration of nuts and governs the nature and the degree of tissue/cell fracture in the oral cavity, with important implications for lipid bioaccessibility, satiety and postprandial lipidaemia (McArthur & Mattes, 2020; Cassady et al., 2009; Grundy, Grassby, et al., 2015; Ellis et al., 2004; Mandalari, Parker, et al., 2018; Paz-Yépez et al., 2019). More specifically, the human mastication of raw and roasted whole almonds crushes and compresses the parenchyma tissues. The resulting boluses consist of a high proportion of large multicellular particles of the almond tissues surrounded by free lipid droplets, some of which coalesce to form larger droplets. These particles contain mostly undamaged cells encapsulating intact OBs with low lipid bioaccessibility (Grundy, Grassby, et al., 2015; Mandalari, Parker, et al., 2018). The cells remain intact in the centre and in layers immediately beneath the fractured particle surfaces with no evidence of cell separation. Only a small amount of lipids within ruptured cells at the particle periphery become either exposed or liberated, hence becoming readily bioaccessible for the early stages of digestion (Ellis et al., 2004; Grundy, Grassby, et al., 2015).

Additionally, roasted almonds are broken down more extensively than raw almonds during mastication, resulting in smaller particles with higher degrees of cell wall degradation. These particles exhibit greater lipid extraction *in vitro* and higher energy absorption (Gebauer et al., 2016). Moreover, it is likely that the compact tissue matrix, comprising tightly packed cells postmastication, may hinder digestive enzymes to gain access to encapsulated nutrients. Having said that, the crushing action of mastication may cause deep fissures running into the core tissue, making it easier for enzymes to penetrate into the particles during subsequent digestion (Grundy, Grassby, et al., 2015; Mandalari et al., 2014).

Gastric digestion of nuts is a critical step that facilitates solid disintegration and the release of intracellular nutrients. Indeed, simulated dynamic gastric digestion of pistachios released the majority of bioactive compounds (i.e., polyphenols, xanthophylls and tocopherols) with more than 90% of the polyphenols becoming readily bioaccessible for absorption in the upper GIT (Mandalari et al., 2013). Using a model dynamic stomach system, Kong and Singh (2009a) demonstrated the breakdown of raw almonds at a slow rate, in addition to their gastric juice absorption and extensive swelling, in the gastric environment. These structural alterations of almonds may partly account for their limited nutrient release/absorption and prolonged satiety. Microscopic examination of digested raw almond tissues revealed most of the cells in the fractured peripheral layer and some internal cells underlying this layer with empty cellular contents and open cracks in the cell walls. Roasting of almonds created channels/voids along the cell walls and intracellular spaces through which steam and oil may have escaped as a result of the heat treatment. These cracks/channels could provide a physical route through which digestive enzymes gain access to lipid and protein bodies. Roasting could also increase the disintegration rate and reduce the swelling capacity of almonds in the stomach, thus facilitating effective digestion and absorption of nutrients (Kong & Singh, 2009a).

Conversely, *in vivo* gastric digestion of almond meals in growing pigs indicated that there were no significant differences in physical property changes between raw almonds and roasted almonds in terms of PSD, rheological properties and textural attributes. This suggested that almonds disintegrated at similar rates in the gastric compartment irrespective of roasting (Bornhorst et al., 2014). Likewise, the gastric emptying rate and the subsequent absorption of lipids were found to be relatively similar between these two almond types. Unlike lipids, the gastric emptying rate and the hydrolysis of the dietary proteins were higher for raw almonds than for roasted almonds, which was possibly linked to structural changes in the almond proteins during roasting (Bornhorst et al., 2016; Bornhorst, Roman, et al., 2013).

Following 12 h of *in vivo* gastric and small intestinal digestion of almonds by ileostomy volunteers, the internal cells underlying the fractured particle surfaces remained largely intact but had lost their intracellular contents. Additionally, the cell walls and the middle lamella increased substantially in thickness after the prolonged incubation in the digestive juices. This implied that the swelling of the cell walls appeared to increase slowly over a longer gut residence time, which, in turn, could lead to the improved bioaccessibility of lipids and proteins over the course of digestion. The cell wall swelling could provide a possible mechanism by which enzymes and bile salts diffuse across the swollen cell walls and hydrolyse lipids/proteins intracellularly, followed by an outflow of lipolytic/proteolytic products, at a very slow rate (Mandalari et al., 2008). The cell wall swelling and the consequent increase in wall porosity have also been observed previously *in vitro* in other nuts, e.g., walnuts and pistachios (McArthur & Mattes, 2020).

Furthermore, the majority of lipid hydrolysis is thought to take place in the duodenal compartment of the small intestine. *In vitro* studies of almonds and hazelnuts have previously revealed that a significant proportion of plant cells remain intact throughout the duodenal stage. The cell walls could retard the rate and the extent of lipolysis, perhaps by acting as physical barriers to lipase access. This underlines the importance of the structural integrity and the porosity of the cell walls in controlling lipolysis (Capuano et al., 2018; Grundy, Carrière, et al., 2016; Grundy, Wilde, et al., 2015). However, recent evidence suggests that the mechanisms for the limited intestinal lipolysis may involve not only cell wall intactness but also additional factors, such as nut structure and internal constituent properties (i.e., lipid and protein bodies) (McArthur & Mattes, 2020).

The structural modifications of OBs during *in vitro* digestion significantly influence lipid digestion and absorption. To illustrate, in almonds, destabilization and flocculation of OBs occur under low pH gastric conditions because of the hydrolysis of interfacial proteins (oleosins) by pepsin. The resulting pepsin-resistant peptides, along with proteins and phospholipids, remain attached to the surface of the OBs and prevent their coalescence throughout gastric digestion. Under intestinal conditions, the displacement of the surface-active peptides by bile salts and the onset of lipolysis by pancreatic lipase disrupt the flocs. Gastric digestion of oleosins facilitates access of lipase to the oil–water interface for efficient lipolysis of the OBs in the duodenum. However, the intact OB protein membrane surrounding the lipid droplets and the accumulation of lipolytic products, mainly long-chain fatty acids, at the interface inhibit the access of lipase/colipase and therefore slow down intestinal lipolysis (Gallier & Singh, 2012). In walnuts, a unique "spontaneous biological multiple-phase emulsion" stabilized by bile salt crystals is formed upon lipolysis of the OB TAGs in the small intestine. The formation of the water-in-oil-in-water emulsions is thought to be driven by the interactions of the polyunsaturated fatty acid (PUFA)-rich fatty acid profile of walnuts, bile salts and some peptides. These self-assembled structures are likely to play a role in the intestinal digestion and absorption of lipids (Gallier et al., 2013). In growing rats, Gallier et al. (2014) showed that the OBs in crushed whole almonds had a higher apparent ileal fatty acid digestibility than almond oil or almond cream. This could be because the OBs undergo coalescence to a lesser extent under gastric conditions, resulting in a smaller droplet size in the stomach.

Hence, the almond OBs may have a larger surface area available for lipase binding, with a corresponding increase in lipid absorption efficiency.

Because of the incomplete digestion of macronutrients in the upper GIT, it is believed that a large proportion of nutrients remain encapsulated by intact cell walls and arrive at the large intestine, where they are fermented by the microbiota and, if undigested in the colon, are excreted in faeces (Mandalari, Parker, et al., 2018). Microscopic examination of human faeces provided direct evidence of an abundance of multicellular particles of almond tissues that had survived enzymatic digestion and microbial fermentation in the GIT (Ellis et al., 2004; Mandalari, Parker, et al., 2018). These particles consisted of intact cell walls that were surrounded by a large number of intestinal bacteria. Bacterial fermentation appeared to have created numerous holes in the cell walls through which bacteria could enter the cells and gain access to their lipid substrates. Bacteria were also detected inside some damaged cells and appeared to have used most of the intracellular nutrients (Ellis et al., 2004). These findings support the epidemiological and clinical evidence that the cell wall encapsulation of lipids enables the ingestion of tree nuts (considered to be energy-dense nutrient-packed foods) to attenuate nutrient absorption. The beneficial metabolic consequences of this include a decrease in postprandial lipidaemia and a reduction in risk factors associated with cardiovascular diseases, diabetes and obesity (Berry et al., 2008; Berryman et al., 2015).

## 11.5 Modification of animal-based food structures in the GIT
### 11.5.1 Meat

Meat is considered to be an important source of protein in the human diet, despite its debatable role in human health. Meat proteins are estimated to have high true ileal digestibility ($>90\%$) (Silvester & Cummings, 1995; Tomé, 2013). Regardless of the species (terrestrial animals or fish), meat proteins (muscle proteins) consist of myofibrillar proteins that are organized in fibrous structures in their natural form but may differ in length, diameter and orientation in different species (Boland et al. (2019). Recently, Boland et al. (2019) thoroughly reviewed the current knowledge of the structures of different muscle proteins.

The digestion of meat proteins in humans may be influenced by many complex factors, such as the postmortem storage (and ageing), the meat tenderization process, the heat treatment (applied during different cooking processes) or other processing treatments, such as sous-vide, high-pressure processing (Boland et al., 2019). Other factors, such as the type of meat (sources, species, muscle cut), the age of the animal, the ultimate pH, the mincing/particle size and other foods eaten with the meat (such as vegetables rich in proteolytic enzymes, starchy foods), can also have an impact on meat protein digestion (Boland et al., 2019; Farouk et al., 2019). Such processes are known to modify the structure of muscle proteins and consequently influence their digestion profiles. There have been limited studies on understanding the impact of structural changes in meat on the kinetics of digestion and the absorption of amino acids. Here we discuss recent studies on the impact of common tenderization and cooking-related processing conditions on the structural and digestion properties of meat proteins.

The tenderization of raw meat before cooking is a common process that is used to improve its sensorial and textural attributes. Zhu et al. (2018) studied the effect of actinidin pretreatment (tenderization) followed by sous-vide cooking on the microstructure and the static *in vitro* oral–gastric protein digestion of beef brisket steaks (meat cut from the breast of the cow). They reported that, compared with raw beef brisket, actinidin-tenderized meat (or actinidin-treated meat) required a shorter sous-vide cooking time and had improved tenderness and juiciness. This was considered to be due to the breakdown of the myofibrillar structure of muscle proteins by the actinidin pretreatment, making the tough muscle meat softer. This change in the structure of the muscle proteins also increased the rate of muscle protein breakdown (during the first 60 min of gastric digestion) under simulated static *in vitro* oral–gastric conditions (as observed using sodium dodecyl sulphate polyacrylamide gel electrophoresis). Previous studies have shown greater hydrolysis or breakdown of beef muscle proteins in the presence of actinidin or actinidin-rich extracts (Kaur et al., 2010; Kaur & Boland, 2013; Rutherfurd et al., 2011).

Postmortem tenderization (postmortem ageing or maturation or conditioning) is also considered to be important for meat tenderization. The inherent sarcoplasmic proteases, such as cathepsins and calpains, in meat play a crucial role in meat tenderization (Kaur et al., 2020). These proteolytic enzymes improve meat (beef) muscle tenderness and lead to postmortem muscle softening during postmortem storage or the low-temperature processing that is optimum for these enzyme activities (such as at sous-vide temperatures of 50 °C for 24 h, 55 °C for 5 h, and 60 °C and 70 °C for 1 h) (Kaur et al., 2020). Such inherent proteolytic activities during postmortem tenderization may also affect gastrointestinal protein digestion. Zou et al. (2020) recently studied the impact of postmortem ageing on the structural changes and protein digestion of vacuum-packed pork longissimus dorsi muscles stored at 4 °C for 3 days. Using Raman spectroscopy, they found that the ordered meat proteins were degraded, denatured and unfolded during postmortem ageing. These structural changes resulted in their high susceptibility to digestive enzymes (mainly to intestinal enzymes) and, thus, increased proteolytic degradation after simulated static *in vitro* gastrointestinal digestion (Zou et al., 2020). During postmortem ageing, the endogenous enzymes such as calpains and cathepsins can catalyze the breakdown of meat proteins, leading to the increased exposure of catalytic sites to digestive enzymes (Boland et al., 2019; Kaur et al., 2020; Zou et al., 2020).

Raw meat is generally cooked (or thermally processed) before consumption to avoid any health risk associated with pathogenic microorganisms and to improve its sensorial characteristics (or acceptability). Raw meat is structurally a complex matrix. Cooking or thermal processing of meat can lead to many microstructural and molecular changes, such as meat lipid oxidation, changes in antioxidant properties, colour, flavour and odour, and myofibrillar protein oxidation, modification and denaturation, which can vary with different temperatures and heating rates (Kaur et al., 2016; Santé-Lhoutellier et al., 2008). The impact of cooking on meat protein digestion can differ based on the extent of the oxidation and the thermal denaturation of proteins as these changes could affect the accessibility of proteins by proteases (Oberli et al., 2016; Santé-Lhoutellier et al., 2008). For example, Santé-Lhoutellier et al. (2008) studied the effect of cooking bovine M. rectus abdominis muscle (100 °C for 5, 15, 30 and 45 min and 270 °C for 1 min) on static *in vitro* gastrointestinal digestion. They found a significant negative correlation between pepsin activity (digestion) and the extent of oxidative modifications (i.e., carbonyl group formation and myofibrillar protein aggregation) during thermal processing when compared with raw meat. However, they did not observe any such correlations between any oxidative parameter during thermal processing and the activity of pancreatic proteases (Santé-Lhoutellier et al., 2008).

The cooking temperature has been considered to be key for determining the rates of digestion of meat proteins *in vitro*. Bax et al. (2012) proposed a possible mechanism for the action of enzyme

## 11.5 Modification of animal-based food structures in the GIT

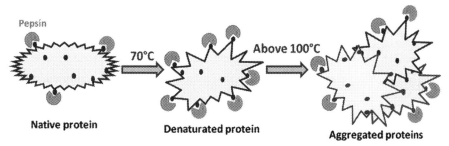

**FIG. 11.5**

Proposed mechanism of pepsin action on meat proteins during heating.

*Adapted from Bax, M.-L., Aubry, L., Ferreira, C., Daudin, J.-D., Gatellier, P., Rémond, D. & Santé-Lhoutellier, V. (2012). Cooking temperature is a key determinant of in vitro meat protein digestion rate: investigation of underlying mechanisms. Journal of Agricultural and Food Chemistry, 60, 10, 2569–2576, with permission from American Chemical Society (ACS) Publications.*

(pepsin) on meat proteins cooked at different temperatures (Fig. 11.5). They suggested that cooking at moderate temperatures (say 70 °C) leads to partial denaturation (unfolding) of proteins, which increases their accessibility to pepsin by exposing more cleavage sites, enhancing the rate of pepsin hydrolysis (digestion). However, at high-temperature cooking (>100 °C), protein aggregation reduces the accessibility of pepsin because of the hiding of cleavage sites, leading to a decrease in protein hydrolysis (Bax et al., 2012). This proposed mechanism was confirmed by Bax et al. (2013), who investigated the effect of different cooking temperatures (60, 75 and 95 °C for 30 min) on the rates of digestion of a minced-meat (veal) protein-based meal in minipigs. They found that the rate of protein digestion (based on the kinetics of the appearance of amino acids in the blood) was higher for the veal meat cooked at 75 °C, followed by 60 °C or 95 °C, within the first 3 h of digestion. They indicated that this could have been due to the progressive protein denaturation at around 75 °C. As expected, oxidation-related protein aggregation slowed pepsin digestion at higher temperatures. They also found that, despite the differences in protein digestion rates, there were no differences in the terminal ileal digestibilities of meats cooked at different temperatures, indicating that cooking temperature did not affect the efficiency of small intestinal and terminal ileal digestibility (Bax et al., 2013).

He et al. (2018) reported that cooking pork meat at relatively low temperatures leads to protein denaturation, but that high-temperature and long-time cooking leads to protein aggregation because of disulphide bonding between protein molecules. Such structural changes in the proteins can alter their interactions with proteolytic enzymes and impact gastrointestinal digestion (He et al., 2018). Wen et al. (2015) also reported a decrease in pork meat protein digestion (in an *in vitro* gastrointestinal digestion model containing both pepsin and trypsin) when the meat was cooked at 100 °C for 3 h compared with 60 °C for 12–16 min. Similarly, Zhang et al. (2020) studied the impact of different heating conditions (60, 70 and 80 °C for 0.5–2.5 h) on the structural changes and *in vitro* pepsin digestion of type I collagen (a major component of connective tissue in meat) from bovine Achilles tendon. They found that increased heating intensity increased the susceptibility of collagen to pepsin (60–70 °C) but that overheating (80 °C) reduced its enzymatic susceptibility. This was because moderate heating led to a reduction in the conformational stability of the type I collagen and the exposure of aromatic residues, leading to greater accessibility to the active sites of pepsin. In contrast, heating at a higher temperature enhanced the surface hydrophobicity of type I collagen, which may have caused protein oxidation and modified the pepsin recognition sites of the protein, reducing digestion (Zhang et al., 2020). Oberli et al. (2016) studied the impact of different cooking processes (raw, boiled, barbecued, grilled and roasted) on the *in vivo* faecal digestibility of bovine meat proteins in male Wistar adult rats for 3 weeks. They found that the true faecal meat protein digestibility was significantly lower in rats fed boiled (94.5%) meat (100°C for 3h) and to some extent roasted meat (96.9%), when measured using the $^{15}$N labelling of meat proteins, compared with raw meat and other cooking processes (97.5%). In addition, they observed that the surface hydrophobicity of meat protein was increased to the same extent in all cooking methods. However, the higher carbonyl content of the boiled and roasted meat proteins compared with the raw meat proteins may have been the reason for their reduced digestibility and consequently would have reduced their proteolytic susceptibility (Oberli et al., 2016).

Conversely, Sayd et al. (2016) did not find any particular trends for the effect of cooking on meat protein digestion when they studied the peptides released after the static *in vitro* gastrointestinal digestion of bull semimembranosus muscle meat cooked at different temperatures (55°C for 5 min, 70°C for 30 min and 90°C for 30 min). Similarly, Oberli et al. (2015) reported that cooking conditions (90°C for 30 min and 55°C for 5 min) did not affect the rates of protein digestion (kinetics of $^{15}$N plasma proteins and appearance of amino acids) of minced bovine meat in 16 human volunteers. They observed similar postprandial utilization of nitrogen between the groups. However, they also reported a moderate decrease in protein digestibility for cooked bovine meat at 90°C for 30 min (high-temperature long-time cooking) compared with at 55°C for 5 min (low-temperature short-time cooking) in 16 human volunteers (5 women and 11 men; aged 28 ± 8 years).

Kaur et al. (2014) studied the effect of high-temperature cooking conditions (used for curries and stews, i.e., approximately 100°C for 10 and 30 min) on the microstructure and static *in vitro* gastrointestinal digestion of beef proteins, compared with raw beef meat. They found that high-temperature cooking resulted in a compact meat structure (observed using transmission electron microscopy) because of the exposure of hydrophobic areas in the denatured myofibrillar proteins, which led to the formation of intermolecular cross-links. The compact structure of the cooked meat was expected to be less susceptible to gastrointestinal enzymes than the raw meat structure. However, they found that raw meat myofibrils formed a fused mass during gastric digestion, which was also resistant to pepsin digestion. Thus, they observed faster digestion of cooked meat proteins in the gastric step because of protein denaturation. However, they found that, compared with raw meat, high-temperature prolonged cooking also led to the generation of "limit peptides", which were not further broken down into free amino acids by digestive enzymes and therefore may have reduced the overall digestibility of the cooked meat (Kaur et al., 2014).

Various nonthermal food processing technologies such as high-pressure processing (Kaur et al. (2016), ultrasound (Zhang et al., 2018) and pulsed electric fields (PEF) (Bhat, Morton, Mason, & Bekhit, 2019; Bhat, Morton, Mason, Jayawardena, & Bekhit, 2019; Chian et al., 2019, 2021) have gained research interest and have been reported to influence the *in vitro* gastrointestinal digestion of meat proteins. Kaur et al. (2016) found that high-pressure processing (at 175 and 600 MPa) of bovine longissimus dorsi muscle meat led to denaturation and aggregation of myofibrillar meat proteins, which enhanced the *in vitro* gastric protein digestion, suggesting increased susceptibility and breakdown of high-pressure-processing-treated meats. Zhang et al. (2018) observed that high-intensity ultrasound (power: 100–800 W, 15 min) degraded and loosened the structure of shrimp (*Exopalaemon modestus*) meat protein (tropomyosin) to generate protein fragments that contributed to higher *in vitro* gastrointestinal digestion and reduced the allergenicity of tropomyosin. In recent years, researchers have used PEF as one of the non-thermal processing options to modulate the kinetics of digestion of nutrients from the meat matrix (Bhat, Morton, Mason, Jayawardena, & Bekhit, 2019). Bhat, Morton, Mason, and Bekhit (2019) studied the effect of PEF processing [T1, 5.0 kV (90 Hz); T2, 10 kV (20 Hz)] on the digestive kinetics of beef semimembranosus muscle during *in vitro* gastrointestinal digestion. PEF treatment led to greater protein digestion, percentage of soluble protein and free amino acids. This was due to increased membrane permeability (electroporation) and structural changes induced in the proteins by the PEF that positively influenced the enzymatic degradation of the beef meat proteins during *in vitro* gastrointestinal digestion (Bhat, Morton, Mason, & Bekhit, 2019). Bhat, Morton, Mason, Jayawardena, and Bekhit (2019) further investigated the impact of PEF treatment (T1, 10 kV, 20 Hz, 20 μs) followed by cooking (core temperature of 75 °C) on the simulated *in vitro* gastrointestinal digestion of beef semimembranosus muscles. They again found that PEF treatment led to a significantly greater protein digestion and percentage of soluble protein during digestion. However, the release of free amino acids was similar to that for the control (Bhat, Morton, Mason, Jayawardena, & Bekhit, 2019). Recently, for the deep and superficial pectoral muscles in beef brisket, Chian et al. (2021) found that a combination of PEF (specific energy of $99 \pm 5$ kJ/kg) and sous-vide cooking (60 °C for 24 h) enhanced the digestive proteolysis of the meat proteins compared with the control sous-vide-cooked meat during *in vitro* gastrointestinal digestion. Although they detected similar muscle micro- and ultrastructural changes between the control sous-vide-cooked pectoral muscles and the PEF-treated sous-vide-cooked pectoral muscles during digestion, the PEF-treated sous-vide-cooked pectoral muscles had more swollen muscle cells and more damaged muscle microstructure and ultrastructure because of the perforation of muscle cell membranes as a result of the electroporation effect of PEF. This would have enhanced the accessibility and penetration of pepsin and gastric juices to substrates, enhancing digestion (Chian et al., 2021).

Most studies have focused on the impact of different structural changes on the proteolysis of meat proteins. However, there is scarce information on assessing the impact of the structural changes in meat on lipid digestion. Asensio-Grau et al. (2019) estimated the impact of matrix structure on *in vitro* lipid digestion in natural meat matrices (beef steak, chicken drumstick, pork lion, cured ham and cooked ham) and processed meat products (hamburger, luncheon meat, pate and sausage). Matrix degradation and lipolysis (free fatty acid release) were found to be higher in processed (unstructured) meat structures than in naturally structured meat matrices. The lower matrix degradation and extent of lipolysis of the natural meat structures were suggested to be due to the encapsulation of lipids in the muscle protein fibres in the natural state. However, in the processed meat products, some of the natural structures, which encapsulate lipid, are disrupted to different extents depending on the type of processing (Asensio-Grau et al., 2019).

A few studies have also focused on the impact of different processing conditions and differently structured fish matrices on the gastrointestinal digestion of fish nutrients. Asensio-Grau et al. (2021) studied the impact of salmon processing (raw, marinating and microwave cooking) on *in vitro* gastrointestinal protein and lipid digestions. They reported that these processing conditions did not affect proteolysis of the fish proteins, even though the matrix degradation index of the microwave-cooked salmon was the least (probably because of protein denaturation and aggregation at high microwave cooking temperatures). However, lipolysis was higher in the microwave-cooked salmon matrix (67%), followed by the raw (57%) and marinated salmon matrices (46%). Higher lipolysis in the microwave-cooked salmon matrix was expected to be due to the release of lipids during cooking, i.e., exudation, leading to higher extractability and subsequent lipolysis during gastrointestinal digestion. The lipids in raw salmon would be expected to be firmly bound to the tissue matrix or entrapped in the fibrous structure of the protein, leading to their slow release during gastrointestinal digestion. The lower lipolysis of the marinated salmon was explained by the release from the salmon meat of the sugar and salt used during marination into the digestion mixture, leading to high ionic strength and affecting the formation of lipid micelles by affecting the interfacial activity of bile salts at the surface of the droplets (Asensio-Grau et al., 2021).

Nasef et al. (2021) investigated the impact of different fish (salmon) matrices, i.e., intact salmon (intact structure), minced salmon (some structure) and defatted salmon+oil (no structure), with identical macronutrient compositions on the absorption of long-chain omega-3 polyunsaturated fatty acids (LCΩ-3PUFAs) in healthy human females (measured using the plasma concentrations of eicosapentaenoic acid and docosahexaenoic acid in the blood at regular time points for 6h postmeal consumption). They found the highest concentration of eicosapentaenoic acid and docosahexaenoic acid in the plasma of human volunteers after the consumption of intact salmon, compared with the other structural forms. Furthermore, they conducted dynamic *in vitro* gastric digestion with the same matrices to understand the mechanisms. They suggested that the encapsulated salmon oil in the salmon muscle protein of the natural salmon matrix would have taken longer to be hydrolysed by pepsin and would have provided a delayed and sustained release of salmon lipids, and consequently a slower gastric emptying of LCΩ-3PUFAs. Conversely, the faster gastric emptying of lipid in the defatted salmon + oil meal was probably due to phase separation, resulting in a separate layer on top and faster gastric emptying of lipid at the end of digestion (Nasef et al., 2021).

Only a small number of *in vivo* studies have been conducted, making it difficult to draw definite conclusions on the effect of cooking conditions on the rates of meat protein digestion and the overall protein digestibility. Also, it has to be noted that, in most of the *in vivo* studies, the meat was minced to reduce the effect of individual variations in chewing efficiency, which may have had an effect on the digestion rate and the postprandial protein metabolism (Rémond et al. (2007). Also, the dynamic digestion process in humans may not be accurately predicted and simulated by static *in vitro* gastrointestinal digestion models. However, it appears that high-temperature and long-time cooking combinations lead to a reduction in the gastrointestinal hydrolysis of meat proteins (Boland et al., 2019). Also, different processing conditions may induce structural changes in the meat matrix and may affect the digestion rates of meat proteins and lipids in the gastrointestinal tract.

### 11.5.2 Milk

Milk is an integral part of a balanced diet for all age groups to meet nutritional and energy requirements. It is also nature's most complex oil-in-water emulsion system (Singh & Gallier, 2017), partly because of its unique production process in the mammary glands, partly because of the unique structure of its naturally assembled nutrients (fat globules and colloidal casein proteins) and partly because of its coagulation behaviour in the stomach, compared with artificially engineered emulsions. Milk proteins are estimated to have high true ileal digestibility (approximately 95%) (Gaudichon et al., 2002; Tomé, 2013). Recent interest in understanding the role of the food structure and the food matrix in digestion and the absorption of nutrients has led to studies on understanding the structural changes in milk during digestion (Thorning et al., 2017).

The coagulation of milk during the gastric digestion process has been suggested to be the main driver in influencing the digestion, delivery and absorption of its nutrients (Dupont & Tomé, 2014; Mulet-Cabero et al., 2019; Ye, 2021; Ye et al., 2020). This has led to the widely accepted concept of slow (i.e., caseins) and fast (i.e., whey proteins) emptying milk proteins with relevance to human digestion (Boirie et al., 1997; Boutrou et al., 2013; Mahe et al., 1996). Whey proteins are soluble globular proteins in milk (Zadow, 2003), whereas casein proteins self-assemble into complex colloidal particles in milk known as casein micelles (McMahon & Brown, 1984). Caseins are regarded as slow-emptying proteins because of the coagulation (or clotting) of casein micelles in the stomach. This leads to relatively slow hydrolysis of caseins by pepsin and slow delivery to the small intestine and, thus, a more gradual postprandial rate of appearance of amino acids in the bloodstream in humans (Boirie et al., 1997; Mahe et al., 1996). In contrast, whey proteins remain soluble, are emptied rapidly from the stomach and provide a fast postprandial rate of release of amino acids into the bloodstream (Boirie et al., 1997; Mahe et al., 1996).

Evidence of milk coagulation has been observed previously in human infants, human adults and animal models, as reviewed by Huppertz and Lambers (2020). In this part of the chapter, we focus on recent studies on understanding the gastrointestinal digestion and absorption of nutrients from naturally structured milk (i.e., liquid raw milk) and differently processed liquid milks. In addition, dynamic and semidynamic *in vitro* and *in vivo* animal digestion studies have been conducted, as such models provide a better understanding of the digestion mechanism of complex structured fluids such as milk.

Ye et al. (2016b) studied the impact of heat treatment (90 °C for 20 min) on the dynamic *in vitro* gastric digestion behaviour (using a human gastric simulator) of raw cow skim milk. They found that raw cow skim milk formed a dense and tight clot network, whereas heated milk formed a loose and open clot network. These structural differences led to a slow rate of casein hydrolysis for the raw skim milk because of the tight structure of the clot, whereas a faster rate of casein hydrolysis was observed for the heated milk because of its open structure (when observed using qualitative sodium dodecyl sulphate polyacrylamide gel electrophoresis). The formation of the clot was due to the hydrolysis of κ-casein by pepsin, leading to destabilization and aggregation of the casein micelles under the simulated conditions of the stomach. The loose structure of the heated milk clot was due to the association of whey proteins with casein micelles during heat treatment, hindering the formation of a strong network (Fig. 11.6). It was also shown that whey proteins were degraded more rapidly in heated milk and were mostly emptied in the form of peptides, which was due to the increased susceptibility of the heated whey proteins to hydrolysis compared with the native whey proteins in raw milk (Ye et al., 2016b).

Ye et al. (2016a) studied the dynamic *in vitro* gastric digestion behaviour of raw and heated (90 °C for 20 min) whole milks. They observed similar clotting behaviour of these milks to those reported previously in their skim milk study (Ye et al., 2016a) and, in addition, reported that the milk fat globules became physically entrapped within the clot network. The release of the fat globules was found to be closely correlated with the structure of the clots, suggesting that the release of the fat globules from the clot was dependent on clot breakdown. This indicated that the clot's structure not only influenced the hydrolysis of the proteins but also influenced the release of the fat globules, which may have implications for lipid digestion. In a further extension of the study, Ye et al. (2017) studied the impact of homogenization treatment (20/5 MPa at 20 °C using a two-stage valve homogenizer) on raw and heated (90 °C for 20 min) milk. They found that raw (unheated) non-homogenized milk formed a firm clot, unheated homogenized milk formed a more porous clot and heated homogenized milk formed a

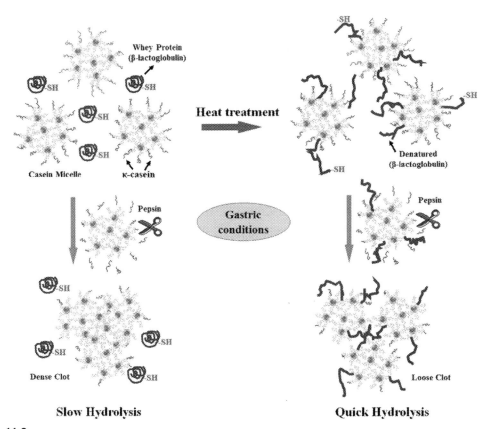

**FIG. 11.6**

Schematic diagram of the possible mechanisms during the formation of protein curds from raw (unheated) milk and heated milk under simulated dynamic *in vitro* gastric digestion conditions.

Adapted from Ye, A. (2021). Gastric colloidal behaviour of milk protein as a tool for manipulating nutrient digestion in dairy products and protein emulsions. Food Hydrocolloids, p. 106599. and Ye, A., Liu, W., Cui, J., Kong, X., Roy, D., Kong, Y., Han, J. & Singh, H. (2019). Coagulation behaviour of milk under gastric digestion: effect of pasteurization and ultra-high temperature treatment. Food Chemistry, 286, 216–225. Available from: https://doi.org/10.1016/j.foodchem.2019.02.010, with permission from Elsevier Inc.

much more fragmented and crumbly structured clot. These differences resulted in a higher degree of hydrolysis of the proteins in the unheated homogenized milk and the heated homogenized milk because of faster diffusion of pepsin into their clots compared with the raw (unheated) non-homogenized milk. The rate of release of the fat globules was highly correlated with the hydrolysis of the clot network, but, surprisingly, the release of the fat globules was not dependent on how the fat globules were associated or incorporated within the clot matrix.

Ye, Liu, et al. (2019) further studied the impact of commercial processing treatments [homogenized raw milk (20/4 MPa and 50 °C using a two-stage valve homogenizer), homogenized pasteurized milk (85 °C for 15 s) and homogenized UHT milk (140 °C for 4 s)] on cow whole milk digestion in an *in vivo* (adult rats) system. They reported that homogenized raw milk formed a firm clot in the rat's stomach, whereas homogenized heated milk formed a soft curd; homogenized UHT milk formed crumbled unstructured curds. This confirmed their previous *in vitro* findings. Similar observations have been reported by Mulet-Cabero et al. (2019) after studying the semidynamic *in vitro* gastric digestion of differently processed whole milks. They reported that raw milk formed a firm coagulum, whereas heated milk formed a fragmented coagulum, which was more evident for homogenized UHT milk. Such structural changes resulted in differences in protein hydrolysis (Fig. 11.7) and nutrient emptying for raw and UHT milks (Mulet-Cabero et al., 2019; Ye, Liu, et al., 2019).

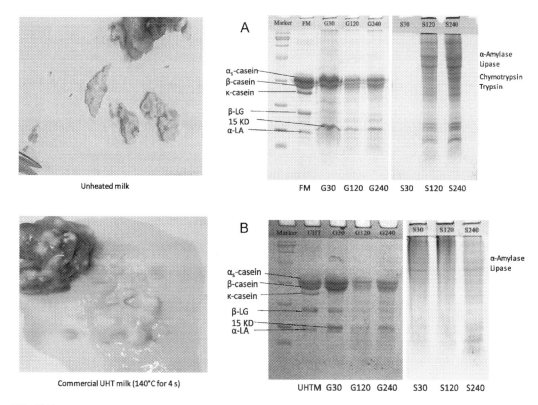

**FIG. 11.7**

Milk clots/curds (photos) obtained from unheated homogenized milk and UHT homogenized milk (140°C for 4s) at 30min in rat stomach after ingestion. SDS-PAGE patterns under reducing conditions of the curds obtained during the digestion in rat stomach and small intestine of unheated homogenized milk (A) and UHT homogenized milk (140°C for 4s) (B) at different times. FM, unheated milk; UHTM, UHT milk; G30, 30min at stomach; G120, 120min at stomach; G240, 240min at stomach; S30, 30min at intestine. S120, 120min at intestine; S240, 240min at intestine.

*Adapted from Ye, A., Liu, W., Cui, J., Kong, X., Roy, D., Kong, Y., Han, J. & Singh, H. (2019). Coagulation behaviour of milk under gastric digestion: effect of pasteurization and ultra-high temperature treatment. Food Chemistry, 286, 216–225. Available from: https://doi.org/10.1016/j.foodchem.2019.02.010, with permission from Elsevier Inc.*

Kaufmann (1984) studied the effect of different processing treatments on milk digestion and absorption in minipigs. They reported that the amino acid concentrations in the blood serum were higher for UHT milk than for other treated milks (raw or pasteurized), which was suggested to be linked to the faster passage through the gastrointestinal tract of UHT milk than other treated milks because of the formation of very soft curds. A study conducted by Lacroix et al. (2008) monitored the postprandial appearance of nutrients in the bloodstream in a group of human volunteers who were given microfiltered raw milk, microfiltered pasteurized milk (72 °C for 20 s) and microfiltered UHT milk (140 °C for 5 s). They reported that the rate of transfer of dietary nitrogen to blood serum amino acids and protein and body urea was much greater in the UHT milk group than in the other milk groups. They suggested that the differences observed were probably due to softer curds and rapid enzymatic hydrolysis of the UHT milk proteins during gastrointestinal digestion. Recently, Fatih et al. (2021) conducted a review to identify scientific studies that assessed the impacts of different heat treatments of milk on postprandial nutrient responses (protein and fat digestion and metabolism) in healthy human adults. They found only little information to suggest that the heat treatment of milk may affect the dynamics of nutrient digestion (Lacroix et al., 2008; Ljungqvist et al., 1979; Nuora, Tupasela, Jokioja, et al., 2018; Nuora, Tupasela, Tahvonen, et al., 2018). However, it was difficult to draw any definite conclusions about the impact of the heat treatment of milk on nutrient digestion and metabolism because of the variability among the different studies (Fatih et al., 2021). This indicated that further human clinical studies are required to determine whether the structural and physicochemical changes observed in *in vitro* or *in vivo* animal studies are relevant to human studies (Fatih et al., 2021).

The true digestibilities of milk are expected to be similar regardless of the source or the type of milk. However, milks from different mammalian species vary in composition and physicochemical properties (protein composition, casein micelle size, fat globule size), which may have implications for the digestion rates of nutrients from different milk species in the GIT (Roy et al., 2020a, 2020b).

Recent studies have also reported the dynamic *in vitro* gastric digestion behaviour of goat and sheep skim and whole milks, compared with cow milk, using a human gastric simulator (Roy et al., 2021a, 2021b). Similar to cow milk, goat and sheep milks separate into a curd (coagulated caseins) and a liquid part (soluble nutrients such as whey proteins) during gastric digestion. The curd is broken down slowly, whereas the soluble whey proteins are transferred rapidly to the small intestine. The curd formed entraps the majority of the fat globules, which are gradually released by the breakdown and hydrolysis of the curd protein network by pepsin and mechanical shearing during digestion. These studies further confirm that, regardless of the species, the rate of fat release from the curd is directly proportional to the breakdown of the protein network of the clots (Roy et al., 2021a, 2021b). Although these studies found that the mechanisms of gastric digestion were similar for cow, goat and sheep milks, the relative amount of curds formed from the milks from different species was dependent on their casein content, i.e., the higher was the casein content, the higher was the amount of curd formed and remaining after gastric digestion (Roy et al., 2021a, 2021b). Other *in vitro* (Mulet-Cabero, Torcello-Gómez, et al., 2020; Phosanam et al., 2021; Ye, Cui, et al., 2019) and *in vivo* (Tari et al., 2018) studies have reported the impact of different casein-to-whey-protein ratios on gastric coagulation and nutrient release in different systems. For example, Phosanam et al. (2021) found that the degree of static *in vitro* gastrointestinal digestion of proteins and lipids was higher in infant formulae in which the casein-to-whey-protein ratio was 40:60 (close to that present in human milk) than in infant formulae with 60:40 and 80:20 casein-to-whey-protein ratios. At the 40:60 ratio, there was little coagulation of the caseins in the gastric phase, and consequently there was greater digestion of proteins and lipids. Blakeborough et al. (1986) also reported higher zinc bioaccessibility from human milk than from cow-milk-based baby foods in piglets. This was due to the softer curds formed from human milk than the hard curds formed from cow milk in the gastrointestinal tract. The different coagulation behaviour of human milk is expected to be due to its lower casein-to-whey-protein ratio and different casein micelle composition. These studies suggest that, apart from the processing treatments, differences in the physicochemical composition of milk may also influence the gastrointestinal digestion and absorption of nutrients.

The impact of structural changes in milk on the kinetics of protein digestion has been well explored, especially during *in vitro* gastrointestinal digestion, as explained previously in this section. However, information on the structural changes in milk fat during gastrointestinal digestion and its impact on the kinetics of fat digestion is limited and still needs to be investigated in detail. The TAG core of the milk fat globules is surrounded, stabilized and protected by a unique trilayer of phospholipids [and proteins, called the milk fat globule membrane (Singh & Gallier, 2017)]. Gastric lipase plays an initial crucial role in facilitating milk fat digestion in human adults (and infants) and facilitates the subsequent hydrolysis of milk fat by small intestinal enzymes (Bernbäck et al., 1990; Brodkorb et al., 2019; Hamosh et al., 1985; Pafumi et al., 2002). However, most of the previous studies focused on intestinal digestion and did not include a suitable alternative to human gastric lipase during *in vitro* adult gastric digestion studies, given that gastric lipolysis accounts for only 10–25% of the overall lipid digestion in adults. There has been a lack of availability of a suitable alternative to human gastric lipase until recently (Brodkorb et al., 2019; Minekus et al., 2014; Mulet-Cabero, Egger, et al., 2020). Moreover, the presence of alkaline sphingomyelinase in the human intestinal tract and in human bile is considered to play a crucial role in sphingomyelin digestion and cholesterol intake in milk (Duan, 2006; Duan et al., 1996), which is not included during the simulated *in vitro* gastrointestinal digestion of milk (Brodkorb et al., 2019; Mulet-Cabero, Egger, et al., 2020). Previous static or dynamic *in vitro* gastrointestinal studies on milk digestion have revealed that the milk fat globules undergo structural changes such as flocculation and coalescence during gastric digestion because of the proteolysis of milk fat globule membrane proteins by pepsin (Gallier et al., 2012; Roy et al., 2021a; Ye et al., 2011). Also, some studies have reported breakthrough discoveries on the formation of ordered geometric nanostructures of milk lipids during the simulated *in vitro* intestinal digestion of milk when observed using small-angle X-ray scattering and cryogenic transmission electron microscopy (Salentinig et al., 2013). However, further studies using *in vitro* models that contain gastric lipase and alkaline sphingomyelinase (and other gastrointestinal lipases and proteases) are needed, to provide a comprehensive understanding of the structural changes in the fat globules in the milk matrix and their relation to the gastrointestinal digestion and absorption of lipids.

Overall, the current literature supports that milk undergoes significant changes in its physicochemical and structural properties during gastric digestion (Mulet-Cabero, Mackie, et al., 2020; Ye et al., 2020). Such changes have an impact on the gastric emptying rates of different nutrients (such as proteins and fats), which may influence their subsequent rates of absorption in the small intestine (Acevedo-Fani & Singh, 2021; Huppertz & Chia, 2021; Mulet-Cabero, Mackie, et al., 2020; Ye, 2021). However, further *in vivo* studies (especially human studies) are required to validate the findings from *in vitro* studies and to draw definite conclusions on the impact of structural changes in the milk matrix on the digestion and absorption of proteins and lipids.

## 11.6 Conclusions

Foods are complex systems in which macronutrients interact in different ways to give rise to complex structures and matrices. These food structures and matrices may be of natural origin (i.e., meat, fruits and vegetables) or may be produced by processing via controlled assemblies and interactions of macromolecules (proteins, lipids and carbohydrates). The structures and chemical compositions of raw natural foods differ widely depending on botanical origin, genetic breeding, animal species and agricultural conditions. Before consumption, most natural foods are submitted to some form of processing, to make them safe to eat, to improve palatability and, in some cases, to disrupt the food matrices to improve the bioavailability of nutrients. It is now recognized that the postprandial behaviour of the nutrients from whole natural foods is different from when individual components of the same food are consumed. These differences are due to the complexity of the interactions between nutrients and the food structure parameters, such as matrix density, particle size and mechanical rigidity. Therefore, understanding the structures in natural foods and how they change during processing and digestion is critically important in delivering optimal nutrition and health benefits.

This chapter has provided recent information on how various technological processes influence the ultrastructures of food materials of botanical or animal species origins. One of the key challenges is to explore the quantitative relationships among the nature of the technological processes applied, the digestion processes and the delivery of nutrients. Much more work needs to be undertaken to develop technological processes that can preserve the natural matrix structure and chemical composition without compromising food safety. This knowledge will in turn redefine processing options for different food groups to achieve a delicate balance between palatability, sensory experience and optimal nutrition.

In addition to technological processes, the processing of foods within the gastrointestinal tract is complex, involving many different mechanical, physical and biochemical processes. Our understanding of how various foods with different compositions, physical properties and structures are affected during digestion has improved over the last decade, largely because of the development of *in vitro* digestion models. However, much of the information obtained in the *in vitro* models needs to be validated in *in vivo* animal/human studies to further understand how different food structures are processed in the human gut. In addition, more sophisticated techniques need to be developed to study the nature of the food matrices and the location and release of nutrients under physiological conditions.

## Acknowledgments

The authors thank the Riddet Institute Centre of Research Excellence (CoRE) and the Tertiary Education Commission, New Zealand, for providing funding for this contribution. Alejandra Acevedo-Fani also thanks the funding received from the European Union's Horizon 2020 Research and Innovation Programme, under the 3D-NANOFOOD project, grant agreement WF-IF-EF 867472.

## References

Acevedo-Fani, A. & Singh, H. 2021, Biopolymer interactions during gastric digestion: Implications for nutrient delivery, Food Hydrocolloids, 116, pp. 106644. Available from: doi:https://doi.org/10.1016/j.foodhyd.2021.106644.

Aguilera, J.M. 2005, Why food microstructure?, Journal of Food Engineering, 67, 1–2, pp. 3-11. Available from: https://doi.org/10.1016/j.jfoodeng.2004.05.050.

Aguilera, J. M. (2006). Food product engineering: Building the right structures. *Journal of the Science of Food and Agriculture*, *86*(8), 1147–1155. Available from: https://doi.org/10.1002/jsfa.2468.

Ahima, R. S., & Antwi, D. A. (2008). Brain regulation of appetite and satiety. *Endocrinology and Metabolism Clinics of North America*, *37*(4), 811–823. Available from: https://doi.org/10.1016/j.ecl.2008.08.005.

Al-Rabadi, G. J., Gilbert, R. G., & Gidley, M. J. (2009). Effect of particle size on kinetics of starch digestion in milled barley and sorghum grains by porcine alpha-amylase. *Journal of Cereal Science*, *50*(2), 198–204.

Altan, A., McCarthy, K. L., Tikekar, R., McCarthy, M. J., & Nitin, N. (2011). Image analysis of microstructural changes in almond cotyledon as a result of processing. *Journal of Food Science*, *76*(2), E212–E221.

Armand, M. (2007). Lipases and lipolysis in the human digestive tract: where do we stand? *Current Opinion in Clinical Nutrition and Metabolic Care*, *10*(2), 156–164. Available from: https://doi.org/10.1097/MCO.0b013e3280177687.

Armand, M., Borel, P., Pasquier, B., Dubois, C., Senft, M., Andre, M., Peyrot, J., Salducci, J. & Lairon, D. 1996, Physicochemical characteristics of emulsions during fat digestion in human stomach and duodenum, The American Journal of Physiology, 271, 1, pp. G172-183. Available from: https://doi.org/10.1152/ajpgi.1996.271.1.G172.

Asensio-Grau, A., Calvo-Lerma, J., Heredia, A., & Andrés, A. (2019). Fat digestibility in meat products: influence of food structure and gastrointestinal conditions. *International Journal of Food Sciences and Nutrition*, *70*(5), 530–539.

Asensio-Grau, A., Calvo-Lerma, J., Heredia, A., & Andrés, A. (2021). In vitro digestion of salmon: Influence of processing and intestinal conditions on macronutrients digestibility. *Food Chemistry*, *342*, 128387. Available from: https://doi.org/10.1016/j.foodchem.2020.128387.

Bach, A. C., & Babayan, V. K. (1982). Medium-chain triglycerides: an update. *The American Journal of Clinical Nutrition*, *36*(5), 950–962. Available from: https://doi.org/10.1093/ajcn/36.5.950.

Bax, M.-L., Aubry, L., Ferreira, C., Daudin, J.-D., Gatellier, P., Rémond, D., & Santé-Lhoutellier, V. (2012). Cooking temperature is a key determinant of in vitro meat protein digestion rate: investigation of underlying mechanisms. *Journal of Agricultural and Food Chemistry*, *60*(10), 2569–2576.

Bax, M.-L., Buffière, C., Hafnaoui, N., Gaudichon, C., Savary-Auzeloux, I., Dardevet, D., Santé-Lhoutellier, V., & Rémond, D. (2013). Effects of meat cooking, and of ingested amount, on protein digestion speed and entry of residual proteins into the colon: a study in minipigs. *PLoS One*, *8*(4), e61252.

Berg, T., Singh, J., Hardacre, A., & Boland, M. J. (2012). The role of cotyledon cell structure during in vitro digestion of starch in navy beans. *Carbohydrate Polymers*, *87*(2), 1678–1688.

Bergstrom, K., Fu, J., Johansson, M.E.V., Liu, X., Gao, N., Wu, Q., Song, J., McDaniel, J.M., McGee, S., Chen, W., Braun, J., Hansson, G.C., Xia, L. 2017, Core 1- and 3-derived O-glycans collectively maintain the colonic mucus barrier and protect against spontaneous colitis in mice, Mucosal Immunology, 10, 1, pp. 91-103. Available from: https://doi.org/10.1038/mi.2016.45.

Bernbäck, S., Bläckberg, L., & Hernell, O. (1990). The complete digestion of human milk triacylglycerol in vitro requires gastric lipase, pancreatic colipase-dependent lipase, and bile salt-stimulated lipase. *The Journal of Clinical Investigation*, *85*(4), 1221–1226.

Berry, S. E., Tydeman, E. A., Lewis, H. B., Phalora, R., Rosborough, J., Picout, D. R., & Ellis, P. R. (2008). Manipulation of lipid bioaccessibility of almond seeds influences postprandial lipemia in healthy human subjects. *The American Journal of Clinical Nutrition*, *88*(4), 922–929.

Berryman, C. E., West, S. G., Fleming, J. A., Bordi, P. L., & Kris-Etherton, P. M. (2015). Effects of daily almond consumption on cardiometabolic risk and abdominal adiposity in healthy adults with elevated LDL-cholesterol: a randomized controlled trial. *Journal of the American Heart Association*, *4*(1), e000993.

Bhat, Z., Morton, J. D., Mason, S. L., & Bekhit, A. E.-D. A. (2019). Pulsed electric field improved protein digestion of beef during in-vitro gastrointestinal simulation. *LWT*, *102*, 45–51.

Bhat, Z., Morton, J. D., Mason, S. L., Jayawardena, S. R., & Bekhit, A. E.-D. A. (2019). Pulsed electric field: A new way to improve digestibility of cooked beef. *Meat Science*, *155*, 79–84.

Bhattarai, R. R., Dhital, S., Mense, A., Gidley, M. J., & Shi, Y.-C. (2018). Intact cellular structure in cereal endosperm limits starch digestion in vitro. *Food Hydrocolloids*, *81*, 139–148.

Blakeborough, P., Gurr, M. I., & Salter, D. N. (1986). Digestion of the zinc in human milk, cow's milk and a commercial babyfood: some implications for human infant nutrition. *British Journal of Nutrition*, *55*(2), 209–217.

Boirie, Y., Dangin, M., Gachon, P., Vasson, M.-P., Maubois, J.-L., & Beaufrère, B. (1997). Slow and fast dietary proteins differently modulate postprandial protein accretion. *Proceedings of the National Academy of Sciences*, *94*(26), 14930–14935. Available from: https://doi.org/10.1073/pnas.94.26.14930.

Boland, M., Kaur, L., Chian, F.M. & Astruc, T. 2019. Muscle Proteins. In: Melton, L., Shahidi, F. & Varelis, P. (eds.) Encyclopedia of Food Chemistry. Oxford: Academic Press. Available: doi:https://doi.org/10.1016/B978-0-08-100596-5.21602-8.

Bordoloi, A., Kaur, L., & Singh, J. (2012). Parenchyma cell microstructure and textural characteristics of raw and cooked potatoes. *Food Chemistry*, *133*(4), 1092–1100.

Bordoloi, A., Singh, J., & Kaur, L. (2012). In vitro digestibility of starch in cooked potatoes as affected by guar gum: Microstructural and rheological characteristics. *Food Chemistry*, *133*(4), 1206–1213.

Bornhorst, G. M., Chang, L. Q., Rutherfurd, S. M., Moughan, P. J., & Singh, R. P. (2013). Gastric emptying rate and chyme characteristics for cooked brown and white rice meals in vivo. *Journal of the Science of Food and Agriculture*, *93*(12), 2900–2908.

Bornhorst, G. M., Drechsler, K. C., Montoya, C. A., Rutherfurd, S. M., Moughan, P. J., & Singh, R. P. (2016). Gastric protein hydrolysis of raw and roasted almonds in the growing pig. *Food Chemistry*, *211*, 502–508.

Bornhorst, G. M., Roman, M. J., Dreschler, K. C., & Singh, R. P. (2014). Physical property changes in raw and roasted almonds during gastric digestion in vivo and in vitro. *Food Biophysics*, *9*(1), 39–48.

Bornhorst, G. M., Roman, M. J., Rutherfurd, S. M., Burri, B. J., Moughan, P. J., & Singh, R. P. (2013). Gastric digestion of raw and roasted almonds in vivo. *Journal of Food Science*, *78*(11), H1807–H1813.

Bornhorst, G., & Singh, R. P. (2014). Gastric digestion in vivo and in vitro: how the structural aspects of food influence the digestion process. *Annual Review of Food Science and Technology*, *5*(5), 111–132. Available from https://doi.org/10.1146/annurev-food-030713-092346.

Boutrou, R., Gaudichon, C., Dupont, D., Jardin, J., Airinei, G., Marsset-Baglieri, A., Benamouzig, R., Tome, D., & Leonil, J. (2013). Sequential release of milk protein-derived bioactive peptides in the jejunum in healthy humans. *The American Journal of Clinical Nutrition*, *97*(6), 1314–1323.

Brett, C. T., & Waldron, K. W. (1996). *Physiology and biochemistry of plant cell walls*. Springer Science & Business Media.

Brodkorb, A., Egger, L., Alminger, M., Alvito, P., Assunção, R., Ballance, S., Bohn, T., Bourlieu-Lacanal, C., Boutrou, R., & Carrière, F. (2019). INFOGEST static in vitro simulation of gastrointestinal food digestion. *Nature Protocols*, *1*.

Brownlee, I. A., Gill, S., Wilcox, M. D., Pearson, J. P., & Chater, P. I. (2018). Starch digestion in the upper gastrointestinal tract of humans. *Starch - Stärke*, *70*(9–10), 1700111. Available from: https://doi.org/10.1002/star.201700111.

Brummer, Y., Kaviani, M., & Tosh, S. M. (2015). Structural and functional characteristics of dietary fibre in beans, lentils, peas and chickpeas. *Food Research International*, *67*, 117–125.

Capuano, E., Pellegrini, N., Ntone, E., & Nikiforidis, C. V. (2018). In vitro lipid digestion in raw and roasted hazelnut particles and oil bodies. *Food & Function*, *9*(4), 2508–2516.

Carrière, F., Grandval, P., Renou, C., Palomba, A., Priéri, F., Giallo, J., Henniges, F., Sander-Struckmeier, S., & Laugier, R. (2005). Quantitative study of digestive enzyme secretion and gastrointestinal lipolysis in chronic pancreatitis. *Clinical Gastroenterology and Hepatology*, *3*(1), 28–38. Available from: https://doi.org/10.1016/S1542-3565(04)00601-9.

Cassady, B. A., Hollis, J. H., Fulford, A. D., Considine, R. V., & Mattes, R. D. (2009). Mastication of almonds: effects of lipid bioaccessibility, appetite, and hormone response. *The American Journal of Clinical Nutrition*, *89*(3), 794–800.

Chen, Z., Huang, Q., Xia, Q., Zha, B., Sun, J., Xu, B., & Shi, Y.-C. (2020). Intact endosperm cells in buckwheat flour limit starch gelatinization and digestibility in vitro. *Food Chemistry*, *330*, 127318.

Chen, L., Tuo, B., & Dong, H. (2016). Regulation of intestinal glucose absorption by ion channels and transporters. *Nutrients*, *8*(1), 43. Available from: https://doi.org/10.3390/nu8010043.

Chian, F. M., Kaur, L., Oey, I., Astruc, T., Hodgkinson, S., & Boland, M. (2019). Effect of Pulsed Electric Fields (PEF) on the ultrastructure and in vitro protein digestibility of bovine longissimus thoracis. *LWT*, *103*, 253–259.

Chian, F. M., Kaur, L., Oey, I., Astruc, T., Hodgkinson, S., & Boland, M. (2021). Effects of pulsed electric field processing and sous vide cooking on muscle structure and in vitro protein digestibility of beef brisket. *Foods*, *10*(3), 512.

Dhital, S., Bhattarai, R. R., Gorham, J., & Gidley, M. J. (2016). Intactness of cell wall structure controls the in vitro digestion of starch in legumes. *Food & Function*, *7*(3), 1367–1379.

Do, D. T. (2020). *Microstructural analysis of edible plants: the possibility of designing low glycaemic biomimetic plant foods*. A dissertation presented in partial fulfilment of the requirements for the degree of Doctor of Philosophy in Food Technology at Massey University, Palmerston North, Manawatū, New Zealand. Massey University.

Do, D. T., Singh, J., Oey, I., & Singh, H. (2018). Biomimetic plant foods: Structural design and functionality. *Trends in Food Science & Technology*, *82*, 46–59. Available from: https://doi.org/10.1016/j.tifs.2018.09.010.

Do, D. T., Singh, J., Oey, I., & Singh, H. (2019). Modulating effect of cotyledon cell microstructure on in vitro digestion of starch in legumes. *Food Hydrocolloids*, *96*, 112–122.

Do, D. T., Singh, J., Oey, I., & Singh, H. (2020). Isolated potato parenchyma cells: Physico-chemical characteristics and gastro-small intestinal digestion in vitro. *Food Hydrocolloids*, *108*, 105972.

Do, D. T., Singh, J., Oey, I., Singh, H., Yada, R. Y., & Frostad, J. M. (2020). A novel apparatus for time-lapse optical microscopy of gelatinisation and digestion of starch inside plant cells. *Food Hydrocolloids*, *104*, 105551.

Duan, R. D. (2006). Alkaline sphingomyelinase: an old enzyme with novel implications. *Biochimica et Biophysica Acta*, *1761*(3), 281–291. Available from: https://doi.org/10.1016/j.bbalip.2006.03.007.

Duan, R. D., Hertervig, E., Nyberg, L., Hauge, T., Sternby, B., Lillienau, J., Farooqi, A., & Nilsson, A. (1996). Distribution of alkaline sphingomyelinase activity in human beings and animals. Tissue and species differences. *Digestive Diseases and Sciences*, *41*(9), 1801–1806. Available from: https://doi.org/10.1007/bf02088748.

Dupont, D., & Tomé, D. (2014). *Milk proteins: digestion and absorption in the gastrointestinal tract*. Milk Proteins: Elsevier.

Edwards, C. H., Grundy, M. M., Grassby, T., Vasilopoulou, D., Frost, G. S., Butterworth, P. J., Berry, S. E., Sanderson, J., & Ellis, P. R. (2015). Manipulation of starch bioaccessibility in wheat endosperm to regulate starch digestion, postprandial glycemia, insulinemia, and gut hormone responses: a randomized controlled trial in healthy ileostomy participants. *The American Journal of Clinical Nutrition*, *102*(4), 791–800.

Ellis, P. R., Kendall, C. W., Ren, Y., Parker, C., Pacy, J. F., Waldron, K. W., & Jenkins, D. J. (2004). Role of cell walls in the bioaccessibility of lipids in almond seeds. *The American Journal of Clinical Nutrition*, *80*(3), 604–613.

Emery, P. W. (2015). Basic metabolism: protein. *Surgery (Oxford)*, *33*(4), 143–147. Available from: https://doi.org/10.1016/j.mpsur.2015.01.008.

Farooq, A. M., Li, C., Chen, S., Fu, X., Zhang, B., & Huang, Q. (2018). Particle size affects structural and in vitro digestion properties of cooked rice flours. *International Journal of Biological Macromolecules, 118*, 160–167.

Farouk, M. M., Wu, G., Frost, D. A., Staincliffe, M., & Knowles, S. O. (2019). Factors affecting the digestibility of beef and consequences for designing meat-centric meals. *Journal of Food Quality*.

Fatih, M., Barnett, M. P., Gillies, N. A., & Milan, A. M. (2021). Heat treatment of milk: a rapid review of the impacts on postprandial protein and lipid kinetics in human adults. *Frontiers in Nutrition, 8*.

Feher, J. 2017. The stomach. In: Feher, J. (ed.) Quantitative human physiology. Boston: Academic Press. Available: https://doi.org/10.1016/b978-0-12-800883-6.00078-1.

Gallier, S., Rutherfurd, S. M., Moughan, P. J., & Singh, H. (2014). Effect of food matrix microstructure on stomach emptying rate and apparent ileal fatty acid digestibility of almond lipids. *Food & Function, 5*(10), 2410–2419.

Gallier, S., & Singh, H. (2012). Behavior of almond oil bodies during in vitro gastric and intestinal digestion. *Food & Function, 3*(5), 547–555. Available from: https://doi.org/10.1039/C2FO10259E.

Gallier, S., Tate, H., & Singh, H. (2013). In vitro gastric and intestinal digestion of a walnut oil body dispersion. *Journal of Agricultural and Food Chemistry, 61*(2), 410–417. Available from: https://doi.org/10.1021/jf303456a.

Gallier, S., Ye, A., & Singh, H. (2012). Structural changes of bovine milk fat globules during in vitro digestion. *Journal of Dairy Science, 95*(7), 3579–3592. Available from: https://doi.org/10.3168/jds.2011-5223.

Gaudichon, C., Bos, C., Morens, C., Petzke, K. J., Mariotti, F., Everwand, J., Benamouzig, R., Daré, S., Tomé, D., & Metges, C. C. (2002). Ileal losses of nitrogen and amino acids in humans and their importance to the assessment of amino acid requirements. *Gastroenterology, 123*(1), 50–59. Available from: https://doi.org/10.1053/gast.2002.34233.

Gebauer, S. K., Novotny, J. A., Bornhorst, G. M., & Baer, D. J. (2016). Food processing and structure impact the metabolizable energy of almonds. *Food & Function, 7*(10), 4231–4238.

Goebel, J. T. S., Kaur, L., Colussi, R., Elias, M. C., & Singh, J. (2019). Microstructure of indica and japonica rice influences their starch digestibility: A study using a human digestion simulator. *Food Hydrocolloids, 94*, 191–198.

Goyal, R. K., Guo, Y., & Mashimo, H. (2019). Advances in the physiology of gastric emptying. *Neurogastroenterology and Motility: the Official Journal of the European Gastrointestinal Motility Society, 31*(4), e13546. Available from: https://doi.org/10.1111/nmo.13546.

Grundy, M., Carrière, F., Mackie, A. R., Gray, D. A., Butterworth, P. J., & Ellis, P. R. (2016). The role of plant cell wall encapsulation and porosity in regulating lipolysis during the digestion of almond seeds. *Food & Function, 7*(1), 69–78.

Grundy, M., Grassby, T., Mandalari, G., Waldron, K. W., Butterworth, P. J., Berry, S. E., & Ellis, P. R. (2015). Effect of mastication on lipid bioaccessibility of almonds in a randomized human study and its implications for digestion kinetics, metabolizable energy, and postprandial lipemia. *The American Journal of Clinical Nutrition, 101*(1), 25–33.

Grundy, M., Lapsley, K., & Ellis, P. R. (2016). A review of the impact of processing on nutrient bioaccessibility and digestion of almonds. *International Journal of Food Science & Technology, 51*(9), 1937–1946.

Grundy, M., Wilde, P. J., Butterworth, P. J., Gray, R., & Ellis, P. R. (2015). Impact of cell wall encapsulation of almonds on in vitro duodenal lipolysis. *Food Chemistry, 185*, 405–412.

Guo, P., Yu, J., Wang, S., Wang, S., & Copeland, L. (2018). Effects of particle size and water content during cooking on the physicochemical properties and in vitro starch digestibility of milled durum wheat grains. *Food Hydrocolloids, 77*, 445–453.

Hamosh, M., Bitman, J., Wood, D. L., Hamosh, P., & Mehta, N. (1985). Lipids in milk and the first steps in their digestion. *Pediatrics, 75*(1), 146–150.

He, J., Zhou, G., Bai, Y., Wang, C., Zhu, S., Xu, X., & Li, C. (2018). The effect of meat processing methods on changes in disulfide bonding and alteration of protein structures: impact on protein digestion products. *RSC Advances, 8*(31), 17595–17605.

Holmes, R. (1971). Carbohydrate digestion and absorption. *Journal of Clinical Pathology. Supplement (Royal College of Pathologists)*, 5, 10–13.

Huppertz, T., & Chia, L. W. (2021). Milk protein coagulation under gastric conditions: A review. *International Dairy Journal*, 113, 104882. Available from: https://doi.org/10.1016/j.idairyj.2020.104882.

Huppertz, T., & Lambers, T. T. (2020). Influence of micellar calcium phosphate on in vitro gastric coagulation and digestion of milk proteins in infant formula model systems. *International Dairy Journal*, 107, 104717. Available from: https://doi.org/10.1016/j.idairyj.2020.104717.

Iqbal, J., & Hussain, M. M. (2009). Intestinal lipid absorption. *American Journal of Physiology. Endocrinology and Metabolism*, 296(6), E1183–E1194. Available from: https://doi.org/10.1152/ajpendo.90899.2008.

Kaufmann, W. (1984). Influences of different technological treatments of milk on the digestion in the stomach. VI. Estimation of amino acid and urea concentrations in the blood: Conclusions regarding the nutritional evaluation. *Milchwissenschaft*, 39(5), 281–284.

Kaur, L., Astruc, T., Vénien, A., Loison, O., Cui, J., Irastorza, M., & Boland, M. (2016). High pressure processing of meat: Effects on ultrastructure and protein digestibility. *Food & Function*, 7(5), 2389–2397.

Kaur, L., & Boland, M. (2013). Influence of kiwifruit on protein digestion. *Advances in Food and Nutrition Research*, 68, 149–167.

Kaur, L., Hui, S. X., & Boland, M. (2020). Changes in cathepsin activity during low-temperature storage and sous vide processing of beef brisket. *Food Science of Animal Resources*, 40(3), 415.

Kaur, L., Maudens, E., Haisman, D. R., Boland, M. J., & Singh, H. (2014). Microstructure and protein digestibility of beef: The effect of cooking conditions as used in stews and curries. *LWT- Food Science and Technology*, 55(2), 612–620.

Kaur, L., Rutherfurd, S. M., Moughan, P. J., Drummond, L., & Boland, M. J. (2010). Actinidin enhances gastric protein digestion as assessed using an in vitro gastric digestion model. *Journal of Agricultural and Food Chemistry*, 58(8), 5068–5073.

Kong, F., Oztop, M. H., Singh, R. P., & McCarthy, M. J. (2011). Physical changes in white and brown rice during simulated gastric digestion. *Journal of Food Science*, 76(6), E450–E457.

Kong, F., & Singh, R. (2008). A model stomach system to investigate disintegration kinetics of solid foods during gastric digestion. *Journal of Food Science*, 73(5), E202–E210.

Kong, F., & Singh, R. P. (2009a). Digestion of raw and roasted almonds in simulated gastric environment. *Food Biophysics*, 4(4), 365–377.

Kong, F., & Singh, R. P. (2009b). Modes of disintegration of solid foods in simulated gastric environment. *Food Biophysics*, 4(3), 180–190.

Kong, F., & Singh, R. P. (2011). Solid loss of carrots during simulated gastric digestion. *Food Biophysics*, 6(1), 84–93.

Korompokis, K., De Brier, N., & Delcour, J. A. (2019). Differences in endosperm cell wall integrity in wheat (*Triticum aestivum* L.) milling fractions impact on the way starch responds to gelatinization and pasting treatments and its subsequent enzymatic in vitro digestibility. *Food & Function*, 10(8), 4674–4684.

Lacroix, M., Bon, C., Bos, C., Léonil, J., Benamouzig, R., Luengo, C., Fauquant, J., Tomé, D., & Gaudichon, C. (2008). Ultra high temperature treatment, but not pasteurization, affects the postprandial kinetics of milk proteins in humans. *The Journal of Nutrition*, 138(12), 2342–2347.

Lemmens, L., Van Buggenhout, S., Van Loey, A. M., & Hendrickx, M. E. (2010). Particle size reduction leading to cell wall rupture is more important for the β-carotene bioaccessibility of raw compared to thermally processed carrots. *Journal of Agricultural and Food Chemistry*, 58(24), 12769–12776.

Li, H.-T., Chen, S.-Q., Bui, A. T., Xu, B., & Dhital, S. (2021). Natural 'capsule' in food plants: cell wall porosity controls starch digestion and fermentation. *Food Hydrocolloids*, 106657.

Liu, D., Dhital, S., Wu, P., Chen, X.-D., & Gidley, M. J. (2019). In vitro digestion of apple tissue using a dynamic stomach model: Grinding and crushing effects on polyphenol bioaccessibility. *Journal of Agricultural and Food Chemistry*, 68(2), 574–583.

Liu, D., Lopez-Sanchez, P., & Gidley, M. J. (2019). Cellular barriers in apple tissue regulate polyphenol release under different food processing and in vitro digestion conditions. *Food & Function*, *10*(5), 3008–3017.

Ljungqvist, B., Blomstrand, E., Hellström, Å., Lindell, I., Olsson, M., & Svanberg, U.-O. (1979). Plasma amino acid response to single test meals in humans. *Research in Experimental Medicine*, *174*(3), 209–219.

Low, D. Y., D'Arcy, B., & Gidley, M. J. (2015). Mastication effects on carotenoid bioaccessibility from mango fruit tissue. *Food Research International*, *67*, 238–246.

Mahe, S., Roos, N., Benamouzig, R., Davin, L., Luengo, C., Gagnon, L., Gaussgers, N., Rautureau, J., & Tomé, D. (1996). Gastrojejunal kinetics and the digestion of [15N] beta-lactoglobulin and casein in humans: the influence of the nature and quantity of the protein. *The American Journal of Clinical Nutrition*, *63*(4), 546–552.

Mandalari, G., Bisignano, C., Filocamo, A., Chessa, S., Sarò, M., Torre, G., Faulks, R. M., & Dugo, P. (2013). Bioaccessibility of pistachio polyphenols, xanthophylls, and tocopherols during simulated human digestion. *Nutrition*, *29*(1), 338–344.

Mandalari, G., Faulks, R. M., Rich, G. T., Lo Turco, V., Picout, D. R., Lo Curto, R. B., Bisignano, G., Dugo, P., Dugo, G., & Waldron, K. W. (2008). Release of protein, lipid, and vitamin E from almond seeds during digestion. *Journal of Agricultural and Food Chemistry*, *56*(9), 3409–3416.

Mandalari, G., Grundy, M. M.-L., Grassby, T., Parker, M. L., Cross, K. L., Chessa, S., Bisignano, C., Barreca, D., Bellocco, E., & Lagana, G. (2014). The effects of processing and mastication on almond lipid bioaccessibility using novel methods of in vitro digestion modelling and micro-structural analysis. *British Journal of Nutrition*, *112*(9), 1521–1529.

Mandalari, G., Merali, Z., Ryden, P., Chessa, S., Bisignano, C., Barreca, D., Bellocco, E., Laganà, G., Faulks, R. M., & Waldron, K. W. (2018). Durum wheat particle size affects starch and protein digestion in vitro. *European Journal of Nutrition*, *57*(1), 319–325.

Mandalari, G., Parker, M. L., Grundy, M. M.-L., Grassby, T., Smeriglio, A., Bisignano, C., Raciti, R., Trombetta, D., Baer, D. J., & Wilde, P. J. (2018). Understanding the effect of particle size and processing on almond lipid bioaccessibility through microstructural analysis: From mastication to faecal collection. *Nutrients*, *10*(2), 213.

McArthur, B., & Mattes, R. (2020). Energy extraction from nuts: walnuts, almonds and pistachios. *British Journal of Nutrition*, *123*(4), 361–371.

McMahon, D. J., & Brown, R. J. (1984). Composition, structure, and integrity of Casein Micelles: A review. *Journal of Dairy Science*, *67, 3*, 499–512. Available from: https://doi.org/10.3168/jds.S0022-0302(84)81332-6.

Minekus, M., Alminger, M., Alvito, P., Ballance, S., Bohn, T., Bourlieu, C., Carriere, F., Boutrou, R., Corredig, M., & Dupont, D. (2014). A standardised static in vitro digestion method suitable for food—An international consensus. *Food & Function*, *5*(6), 1113–1124.

Miquel-Kergoat, S., Azais-Braesco, V., Burton-Freeman, B., & Hetherington, M. M. (2015). Effects of chewing on appetite, food intake and gut hormones: A systematic review and meta-analysis. *Physiology & Behavior*, *151*, 88–96. Available from: https://doi.org/10.1016/j.physbeh.2015.07.017.

Moughan, P.J. (2009). Digestion and absorption of proteins and peptides. In D. J. McClements, & E. A. Decker (Eds.), *Designing functional foods* Woodhead Publishing. https://doi.org/10.1533/9781845696603.1.148. Available:.

Moughan, P. J. (2020). Holistic properties of foods: a changing paradigm in human nutrition. *Journal of the Science of Food and Agriculture*, *100*(14), 5056–5063. Available from: https://doi.org/10.1002/jsfa.8997.

Mu, H., & Hoy, C. E. (2004). The digestion of dietary triacylglycerols. *Progress in Lipid Research*, *43*(2), 105–133. Available from: https://doi.org/10.1016/s0163-7827(03)00050-x.

Mulet-Cabero, A.-I., Egger, L., Portmann, R., Ménard, O., Marze, S., Minekus, M., Le Feunteun, S., Sarkar, A., Grundy, M. M. L., Carrière, F., Golding, M., Dupont, D., Recio, I., Brodkorb, A., & Mackie, A. (2020). A standardised semi-dynamic in vitro digestion method suitable for food—An international consensus. *Food & Function*, *11*(2), 1702–1720. Available from: https://doi.org/10.1039/C9FO01293A.

Mulet-Cabero, A.-I., Mackie, A. R., Brodkorb, A., & Wilde, P. J. (2020). Dairy structures and physiological responses: A matter of gastric digestion. *Critical Reviews in Food Science and Nutrition, 60*(22), 3737–3752. Available from: https://doi.org/10.1080/10408398.2019.1707159.

Mulet-Cabero, A.-I., Mackie, A. R., Wilde, P. J., Fenelon, M. A., & Brodkorb, A. (2019). Structural mechanism and kinetics of in vitro gastric digestion are affected by process-induced changes in bovine milk. *Food Hydrocolloids, 86*, 172–183. Available from: https://doi.org/10.1016/j.foodhyd.2018.03.035.

Mulet-Cabero, A.-I., Torcello-Gómez, A., Saha, S., Mackie, A. R., Wilde, P. J., & Brodkorb, A. (2020). Impact of caseins and whey proteins ratio and lipid content on in vitro digestion and ex vivo absorption. *Food Chemistry, 319*, 126514. Available from: https://doi.org/10.1016/j.foodchem.2020.126514.

Muniz, C. R., Freire, F. C., Soares, A. A., Cooke, P. H., & Guedes, M. I. (2013). The ultrastructure of shelled and unshelled cashew nuts. *Micron, 54*, 52–56.

Nasef, N. A., Zhu, P., Golding, M., Dave, A., Ali, A., Singh, H., & Garg, M. (2021). Salmon food matrix influences digestion and bioavailability of long-chain omega-3 polyunsaturated fatty acids. *Food & Function, 12*(14), 6588–6602.

Nichols, B. L., Avery, S., Sen, P., Swallow, D. M., Hahn, D., & Sterchi, E. (2003). The maltase-glucoamylase gene: common ancestry to sucrase-isomaltase with complementary starch digestion activities. *Proceedings of the National Academy of Sciences of the United States of America, 100*(3), 1432–1437. Available from: https://doi.org/10.1073/pnas.0237170100.

Noah, L., Guillon, F., Bouchet, B., Buleon, A., Molis, C., Gratas, M., & Champ, M. (1998). Digestion of carbohydrate from white beans (*Phaseolus vulgaris* L.) in healthy humans. *The Journal of Nutrition, 128*(6), 977–985.

Norton, J. E., Wallis, G. A., Spyropoulos, F., Lillford, P. J., & Norton, I. T. (2014). Designing food structures for nutrition and health benefits. *Annual Review of Food Science and Technology, 5*, 177–195. Available from: https://doi.org/10.1146/annurev-food-030713-092315.

Nuora, A., Tupasela, T., Jokioja, J., Tahvonen, R., Kallio, H., Yang, B., Viitanen, M., & Linderborg, K. (2018). The effect of heat treatments and homogenisation of cows' milk on gastrointestinal symptoms, inflammation markers and postprandial lipid metabolism. *International Dairy Journal, 85*, 184–190.

Nuora, A., Tupasela, T., Tahvonen, R., Rokka, S., Marnila, P., Viitanen, M., Mäkelä, P., Pohjankukka, J., Pahikkala, T., Yang, B., Kallio, H., & Linderborg, K. (2018). Effect of homogenised and pasteurised versus native cows' milk on gastrointestinal symptoms, intestinal pressure and postprandial lipid metabolism. *International Dairy Journal, 79*, 15–23. Available from: https://doi.org/10.1016/j.idairyj.2017.11.011.

Oberli, M., Lan, A., Khodorova, N., Santé-Lhoutellier, V., Walker, F., Piedcoq, J., Davila, A.-M., Blachier, F., Tomé, D., & Fromentin, G. (2016). Compared with raw bovine meat, boiling but not grilling, barbecuing, or roasting decreases protein digestibility without any major consequences for intestinal mucosa in rats, although the daily ingestion of bovine meat induces histologic modifications in the colon. *The Journal of Nutrition, 146*(8), 1506–1513.

Oberli, M., Marsset-Baglieri, A., Airinei, G., Santé-Lhoutellier, V., Khodorova, N., Rémond, D., Foucault-Simonin, A., Piedcoq, J., Tomé, D., & Fromentin, G. (2015). High true ileal digestibility but not postprandial utilization of nitrogen from bovine meat protein in humans is moderately decreased by high-temperature, long-duration cooking. *The Journal of Nutrition, 145*(10), 2221–2228.

Ormerod, A., Ralfs, J., Jobling, S., & Gidley, M. (2002). The influence of starch swelling on the material properties of cooked potatoes. *Journal of Materials Science, 37*(8), 1667–1673.

Pafumi, Y., Lairon, D., De La Porte, P. L., Juhel, C., Storch, J., Hamosh, M., & Armand, M. (2002). Mechanisms of inhibition of triacylglycerol hydrolysis by human gastric lipase. *Journal of Biological Chemistry, 277*(31), 28070–28079.

Pallares, A. P., Loosveldt, B., Karimi, S. N., Hendrickx, M., & Grauwet, T. (2019). Effect of process-induced common bean hardness on structural properties of in vivo generated boluses and consequences for in vitro starch digestion kinetics. *British Journal of Nutrition, 122*(4), 388–399.

Pallares, A. P., Miranda, B. A., Truong, N. Q. A., Kyomugasho, C., Chigwedere, C. M., Hendrickx, M., & Grauwet, T. (2018). Process-induced cell wall permeability modulates the in vitro starch digestion kinetics of common bean cotyledon cells. *Food & Function*, *9*(12), 6544–6554.

Patel, H., Royall, P. G., Gaisford, S., Williams, G. R., Edwards, C. H., Warren, F. J., Flanagan, B. M., Ellis, P. R., & Butterworth, P. J. (2017). Structural and enzyme kinetic studies of retrograded starch: Inhibition of α-amylase and consequences for intestinal digestion of starch. *Carbohydrate Polymers*, *164*, 154–161. Available from: https://doi.org/10.1016/j.carbpol.2017.01.040.

Paz-Yépez, C., Peinado, I., Heredia, A., & Andrés, A. (2019). Influence of particle size and intestinal conditions on in vitro lipid and protein digestibility of walnuts and peanuts. *Food Research International*, *119*, 951–959.

Phosanam, A., Chandrapala, J., Huppertz, T., Adhikari, B., & Zisu, B. (2021). In vitro digestion of infant formula model systems: Influence of casein to whey protein ratio. *International Dairy Journal*, *117*, 105008.

Pletsch, E. A., & Hamaker, B. R. (2018). Brown rice compared to white rice slows gastric emptying in humans. *European Journal of Clinical Nutrition*, *72*(3), 367–373.

Ranawana, V., Leow, M. K., & Henry, C. (2014). Mastication effects on the glycaemic index: impact on variability and practical implications. *European Journal of Clinical Nutrition*, *68*(1), 137–139.

Ranawana, V., Monro, J. A., Mishra, S., & Henry, C. J. K. (2010). Degree of particle size breakdown during mastication may be a possible cause of interindividual glycemic variability. *Nutrition Research*, *30*(4), 246–254.

Rémond, D., Machebeuf, M., Yven, C., Buffière, C., Mioche, L., Mosoni, L., & Mirand, P. P. (2007). Postprandial whole-body protein metabolism after a meat meal is influenced by chewing efficiency in elderly subjects. *The American Journal of Clinical Nutrition*, *85*(5), 1286–1292. Available from: https://doi.org/10.1093/ajcn/85.5.1286.

Rogers, K. (2010). *The digestive system*. Britannica Educational Publishing.

Roman, L., Gomez, M., Li, C., Hamaker, B. R., & Martinez, M. M. (2017). Biophysical features of cereal endosperm that decrease starch digestibility. *Carbohydrate Polymers*, *165*, 180–188.

Rovalino-Córdova, A. M., Fogliano, V., & Capuano, E. (2019). The effect of cell wall encapsulation on macronutrients digestion: A case study in kidney beans. *Food Chemistry*, *286*, 557–566.

Roy, D., Ye, A., Moughan, P. J., & Singh, H. (2020a). Composition, structure, and digestive dynamics of milk from different species—A review. *Frontiers in Nutrition*, *7*, 195. Available from: https://doi.org/10.3389/fnut.2020.577759.

Roy, D., Ye, A., Moughan, P. J., & Singh, H. (2020b). Gelation of milks of different species (dairy cattle, goat, sheep, red deer, and water buffalo) using glucono-δ-lactone and pepsin. *Journal of Dairy Science*, *103*(7), 5844–5862.

Roy, D., Ye, A., Moughan, P. J., & Singh, H. (2021a). Impact of gastric coagulation on the kinetics of release of fat globules from milk of different species. *Food & Function*, *12*(4), 1783–1802.

Roy, D., Ye, A., Moughan, P. J., & Singh, H. (2021b). Structural changes in cow, goat and sheep skim milk during dynamic in vitro gastric digestion. *Journal of Dairy Science*, *104*(2), 1394–1411.

Rutherfurd, S. M., Montoya, C. A., Zou, M. L., Moughan, P. J., Drummond, L. N., & Boland, M. J. (2011). Effect of actinidin from kiwifruit (Actinidia deliciosa cv. Hayward) on the digestion of food proteins determined in the growing rat. *Food Chemistry*, *129*(4), 1681–1689.

Salentinig, S., Phan, S., Khan, J., Hawley, A., & Boyd, B. J. (2013). Formation of highly organized nanostructures during the digestion of milk. *ACS Nano*, *7*(12), 10904–10911.

Santé-Lhoutellier, V., Astruc, T., Marinova, P., Greve, E., & Gatellier, P. (2008). Effect of meat cooking on physicochemical state and in vitro digestibility of myofibrillar proteins. *Journal of Agricultural and Food Chemistry*, *56*(4), 1488–1494.

Sayd, T., Chambon, C., & Santé-Lhoutellier, V. (2016). Quantification of peptides released during in vitro digestion of cooked meat. *Food Chemistry*, *197*, 1311–1323.

Sensoy, I. (2021). A review on the food digestion in the digestive tract and the used in vitro models. *Current Research in Food Science*, 4, 308–319. Available from: https://doi.org/10.1016/j.crfs.2021.04.004.

Silvester, K. R., & Cummings, J. H. (1995). Does digestibility of meat protein help explain large bowel cancer risk? *Nutrition and Cancer*, 24(3), 279–288.

Sim, L., Quezada-Calvillo, R., Sterchi, E. E., Nichols, B. L., & Rose, D. R. (2008). Human intestinal maltase–glucoamylase: Crystal structure of the N-terminal catalytic subunit and basis of inhibition and substrate specificity. *Journal of Molecular Biology*, 375(3), 782–792. Available from: https://doi.org/10.1016/j.jmb.2007.10.069.

Simonian, H. P., Vo, L., Doma, S., Fisher, R. S., & Parkman, H. P. (2005). Regional postprandial differences in pH within the stomach and gastroesophageal junction. *Digestive Diseases and Sciences*, 50(12), 2276–2285. Available from: https://doi.org/10.1007/s10620-005-3048-0.

Singh, H., & Gallier, S. (2014). Processing of food structures in the gastrointestinal tract and physiological responses. In M. Boland, M. Golding, & H. Singh (Eds.), *Food structures, digestion and health*. San Diego: Academic Press. https://doi.org/10.1016/b978-0-12-404610-8.00002-5. Available:.

Singh, H., & Gallier, S. (2017). Nature's complex emulsion: The fat globules of milk. *Food Hydrocolloids*, 68, 81–89.

Singh, H., Ye, A. Q., & Ferrua, M. J. (2015). Aspects of food structures in the digestive tract. *Current Opinion in Food Science*, 3, 85–93. Available from: https://doi.org/10.1016/j.cofs.2015.06.007.

Singh, H., Ye, A., & Horne, D. (2009). Structuring food emulsions in the gastrointestinal tract to modify lipid digestion. *Progress in Lipid Research*, 48(2), 92–100. Available from: https://doi.org/10.1016/j.plipres.2008.12.001.

Sivakamasundari, S. K., Moses, J. A., & Anandharamakrishnan, C. (2021). Chewing cycle during mastication influences the in vitro starch digestibility of rice. *Pharma Innovation Journal*, 10, 734–738.

Smith, M. M., & Morton, D. G. (2010). *The Digestive System*. Edinburgh: Elsevier.

Somaratne, G., Ye, A., Nau, F., Ferrua, M. J., Dupont, D., Singh, R. P., & Singh, J. (2020). Role of biochemical and mechanical disintegration on β-carotene release from steamed and fried sweet potatoes during in vitro gastric digestion. *Food Research International*, 136, 109481.

Soybel, D. I. (2005). Anatomy and physiology of the stomach. *Surgical Clinics of North America*, 85(5), 875–894. Available from: https://doi.org/10.1016/j.suc.2005.05.009.

Sullivan, R. J. (2009). *Digestion and nutrition*. Infobase Publishing.

Tari, N. R., Fan, M. Z., Archbold, T., Kristo, E., Guri, A., Arranz, E., & Corredig, M. (2018). Effect of milk protein composition of a model infant formula on the physicochemical properties of in vivo gastric digestates. *Journal of Dairy Science*, 101(4), 2851–2861. Available from: https://doi.org/10.3168/jds.2017-13245.

Thorning, T. K., Bertram, H. C., Bonjour, J.-P., De Groot, L., Dupont, D., Feeney, E., Ipsen, R., Lecerf, J. M., Mackie, A., & McKinley, M. C. (2017). Whole dairy matrix or single nutrients in assessment of health effects: Current evidence and knowledge gaps. *The American Journal of Clinical Nutrition*, 105(5), 1033–1045.

Tomé, D. (2013). Digestibility issues of vegetable versus animal proteins: protein and amino acid requirements—functional aspects. *Food and Nutrition Bulletin*, 34(2), 272–274.

Tortora, G. J., & Derrickson, B. H. (2008). *Principles of Anatomy and Physiology*. John Wiley & Sons.

Tydeman, E. A., Parker, M. L., Faulks, R. M., Cross, K. L., Fillery-Travis, A., Gidley, M. J., Rich, G. T., & Waldron, K. W. (2010). Effect of carrot (Daucus carota) microstructure on carotene bioaccessibility in the upper gastrointestinal tract. 2. In vivo digestions. *Journal of Agricultural and Food Chemistry*, 58(17), 9855–9860.

Tydeman, E. A., Parker, M. L., Wickham, M. S., Rich, G. T., Faulks, R. M., Gidley, M. J., Fillery-Travis, A., & Waldron, K. W. (2010). Effect of carrot (Daucus carota) microstructure on carotene bioaccessibilty in the upper gastrointestinal tract. 1. In vitro simulations of carrot digestion. *Journal of Agricultural and Food Chemistry*, 58(17), 9847–9854.

Waldum, H. L., Hauso, O., & Fossmark, R. (2014). The regulation of gastric acid secretion—Clinical perspectives. *Acta Physiologica (Oxford, England)*, *210*(2), 239–256. Available from: https://doi.org/10.1111/apha.12208.

Wang, Z., Ichikawa, S., Kozu, H., Neves, M. A., Nakajima, M., Uemura, K., & Kobayashi, I. (2015). Direct observation and evaluation of cooked white and brown rice digestion by gastric digestion simulator provided with peristaltic function. *Food Research International*, *71*, 16–22.

Wen, S., Zhou, G., Li, L., Xu, X., Yu, X., Bai, Y., & Li, C. (2015). Effect of cooking on in vitro digestion of pork proteins: A peptidomic perspective. *Journal of Agricultural and Food Chemistry*, *63*(1), 250–261.

Wu, P., Deng, R., Wu, X., Wang, Y., Dong, Z., Dhital, S., & Chen, X. D. (2017). In vitro gastric digestion of cooked white and brown rice using a dynamic rat stomach model. *Food Chemistry*, *237*, 1065–1072.

Ye, A. (2021). Gastric colloidal behaviour of milk protein as a tool for manipulating nutrient digestion in dairy products and protein emulsions. *Food Hydrocolloids*, 106599.

Ye, A., Cui, J., Carpenter, E., Prosser, C., & Singh, H. (2019). Dynamic in vitro gastric digestion of infant formulae made with goat milk and cow milk: influence of protein composition. *International Dairy Journal*, *97*, 76–85. Available from: https://doi.org/10.1016/j.idairyj.2019.06.002.

Ye, A., Cui, J., Dalgleish, D. G., & Singh, H. (2016a). The formation and breakdown of structured clots from whole milk during gastric digestion. *Food & Function*, *7*(10), 4259–4266.

Ye, A., Cui, J., Dalgleish, D. G., & Singh, H. (2016b). Formation of a structured clot during the gastric digestion of milk: impact on the rate of protein hydrolysis. *Food Hydrocolloids*, *52*, 478–486. Available from: https://doi.org/10.1016/j.foodhyd.2015.07.023.

Ye, A., Cui, J., Dalgleish, D. G., & Singh, H. (2017). Effect of homogenization and heat treatment on the behavior of protein and fat globules during gastric digestion of milk. *Journal of Dairy Science*, *100*(1), 36–47.

Ye, A., Cui, J., & Singh, H. (2011). Proteolysis of milk fat globule membrane proteins during in vitro gastric digestion of milk. *Journal of Dairy Science*, *94*(6), 2762–2770. Available from: https://doi.org/10.3168/jds.2010-4099.

Ye, A., Liu, W., Cui, J., Kong, X., Roy, D., Kong, Y., Han, J., & Singh, H. (2019). Coagulation behaviour of milk under gastric digestion: effect of pasteurization and ultra-high temperature treatment. *Food Chemistry*, *286*, 216–225. Available from: https://doi.org/10.1016/j.foodchem.2019.02.010.

Ye, A., Roy, D., & Singh, H. (2020). Structural changes to milk protein products during gastrointestinal digestion. In M. Boland, & H. Singh (Eds.), *Milk proteins* (3rd ed.). Academic Press. https://doi.org/10.1016/B978-0-12-815251-5.00019-0. Available:.

Young, C. T., Schadel, W. E., Pattee, H. E., & Sanders, T. H. (2004). The microstructure of almond (Prunus dulcis (Mill.) DA Webb cv. 'Nonpareil') cotyledon. *LWT- Food Science and Technology*, *37*(3), 317–322.

Zadow, J. G. (2003). In B. Caballero (Ed.), *Encyclopedia of food sciences and nutrition* (2nd ed.). Oxford: Academic Press. https://doi.org/10.1016/B0-12-227055-X/01288-8. Available:.

Zebrowska, T., Low, A., & Zebrowska, H. (1983). Studies on gastric digestion of protein and carbohydrate, gastric secretion and exocrine pancreatic secretion in the growing pig. *British Journal of Nutrition*, *49*(3), 401–410.

Zhang, Z., Zhang, X., Chen, W., & Zhou, P. (2018). Conformation stability, in vitro digestibility and allergenicity of tropomyosin from shrimp (Exopalaemon modestus) as affected by high intensity ultrasound. *Food Chemistry*, *245*, 997–1009. Available from: https://doi.org/10.1016/j.foodchem.2017.11.072.

Zhang, M., Zhao, D., Zhu, S., Nian, Y., Xu, X., Zhou, G., & Li, C. (2020). Overheating induced structural changes of type I collagen and impaired the protein digestibility. *Food Research International*, *134*, 109225.

Zheng, Y., & Wang, Z. (2015). The cereal starch endosperm development and its relationship with other endosperm tissues and embryo. *Protoplasma*, *252*(1), 33–40.

Zhu, X., Kaur, L., Staincliffe, M., & Boland, M. (2018). Actinidin pretreatment and sous vide cooking of beef brisket: Effects on meat microstructure, texture and in vitro protein digestibility. *Meat Science*, *145*, 256–265.

Zou, X., He, J., Zhao, D., Zhang, M., Xie, Y., Dai, C., Wang, C., & Li, C. (2020). Structural changes and evolution of peptides during chill storage of pork. *Frontiers in Nutrition*, *7*, 151.

# CHAPTER 12

# Assessing nutritional behavior of foods through *in vitro* and *in vivo* studies

**Didier Dupont and Olivia Ménard**
*STLO, INRAE—Institut Agro, Rennes, France*

## 12.1 Introduction

The nutritional quality of food cannot be only described by the composition in nutrients, i.e. protein, lipid, carbohydrate and micronutrients. The quality highly depends on the ability of the food to be digested (digestibility) and of the nutrients to be available for use by the organs (bioavailability). The fate of food in the gastrointestinal tract is highly dependent on the way food constituents are organized and interact with each other in the food matrix. The micro and macrostructure of food have been shown to play a key role in the kinetics of food digestion and nutrient release (Fardet et al., 2019; Somaratne et al., 2020; Turgeon & Rioux, 2011). The architecture of food clearly affects the nutrients released in the GI tract (bioaccessibility) and their absorption.

Therefore, understanding the mechanisms of food disintegration in the gastrointestinal tract is of paramount importance to assess the nutritional properties of food. Although digestive physiology has been the subject of numerous studies in the past, the mechanisms of food disintegration in the digestive tract remain unclear. During digestion, chemical, enzymatic and mechanical phenomena occur simultaneously. Understanding the mechanisms linked to digestion remains an ambitious objective and the growing number of studies focusing on digestion over the past ten years bears witness to this. Studying digestion requires the use of *in vitro* or *in vivo* models. The choice of the digestion model depends on many factors especially ethical, technical or financial. The aim of this chapter is to describe the different models existing for digestion studies. In the first part, the models of *in vitro* digestion will be presented. The different stages of digestion will be detailed and the associated parameters to static models will be clarified. To move towards more complex *in vitro* systems and physiologically relevant, different dynamic mono or multi-compartmental models have been developed and will be explained. Since *in vitro* models can hardly perfectly reproduce the biological complexity of the digestive tract, the use of *in vivo* models, animal or human, remains the "gold standard" strategy to study digestion. In the second part of the chapter, *in vivo* models will be described and discussed. This synthesis highlights the great diversity of digestion models. It is important to know the advantages and the limits of these in order to set up the model answering better the objectives of the study to be conducted. Finally, in the third part of the chapter, the main *in vitro* and *ex vivo* models of absorption will be presented.

Digestion is a complex biological process that allows people to obtain nutrients from the food that are essential for their life. Foods, during their passage through the digestive tract, are exposed to mechanical and enzymatic phenomena. Chewing in the mouth, and peristaltic movements of the stomach and intestines lead to the breakdown of food by reducing its size. At the same time, enzymes in the digestive tract, from mouth to intestine, chemically alter food and convert it into nutrients. Absorption of these nutrients begins in the stomach but takes place primarily in the small intestine.

## 12.2 *In vitro* oro-gastro-intestinal digestion models

Compared with *in vivo* models, *in vitro* digestion models have certain advantages: cost reduction, time saving, better repeatability and reproducibility, and no ethical constraints. Several models mimicking the digestive tract have been developed to answer scientific research questions in different fields. Two types of *in vitro* models are used: static and dynamic models.

### 12.2.1 Static *in vitro* digestion models

Static *in vitro* digestion models are simple biochemical models mimicking the digestive physicochemical conditions. *In vitro* models can mimic three major stages of digestion: the mouth, stomach and small intestine (Fig. 12.1). These different steps can be considered alone or in combination.

Static *in vitro* digestion models consist in recreating in a series of reactors the conditions to which the food will be subjected as it passes through the digestive tract. They are widely used in order to predict the behavior of food during digestion.

At each step, the food is incubated at 37°C. It is mixed with fluids (salivary, gastric or intestinal) for a defined time. The pH is adapted for each step of the digestion and maintained at a fixed pH. Likewise, for each step, the enzymes are added in a single addition and the enzyme/substrate ratio is constant.

The digestive process can be divided into three main phases:

#### 12.2.1.1 The oral phase

This step corresponds to the physical breakdown of food during mastication and the lubrification of the food by saliva in order to form the food bolus. The saliva that is viscous due to its high content in mucins will cover the bolus enabling it to go through the oesophagus without being blocked. This phase will occur at a pH of around 7 and α-amylase is the main enzyme to be involved and will initiate the digestion of carbohydrates in the mouth and the stomach, as long as the pH remains higher than 5. A lipolytic activity has been measured in saliva by different research groups but it is commonly accepted that no lingual lipase is produced in humans compared to rats. The oral step has been shown to have a strong impact on food digestion and nutrient absorption especially when comparing elderly people with impaired dentition to normal ones (Rémond et al., 2007). The duration of this step is short and around 2 min.

#### 12.2.1.2 The gastric phase

The function of the stomach is to store food, crush it and break it down. In the upper part of the stomach (fundus), food is stored and in the lower part (antrum), food is mixed by stronger contractions with gastric enzymes, hydrochloric acid, and the pylorus opening allows the chyme to be transferred into the duodenum. Different parameters must be considered to mimic the gastric phase: the quantity and

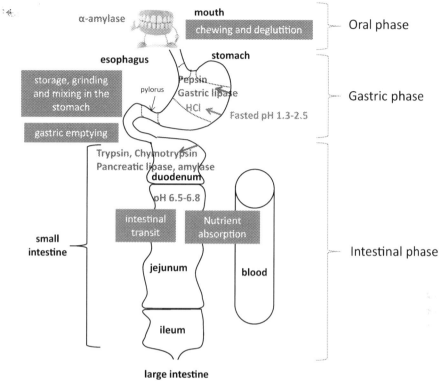

**FIG. 12.1**

The Human gastrointestinal tract and the different steps of the digestive process.

nature of the enzymes (pepsin and lipase), the pH, the mechanical force (agitation), the dilution with the gastric fluid, and the residence time. The duration of this step can vary between 30 min to 2 h depending on the macrostructure of the food (liquid/gel/solid).

### 12.2.1.3 The intestinal phase

The intestinal phase is very important since this is where nutrient absorption takes place. After emptying the gastric content into the duodenum, the pH of the chyme is neutralized by the bicarbonate and mixed with pancreatic enzymes and bile. The key parameters of the intestinal phase are: pH, level of pancreatic enzymes (proteases, amylase and lipases), bile, dilution with intestinal fluid, and residence time. Static digestion *in vitro* models have been developed and differ according to the type of food, the objective of the study, the number of reproduced steps and the physiological stage (infant, adult or elderly). They have been recently reviewed (Hur et al., 2011) and some examples are presented in Table 12.1.

A literature search shows that more than 90 different *in vitro* digestion models have been proposed by different research groups. Those models mainly differ by:

The number of steps:

**Table 12.1** Example of variability of static *in vitro* models applied to the study of digestion of dairy products.

| Objective of the study | Type of enzymes | Digestion time (min) | pH | Reference |
|---|---|---|---|---|
| Digestion of cow's/goat milk | Human gastric juice from aspirates | 30 | 2.5 | Almaas et al. (2006) |
| | Human duodenal juice | 30 | 7.5 | |
| Digestion of milk proteins and human milk | Porcine pepsin | 60 | Decrease from pH=6.5 to pH=2 in 60 min | Chatterton et al. (2004) |
| | Or infant gastric juice | 60 | | |
| β-lactoglobulin digestion | Porcine pepsin | 120 | 2.5 | Mandalari et al. (2009) |
| | Porcine trypsin and chymotrypsin and bile salts | 15 | 6.5 | |
| Milk protein digestion (β-lactoglobulin) (β-casein) | Porcine pepsin | 60 | pH=3 (infant model) pH=2.5 (adult model) | Dupont et al. (2019) |
| | Porcine trypsin, chymotrypsin and bile salts | 30 | pH=6.5 | |

Depending on the type of matrix studied, static *in vitro* models include one, two or three steps. For liquid type matrices the oral phase is generally neglected. On the other hand, most static *in vitro* digestion models include two successive phases (gastric then intestinal). This succession of steps does not reflect the physiological reality because these phenomena take place simultaneously *in vivo* (part of the meal has already been transferred to the small intestine while the other part is still in the stomach).

The duration of the steps:

The oral phase is very quick, less than 2 min and is strongly related to the type of food ingested. The gastric phase lasts from 30 min to 2 h depending on the studies. Regarding the intestinal phase, it also varies between 30 min and 6 h and depends in particular on the part of the intestine that needs to be mimicked (duodenum, or the whole small intestine).

The fluids and enzymes used:

The fluids, mimicking saliva, gastric and intestinal secretions, can contain a varied mineral composition (Oomen et al., 2003; Versantvoort et al., 2005), or just sodium chloride (Dupont et al., 2010). They aim to simulate the ionic strength of the secretions. The pH is adjusted according to the digestive phase considered. The dilution of the meal during its passage through the digestive tract generally respects these meal/salivary fluid/gastric fluid/intestinal fluid proportions: 2/1/2/3 (v/v) and has been well described by Versantvoort et al. (2005).

The most frequently used enzymes in *in vitro* digestion systems are pepsin, chymotrypsin, trypsin, lipase and amylase, but bile salts are also added. These enzymes and bile salts are most often of animal origin, especially porcine, rabbit or even bovine. Regarding gastric lipase of animal origin, it is only commercialized by a French SME (Lipolytech) and other sources of fungal origin have also been considered but exhibit different specificities. The enzymes are added in purified form or, in the case of intestinal digestion, in the form of an extract of animal pancreatic secretions containing all the digestive enzymes involved in this digestion step. Some studies use enzymes of human origin after the collection of gastric or intestinal juices in adults (Almaas et al., 2006), or infants (Armand et al., 1996; Chatterton et al., 2004). In addition to adding enzymes under physiological conditions in *in vitro* digestion systems, it is essential to characterize and measure their activity under digestive conditions since factors such as concentration, temperature, pH, presence of inhibitors or activators, and incubation time influence enzyme activities.

Different static digestion models have been developed, with adapted and different parameters, which makes the results difficult to compare between studies. For example, depending on the origin of the digestive enzymes used (i.e., porcine, human, fungal), the specificity and activity of the digestive enzymes will be different. Also, differences in pH, ionic strength, nature of minerals added to fluids, and digestion length have been observed between the different models. These variations in the parameters of the *in vitro* digestion system will directly impact enzymes activities, and consequently, the digestion result. To overcome the problem, namely the impossibility of comparing the results between different studies and the need to harmonize a digestion protocol that the entire scientific community can use, the network of European researchers INFOGEST, whose objective is to bring together a community of scientists in the field of digestion, has established a consensus around a static digestion protocol (Minekus et al., 2014). Since then, the model has been extensively used to assess food digestibility, nutrient bioaccessibility, food matrix effect, allergen persistence in the GI tract, etc. The model is now used all around the world and is about to be recognized as a reference method by International Organization for Standardization (ISO) and International Dairy Federation (IDF).

Although not fully physiologically relevant (i.e., constant pH, absence of transit or emptying between the different compartments, absence of absorption of nutrients, etc.), the static *in vitro* digestion models have many advantages. They are simple and inexpensive tools. They are technically easy to set up and require little material. Recently, the advantages and the limits of static *in vitro* digestion models have been reviewed by experts from the INFOGEST network (Bohn et al., 2018). Static digestion models can be used as a pre-screening method, when a large number of tests need to be performed or before moving to more complex systems. Since they simplify the digestive process, they can also allow unravelling mechanisms that occur at a molecular scale. For instance, phospholipids such as phosphatidylcholine released by the gastric mucosa have been shown to interact with globular proteins like β-lactoglobulin to harden its structure and make it more resistant to the action of pepsin (Mandalari et al., 2009). Finally, static *in vitro* digestion models can also be relevant to estimate end-point values such as the glycaemic index, protein digestibility, among others.

### 12.2.2 Dynamic *in vitro* digestion models

Dynamic *in vitro* digestion models, unlike static models, take the evolution during digestion of biochemical and mechanical phenomena that take place in the digestive tract into account. These artificial systems are physiologically closer to human physiology. The flow of secretions and the gastric and intestinal transit time are considered. The evolution of acidification in the gastric phase, the flow of enzymes and secretions, and the gastric and intestinal emptying are mimicked and regulated.

#### 12.2.2.1 Gastric and intestinal emptying modelling

In order to control the emptying of food in the stomach and small intestine, an exponential power equation is used:

$$f = 2^{-(t/t_{1/2})\beta} \tag{12.1}$$

where $f$ represents the fraction of the meal remaining in the stomach (or intestine), $t$ the time (in minutes) after the food ingestion, $t_{1/2}$ emptied half-time, where 50% of the meal is remaining in the stomach and the 50% left emptied and $\beta$ the coefficient describing the shape of the curve as described previously (Elashoff et al., 1982). For each type of food, the emptying half-time and the coefficient $\beta$ must be defined.

### 12.2.2.2 Gastric acidification modelling

In the stomach, the production of hydrochloric acid by the gastric mucosa contributes to acidify the chyme and leads to the chemical breakdown of the food. The low pH of the gastric phase optimizes enzyme activity because pepsin has an optimum activity at pH 2 (DiPalma et al., 1991) and gastric lipase at pH 5.5 (Ville et al., 2002). This acidification is gradual and is highly dependent on the buffering capacity of the food.

The gastric acidification data presented in Fig. 12.2 were obtained in an *in vivo* study on adults (Malagelada et al., 1976) and several *in vitro* studies (Blanquet et al., 2004; Kong & Singh, 2010; Lvova et al., 2012; Minekus et al., 1995; Pitino et al., 2012; Souliman et al., 2006). Fig. 12.2 underlines the importance of the initial pH of the meal and its buffering capacity on the kinetics of gastric acidification. Overall, two hours after ingestion the gastric pH is around 2.

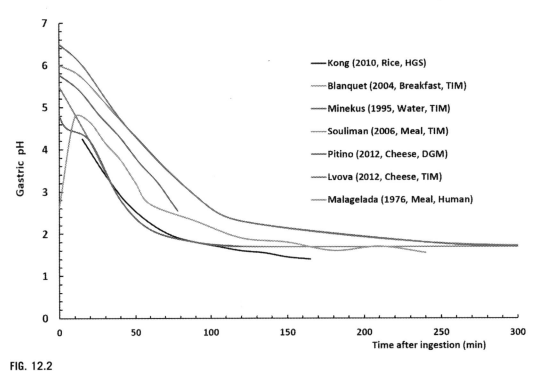

**FIG. 12.2**

Gastric acidification kinetics.

### 12.2.2.3 Dynamic models

Different dynamic *in vitro* digestion systems have been developed. Most of them remain commercially unavailable or very expensive. The different systems are mono or multi-compartmentalized and mimic one or more steps of the digestion process. A review published by the INFOGEST network (Dupont et al., 2019) summarized the ability of *in vitro* models to simulate *in vivo* data.

#### Dynamic stomach models

**The Dynamic Gastric Model.** The Dynamic Gastric Model (DGM) has been developed at the Institute of Food Research (now known as the Quadram Institute) by Mercuri et al. (2011). This gastric model consists of two stages. The first represents the upper part of the stomach (fundus or body of the stomach) and the second the lower part (antrum). The upper part, which represents the body of the stomach, reproduces the mixture of food with gastric secretions. The lower part simulates the shear forces allowing the crushing of food and gastric emptying. Appropriate software controls the regulation of pH, the flow of gastric secretions and gastric emptying. This model has been used to investigate the breakdown of agar gel beads of various fracture strengths in high and low-viscosity meals and compared to *in vivo* data collected on human volunteers (Vardakou et al., 2011). The Bioneer company in Denmark (https://bioneer.dk/product-services/drug-development/) provides services on this system.

**The Human Gastric Simulator.** This model has been developed by Kong and Singh (2010). It was specially designed to mimic the shear forces applied to food in the stomach. Stomach contractions are reproduced by rollers compressing the stomach and simulate the amplitude, frequency and intensity of the gastric contractions. The more intense contractions at the pylorus level allows the emptying of the chyme towards the intestine. This system also simulates gastric acidification and the flow of secretions. This system is reliable to study the food breakdown in the stomach. The model has been validated, by comparison with stomach contraction forces obtained *in vivo*, to simulate stomach shear forces (Kong & Singh, 2010). It has been shown to reproduce the gastric emptying and pH curve in comparison with *in vivo* data on pigs fed with white rice (Bornhorst et al., 2014).

#### Dynamic gastrointestinal models

Gastrointestinal models incorporate, at least, two stages of digestion, the gastric phase where food is crushed and undergoes chemical transformation and the intestinal phase where most of the chemical digestion and absorption of nutrients takes place.

**The TIM model.** This model was developed by Minekus et al. (1995) from TNO in the Netherlands and is considered one of the gastrointestinal simulator references. It is now commercialized by the TIM company in The Netherlands (http://www.tno-pharma.com/TIM_en.html).

TIM-1 is the most complete gastrointestinal model currently available to mimic digestive conditions. This model includes the stomach and all three parts of the small intestine (duodenum, jejunum, and ileum). It allows to mimic the essential parameters of digestion, namely: mixing of chyme, gastric and intestinal emptying, gastric acidification and neutralization of pH in the intestinal phase, sequential addition of digestive secretions and absorption of water and digestive products in the intestinal parts by a dialysis system. This model has been validated by comparison with *in vivo* data.

**The DiDGi system.** This gastrointestinal simulator has been developed by INRAE (Ménard et al., 2014) and is a simple, inexpensive system that can be adapted to different types of food. It was constructed to monitor the disintegration and hydrolysis kinetics of food during simulated digestion. The system reproduces the upper part of the digestive tract, i.e., stomach and intestine. To make it

**FIG. 12.3**

DiDGI dynamic gastrointestinal digestion system.

physiologically relevant, the software monitors the digestive parameters such as gastric and intestinal emptying, the kinetics of gastric acidification, the flow of secretions and the mixing in the two compartments (Fig. 12.3).

For the validation of the tool, the digestion of an infant formula, adapted to the nutritional needs of the piglet, was studied both *in vitro*, using the DiDGI system, and *in vivo*, on piglets. The kinetics of protein digestion, obtained *in vitro* or *in vivo*, were monitored and compared (Ménard et al., 2014) (Fig. 12.4).

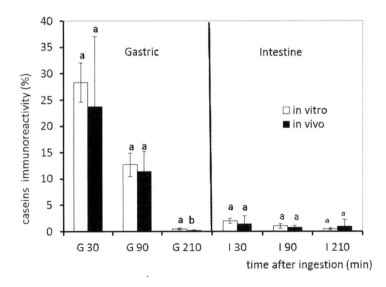

**FIG. 12.4**

Comparison of the % of intact caseins (A) and β-lactoglobulin (B) after 30 min, 90 min and 210 min of digestion in the gastric compartment (G) and intestine (I) between *in vitro* (*black* bars) and *in vivo* (*white* bars).

The digestion kinetics of the two major proteins of milk, caseins and β-lactoglobulin, obtained *in vitro* or *in vivo*, were similar. The validation of this simulator is an essential step and is mandatory to ensure the physiological relevance of the tool. However, validating a model for the digestion of a specific food does not mean that the model is valid for all of them and this validation step needs to be done on, at least, families of food structures (liquid/gel/solid) using representative foods.

Colonic models

The large intestine with the gut microbiota residing there is a complex ecosystem that plays a role in human digestion. To simulate this step, *in vitro* systems recreate the intestinal flora to mimic the colonic part of the human digestive system (Durand et al., 1997). The objective is to maintain a complex microbial population over a long period (several weeks) by adding a nutrient medium simulating the nutritional and biochemical conditions of ileal chyme and at the same time removing an equivalent amount of the fermentation medium. Several colonic models have been proposed recently by different groups (Buckley et al., 2021; Hoshi et al., 2021; Ryan et al., 2021).

**The Simulated Human Intestinal Microbial Ecosystem**. The Simulated Human Intestinal Microbial Ecosystem Model (SHIME) was developed by Molly et al. (Molly et al., 1993). It is by far one of the most well-known digestion models and covers the whole digestive tract. It is now commercialized by the ProDigest company (https://www.prodigest.eu/en/technology/shime-and-m-shime) that also provides services using the system.

This complete digestion system has six compartments and mimics the digestive tract from the stomach to the colon. The colonic part is mimicked by three reactors in series. The entire system is maintained anaerobically and the temperature and pH are computer-controlled.

**The ARCOL system**. ARCOL (Artificial colon) is a one-stage fermentation model that reproduces the colonic environment of humans or animals. This model has been developed by the University of Clermont Auvergne (Clermont-Ferrand, France). One of the advantages of the model is that it allows the maintaining of anaerobiosis inside the fermenter by the sole metabolic activity of the microbiota and not by flushing with $N_2$ or $CO_2$, as usually done in other colonic *in vitro* models. Up to date, ARCOL has been used to reproduce the colon of humans (Blanquet-Diot et al., 2012; Cordonnier et al., 2015; Thévenot et al., 2013, 2015), and pre-ruminant calves (Gerard-Champod et al., 2010).

ARCOL has been validated towards *in vivo* data regarding the composition of the colonic microbiota (main bacterial populations followed by qPCR or plating), its metabolic activity [production of major end products of fermentation, such as short chain fatty acids (SCFA)] and/or the composition of the nutritive medium used to feed the fermenter (Cordonnier et al., 2015; Gerard-Champod et al., 2010; Thévenot et al., 2015). Furthermore, the relevance of the ARCOL model for probiotic studies was also shown as the survival of probiotic yeasts and their influence on SCFA production obtained *in vitro* corroborate the available data in human adult volunteers (Blanquet-Diot et al., 2012; Cordonnier et al., 2015; Thévenot et al., 2015).

## 12.3 Absorption models
### 12.3.1 Cellular models
#### *12.3.1.1 Caco-2 cell monolayer*

The human epithelial cell line Caco-2 has been widely used as a model of the intestinal epithelial barrier. The Caco-2 cell line is originally derived from a colon carcinoma. However, one of its most advantageous properties is its ability to spontaneously differentiate into a monolayer of cells with many properties typical of absorptive enterocytes with brush border layer as found in the small intestine. Towards confluence, Caco-2 cells start to polarize acquiring a characteristic apical brush border with microvilli. Tight junctions are formed between adjacent cells, and they express enzyme activities typical of enterocytes, i.e. lactase, aminopeptidase N, sucrase-isomaltase and dipeptidylpeptidase IV. Several studies have compared Caco-2 permeability coefficients with absorption data in humans and found a high correlation, particularly if the compounds are transported by passive paracellular transport mechanisms (Sun et al., 2008). In comparison with normal intestinal epithelium the Caco-2 cell model has several limitations. First of all, the normal epithelium contains more than one cell type, not only enterocytes, and secondly, when using the Caco-2 cell model, no mucus and unstirred water layer are present.

### *12.3.1.2 Caco-2-HT29-MTX co-cultures*
Even though the Caco-2 cells have been widely employed to measure drug and nutrient transport, this model has been criticized because the permeability of the Caco-2 monolayer to hydrophilic compounds, generally transported by paracellular mechanisms, are poor because of the relatively tight junction that is characteristic of these cells (Artursson et al., 2001). A pure Caco-2 cell model also has an overexpression of P-glycoprotein which may lead to higher secretion rates and consequently lower permeability in the absorptive direction (Anderle et al., 1998). The HT29MTX cell line has less expression of tight junctions. Thus, for instance, the ability of mannitol to penetrate tight junctions in the HT29 monolayer is 50-fold higher than that of Caco-2 monolayers (Wikman et al., 1993). The permeability of a cell layer resulting from co-cultivation between Caco-2 and HT29 are more in resemblance with that of the normal intestine. The permeability of the Caco-2/HT29 co-culture model was correlated with fractions absorbed in humans for selected drugs, and it was found relatively good correlations (Walter et al., 1996).

## 12.3.2 Ussing chambers
The ussing chamber (UC) consists of two halves separated by an epithelium (sheet of intestinal mucosa or monolayer of epithelial cells grown on permeable supports). Epithelia are polar in nature, i.e., they have an apical or mucosal side and a basolateral or serosal side. An UC can isolate the apical side from the basolateral side. The two half chambers are filled with equal amounts of symmetrical Ringer solution to remove chemical, mechanical or electrical driving forces. Ion transport takes place across any epithelium. Transport may be in either direction. This technique has been extensively used to study drug absorption, and several reviews have shown a satisfactory correlation with the absorption rates measured on human volunteers (Herrmann & Turner, 2016; Lomasney & Hyland, 2013), particularly with respect to passively transported compounds (Fagerholm et al., 1999). UC thus may allow investigation into the absorption of several bioactive compounds such as dietary peptides that cannot easily be done *in vivo*. Indeed, the presence of plasma proteases that further hydrolyze peptides reaching the bloodstream and of blood proteins make the untargeted identification of peptides difficult. Tissue from various animals may be used depending on the availability and purpose of the research. For human nutrition, pigs are considered the best non-primate model, since they have a large single-stomach and a comparable gut physiology to humans (Merrifield et al., 2011). However, a literature search revealed only one study (Awati et al., 2009) using the UC model to investigate the permeation of amino

acids from whey proteins. Recently, we published the first paper investigating the passage of whey protein-derived peptides through an intestinal epithelium (proximal jejunum of a 7-week-old piglet) by using mass spectrometry for peptide identification (Ozorio et al., 2020). We observed that the brush border peptidase activity was intense in the first 10 min of the experiment, producing several new peptides in the apical compartment. Then, 286 peptides originating from β-lactoglobulin (60%) and casein (20%) were shown to pass through the intestinal epithelium as they were detected in the basolateral compartment after 120 min of incubation. Nevertheless, only a small percentage of a peptide (from 0.6% to 3.35%) is able to cross the epithelial barrier. This study was a pioneer work for demonstrating transepithelial jejunum absorption of whey oligopeptides in an *ex vivo* model. It also confirmed the proteolytic activity of brush border enzymes on these oligopeptides, giving birth to a myriad of new bioactive peptides available for absorption.

### 12.3.3 Organoids

Organoids are three-dimensional cellular structures that have similar architectures and functionality to the organs from which they are derived. They are obtained from stem cells through a self-organization process, promoted by a culture medium containing the appropriate growth and differentiation factors. Developed in the 2010s, these methodologies currently make it possible to obtain organoids from several organs: intestine, retina, mammary glands, liver, kidney, lung and even brain.

Organoids make it possible to mimic what happens in the whole organ and their applications are numerous: for example, to study diet-microbiome-host interactions (Rubert et al., 2020), to decipher physiological processes (Capeling et al., 2020), or to investigate intestinal pathologies (Angus et al., 2020). Intestinal organoids open a new world of possibilities for GI culture methods, by manipulation of these cells. The great potential of these now self-sustaining 'mini-guts' in the field of transplantation and regenerative medicine is evident.

## 12.4 *In vivo* models
### 12.4.1 Animal models

Several animal species have increased scientific knowledge in the field of nutrition, in particular rats and pigs, which are the most frequently used. Numerous studies have been performed on rats for assessing food intake (Opara et al., 1996), muscle protein synthesis (Dardevet et al., 2000), and the effect of food constituents on gut microbiota interaction (Coklo et al., 2020) or food-related pathologies (Rodriguez-Correa et al., 2020). The rat is a small omnivorous and monogastric animal and has a digestive tract whose general organization is similar to that of humans: the small intestine represents 81% of the total intestine in humans and 83% in rats (DeSesso & Jacobson, 2001). Anatomical differences should be noted, however, the rat stomach has two parts, one for bacterial digestion and the other for acid and enzymatic secretions. The rat continuously produces bile in the intestine. Its large intestine is mainly represented by a large cecum. In addition, it is a rodent and does not have the same eating habits as humans because it breaks up its meals throughout the day. This animal model has the advantage of being inexpensive. After euthanasia of the animals, samples can be taken from the various compartments of the digestive tract, but small quantities of effluents are recovered. Usually, these studies require the sacrifice of a large number of animals.

The animal model of choice currently used for nutritional studies is the pig, whose digestive tract physiology is similar to humans in terms of transit and digestive efficiency (Guilloteau et al., 2010). It can ingest a wide variety of foods. Under surgery, the cannulas are placed at different levels of the digestive tract and catheters allow to sample effluents and blood. These cannulas and catheters allow the study of food hydrolysis and absorption. Nutrient absorption is evaluated by blood samples characterization. Unlike rats, cannulated pigs are not euthanized following ingestion of the food and can be used in several studies. Due to its small size, weight stability and ease of handling, the mini-pig is now increasingly used for nutritional studies. As with conventional pigs, catheters and cannulas can be implanted at different levels of the digestive tract. This model was used in a study to compare the kinetics of protein digestion according to the structure of the food (Barbé et al., 2013, 2014). The piglets are also used as a model for the digestion of the newborn. They have been used to evaluate the gastric emptying, the gastric acidification of meals based on milk and/or soy proteins (Moughan et al., 1992), but also to study the kinetics of infant formula digestion (Lemaire et al., 2021).

The pig model requires the use of qualified personnel, particularly for surgical procedures and for monitoring animals, and the cost of studies remains higher than for studies carried out in rats. These models also present strong constraints in terms of ethics. Depending on the objectives of the study, the animal model to be used should be considered with regard to the constraints and advantages associated with each animal model. Animal studies are technically cumbersome to conduct, variability between individuals is a major drawback and a large number of animals must be included to overcome this variability.

### 12.4.2 Human subjects

Human digestion: direct or indirect methods are used.

#### 12.4.2.1 Direct methods

The digestive contents should be collected by aspiration of the chyme using a nasogastric or nasal intestinal tube. This method is invasive because a tube is inserted into the nose of the subject, healthy volunteer or patient to the desired digestive target. The evaluation of stomach pH, or residual stomach volume can be done through access to gastric contents during digestion. Access to ileal contents can also be achieved in ileostomized patients, for whom the intestine has undergone a bypass and the intestinal effluents are then collected in an external bag. However, it should be noted that these patients suffer from inflammatory digestive pathologies, but studies have reported a level of absorption similar to that of healthy subjects.

The kinetic samples after ingestion allow to evaluate the hydrolysis of macronutrients. Depending on the composition of the food, the amount of intact proteins or residual triglycerides are measured at different post-meal times but also the associated hydrolysis products (peptides, amino acids, or free fatty acids, mono or diglycerides) (Armand et al., 1996; Gaudichon et al., 1995; Roman et al., 2007). The use of a non-absorbable marker allows to dissociate the food part from the secretory part, the dilution of the meal by the secretions can thus be evaluated, as well as the gastric emptying.

Digestibility can be measured either in the ileum or in the feces. It is estimated from the amount of proteins, sugars or lipids found in the ileum or in the feces compared to the amount present in the food before ingestion. For proteins, the use of labelling techniques is necessary to differentiate dietary proteins from endogenous proteins.

To evaluate gastric emptying, scintigraphy is the most frequently imaging technique used to observe the gastric chyme and evaluate the gastric emptying. This method, described by Griffith et al. (1966), involves labelling the food with a radioisotope, such as Technicium 99, which binds to the solid phase of the food. By gamma ray camera, radioactivity is detected and measured. The use of radioactive markers limits these studies. Scintigraphy is also used in pigs to assess gastric emptying depending on the nature of the food ingested (Ménard et al., 2018). By modelling the data obtained, the fraction emptied over time can be plotted.

#### 12.4.2.2 Indirect methods

To measure bioavailability, i.e., the amount of nutrients digested and absorbed blood samples are withdrawn from subjects at different postprandial times. The kinetics of digestion by measuring amino acids, triglycerides and glucose in the blood plasma are evaluated. The difficulty lies in the differentiation and evaluation of exogenous (food origin) and endogenous nutrients (Deglaire et al., 2008).

Access to effluents by gastrointestinal tubes in humans remains the "gold standard" method for nutrition studies, but these invasive investigations remain rare due to major technical and ethical constraints. The rat and the pig are the animal models most commonly used today, but they become more and more difficult to use due to ethical concerns. The high individual variability observed during *in vivo* studies requires a large number of subjects (animals or humans) and makes the study cumbersome to conduct. The choice of the *in vivo* model depends on the research question asked and on the feasibility. It is essential to understand the pros and the cons of the different models in order to consider the best option.

## 12.5 Conclusion

For understanding the effect of food on human health it is critical to better understand the fate of food in the gastrointestinal tract and identify the nutrients and bioactive molecules that are released in the lumen and absorbed. In order to do so, a large range of *in vitro* digestion and absorption models are available. The choice of the most relevant model to use must be driven by the research question that needs to be solved. For comparing foods in identical conditions or following the evolution of compounds in digestive conditions, *in vitro* models appear relevant. Their low cost and simplicity to perform, make them a good alternative to *in vivo* models. However, when a physiological effect of food on the host is being investigated, nothing can really replace *in vivo* models. Nevertheless, in the near future, new *in vitro* models more and more sophisticated will be available to replace *in vivo* studies. Models including gastric anatomical structures and motility, particularly the gastric emptying process and pylorus sieving effect have a great potential to replace human studies (Peng et al., 2021). Similarly, gut-on-a-chip systems covering partly or totally the different compartments of the gastrointestinal tract have a strong potential for studies where high throughput will be needed (Ashammakhi et al., 2020; Puschhof et al., 2021; Xiang et al., 2020).

## References

Almaas, H., Cases, A. L., Devold, T. G., Holm, H., Langsrud, T., Aabakken, L., Aadnoey, T., & Vegarud, G. E. (2006). In vitro digestion of bovine and caprine milk by human gastric and duodenal enzymes. *International Dairy Journal*, *16*(9), 961–968. https://doi.org/10.1016/j.idairyj.2005.10.029.

Anderle, P., Niederer, E., Rubas, W., Hilgendorf, C., Spahn-Langguth, H., Wunderli-Allenspach, H., Merkle, H. P., & Langguth, P. (1998). P-glycoprotein (P-gp) mediated efflux in Caco-2 cell monolayers: The influence of culturing conditions and drug exposure on P-gp expression levels. *Journal of Pharmaceutical Sciences*, *87*(6), 757–762. https://doi.org/10.1021/js970372e.

Angus, H. C. K., Butt, A. G., Schultz, M., & Kemp, R. A. (2020). Intestinal organoids as a tool for inflammatory bowel disease research. *Frontiers in Medicine*, *6*. https://doi.org/10.3389/fmed.2019.00334.

Armand, M., Hamosh, M., Mehta, N. R., Angelus, P. A., Philpott, J. R., Henderson, T. R., Dwyer, N. K., Lairon, D., & Hamosh, P. (1996). Effect of human milk or formula on gastric function and fat digestion in the premature infant. *Pediatric Research*, *40*(3), 429–437. https://doi.org/10.1203/00006450-199609000-00011.

Artursson, P., Palm, K., & Luthman, K. (2001). Caco-2 monolayers in experimental and theoretical predictions of drug transport. *Advanced Drug Delivery Reviews*, *46*(1–3), 27–43. https://doi.org/10.1016/S0169-409X(00)00128-9.

Ashammakhi, N., Nasiri, R., Barros, N. R., Tebon, P., Thakor, J., Goudie, M., Shamloo, A., Martin, M. G., & Khademhosseni, A. (2020). Gut-on-a-chip: Current progress and future opportunities. *Biomaterials*, *255*. https://doi.org/10.1016/j.biomaterials.2020.120196.

Awati, A., Rutherfurd, S. M., Plugge, W., Reynolds, G. W., Marrant, H., Kies, A. K., & Moughan, P. J. (2009). Ussing chamber results for amino acid absorption of protein hydrolysates in porcine jejunum must be corrected for endogenous protein. *Journal of the Science of Food and Agriculture*, *89*(11), 1857–1861. https://doi.org/10.1002/jsfa.3662.

Barbé, F., Ménard, O., Gouar, Y. L., Buffière, C., Famelart, M. H., Laroche, B., Feunteun, S. L., Rémond, D., & Dupont, D. (2014). Acid and rennet gels exhibit strong differences in the kinetics of milk protein digestion and amino acid bioavailability. *Food Chemistry*, *143*, 1–8. https://doi.org/10.1016/j.foodchem.2013.07.100.

Barbé, F., Ménard, O., Le Gouar, Y., Buffière, C., Famelart, M. H., Laroche, B., Le Feunteun, S., Dupont, D., & Rémond, D. (2013). The heat treatment and the gelation are strong determinants of the kinetics of milk proteins digestion and of the peripheral availability of amino acids. *Food Chemistry*, *136*(3–4), 1203–1212. https://doi.org/10.1016/j.foodchem.2012.09.022.

Blanquet, S., Zeijdner, E., Beyssac, E., Meunier, J. P., Denis, S., Havenaar, R., & Alric, M. (2004). A dynamic artificial gastrointestinal system for studying the behavior of orally administered drug dosage forms under various physiological conditions. *Pharmaceutical Research*, *21*(4), 585–591. https://doi.org/10.1023/B:PHAM.0000022404.70478.4b.

Blanquet-Diot, S., Denis, S., Chalancon, S., Chaira, F., Cardot, J. M., & Alric, M. (2012). Use of artificial digestive systems to investigate the biopharmaceutical factors influencing the survival of probiotic yeast during gastrointestinal transit in humans. *Pharmaceutical Research*, *29*(6), 1444–1453. https://doi.org/10.1007/s11095-011-0620-5.

Bohn, T., Carriere, F., Day, L., Deglaire, A., Egger, L., Freitas, D., Golding, M., Le Feunteun, S., Macierzanka, A., Menard, O., Miralles, B., Moscovici, A., Portmann, R., Recio, I., Rémond, D., Santé-Lhoutelier, V., Wooster, T. J., Lesmes, U., Mackie, A. R., & Dupont, D. (2018). Correlation between in vitro and in vivo data on food digestion. What can we predict with static in vitro digestion models? *Critical Reviews in Food Science and Nutrition*, 2239–2261. https://doi.org/10.1080/10408398.2017.1315362.

Bornhorst, G. M., Rutherfurd, S. M., Roman, M. J., Burri, B. J., Moughan, P. J., & Singh, R. P. (2014). Gastric pH distribution and mixing of soft and rigid food particles in the stomach using a dual-marker technique. *Food Biophysics*, *9*(3), 292–300. https://doi.org/10.1007/s11483-014-9354-3.

Buckley, A. M., Moura, I. B., Arai, N., Spittal, W., Clark, E., Nishida, Y., Harris, H. C., Bentley, K., Davis, G., Wang, D., Mitra, S., Higashiyama, T., & Wilcox, M. H. (2021). Trehalose-induced remodelling of the human microbiota affects clostridioides difficile infection outcome in an in vitro colonic model: A pilot study. *Frontiers in Cellular and Infection Microbiology, 11*. https://doi.org/10.3389/fcimb.2021.670935.

Capeling, M., Huang, S., Mulero-Russe, A., Cieza, R., Tsai, Y. H., Garcia, A., & Hill, D. R. (2020). Generation of small intestinal organoids for experimental intestinal physiology. In *Vol. 159. Methods in cell biology* (pp. 143–174). Academic Press Inc. https://doi.org/10.1016/bs.mcb.2020.03.007.

Chatterton, D. E. W., Rasmussen, J. T., Heegaard, C. W., Sørensen, E. S., & Petersen, T. E. (2004). In vitro digestion of novel milk protein ingredients for use in infant formulas: Research on biological functions. *Trends in Food Science and Technology, 15*(7–8), 373–383. https://doi.org/10.1016/j.tifs.2003.12.004.

Coklo, M., Maslov, D., & Kraljevic Pavelic, S. (2020). Modulation of gut microbiota in healthy rats after exposure to nutritional supplements. *Gut Microbes, 12*.

Cordonnier, C., Thévenot, J., Etienne-Mesmin, L., Denis, S., Alric, M., Livrelli, V., & Blanquet-Diot, S. (2015). Dynamic in vitro models of the human gastrointestinal tract as relevant tools to assess the survival of probiotic strains and their interactions with gut microbiota. *Microorganisms, 3*(4), 725–745. https://doi.org/10.3390/microorganisms3040725.

Dardevet, D., Sornet, C., Balage, M., & Grizard, J. (2000). Stimulation of in vitro rat muscle protein synthesis by leucine decreases with age. *Journal of Nutrition, 130*(11), 2630–2635. https://doi.org/10.1093/jn/130.11.2630.

Deglaire, A., Moughan, P. J., Bos, C., Petzke, K., Rutherfurd, S. M., & Tomé, D. (2008). A casein hydrolysate does not enhance gut endogenous protein flows compared with intact casein when fed to growing rats. *Journal of Nutrition, 138*(3), 556–561. https://doi.org/10.1093/jn/138.3.556.

DeSesso, J. M., & Jacobson, C. F. (2001). Anatomical and physiological parameters affecting gastrointestinal absorption in humans and rats. *Food and Chemical Toxicology, 39*(3), 209–228. https://doi.org/10.1016/S0278-6915(00)00136-8.

DiPalma, J., Kirk, C. L., Hamosh, M., Colon, A. R., Benjamin, S. B., & Hamosh, P. (1991). Lipase and pepsin activity in the gastric mucosa of infants, children, and adults. *Gastroenterology, 101*(1), 116–121. https://doi.org/10.1016/0016-5085(91)90467-Y.

Dupont, D., Alric, M., Blanquet-Diot, S., Bornhorst, G., Cueva, C., Deglaire, A., Denis, S., Ferrua, M., Havenaar, R., Lelieveld, J., Mackie, A. R., Marzorati, M., Menard, O., Minekus, M., Miralles, B., Recio, I., & Van den Abbeele, P. (2019). Can dynamic in vitro digestion systems mimic the physiological reality? *Critical Reviews in Food Science and Nutrition, 59*(10), 1546–1562. https://doi.org/10.1080/10408398.2017.1421900.

Dupont, D., Mandalari, G., Molle, D., Jardin, J., Léonil, J., Faulks, R. M., Wickham, M. S. J., Clare Mills, E. N., & Mackie, A. R. (2010). Comparative resistance of food proteins to adult and infant in vitro digestion models. *Molecular Nutrition & Food Research, 54*(6), 767–780. https://doi.org/10.1002/mnfr.200900142.

Durand, M., Beaumatin, P., Bulman, B., Bernalier, A., Grivet, J. P., Serezat, M., Gramet, G., & Lahaye, M. (1997). Fermentation of green alga sea-lettuce (Ulva sp) and metabolism of its sulphate by human colonic microbiota in a semi-continuous culture system. *Reproduction Nutrition Development, 37*(3), 267–283. https://doi.org/10.1051/rnd:19970303.

Elashoff, J. D., Reedy, T. J., & Meyer, J. H. (1982). Analysis of gastric emptying data. *Gastroenterology, 83*(6), 1306–1312. https://doi.org/10.1016/S0016-5085(82)80145-5.

Fagerholm, U., Nilsson, D., Knutson, L., & Lennernäs, H. (1999). Jejunal permeability in humans in vivo and rats in situ: Investigation of molecular size selectivity and solvent drag. *Acta Physiologica Scandinavica, 165*(3), 315–324. https://doi.org/10.1046/j.1365-201X.1999.00510.x.

Fardet, A., Dupont, D., Rioux, L.-E., & Turgeon, S. L. (2019). Influence of food structure on dairy protein, lipid and calcium bioavailability: A narrative review of evidence. *Critical Reviews in Food Science and Nutrition*, 1987–2010. https://doi.org/10.1080/10408398.2018.1435503.

Gaudichon, C., Mahé, S., Roos, N., Benamouzig, R., Luengo, C., Huneau, J.-F., Sick, H., Bouley, C., Rautureau, J., & Tome, D. (1995). Exogenous and endogenous nitrogen flow rates and level of protein hydrolysis in the human jejunum after [15N]milk and [15N]yoghurt ingestion. *British Journal of Nutrition*, *74*(2), 251–260. https://doi.org/10.1079/bjn19950128.

Gerard-Champod, M., Blanquet-Diot, S., Cardot, J.-M., Bravo, D., & Alric, M. (2010). Development and validation of a continuous in vitro system reproducing some biotic and abiotic factors of the veal calf intestine. *Applied and Environmental Microbiology*, *76*(16), 5592–5600. https://doi.org/10.1128/AEM.00524-10.

Griffith, G. H., Owen, G. M., Kirkman, S., & Shields, R. (1966). Measurement of rate of gastric emptying using chromium-51. *Lancet*, *287*(7449), 1244–1245. https://doi.org/10.1016/s0140-6736(66)90247-9.

Guilloteau, P., Zabielski, R., Hammon, H. M., & Metges, C. C. (2010). Nutritional programming of gastrointestinal tract development. Is the pig a good model for man? *Nutrition Research Reviews*, *23*(1), 4–22. https://doi.org/10.1017/S0954422410000077.

Herrmann, J. R., & Turner, J. R. (2016). Beyond Ussing's chambers: Contemporary thoughts on integration of transepithelial transport. *American Journal of Physiology - Cell Physiology*, *310*(6), C423–C431. https://doi.org/10.1152/ajpcell.00348.2015.

Hoshi, N., Inoue, J., Sasaki, D., & Sasaki, K. (2021). The Kobe University human intestinal microbiota model for gut intervention studies. *Applied Microbiology and Biotechnology*, *105*(7), 2625–2632. https://doi.org/10.1007/s00253-021-11217-x.

Hur, S. J., Lim, B. O., Decker, E. A., & McClements, D. J. (2011). In vitro human digestion models for food applications. *Food Chemistry*, *125*(1), 1–12. https://doi.org/10.1016/j.foodchem.2010.08.036.

Kong, F., & Singh, R. P. (2010). A human gastric simulator (HGS) to study food digestion in human stomach. *Journal of Food Science*, *75*(9), E627–E635. https://doi.org/10.1111/j.1750-3841.2010.01856.x.

Lemaire, M., Ménard, O., Cahu, A., Nogret, I., Briard-Bion, V., Cudennec, B., Cuinet, I., Le Ruyet, P., Baudry, C., Dupont, D., Blat, S., Deglaire, A., & Le Huërou-Luron, I. (2021). Addition of dairy lipids and probiotic lactobacillus fermentum in infant formulas modulates proteolysis and lipolysis with moderate consequences on gut physiology and metabolism in Yucatan piglets. *Frontiers in Nutrition*, *8*. https://doi.org/10.3389/fnut.2021.615248.

Lomasney, K. W., & Hyland, N. P. (2013). The application of Ussing chambers for determining the impact of microbes and probiotics on intestinal ion transport. *Canadian Journal of Physiology and Pharmacology*, *91*(9), 663–670. https://doi.org/10.1139/cjpp-2013-0027.

Lvova, L., Denis, S., Barra, A., Mielle, P., Salles, C., Vergoignan, C., Di Natale, C., Paolesse, R., Temple-Boyer, P., & Feron, G. (2012). Salt release monitoring with specific sensors in in vitro oral and digestive environments from soft cheeses. *Talanta*, *97*, 171–180. https://doi.org/10.1016/j.talanta.2012.04.013.

Malagelada, J.-R., Longstreth, G. F., Summerskill, W. H. J., & Go, V. L. W. (1976). Measurement of gastric functions during digestion of ordinary solid meals in man. *Gastroenterology*, *70*(2), 203–210. https://doi.org/10.1016/S0016-5085(76)80010-8.

Mandalari, G., Mackie, A. M., Rigby, N. M., Wickham, M. S. J., & Mills, E. N. C. (2009). Physiological phosphatidylcholine protects bovine β-lactoglobulin from simulated gastrointestinal proteolysis. *Molecular Nutrition & Food Research*, *53*(1), S131–S139. https://doi.org/10.1002/mnfr.200800321.

Ménard, O., Cattenoz, T., Guillemin, H., Souchon, I., Deglaire, A., Dupont, D., & Picque, D. (2014). Validation of a new in vitro dynamic system to simulate infant digestion. *Food Chemistry*, *145*, 1039–1045. https://doi.org/10.1016/j.foodchem.2013.09.036.

Ménard, O., Famelart, M. H., Deglaire, A., Le Gouar, Y., Guérin, S., Malbert, C. H., & Dupont, D. (2018). Gastric emptying and dynamic in vitro digestion of drinkable yogurts: Effect of viscosity and composition. *Nutrients*, *10*(9). https://doi.org/10.3390/nu10091308.

Mercuri, A., Passalacqua, A., Wickham, M. S. J., Faulks, R. M., Craig, D. Q. M., & Barker, S. A. (2011). The effect of composition and gastric conditions on the self-emulsification process of ibuprofen-loaded self-emulsifying drug delivery systems: A microscopic and dynamic gastric model study. *Pharmaceutical Research*, 28(7), 1540–1551. https://doi.org/10.1007/s11095-011-0387-8.

Merrifield, C. A., Lewis, M., Claus, S. P., Beckonert, O. P., Dumas, M. E., Duncker, S., Kochhar, S., Rezzi, S., Lindon, J. C., Bailey, M., Holmes, E., & Nicholson, J. K. (2011). A metabolic system-wide characterisation of the pig: A model for human physiology. *Molecular BioSystems*, 7(9), 2577–2588. https://doi.org/10.1039/c1mb05023k.

Minekus, M., Alminger, M., Alvito, P., Ballance, S., Bohn, T., Bourlieu, C., Carrière, F., Boutrou, R., Corredig, M., Dupont, D., Dufour, C., Egger, L., Golding, M., Karakaya, S., Kirkhus, B., Le Feunteun, S., Lesmes, U., Macierzanka, A., Mackie, A., & Brodkorb, A. (2014). A standardised static in vitro digestion method suitable for food—An international consensus. *Food & Function*, 5(6), 1113–1124. https://doi.org/10.1039/C3FO60702J.

Minekus, M., Marteau, P., Havenaar, R., & Veld, J. H. (1995). A multicompartmental dynamic computer-controlled model simulating the stomach and small intestine. *Alternatives to Laboratory Animals*, 23(2), 197–209. https://doi.org/10.1177/026119299502300205.

Molly, K., Vande Woestyne, M., & Verstraete, W. (1993). Development of a 5-step multi-chamber reactor as a simulation of the human intestinal microbial ecosystem. *Applied Microbiology and Biotechnology*, 39(2), 254–258. https://doi.org/10.1007/BF00228615.

Moughan, P. J., Birtles, M. J., Cranwell, P. D., Smith, W. C., & Pedraza, M. (1992). The piglet as a model animal for studying aspects of digestion and absorption in milk-fed human infants. *World Review of Nutrition and Dietetics*, 67, 40–113.

Oomen, A. G., Rompelberg, C. J. M., Bruil, M. A., Dobbe, C. J. G., Pereboom, D. P. K. H., & Sips, A. J. (2003). Development of an in vitro digestion model for estimating the bioaccessibility of soil contaminants. *Archives of Environmental Contamination and Toxicology*, 44(3), 281–287. https://doi.org/10.1007/s00244-002-1278-0.

Opara, E. I., Meguid, M. M., Zhong-Jin, Y., & Hammond, W. G. (1996). Studies on the regulation of food intake using rat total parenteral nutrition as a model. *Neuroscience and Biobehavioral Reviews*, 20(3), 413–443. https://doi.org/10.1016/0149-7634(95)00027-5.

Ozorio, L., Mellinger-Silva, C., Cabral, L. M. C., Jardin, J., Boudry, G., & Dupont, D. (2020). The influence of peptidases in intestinal brush border membranes on the absorption of oligopeptides from whey protein hydrolysate: An ex vivo study using an Ussing chamber. *Foods*, 9(10). https://doi.org/10.3390/foods9101415.

Peng, Z., Wu, P., Wang, J., Dupont, D., Menard, O., Jeantet, R., & Chen, X. D. (2021). Achieving realistic gastric emptying curve in an advanced dynamic in vitro human digestion system: experiences with cheese—A difficult to empty material. *Food & Function*, 12(9), 3965–3977. https://doi.org/10.1039/d0fo03364b.

Pitino, I., Randazzo, C. L., Cross, K. L., Parker, M. L., Bisignano, C., Wickham, M. S. J., Mandalari, G., & Caggia, C. (2012). Survival of Lactobacillus rhamnosus strains inoculated in cheese matrix during simulated human digestion. *Food Microbiology*, 31(1), 57–63. https://doi.org/10.1016/j.fm.2012.02.013.

Puschhof, J., Pleguezuelos-Manzano, C., & Clevers, H. (2021). Organoids and organs-on-chips: Insights into human gut-microbe interactions. *Cell Host & Microbe*, 29(6), 867–878. https://doi.org/10.1016/j.chom.2021.04.002.

Rémond, D., Machebeuf, M., Yven, C., Buffière, C., Mioche, L., Mosoni, L., & Mirand, P. P. (2007). Postprandial whole-body protein metabolism after a meat meal is influenced by chewing efficiency in elderly subjects. *American Journal of Clinical Nutrition*, 85(5), 1286–1292. https://doi.org/10.1093/ajcn/85.5.1286.

Rodriguez-Correa, E., Gonzalez-Perez, I., Clavel-Perez, P., Contreras-Vargas, Y., & Carvajal, K. (2020). Biochemical and nutritional overview of diet-induced metabolic syndrome models in rats: What is the best choice? *Nutrition & Diabetes*, 10.

Roman, C., Carriere, F., Villeneuve, P., Pina, M., Millet, V., Simeoni, U., & Sarles, J. (2007). Quantitative and qualitative study of gastric lipolysis in premature infants: Do MCT-enriched infant formulas improve fat digestion? *Pediatric Research*, 61(1), 83–88. https://doi.org/10.1203/01.pdr.0000250199.24107.fb.

Rubert, J., Schweiger, P. J., Mattivi, F., Tuohy, K., Jensen, K. B., & Lunardi, A. (2020). Intestinal organoids: A tool for modelling diet-microbiome-host interactions. *Trends in Endocrinology and Metabolism*, *31*(11), 848–858. https://doi.org/10.1016/j.tem.2020.02.004.

Ryan, J. J., Monteagudo-Mera, A., Contractor, N., & Gibson, G. R. (2021). Impact of 2′-fucosyllactose on gut microbiota composition in adults with chronic gastrointestinal conditions: Batch culture fermentation model and pilot clinical trial findings. *Nutrients*, *13*(3), 1–16. https://doi.org/10.3390/nu13030938.

Somaratne, G., Ferrua, M. J., Ye, A., Nau, F., Floury, J., Dupont, D., & Singh, J. (2020). Food material properties as determining factors in nutrient release during human gastric digestion: A review. *Critical Reviews in Food Science and Nutrition*, *60*(22), 3753–3769. https://doi.org/10.1080/10408398.2019.1707770.

Souliman, S., Blanquet, S., Beyssac, E., & Cardot, J. M. (2006). A level A in vitro/in vivo correlation in fasted and fed states using different methods: Applied to solid immediate release oral dosage form. *European Journal of Pharmaceutical Sciences*, *27*(1), 72–79. https://doi.org/10.1016/j.ejps.2005.08.006.

Sun, H., Chow, E. C., Liu, S., Du, Y., & Pang, K. S. (2008). The Caco-2 cell monolayer: Usefulness and limitations. *Expert Opinion on Drug Metabolism & Toxicology*, *4*(4), 395–411. https://doi.org/10.1517/17425255.4.4.395.

Thévenot, J., Cordonnier, C., Rougeron, A., Le Goff, O., Nguyen, H. T. T., Denis, S., Alric, M., Livrelli, V., & Blanquet-Diot, S. (2015). Enterohemorrhagic Escherichia coli infection has donor-dependent effect on human gut microbiota and may be antagonized by probiotic yeast during interaction with Peyer's patches. *Applied Microbiology and Biotechnology*, *99*(21), 9097–9110. https://doi.org/10.1007/s00253-015-6704-0.

Thévenot, J., Etienne-Mesmin, L., Denis, S., Chalancon, S., Alric, M., Livrelli, V., & Blanquet-Diot, S. (2013). Enterohemorrhagic Escherichia coli O157:H7 survival in an in vitro model of the human large intestine and interactions with probiotic yeasts and resident microbiota. *Applied and Environmental Microbiology*, *79*(3), 1058–1064. https://doi.org/10.1128/aem.03303-12.

Turgeon, S. L., & Rioux, L. E. (2011). Food matrix impact on macronutrients nutritional properties. *Food Hydrocolloids*, *25*(8), 1915–1924. https://doi.org/10.1016/j.foodhyd.2011.02.026.

Vardakou, M., Mercuri, A., Barker, S. A., Craig, D. Q. M., Faulks, R. M., & Wickham, M. S. J. (2011). Achieving antral grinding forces in biorelevant in vitro models: Comparing the USP dissolution apparatus II and the dynamic gastric model with human in vivo data. *AAPS PharmSciTech*, *12*(2), 620–626. https://doi.org/10.1208/s12249-011-9616-z.

Versantvoort, C. H. M., Oomen, A. G., Van De Kamp, E., Rompelberg, C. J. M., & Sips, A. J. A. M. (2005). Applicability of an in vitro digestion model in assessing the bioaccessibility of mycotoxins from food. *Food and Chemical Toxicology*, *43*(1), 31–40. https://doi.org/10.1016/j.fct.2004.08.007.

Ville, E., Carrière, F., Renou, C., & Laugier, R. (2002). Physiological study of pH stability and sensitivity to pepsin of human gastric lipase. *Digestion*, *65*(2), 73–81. https://doi.org/10.1159/000057708.

Walter, E., Janich, S., Roessler, B. J., Hilfinger, J. M., & Amidon, G. L. (1996). HT29-MTX/Caco-2 cocultures as an in vitro model for the intestinal epithelium: In vitro-in vivo correlation with permeability data from rats and humans. *Journal of Pharmaceutical Sciences*, *85*(10), 1070–1076. https://doi.org/10.1021/js960110x.

Wikman, A., Karlsson, J., Carlstedt, I., & Artursson, P. (1993). A drug absorption model based on the mucus layer producing human intestinal goblet cell line HT29-H. *Pharmaceutical Research: An Official Journal of the American Association of Pharmaceutical Scientists*, *10*(6), 843–852. https://doi.org/10.1023/A:1018905109971.

Xiang, Y., Wen, H., Yu, Y., Li, M., Fu, X., & Huang, S. (2020). Gut-on-chip: Recreating human intestine in vitro. *Journal of Tissue Engineering*, *11*. https://doi.org/10.1177/2041731420965318.

# CHAPTER 13

# Application of artificial neural networks (ANN) for predicting the effect of processing on the digestibility of foods

L.A. Espinosa Sandoval[a,b], A.M. Polanía Rivera[a,c], L. Castañeda Florez[c], and A. García Figueroa[c]

[a]*GIPAB Group, School of Food Engineering, Faculty of Engineering, Universidad del Valle, Cali, Valle del Cauca, Colombia,* [b]*GIFTEX Group, Theoretical and Experimental Physical-Chemistry Interdisciplinary Group, Universidad Industrial de Santander, Piedecuesta, Santander, Colombia,* [c]*School of Food Engineering, Faculty of Engineering, Universidad del Valle, Tuluá, Valle del Cauca, Colombia*

## 13.1 Introduction

Due to consumers' demand for food products of better nutritional quality and their interest in their health and wellness benefits, predictive models that employ artificial intelligence are a useful tool to achieve these objectives. Artificial neural networks (ANN) are computational systems that mimic neurological behavior; the models established through this methodology are easily adaptable and with a high degree of accuracy. These networks have been used in different fields such as medicine, where they have been used as clinical decision support systems; computing, used for classification, grouping, pattern recognition, anomaly detection, and pharmaceuticals employed to the identification of new drugs, laboratory synthesis, animal studies or clinical phases. In recent years, research has been conducted on the potential of ANN as models that allow finding the best combination of parameters involved in an experimental design and lead to the prediction of results (response variables of interest). Also allow to optimize food processes, product development, shelf-life estimation, as well as for the evaluation of digestibility and physiological response associated with the consumption of food products.

Therefore, this chapter will discuss the different applications of ANN for optimizing different food processing operations, such as extraction of bioactive compounds, drying, extrusion, filtration, shelf-life estimation, rheology, fermentation and encapsulation, showing the benefits of the application of these models. It also describes the models used to simulate the *in vitro* gastrointestinal process, such as static, semi-dynamic and dynamic models, being those the experimental basis for the application of ANN that allows the rapid evaluation of ingredients and food products based on their digestibility and the prediction of physiological responses generated by their ingestion, such as the glycemic index and satiety sensation.

## 13.2 Artificial intelligence in food processing

The term artificial intelligence (AI) refers to the artificial creation of intelligence equivalent to human intelligence, i.e., that is capable of reasoning, planning and processing natural language (Sebastin, 2018). Through it, it is possible to develop computer systems that perform tasks that would traditionally require human intelligence to be performed. AI was defined by John McCarthy as "the science and engineering of making intelligent machines, especially intelligent computer programs" (Singh, 2019). Models developed from AI learn from data and are applied by individuals, organizations or companies to make predictions. Currently, machine learning models are being developed to reduce the complexity and variety of data presented in the food industry (Chidinma-Mary-Agbai., 2020).

Due to the demands of consumers in terms of quality and product development, it has generated the deployment of tools that use AI to achieve better predictions and thus meet the needs of customers (Guiné, 2019). One of these tools has been the use of ANN. These networks are a set of statistical methods and nonlinear discrimination that have predictive capabilities and have been widely analyzed. ANN are computational systems that mimic biological neurons. Methods based on ANN use computer programs that are very useful because they have advantages such as: adaptability, model independence, ease of use and high accuracy (Song et al., 2017; Torkashvand et al., 2017).

An ANN performs a numerical estimation that manages to simulate the learning and memorization processes. Moreover, it is a tool that works based on the experimental input variables and finds its operating rules in the factors (Lai et al., 2016). To better understand ANN it is necessary to know three concepts, the first is "*neuron*", which can be defined as the basic computational unit of the network, second "*architecture*", which represents the structure, i.e., how the neurons are connected and finally "*learning*", the process that manages to adapt the network to perform a task or calculate a desired function (Fig. 13.1). "The structure of an ANN comprises an input layer, one or more hidden layers and an output layer of neurons". The coefficients used to establish the relevance of the different inputs represent the connection weights of the ANN, and it is possible to establish various equations to determine the relative importance as a function of the magnitude of the weights (Giwa et al., 2016; Zhang & Zimba, 2017). These networks use known data, such as previously solved exercises, thus recognizing complex patterns between inputs and outputs, and applying that knowledge to new input data. With prior training, ANN can predict results with excellent accuracy compared to conventional regression

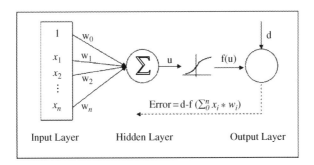

**FIG. 13.1**

General layout of an artificial neural network (ANN).

or classification analysis. These networks have been successfully employed in fields such as computing, medicine, food, product development, among others to predict response variables optimizing processes and using less experimental trials to obtain trustworthy results (Disse et al., 2018).

### 13.2.1 Artificial neural network application in food processing operations

Bhagya Raj and Dash (2020b) posited that ANN is a simplified model of the biological nervous system consisting of nerve cells or neurons and can be applied to food process engineering. They have been used in food quality and safety analysis, food image analysis, and modeling of various thermal and nonthermal food processing operations by means of simple non-linear relationships and predict responses without knowing the mechanism involved in the process. Each neural network uses data by assigning connection weights representing input lines to neurons in the hidden layer and weights assigned to neurons in the hidden layer to the output (Espinosa-Sandoval et al., 2020). This is of great importance in predicting the output data. Different ANN in different unit operations in food processing are presented based on theoretical developments related to adaptability, machine learning, classification and prediction. An ANN gives fast responses which makes it suitable for its application. Due to the easy adaptability of these methods, they have been successfully used in numerous investigations to model or predict processes in the food industry, food properties, food engineering or quality control. Applications in different unit operations of food processing such as extraction, drying, extrusion, and filtration, are presented below.

#### 13.2.1.1 Extraction techniques

Traditionally, the optimization of some extraction processes have used statistical tools such as response surface methodology (RSM) (Mohamed & Mohamed, 2015). During the last two decades ANN have proven to be an effective methodology for the modeling of these processes and their optimization, by using in nonlinear equations (Alara et al., 2018).

In order to determine the optimal conditions for the extraction process of total phenolic compounds (TPC) in mangosteen (*Garcinia mangostana* L.) peel powder, two modeling methodologies were compared: response surface methodology (RSM) and ANN (Cheok et al., 2012). When analyzing the results of both tools, it was possible to observe that a better coefficient of correlation ($R^2$) was obtained with ANN (0.945) compared to RSM (0.897) evidencing a good fit of the model to the experimental data; likewise, the average absolute deviation values (ADD) were 4.01% in ANN versus 5.37% for RSM. These data indicate that the ANN technique is a better modeling tool for both data fitting and estimation capabilities.

Microwave assisted extraction (MAE) was employed by Sinha et al. (2013) for the extraction of bioactive compounds from *Bixa orellana* (*annatto*) seeds and was modeled by RSM and ANN to predict the responses. The input parameters or input neurons used in the process included pH, extraction time, and amount of *annatto* seeds. The results showed that the solvent pH of the input neurons and extraction time showed a significant effect (positive effect) on pigment yield. This extraction study revealed that ANN had a higher $R^2$ (0.95) and lower Root Mean Square Error (RMSE) and, therefore, ANN had a higher predictive ability than the RSM model. In order to model and optimize the microwave-assisted extraction of total phenolic content in choke berries (*Aronia melanocarpa*), the RSM and ANN methodologies were used (Simić et al., 2016) to evaluate the effect of power, ethanol concentration and processing time on the Total Phenolic Capacity (TPC). According to the statistical indicators evaluated ($R^2$, RMSE and mean absolute error), it was demonstrated that the ANN tool provide better adjustment; in addition, the conditions yielded by ANN provided the highest yield for the TPC and could be used for scaling up, which were: 300 W of power, 53.6% ethanol and 5 min, obtaining a yield of TPC expressed as galic acid equivalent 420.1 mg GAE/100 g of fresh plant material. In the study carried out by Alara et al. (2018), the results of the RSM and ANN methodologies were compared in the optimization of the extraction yield and antioxidant capacity of extracts obtained by MAE in leaf samples of *Vernonia cinerea*. In order to evaluate the two methodologies, the parameters of coefficient of correlation ($R^2$), root mean square error (RMSE) and mean absolute deviation (MAD) were established to conclude which was better. The ANN methodology presented a higher prediction potential compared to RSM because the predicted model is able to predict the experimental data with good fit parameters, for example, higher $R^2$ 0.9912, 0.9928 and 0.9944 were obtained for extraction yield and antioxidant activities by DPPH and ABTS, respectively. Therefore, this methodology could be a better alternative in data fitting by MAE of antioxidant compounds obtained from this substrate.

The Supercritical fluid extraction (SCFE) is a process that has been modeled in various aspects such as heat transfer phenomena-based models (HTPBM's), mass transfer phenomenon-based models (MTPBM's) and empirical models (EM's). SCFE allows the extraction and fraction of fats, oils, essences, spices, pigments, essential oils (Suryawanshi & Mohanty, 2018). In recent years, ANN has been applied to this process, due to the high sensitivity of the mathematical models to small variations in the process variables. ANN have started to be used as predictive methodologies to optimize the operating parameters in the SCFE process applied to different substrates (Khajeh et al., 2012; Lashkarbolooki et al., 2013). The first reports of ANN applied to SFE systems were made by Ghoreishi and Heidari (2013) and Pilkington et al. (2014), who compared this methodology with RSM in extracts obtained from green tea and Artemisinin from *Artemisia annua* respectively, observing a high accuracy in ANN. This methodology has arisen the interest of researchers due to its high reliability between input and output data of complex systems as well as nonlinear ones. The extraction of essential oils from *Diplotaenia cachrydifolia* by SFE was predicted using a three-layer ANN model (Khajeh et al., 2012). A multilayer feed-forward neural network, with a backpropagation learning algorithm, was implemented to model the extraction process. Training data were used to calculate the network parameters; validation data were used to ensure the robustness of the network parameters and test data were used to evaluate the predictive ability of the generated model. A network with five hidden neurons was ideal for predicting the extraction yield of these oils with great accurately. Suryawanshi and Mohanty (2018) developed a three-layer ANN model with the objective of optimizing the cumulative extraction yield (CEY) of *Argemone mexicana* (L.) (*A. Mexicana*) seed oil; the objective of this network was to simulate the oil extraction process by means of SFE, using supercritical $CO_2$ as the solvent. According to the ANN models evaluated, those with the best performance were selected and an equation was established that could be used to predict the CEY. The most suitable network was the one developed with the Levenberg-Marquardt algorithm, and which included six neurons in the hidden layer. The selected model was ANN-FFBP (Feed Forward Back-Propagation) [5-6-1] since it allowed the CEY to be adequately predicted, making this model a viable tool for the prediction of yields in this type of substrates.

Pavlić et al. (2020) applied SFE on raspberry seed samples to obtain oil, and aiming to optimize the process conditions they used kinetic modeling and ANN in the extraction stage. In the case of ANN, the initial mass transfer rate data determined from the experimental data were used. The results of mass transfer in the extraction curves were adjusted to an ANN model, using the $R^2$ parameter to check the fit, obtaining a value of 0.917, likewise, a low error term ($2.890 \times 10^{-5}$) was obtained, and the accuracy of the model was corroborated with the high goodness-of-fit tests: Square Sum Error - SSE (0.001) and AARD (0.014). It is possible to conclude that ANN are an excellent tool to establish the optimum conditions that can provide a higher initial mass transfer rate, ideal in this type of process.

Ultrasound-assisted extraction method (UAE) is another method widely employed for extraction of biocompounds. This method was applied by Bhagya Raj and Dash (2020a) to recover compounds from dragon fruit peel and were used ANN and genetic algorithm (GA) (algorithms inspired from the biological evolutionary process) to model and optimize the process (Katoch et al., 2021). The ANN model was adequately adjusted to the experimental data and the model output was applied to perform the optimization together with the genetic algorithm. In this way, the optimized conditions to achieve maximum yield were established, which were: liquid/solid ratio of 25:1 mL/g, 60% of solvent concentration, 60 °C temperature and 20 min of extraction time. In the study conducted by Kashyap et al. (2020), they used UAE to obtain polyphenols from *Meghalayan cherry*, the authors used RSM and ANN methodologies to evaluate the effect of solvent to solid ratio, solvent concentration, extraction time and amplitude on the TPC of the extracts. The results evidenced that these methodologies are adequate to describe the experimental data, being the ANN more accurate in the predictions because it presents higher $R^2$ and lower MSE. Therefore, ANN is considered as a viable tool to predict the extraction efficiency of this type of compounds.

### *13.2.1.2 Drying*
The objective of applying ANN modeling in drying is to produce a dry product of desired quality at minimum cost and maximum yield. ANN has been implemented to model the various drying processes such as tray drying, vacuum drying, microwave drying, osmotic dehydration, among others, drying by analyzing the effect of drying process parameters on product quality, process efficiency and for modelling and prediction in energy-engineering systems (Guiné, 2019).

Azadbakht et al. (2017) used a fluidized bed dryer to dehydrate potato samples, they used the ANN methodology to model the energy and exergy parameters during the process. They concluded that the proposed model was effective in simulating the thermodynamics of the process. Khazaei et al. (2013) employed applied machine vision and ANN for modeling the grape drying process predicting moisture content in real time. The best ANN was obtained with three layers: one hidden layer with 5 nodes, one output layer with 1 node and 4 nodes (units).

Motavali et al. (2013) employed vacuum microwave drying to dehydrate sour cherry samples by evaluating different powers. In order to predict the moisture content and drying rate as a function of the input variables, the researchers developed an ANN using the backward propagation algorithm, showing an excellent correlation between the predicted and experimental data, as well as $R^2$ values close to 1, thus proving that ANN are a reliable alternative due to simplicity and generality.

Bhagya Raj and Dash (2020a) applied feed-forward back-propagation ANN to model the vacuum microwave drying process in dragon fruit (*Hylocereus undatus*) samples. This model allowed obtaining the optimum conditions of the process, and the predicted values coincided with the experimental data, presenting a low deviation (1.227–2.936) %. Rasooli Sharabiani et al. (2021) applied two drying methods to dehydrate apple samples [convective (CD) and microwave (MD)], employed ANN to evaluate the performance of the methods with respect to moisture ratio (MR). The results showed that ANN presented a good fit since the $R^2$ obtained were 0.9993 and 0.9991 in CD and MD, respectively. In the study conducted by (Sarkar et al., 2020) different drying methods for pineapple slices were evaluated, namely hot air oven, microwave, microwave convective and freeze drying, using ANN to model the methods analyzed. The correlation coefficients obtained for all models were close to 1, concluding that ANN adequately predict the dehydration kinetics of the processes evaluated; however, the freeze-dried product presented better reconstitution properties.

ANN were also used to evaluate whether they were feasible to predict the osmotic dehydration properties of courgette rings (Mokhtarian & Garmakhany, 2017). The input variables were immersion time, temperature, solution concentration and treatment, and the output variables were water loss and solids gain. The best activation function for the ANN was tanh with 46 neurons in the first and second hidden layers, this network was able to adequately predict water loss and solids gain with $R^2$ values of 0.985 and 0.938, respectively. (Prakash Maran et al., 2013) also used ANN to model the mass transfer process during osmotic dehydration of papaya (*Carica papaya L.*) samples, compared to the RSM method. According to the statistical parameters used RMSE, MAE, $\chi^2$, $R^2$ and others; it was possible to conclude that the ANN model was more accurate in prediction as compared to RSM model.

### 13.2.1.3 Filtration

One of the advantages of neural networks is that they can learn the complex transport processes of a system from observed inputs and outputs, being a suitable tool to approximate functions without the need to use mathematical models, since this methodology can learn from examples and recognize patterns in a series of input and output data. Therefore, it is possible to find ANN applications in filtration processes (Karimi et al., 2012).

Permeate flux of red plum juice during membrane processing was modeled and predicted using ANN (Nourbakhsh et al., 2014). A multilayer direct feed neural network was designed with the input parameters: transmembrane pressure (TMP), feed temperature, feed cross-flow velocity, membrane pore size, and processing time for filtration. The ANN developed adequately models the dynamic behavior of the permeate flow with the parameters evaluated during juice clarification.

Ibrahim et al. (2020) employed ANN for the prediction of palm oil mill effluent permeate flux (POME) in membrane bioreactor (MBR) filtration also using RSM. The developed network correlated the output (permeate flux) with four input variables (air flow rate, transmembrane pressure, permeate pump and aeration pump). The results obtained showed that the combination of RSM and ANN was ideal for predicting permeate fluxes in the MBR system.

### 13.2.1.4 Extrusion

Extrusion technologies play an essential role in the food industry to produce nutritious food varieties; applying ANN in these processes has aroused the interest of researchers as it would allow predicting the behavior of the variables involved in the process as well as the characteristics of the final product. (Fan et al., 2013) used ANN to predict texture characteristics from the external color of rice flour mixed with extruded cheese product. To model the extrusion process, a two-hidden-layer ANN model with 10 and 30 neurons in each hidden layer was used. The model presented a good fit, which was corroborated by the correlation coefficient values obtained for hardness and gumminess (0.967 and 0.986). (Cubeddu et al., 2014) also used ANN to configure and monitor the extrusion process. The objective was to optimize the process variables (water content and screw speed) considering the desired characteristics of the final product (outputs). They obtained a maximum error of 4.1% for screw speed and 1.6% for water content; however, trying to regulate high perturbations in water content may result in a divergence of the calculation. Therefore, the authors suggest that future work should include a larger number of training patterns, including a wider range of water content.

### 13.2.1.5 Shelf-life estimation

The sensorial quality, as well as the physicochemical characteristics of a product depend on many factors, the main ones being the raw materials with which they are elaborated, the processing, their microbiological state, as well as the packaging. During storage, it is possible to observe some important changes in color, flavor, aroma or consistency, depending on whether the environmental conditions are altered, directly influencing the reactions that may be generated in the product. For these reasons, it is relevant to have models that allow to predict the shelf life of products and to have a much more accurate estimation (Zhi et al., 2018). Some predictive models developed to evaluate shelf life are often expensive and complex, including electronic detectors for rapid food quality diagnosis (Gómez et al., 2008). However, in recent years, more accurate, efficient and faster mathematical models have been developed and applied, such as the Weibull model, used to estimate shelf life in pickles (Keklik et al., 2017), or the Q10 model, which has been used to estimate shelf life in products such as frozen shrimp, *Pangasius* fillets, refrigerated pork, tomato sauce, juices, among others (Dong et al., 2015; Mai & Huynh, 2017; Nguyen et al., 2016). Multiple linear regression (MLR) is another predictive model of great help in predicting shelf life in foods, its main advantage being its ability to show relationships between variables, as well as its ease of use and applicability (Weiss et al., 2018). ANN have also been used for shelf-life prediction in food, some of the advantages being fault tolerance, ability to learn from examples, real-time operation and forecasting of nonlinear data (Carvalho et al., 2013). ANN have also been applied for this purpose due to their reliability and accuracy. ANN have been used to define shelf-life for predicting food quality, e.g., Stangierski et al. (2019) compared multiple linear regression (MLR) and ANN models to predict shelf life at different temperatures in Gouda cheese. When analyzing the results, they concluded that the ANN models (multilayer perceptron type) were better for the three temperatures evaluated, since they presented higher correlation coefficients $R^2 = 0.99$, 0.96 and 0.96 with respect to the MRL model $R^2 = 0.99$, 0.87 and 0.87; in addition, the RMSE values were lower for ANN, showing greater precision. The curves of predicted values of general desirability using the MLR model and the ANN model had a similar behavior. However, a better fit was observed for the ANN model. Finally, although both models could be used for the prediction of shelf life in this product, ANN presented a better fit according to the statistical parameters considered. In addition, Sánchez-González and Oblitas-Cruz (2017) evaluated the predictive ability of ANN on the shelf life and acidity of vacuum-packed fresh cheese. Samples were stored at intervals of two to four days at three different temperatures and constant relative humidity. During this period, acidity (AC) and sensory acceptability were determined, which was used to determine shelf life time (SLT) through the modified Weibull sensory risk method. For the neural network, temperature, failure possibility and maturation time were used as inputs, and AC and SLT were used as outputs. The selected networks were those with the lowest mean square error (MSE) and the best $R^2$. Compared to the regression models, ANN presented good accuracy and higher $R^2$ for both SLT and AC, 0.996 and 0.6897, respectively. According to these results, it is possible to employ ANN to adequately model the shelf life in this product.

X. Liu et al. (2015) compared Arrhenius models and ANN to predict quality changes in rainbow trout fillets at different temperatures, evaluating total aerobic counts (TAC), electrical conductivity (EC), K-value and sensory evaluation (SA), correlating these variables with product shelf life. The Arrhenius model was viable for TAC and EC variables but presented poor results for SA and K-value. On the other hand, ANN was effective in predicting changes in all the variables analyzed in the range of temperatures evaluated (270–285 K); moreover, it presented relative errors below 10%. This indicates that it is a viable tool for modeling these quality variables that affect the shelf life of this product.

Other applications of ANN are presented in Table 13.1 (Bhagya Raj & Dash, 2020c).

Table 13.1 ANN applications in different unit operations in food processing such as rheology, fermentation, and encapsulation.

| Process | Description | Type of ANN | N° of neurons, inputs and outputs | Independent variables | Depend variables | Results | References |
|---|---|---|---|---|---|---|---|
| Rheology | Two ANN models used to model honey rheological behavior | Genetic algorithm—ANN (GA-ANN) and adaptive neuro-fuzzy inference system (ANFIS) | Model of 3 input neurons, 1 output neuron and 11 neurons in the hidden layer (GA-ANN) and 3 membership functions of Gaussian type for all input variables and linear for the output (ANFIS) | Porosity, moisture content and formulation components | Honey viscosity (1.7–270.48 Pa s) | Both GA-ANN and ANFIS models predictions agreed well with testing data sets and could be useful for understanding and controlling factors affecting viscosity of honey | Ramzi et al. (2015) |
| | Prediction of nanofluid viscosity using a multilayer perceptron artificial neural network (MLP-ANN) | Multilayer perceptual ANN having two hidden layers trained with the Levenberg-Marquardt algorithm | Model of 24 input neurons, 1 output neuron and 5 neurons in the visible layer and 14 neurons in the hidden layer | Protein content, wet gluten, sedimentation value and falling number | Relative viscosity | The theoretical models developed presented a high prediction capability, in addition the comparison of the predicted values with the experimental data had low values of absolute average relative deviation | Heidari et al. (2016) |

| | | | | | | |
|---|---|---|---|---|---|---|
| Fermentation | ANN modeling and optimization of bioethanol production from bread starch hydrolysate | ANN-GA backpropagation with LM and integrated algorithm. | Model of 3 input neurons, 1 output neuron, and 3 neurons in the hidden layer | Inch size), initial temperature (30–60 °C), thermal diffusivity microorganism sensitivity indicator (10–25 °C) and food quality sensitivity indicator (15–45 °C) | Bioethanol yield (1.63% - 3.82%) | 0.41% and maximum average relative deviation 6.44% corroborating the good fit of the model to the data Bioethanol yield from the RSM and ANN models were 4.10 and 4.22% vol, respectively. Also, ANN was superior to RSM in both predictability and data fittings | Betiku and Taiwo (2015) |
| | Prediction of the fermentation rate of cocoa beans (Theobroma cacao L.) using ANN | Multi-layer perceptron with feedforward algorithm. | The number of neurons in the hidden layer was set to a range from 3 to 10, network to train = 20 and network to retrain = 5 | Breadfruit starch hydrolysate (80–160 g L$^{-1}$), time (12 to 36 h), and pH (4.5 to 5.5) | Fermentation rate of fermented cocoa beans | The results of the present work demonstrate that the proposed ANN model can be adopted as a low-cost and in situ procedure to predict FI in fermented cocoa beans through apps developed for mobile device | León-Roque et al. (2016) |
| | Modeling of wheat germ fermentation process for flavonoids and | Multilayer perceptron with backpropagation and extended Kalman filter | Model of 4 input neurons, 2 output neurons, and 8 neurons in the | Measurement of RGB color of fermented cocoa surface and R/G ratio in | Flavonoid production and methoxy-r-benzoquinone +2.6-dimethoxy- | The regression coefficients ($R^2$) between experimental and predicted | Zheng et al. (2016) |

Continued

Table 13.1 ANN applications in different unit operations in food processing such as rheology, fermentation, and encapsulation—cont'd

| Process | Description | Type of ANN | N° of neurons, inputs and outputs | Independent variables | Depend variables | Results | References |
|---|---|---|---|---|---|---|---|
| | two benzoquinones using ANNs | (EKF) gain algorithms | hidden layer - Model of 3 input neurons, 2 output neurons and 8 neurons in the hidden layer | cocoa beans from alkaline extracts | r-benzoquinone content of wheat germ | values indicated that EKF-ANN models had better accuracy | |
| Encapsulation | Modeling to predict particle size (PS), polydispersity index (PDI) and encapsulation efficiency (EE) of ultrasound assisted nanoemulsions | Multilayer feedforward neural network | Model of a hidden layer with three, four and two neurons in the hidden layer for PS, PDI and EE, respectively | Phenolic compound mass; encapsulating copolymer concentration, ratio and mass; solvent volume; surfactant concentration; emulsion volume; and ultrasound time and power | Particle size (PS), polydispersity index (PDI) and encapsulation efficiency (EE) | The models allowed us to predict PS, PDI and EE for a wide range of factors. Mean square errors of 0.0538, 0.0337, and 0.0198 and correlation coefficients of 0.9139, 0.9064, and 0.8472 for PS, PDI, and EE, respectively, were obtained during training. Furthermore, mean square errors of 0.0408, 0.0224, and 0.0117 and | Espinosa-Sandoval et al. (2019) |

| Process | Model | Architecture | Input parameters | Output parameters | Results | Reference |
|---|---|---|---|---|---|---|
| Development of piceid ultrasound-assisted liposome encapsulation | Multilayer feedforward neural network model | Model of 3 input neurons, 2 output neurons and 6 neurons in the hidden layer - Model of 3 input neurons, 2 output neuron and 6 neurons in the hidden layer | Stirring speed (40–200 rpm), initial pH (4–8), fermentation temperature (24–40 °C), and fermentation time (5–45 h) | Encapsulation efficiency (%) and absolute lipid loading (%) | correlation coefficients of 0.9138, 0.9115, and 0.8955 for PS, PDI, and EE, respectively, were achieved during verification. The results indicated that the prediction accuracy of ANN was better than that of RSM. Likewise, ANN permit predicting and optimizing encapsulation | Huang et al. (2017) |
| Modeling of the emulsification process using an evolving hybrid ANN-GA framework assisted by RSM | Multilayer feedforward neural network model | Model of four input neurons, one output neuron, and twenty-four neurons in the hidden layer | Lipid content (120–180 mg), ultrasound power (90–150 W) and ultrasound time (10–50 min) | Emulsion stability index | The proposed hybrid GA model was found to be useful for the optimization of process parameters for emulsion formation and stability analysis | Kundu et al. (2015) |

Adapted from Bhagya Raj, G. V. S., & Dash, K. K. (2020). Comprehensive study on applications of artificial neural network in food process modeling. *Critical Reviews in Food Science and Nutrition, 62*(10), 2756–2783. https://doi.org/10.1080/10408398.2020.1858398, with modifications.

## 13.3 ANN in digestion modelling

As previously observed, ANN can be used as a powerful tool for the design, optimization and prediction of results of multiple food processes; these capabilities can also be applied for the simulation of human digestion and prediction of the absorption and bioavailability of nutrients, as well as physiological responses to foods such as the glycemic index and the feeling of satiety by extrapolating existing experimental data. In this sense, ANN are used to produce a model, a set of nonlinear equations, capable of predicting a response with great precision from input variables obtained from experimental conditions of digestion models (proximal composition of the food or concentration of the substance of interest, for example). An adequate optimization of the parameters of the neural network must be carried out in order for the model to predict correctly outside calibration set, this predictive ability is evaluated through cross-validation using parameters such as $R^2$ and cross-validated $R^2$ ($CVR^2$) in order to determine the reliability of the selected model.

Commonly, the digestion process has been modeled through *in vivo* models, however these can be expensive, require long time and present high variability due to test subjects. Considering this, *in vitro* models combined whit ANN could be used to reduce the number of experiments and accurately predict physiological processes' responses.

This section provides a review of the status of ANN in modeling human digestion. Initially, the main *in vitro* models for digestion will be mentioned, which are used to provide the essential experimental data for the construction of artificial neural networks. The use of ANN for the prediction of human digestion will be described later.

### 13.3.1 *In vitro* models

*In vitro* digestion models simulate the physiological processes of *in vivo* digestion and are very useful to study and understand the interactions, changes and bioaccessibility of nutrients, drugs and compounds. This tool is used in diverse areas such as chemistry, pharmacology, nutrition, among others Espinosa-Sandoval et al., 2021; Lucas-González et al., 2018).

In recent years, researchers around the world have used different *in vitro* digestion models to analyze chemical and structural changes that occur in foods or compounds when subjected to various simulated gastrointestinal conditions. The most used models and their applications are presented below.

#### 13.3.1.1 *Static in vitro digestion models*

Static digestion models are easy to perform and relatively inexpensive. In fact, a large part of the *in vitro* digestion researches carried out (~90%) have used static models (Mackie et al., 2020; Lucas-González et al., 2018). These systems are capable of simulate the gastric phase using monocompartmental test or the intestinal phase with multicompartmental protocols where the oral, gastric and small intestine phases are simulated in sequence (Espinosa-Sandoval et al., 2021; Minekus et al., 2014). A series of initial biochemical parameters are established (pH, temperature, concentration of bile salts and enzymes) to replicate these sections gastrointestinal tract and that remain fixed in the process. The static models are frequently used for mechanistic studies and to elaborate hypotheses with specific applications for screening purposes (Li et al., 2020). Some studies show the use of these models to evaluate the behavior of different delivery systems under the different simulated gastrointestinal conditions. Gao et al. (2021) used a simulated gastrointestinal digestion model having gastric and intestinal phases to investigate the epigallocatechin gallate (EGCG) release from β-lactoglobulin/gum arabic (β-Lg-GA) complexes. Peanparkdee et al. (2018) applied an *in vitro* simulated gastrointestinal static digestion model to evaluate the digestive stability of the capsules containing Riceberry bran extract, which were prepared with two types of gelatins [acidic (type A) and alkaline (type B)]. The extract encapsulated with type A gelatin showed a decrease in bioactive compounds (BC) activity and their antioxidant activity quantified by FRAP assays during simulated *in vitro* gastrointestinal digestion.

Thumann et al. (2020) applied this static model with the objective of analyzing the changes in the metabolic profile in each phase of the digestion of nine extracts obtained from medicinal plants. According to their results, the gastric phase showed a bioaccessibility of 93–118%, which reflects that the compounds were not affected by the digestion process; however, in the intestinal phase there were changes in the metabolic profile.

Although these models have been used to investigate digestion in a wide variety of foods, they are very simple and do not represent an exact simulation of the complex processes that occur in the gastrointestinal tract (de Souza Simões et al., 2017). In addition to this, in the human body the biochemical environment changes regularly and this fact cannot be simulated with static models. For these reasons, dynamic and semi-dynamic *in vitro* digestion models have been developed to overcome these disadvantages (Gonçalves et al., 2021).

### *13.3.1.2 Dynamic in vitro digestion model*

Dynamic model systems are designed to simulate some factors of the gastric tract, such as the viscosity, concentration of enzymes, pH, the transit of chyme and the adequate mixing at each stage due to peristalsis (Cardoso et al., 2015). Due to its characteristics, this model allows to achieve a better approximation of the mechanisms and processes that occur during the digestive process and the changes that occur at the structural level in food, likewise, it allows a more exact evaluation of the bioaccessibility of specific bioactive compounds (Li et al., 2020). The research carried out that relates to the behavior of the systems of controlled delivery of bioactive compounds in dynamic models have used dynamic gastric model (DGM), human gastric simulator (HGS), TNO's gastrointestinal model (TIM) and simulator of the human intestinal microbial ecosystem (SHIME), which are explained below.

#### Dynamic gastric model (DGM)

The DGM was developed at the Institute of Food Research (Norwich, UK) (Wickham et al., 2012). It is a monocompartmental model that is controlled by a computer, consisting of an inverted cone shape, made of latex. This system simulates the fundus (proximal stomach, the upper part) and antrum (distal stomach, the lower part). In the part corresponding to the fundus, the sample is subjected to slight forces because of peristaltic contractions, which are applied by water pressure at 37 °C and is applied with a piston and barrel. Gastric secretions such as enzymes or gastric acid are also supplied in this part through pumps controlled by a computer, being the acid or enzyme rates dependent of the pH response of the sample and the volume of product fed. With respect to the antrum, the digested product is submitted to high shear stress and mixing. To simulate the emptying of the stomach, a valve is assigned. For its part, the residence time is determined by software and is dependent on the composition of the sample, size and caloric content. The emptying is set at pre-defined intervals and after, samples may be subjected to further digestion depending according to the aim of the study (Li et al., 2020; Mackie et al., 2020; Thuenemann et al., 2015).

This model has been used to carry out research on food since it is possible to use complex food matrices such as those used in *in vivo* studies. Therefore, it has been used to analyze structural changes during digestion, mainly to evaluate the rate of hydrolysis and the spread of macronutrients in the upper gastrointestinal tract (Mackie et al., 2020).

### The human gastric simulator (HGS)

This model was developed at the University of California (USA) and replicated at the Riddet Institute (New Zealand) (Kong & Singh, 2010). This model is also a single compartment and was designed to simulate the peristaltic activity of the contraction waves of the antrum and therefore provide a better simulation of the mechanical shear and crushing forces that carry out the disintegration of ingested food (Ferrua & Singh, 2015). The gastric compartment is simulated by a flexible cylindrical latex chamber with an inverted conical shape and a 5.7 L capacity. A peristaltic pump is used for the release of gastric secretion. Inside the chamber, a peristaltic pump is installed and connected to the bottom of the gastric compartment with a mesh bag to perform gastric emptying. It has a temperature control system (37 °C) based on two bulbs and a mini fan Ferrua & Singh, 2015; Li et al., 2020).

Some disadvantages of this model are related to gastric fluid secretion and emptying rates as well as residence time, since this factor does not consider the properties of the food. However, research is observed where this model is applied to evaluate the effect of physical degradation and gastric stability during digestion on the kinetics of nutrient release, which is relevant to understand the physiological response mechanisms of the products (Mackie et al., 2020).

### TNO's gastrointestinal model (TIM)

This model was developed in Zeist (Netherlands) by TNO Nutrition and Food Research Institute. It is composed by four compartments that simulate the stomach, duodenum, jejunum and ileum (Mackie et al., 2020). The temperature is controlled through a water pump (at 37 °C) which is located in the middle of walls of glass jackets which compose these compartments. The inner wall is flexible enough, allowing to imitate peristaltic movements. The transit of chyme through the compartments is performed with peristaltic pumps. Using software that contains a database of *in vivo* studies in animals and humans, it is possible to control digestion conditions such as pH and secretion rates in each compartment (Li et al., 2020; Mackie et al., 2020). Based on the above-mentioned characteristics, this model has been reported to offer a complete mimicry of the GI tract. Therefore, this model has been used to study the digestion of food products (Li et al., 2020).

### Simulator of the human intestinal microbial ecosystem (SHIME)

This is an *in vitro* multicompartment system that includes almost the entire gastrointestinal tract except for the buccal phase. It is composed of six compartments that simulate the stomach, duodenum/jejunum, ileum, cecum/ascending colon, transverse colon and descending colon (de Souza Simões et al., 2017). A magnetic stirrer is used to stir the last five compartments. It should be noted that this type of agitation is different from *in vivo* peristaltic movements that involve the behaviors of mechanistic agitation and digestion. If compared with *in vivo* methodologies, the amount of processed data is greater and it has been used to investigate punctually the interactions of the ingested components with the resident human microbiota (Van de Wiele et al., 2015). The absorption methods, the effects of microbiota colonization, the adhesion of microorganisms to the vessels and tubing does are not included in this model (Van de Wiele et al., 2015). Some research carried out in recent years that have used the dynamic *in vitro* digestion models are presented in Table 13.2.

**Table 13.2 Applications of dynamic *in vitro* digestion model.**

| Type of model | Conditions | Results | References |
|---|---|---|---|
| Dynamic gastric model (DGM) | Food matrices: Muffins containing two types of almonds flour: AP (particle size range 1700–2000 μm) AF (particle size <450 μm)<br>**Mastication**<br>One volunteer<br>Average chewing time: 3 min 22 s (AF) and 6 min 38 s (AP)<br>**Gastric digestion**<br>Porcine gastric mucosa pepsin and a gastric lipase analogue from *Rhizopus oryzae*<br>Pepsine (9000 U/mL)<br>Lipase (60 U/mL)<br>**Duodenal digestion**<br>Bile solution (10.4 mL) Pancreatic enzyme solution (29.2 mL) | • For de AF muffin the total lipid released was of 97.1 ± 1.7% meanwhile for AP was 57.6 ± 1.1%. It indicates that there were significant differences in this factor. A enhance in lipid release was evidenced in AF muffin during the gastric phase compared with the masticated samples.<br>• *In vivo* digestion of these muffins by an ileostomy volunteer (0–10 h) gave similar results with 96.5% and 56.5% lipid digested, respectively<br>• Microstructural analysis showed that some lipid remained encapsulated within the plant tissue throughout digestion | Grassby et al. (2017) |
| | Matrix: Natural (NS) and Blanched (BS) skins<br>In order to establish the effect of a food matrix on polyphenols bioaccessibility, NS and BS were either digested in water (WT) or incorporated into home-made biscuits (HB), crisp-bread (CB) and full-fat milk (FM)<br>**Gastric digestion**<br>The simulated gastric enzyme solution was prepared by dissolving porcine gastric mucosa pepsin and a gastric lipase analogue from *Rhizopus oryzae*<br>**Duodenal digestion**<br>Simulated bile solution (2.5 mL) Pancreatic enzyme solution (7.0 mL) Incubation temperature: 37°C<br>Shaking conditions: 170 rpm for 2 h | • Phenolic acids presented a high release percentage, being 47.1% for NS and 45.3% for BS, when dissolved in water. An increase in the duodenal phase of 68.5% was observed in NS and 64.7% in BS<br>• Regarding flavonones, after gastric incubation a lower percentage of release was observed in BS (29.3%) and after gastric plus duodenal (48.2%) incubation<br>• WT increased the release of flavan-3-ols ($P < 0.05$) and flavonols ($P < 0.05$) from NS after gastric plus duodenal digestion, whereas CB and HB were better vehicles for BS<br>• The release of bioactives from almond skins could explain the beneficial effects associated with almond consumption | Mandalari et al. (2016) |

*Continued*

| Table 13.2 Applications of dynamic *in vitro* digestion model—cont'd | | | |
|---|---|---|---|
| **Type of model** | **Conditions** | **Results** | **References** |
| The human gastric simulator (HGS) | Encapsulated microorganisms: *Bifidobacterium lactis* (HN019) or *Lactobacillus acidophilus* (NCFM®) in different types of chocolates (milk chocolate, and dark chocolate).<br>**Methodology**<br>30.0 mL artificial saliva<br>Agitated time: 5 min<br>Shaking incubator: 15 rpm at 37°C<br>20.0 mL of gastric priming solution<br>The amounts of gastric acid and enzyme secretions added in the DGM were 9.0 mL, and 21.0 mL respectively.<br>Time: 30 min<br>Linear emptying rate of 5 min between samples | • During the simulated passage of the upper gastrointestinal tract, it was observed that chocolate was an excellent carrier for the release of probiotics, due to the survival of these microorganisms, compared to probiotic products found on the market<br>• The viability of *B. lactis* was slightly higher than that of Lb. acidophilus and survival rates were greater than 6.5 log CFU/g in the case of the static model, and greater than 7.0 log CFU/g, in the case of the dynamic gastric model, with milk chocolate being the most protective vehicle | Klindt-Toldam et al. (2016) |
| | Sample: millet and wheat couscous<br>- Digesta was collected from the HGS at 30 min intervals over 180 min<br>- Particle size and percent starch hydrolysis of couscous in the digesta were evaluated at each time point | The lowest starch hydrolysis was present in the millet couscous sample. Slower starch hydrolysis was associated with a smaller amylose chain length for millet (839–963 DP) than for wheat (1225–1563 DP), which can allow denser packing of millet starch molecules that prevents hydrolysis<br><br>A slow gastric emptying rate was also observed in millet couscous in people, which could be associated with the slow hydrolysis property of starch that activates the ileal brake system, regardless of the high rate of decomposition of particles in the stomach | Hayes et al. (2020) |
| | Matrix: sweet potatoes (SSP) and fried sweet potatoes (FSP)<br>Oral phase: 0.5 mL salivary α-amylase solution of 1500 U/mL, 25 µL of 0.3 M $CaCl_2$ $(H_2O)_2$ and 975 µL of water were added and thoroughly mixed.<br>Incubation time: 2 min, T: 37°C at 50 rpm | According to the optical microscopy images, it is possible to conclude that the microstructure of the samples was significantly influenced by the cooking method and by simulated gastric digestion; evidencing that before digestion the cells are complete but after the process there is a break in the cell walls | Somaratne et al. (2020) |

Table 13.2 Applications of dynamic *in vitro* digestion model—cont'd

| Type of model | Conditions | Results | References |
|---|---|---|---|
| | After the oral phase, the sample was mixed with 7.5 mL of previously warmed (37 °C) simulated gastric fluid (SGF) electrolyte stock solution and 5 μL of 0.3 M $CaCl_2$ $(H_2O)_2$.<br><br>Enzymes: 1.6 mL porcine pepsin stock solution (2000 U/mL in the final mixture) and 1.0 mL lipase (120 U/mL in the final mixture) Ph: 3, T: 37 °C and 50 rpm Digestion Time: 4 h | Both SSP and FSP had a similar percentage of total β-carotene loss at the end of the 4 h of gastric digestion (17.6 ± 3.3% and 18.3 ± 0.7%, respectively)<br><br>FSP presented greater softening and collapsed faster during the digestion process compared to SSP, which could be attributed to the rapid breakdown of the cell wall and the release of β-carotene when mechanical forces are applied as in the human gastric simulator (HGS) | |
| | Matrix: Whey protein gels with different geometries<br>- Small, medium, and large cubes with side lengths of 3.1, 5.2, and 10.3 mm<br>- Spheres with a diameter of 6.5 mm<br>Gastric digestion time: 180 min | The size of the particles during digestion showed that the breaking mechanisms in the spheres occurred by erosion and for the cubes there was fragmentation at the beginning of digestion, followed by erosion<br><br>The digestion time, the size of the particles and their interaction significantly influenced the absorption of humidity and the gastric emptying | Mennah-Govela and Bornhorst (2021) |
| TNO's gastrointestinal model (TIM) | The bioaccessibility of polymethoxyflavones (PMFs) loaded in high internal phase emulsions (HIPE, $\Phi oil = 0.82$) stabilized by whey protein isolate (WPI) low methoxy pectin (LMP) complexes was evaluated using *in vitro* lipolysis and dynamic *in vitro* intestinal digestion studies Gastric emptying time: 70 min T: 37 °C Stomach (pH 1.5), duodenum (pH 6.4), jejunum (pH 6.9) and ileum (pH 7.2) Lipase and pepsin were used for the gastric phase Intestinal digestion | Simulating the entire human gastrointestinal (GI) tract, the GI model TIM-1 demonstrated a 5- and 2-fold increase in the total bioaccessibility for two major PMFs encapsulated in HIPE<br><br>The incorporation of a high amount of PMFs into the viscoelastic matrix of HIPE could represent an innovative and effective way to design an oral delivery system<br><br>This system could be used to control and enhance the release of lipophilic bioactive compounds within the digestive tract, | Wijaya et al. (2020) |

*Continued*

## Table 13.2 Applications of dynamic *in vitro* digestion model—cont'd

| Type of model | Conditions | Results | References |
|---|---|---|---|
| | Pancreatin and bile extract (fresh pig bile) | specifically the human upper gastrointestinal tract | |
| | This study was designed understand the binding of fumonisin B (FB) to maize and its major macrocomponents, during the preparation of porridge. The bioaccessibility of fumonisins in the gastrointestinal tract, was determinate using TNO Gastrointestinal Model | The results proved that fumonisins $B_1$, $B_2$ and $B_3$ are released rapidly from stiff porridge prepared from contaminated maize meal<br><br>The low bioavailability of FB is not associated with intestinal content. In fact, the data obtained demonstrate that FB is rapidly released from maize porridge | Plessis et al. (2020) |
| | The goal of this study was investigated to what extent the dAGE-protein(dietary advanced glycation endproducts) binding is affected by human GI digestion<br><br>Matrix: ginger biscuits and apple juice<br>Using the TIM model was analyzed the digests for different free-dAGEs and protein-bound dAGEs by ultrahigh pressure liquid chromatography coupled to triple quadrupole mass spectrometry (UPLC-MS/MS) | • The dAGEs enter the small intestinal and large intestinal tract in predominantly protein-bound and thereby pro-inflammatory state<br>• All food products showed endogenous formation of AGEs in the small intestinal tract, with ginger biscuits showing the highest amount of endogenously formed MG-H1 represented by the recovery of more than 400% | Van der Lugt et al. (2020) |
| Simulator of the human intestinal microbial ecosystem (SHIME) | The authors evaluated the effect of an infant cereal with probiotic (*Bifidobacterium animalis* ssp. *lactis* BB-12®) on infant's intestinal microbiota<br><br>The ascending colon was inoculated with fecal microbiota from three children (2–3 years old) | The production of butyric acid increased by 52% and there was a decrease in $NH_4^+$, at the end of the treatment<br><br>An increase in *Lactobacillaceae* families was also observed, more precisely *L. gasseri* and *L. kefiri*. The first, associated with the prevention of rhinitis in children and the second, prevention of obesity | Salgaço et al. (2021) |
| | The behavior of two mycotoxin detoxifying additives (aflatoxin bentonite clay binder and a fumonisin esterase) in a human child gut model was investigated Aflatoxin B1 ($AFB_1$) and fumonisin $B_1$ ($FB_1$) were added to the SHIME diet during one week. | • The addition of the detoxifiers resulted in a significant decrease in aflatoxin B1 (AFB1) and fumonisin B1 (FB1) concentrations<br>• Enhanced significantly the concentration of HFB1. Concentrations of short-chain | Neckermann et al. (2021) |

Table 13.2 Applications of dynamic *in vitro* digestion model—cont'd

| Type of model | Conditions | Results | References |
|---|---|---|---|
| | After, the two detoxifiers and mycotoxins were added to the system for an additional week Using ultra-performance liquid chromatography-tandem mass spectrometry method the concentrations of AFB1, FB1, hydrolysed FB1 (HFB1), partially hydrolysed FB1a and FB1b, were determined | fatty acid remained generally stable throughout the experiment<br>• No major changes in bacterial composition occurred during the experiment | |
| | The authors compared the upper gastrointestinal digestion and subsequent colonic fermentation of human milk vs. goat and cow milk-based infant formulas (goat IF and cow IF, respectively), without additional oligosaccharides using an *in vitro* model for 3-month-old infants based on the Simulator of the Human Intestinal Microbial Ecosystem (SHIME®) | • Oligosaccharides were identified in human and goat milk IF. Also, all three milk matrices decreased colonic pH by boosting acetate, lactate, and propionate production, which related to increased abundances of acetate/lactate-producing *Bifidobacteriaceae* for human milk (+25.7%) and especially goat IF (33.8%) and cow IF (37.7%)<br>• Overall, goat and cow milk-based formulas without added oligosaccharides impacted gut microbial activity and composition similarly to human milk | Gallier et al. (2020) |

### 13.3.1.3 Semi-dynamic in vitro digestion model

Those models that mimic at least one gastrointestinal dynamic characteristic are called semi-dynamic. These models have been used frequently to perform controlled enzymatic release or to simulate the pH of the chyme. One of the advantages it presents with respect to the dynamic model where several compartments are used is the use of a single simple container. This model has been used regularly to simulate dynamic changes in gastric pH (Mackie et al., 2020).

Almada-Érix et al. (2021) developed a semi-dynamic *in vitro* model that simulated the different human gastric and intestinal phases. The model was used for evaluating the behavior of *Bacillus coagulans* GBI-30 (BC) through the gastrointestinal tract in yogurt and orange juice samples. When the viability of the microorganism was evaluated during the digestion stages, it was quantified and evidenced a high survival in both the juice and the yogurt. Therefore, these products are suitable to provide this probiotic in the diet. Regarding the semi-dynamic system, the authors mention that the assembly was simple as well as its operation, and that the model adjusts to evaluate the resistance of these strains under the simulated intestinal conditions.

Markussen et al. (2021) used a semi-dynamic *in vitro* digestion model in a system composed of milk protein concentrate (MPC) and 1% alginate, guar gum and pectin. In order to evaluate their behavior in the gastrointestinal tract and assess how their structures were related to changes in protein breakdown. According to their results, it was found that the kinetics of proteolysis is influenced by the type of polysaccharide. Through this modeling it is possible to understand how the structure and composition of the protein together with the added polysaccharide affects nutrient absorption, which is useful when developing food products for different needs and lifestyles.

Mulet-Cabero et al. (2019) studied gastric behavior using a semi-dynamic *in vitro* model in pasteurized and ultra-pasteurized milk samples obtained on a pilot scale. This model reproduces some gastric behaviors observed in the stomach and can simulate its main dynamics including gradual acidification, fluid secretion, gastric enzymes and emptying. Similarly, in this study, it was demonstrated that changes induced by milk processes (homogenization and heat treatments) affect gastric digestion *in vitro*, generating separation of the gastric phase, structural changes in proteins, which directly affect nutrient metabolism *in vivo*.

These three methodologies are very important to understand all the modelling aspects of the digestion process. However, it is also essential to consider the effect of the processing variables on the nutritional attributes of food products. Thus, as mentioned before, it is also important to understand the application of the ANN in human digestion since they can allow predicting the behavior of the compounds in each of the evaluated phases.

### 13.3.2 ANN for the prediction of human digestion

One of the applications of ANN in digestive processes is the prediction of the glycemic index generated by the consumption of food products (Magaletta et al., 2010). The glycemic index (GI) is a scale (1 to 100) that describes the rate at which foods and beverages containing available carbohydrates are digested, absorbed, and metabolized to glucose and released into the bloodstream. This parameter is of particular interest for the development of food products especially for people who require strict diets such as diabetics. Traditionally, the determination of GI is developed through *in vivo* studies in human test subjects, where the process of recruitment, selection, and development of the analyzes can take 2 to 3 months and represent high costs per sample.

These factors have led to the search for alternatives that facilitate the analysis and reduce costs and waiting times, so the combination of *in vitro* studies with neural networks represents a valuable tool for the development of a model that allows predicting the glycemic index of a food with sufficient certainty, which is ideal for evaluating different formulations during product development. In the research carried out by (Espinosa-Sandoval et al. (2020); Magaletta et al., 2010), a neural network derived from *in vitro* data, proximal food composition and glycemic index obtained *in vivo* was developed, which allowed a rapid and accurate prediction of the GI, even for products outside the calibration set, which included cookies, juices, dairy products, fruits among others.

In a similar study, (Yousefi & Razavi, 2017) used artificial neural networks to predict the amount of glucose released during gastrointestinal digestion of native and modified wheat starch. Through a neural network methodology combined with a genetic algorithm, an optimization technique, a model was developed that efficiently predicted glucose release under *in vitro* conditions considering as input parameters digestion time, concentration, and volume of the starch gel. These results represent an advantage when controlling and predicting the possible effects of consumption of food with the presence of starch in its composition, allowing the classification and selection of this food ingredient by their degree of digestibility, being classified as rapidly digestible, slowly digestible and resistant starch (RS).

Resistant starches are of particular interest for their multiple health benefits, including an increased sensation of fullness, improved insulin sensitivity, and reduced triglyceride levels. Das et al. (2022) studied the production of RS from green bananas by enzymatic treatment. The amount of resistant starch obtained was determined from *in vitro* digestibility studies of the samples. Response surface methodology and neural networks were used to predict the resulting RS after enzymatic modification of the native starch. The ANN showed a greater predictive ability due to the use of a training algorithm Levenberg-Marquardt Back Propagation (LMBP), which increased the efficiency and accuracy of the model, allowing an adequate prediction of the results obtained through the *in vitro* digestion processes (a static model with sodium maleate buffer (pH 6.0) and pancreatin plus amyloglucosidase enzymes was used). These results allowed the development of functional ingredients for the formulation of foods where glucose release aimed to be controlled. This could be observed in the research of Olawoye et al. (2020), where resistant starch from *Cardaba banana* was used as an ingredient for the production of gluten-free cookies with low GI. The digestibility of the product was evaluated through an *in vitro* analysis (the digestion process of the starches was carried out by a biphasic static model at pH 2.0 and then pH 6.9 with the addition of enzymes) in order to obtain the predicted glycemic index and the glycemic load (GL). From these experimental results, two prediction models of the digestibility of gluten-free biscuits were proposed through neural networks and response surface methodology, considering as input variables the process conditions for the elaboration of the product (time and baking temperature). ANN showed a better prediction of the results of glycemic index, glycemic load and amount of resistant starch of the product, presenting a higher correlation coefficient and lower RMSE. In addition to the advantages of a low glycemic index, the consumption of foods with the presence of resistant starches increases the feeling of satiety, allowing better control over appetite.

ANN can be used as a tool for predicting satiety after food consumption. Satiety is a complex phenomenon that involves the release of hormones from the gastro-intestinal tract as well as neuro-endocrine signals from the stomach, liver, and pancreas to the brain. The evaluation of appetite can be performed by applying the Visual Analog Scale (VAS) method, a 100-point scale where according to the score, desire to eat, hunger, fullness and satiety is evaluated. The extremes of this scale are not at all hungry (0) and extremely hungry (100).

Krishnan et al. (2016) built a neural network to model the relationship between hormones linked to the feeling of satiety and the scores obtained from VAS. *In vivo* studies were carried out as input parameters for the development of ANN, considering the composition of the food as well as the route of administration of these (oral or through enteral nutrition using gastric infusions). The model obtained allowed predicting with an adequate correlation the satiety values for given measurements of gastro-intestinal hormones, finding that the glucagon-like peptide-1 (GLP-1) hormone has a crucial significance for an adequate prediction of satiety. On the other hand, the predicted VAS responses made it possible to differentiate the effects on satiety based on the composition of the food and the route of administration.

The ability to predict the feelings of hunger and satiety, becomes a valuable tool for the development of products and diets that contribute to well-being and health. Under this premise, Bellmann et al. (2019) evaluated the use of *in vitro* models of digestion combined with neural networks for the prediction of satiety. The bioaccessibility of carbohydrates and proteins, obtained from the *in vitro* digestion of different breakfast diets studied, were used as input parameters for the construction of the model used for the prediction of the corresponding VAS, which were calibrated from known data for three of the foods studied. A dynamic three-sectional gastric model, TIMagc, was used for simulation of the digestive process allowing to control motility and pressure forces in the sections corresponding to fundus, proximal antrum, and distal antrum of the stomach. Body temperature, motility patterns, secretion of salivary and gastric juices, gastric pH curve, and gastric emptying were all computer-programmed. The predicted VAS values were compared with those obtained through an *in vivo* clinical study for the same diets. The model obtained presented a good fit and adequate prediction of the degree of satiety generated by the different foods evaluated, which opens up great potential for the development of products with different satiating capacities which adjust to different lifestyles or medical requirements. The combination of an *in vitro* model with neural networks reduces costs and speeds up obtaining results when compared to an *in vivo* clinical study.

Another interesting application for the use of neural networks in the modeling of digestive processes is through the representation of the relationship between nutrients ingested in the diet and the microbiota present in the gastrointestinal tract. The gastrointestinal microbiota is a complex system made up of bacteria, fungi and archaebacteria that significantly affects intestinal and systematic health (circulatory, immune, biliary systems, among others) and the relationship between the gastrointestinal tract and the central nervous system. The prediction of how the intake of certain nutrients affects the distribution and balance of microorganisms, becomes a challenge to develop food products, as well as probiotics, prebiotics and symbiotics (synergistic combination of these) that favor the microbial populations present in the gastrointestinal tract, as well as the production of beneficial metabolites to health by the microbiota. The use of artificial neural networks and machine learning methodologies are used as powerful tools when it comes to understanding and predicting the behavior of the microbiota during digestion processes.

During digestion, microorganisms release enzymes that act on carbohydrates, commonly dietary fiber, recognizing and degrading them; the fermentation of dietary fiber by microorganisms generates metabolites with therapeutic actions, so understanding the relationship between microbial enzymes and substrates will allow a more informed selection of food ingredients (for example, prebiotic fibers) that favor the growth of the microbiota and the release of beneficial metabolites.

Using neural networks S. Liu et al. (2020) developed a model that allowed predicting the affinity between carbohydrate substrate (fructo-oligosaccharides) and bacterial enzyme of specific microorganism species in the human gut microbiome. Fructo-oligosaccharides have important health benefits, so understanding their affinity with enzymes is of great importance for the formulation of food products. Through a Poisson noise-based few-shots learning neural network methodology, a robust and more accurate model was found compared to other machine learning methodologies that allowed predicting the relationships between the active sites of the enzymes and the model fructo-oligosaccharides studied (1-kestose, raffinose, nystose, and stachyose.) Reducing simulation time and improving understanding of the relationship between the structure of the active sites of bacterial enzymes and fructo-oligosaccharides.

## 13.4 Conclusions

Using artificial intelligence models in food processing offers a great possibility to establish predictive scenarios of wide potential for product development or production process improvement. The models that were presented show that they are very useful to validate *in vivo* data. In particular, the combination of *in vitro* models applying artificial neural networks offers a viable alternative to replace animal and human tests in certain areas of research. In addition, these models can predict behavior easily and with a high degree of accuracy, saving costs and time. Artificial neural networks also facilitate the development of functional foods by allowing the rapid evaluation of possible formulations based on food ingredients with different degrees of digestibility that allow the design of products oriented to people requiring special diets.

## References

Alara, O. R., Abdurahman, N. H., Afolabi, H. K., & Olalere, O. A. (2018). Efficient extraction of antioxidants from Vernonia cinerea leaves: Comparing response surface methodology and artificial neural network. *Beni-Suef University Journal of Basic and Applied Sciences, 7*(3), 276–285. https://doi.org/10.1016/j.bjbas.2018.03.007.

Almada-Érix, C. N., Almada, C. N., Souza Pedrosa, G. T., Lollo, P. C., Magnani, M., & Sant'Ana, A. S. (2021). Development of a semi-dynamic in vitro model and its testing using probiotic *Bacillus coagulans* GBI-30, 6086 in orange juice and yogurt. *Journal of Microbiological Methods, 183*, 106187. https://doi.org/10.1016/j.mimet.2021.106187.

Azadbakht, M., Aghili, H., Ziaratban, A., & Torshizi, M. V. (2017). Application of artificial neural network method to exergy and energy analyses of fluidized bed dryer for potato cubes. *Energy, 120*(Supplement C), 947–958.

Bellmann, S., Krishnan, S., de Graaf, A., de Ligt, R. A., Pasman, W. J., Minekus, M., & Havenaar, R. (2019). Appetite ratings of foods are predictable with an in vitro advanced gastrointestinal model in combination with an in silico artificial neural network. *Food Research International, 122*(March), 77–86. https://doi.org/10.1016/j.foodres.2019.03.051.

Betiku, E., & Taiwo, A. E. (2015). Modeling and optimization of bioethanol production from breadfruit starch hydrolyzate vis-à-vis response surface methodology and artificial neural network. *Renewable Energy, 74*, 87–94. https://doi.org/10.1016/j.renene.2014.07.054.

Bhagya Raj, G. V. S., & Dash, K. K. (2020a). Microwave vacuum drying of dragon fruit slice: Artificial neural network modelling, genetic algorithm optimization, and kinetics study. *Computers and Electronics in Agriculture, 178*, 105814. https://doi.org/10.1016/j.compag.2020.105814.

Bhagya Raj, G. V. S., & Dash, K. K. (2020b). Ultrasound-assisted extraction of phytocompounds from dragon fruit peel: Optimization, kinetics and thermodynamic studies. *Ultrasonics Sonochemistry, 68*, 105180.

Bhagya Raj, G. V. S., & Dash, K. K. (2020c). Comprehensive study on applications of artificial neural network in food process modeling. *Critical Reviews in Food Science and Nutrition, 62*(10), 2756–2783. https://doi.org/10.1080/10408398.2020.1858398.

Cardoso, C., Afonso, C., Lourenço, H., Costa, S., & Nunes, M. L. (2015). Bioaccessibility assessment methodologies and their consequences for the risk benefit evaluation of food. *Trend in Food Science & Technology, 41*(1), 5–23. https://doi.org/10.1016/j.tifs.2014.08.008.

Carvalho, N. B., Minim, V. P. R., Silva, R.d.C.d. S. N., della Lucia, S. M., & Minim, L. A. (2013). Artificial neural networks (ANN): prediction of sensory measurements from instrumental data. *Food Science and Technology (Campinas), 33*(4), 722–729. https://doi.org/10.1590/S0101-20612013000400018.

Cheok, C. Y., Chin, N. L., Yusof, Y. A., Talib, R. A., & Law, C. L. (2012). Optimization of total phenolic content extracted from *Garcinia mangostana* Linn. hull using response surface methodology versus artificial neural network. *Industrial Crops and Products, 40*, 247–253. https://doi.org/10.1016/j.indcrop.2012.03.019.

Chidinma-Mary-Agbai. (2020). Application of artificial intelligence (AI) in food industry. *GSC Biological and Pharmaceutical Sciences, 13*(1), 171–178. https://doi.org/10.30574/gscbps.2020.13.1.0320.

Cubeddu, A., Rauh, C., & Delgado, A. (2014). Hybrid artificial neural network for prediction and control of process variables in food extrusion. *Innovative Food Science & Emerging Technologies, 21*, 142–150. https://doi.org/10.1016/j.ifset.2013.10.010.

Das, M., Rajan, N., Biswas, P., & Banerjee, R. (2022). A novel approach for resistant starch production from green banana flour using amylopullulanase. *Lwt, 153*(August 2021), 112391. https://doi.org/10.1016/j.lwt.2021.112391.

de Souza Simões, L., Madalena, D. A., Pinheiro, A. C., Teixeira, J. A., Vicente, A. A., & Ramos, Ó. L. (2017). Micro- and nano bio-based delivery systems for food applications: In vitro behavior. *Advances in Colloid and Interface Science, 243*, 23–45.

Disse, E., Ledoux, S., Bétry, C., Caussy, C., Maitrepierre, C., Coupaye, M., Laville, M., & Simon, C. (2018). An artificial neural network to predict resting energy expenditure in obesity. *Clinical Nutrition (Edinburgh, Scotland), 37*(5), 1661–1669. https://doi.org/10.1016/j.clnu.2017.07.017.

Dong, X., Li, Q., Sun, D., Chen, X., & Yu, X. (2015). Direct FTIR analysis of free fatty acids in edible oils using disposable polyethylene films. *Food Analytical Methods, 8*(4), 857–863. https://doi.org/10.1007/s12161-014-9963-y.

Espinosa-Sandoval, L. A., Cerqueira, M. A., Ochoa-Martínez, C. I., & Ayala-Aponte, A. A. (2019). Phenolic compound–loaded nanosystems: artificial neural network modeling to predict particle size, polydispersity index, and encapsulation efficiency. *Food and Bioprocess Technology, 8*(12), 1395–1408. https://doi.org/10.1007/s11947-019-02298-8.

Espinosa-Sandoval, L. A., Ochoa-Martínez, C. I., & Ayala-Aponte, A. A. (2020). Prediction of in vitro release of nanoencapsulated phenolic compounds using Artificial Neural Networks. *DYNA, 87*(212). https://doi.org/10.15446/dyna.v87n212.72883.

Espinosa-Sandoval, L. A., Cerqueira, M. A., Ochoa-Martínez, C. I., & Ayala-Aponte, A. A. (2021). Polysaccharide-based multilayer nano-emulsions loaded with oregano oil: Production, characterization, and in vitro digestion assessment. *Nanomaterials, 11*(4), 878–890. https://doi.org/10.3390/nano11040878.

Fan, F., Ma, Q., Ge, J., Peng, Q. Y., Roley, W. W., & Tang, S. Z. (2013). Prediction of texture characteristics from extrusion food surface images using a computer vision system and artificial networks. *Journal of Food Engineering, 118*(4), 426–433.

Ferrua, M., & Singh, R. (2015). Human gastric simulator (Riddet Model). In *The impact of food bioactives on health in vitro and ex vivo models* (pp. 61–72). Springer.

Gallier, S., Van den Abbeele, P., & Prosser, C. (2020). Comparison of the bifidogenic effects of goat and cow milk-based infant formulas to human breast milk in an in vitro gut model for 3-month-old infants. *Frontiers in Nutrition, 7*, 1–11. https://doi.org/10.3389/fnut.2020.608495.

Gao, J., Mao, Y., Xiang, C., et al. (2021). Preparation of β-lactoglobulin/gum arabic complex nanoparticles for encapsulation and controlled release of EGCG in simulated gastrointestinal digestion model. *Food Chemistry, 354*, 129516. https://doi.org/10.1016/j.foodchem.2021.129516.

Ghoreishi, S. M., & Heidari, E. (2013). Extraction of epigallocatechin-3-gallate from green tea via supercritical fluid technology: neural network modeling and response surface optimization. *The Journal of Supercritical Fluids, 74*, 128–136.

Giwa, A., Daer, S., Ahmed, I., Marpu, P. R., & Hasan, S. W. (2016). Experimental investigation and artificial neural networks ANN modeling of electrically-enhanced membrane bioreactor for wastewater treatment. *Journal of Water Process Engineering, 11*(Supplement C), 88–97.

Gómez, A. H., Wang, J., Hu, G., & Pereira, A. G. (2008). Monitoring storage shelf life of tomato using electronic nose technique. *Journal of Food Engineering, 85*(4), 625–631. https://doi.org/10.1016/j.jfoodeng.2007.06.039.

Gonçalves, A., Estevinho, B. N., & Rocha, F. (2021). Methodologies for simulation of gastrointestinal digestion of different controlled delivery systems and further uptake of encapsulated bioactive compounds. *Trends in Food Science & Technology, 114*, 510–520. https://doi.org/10.1016/j.tifs.2021.06.007.

Grassby, T., Mandalari, G., Grundy, M. M.-L., Edwards, C. H., Bisignano, C., Trombetta, D., et al. (2017). In vitro and in vivo modeling of lipid bioaccessibility and digestion from almond muffins: The importance of the cell-wall barrier mechanism. *Journal of Functional Foods, 37*, 263–271. https://doi.org/10.1016/j.jff.2017.07.046.

Guiné, R. P. F. (2019). The use of artificial neural networks (ANN) in food process engineering. *ETP International Journal of Food Engineering*, 15–21. https://doi.org/10.18178/ijfe.5.1.15-21.

Hayes, A. M. R., Swackhamer, C., Mennah-Govela, Y. A., Martinez, M. M., Diatta, A., Bornhorst, G. M., & Hamaker, B. R. (2020). Pearl millet (*Pennisetum glaucum*) couscous breaks down faster than wheat couscous in the Human Gastric Simulator, though has slower starch hydrolysis. *Food & Function, 11*(1), 111–122. https://doi.org/10.1039/C9FO01461F.

Heidari, E., Sobati, M. A., & Movahedirad, S. (2016). Accurate prediction of nanofluid viscosity using a multilayer perceptron artificial neural network (MLP-ANN). *Chemometrics and Intelligent Laboratory Systems, 155*, 73–85. https://doi.org/10.1016/j.chemolab.2016.03.031.

Huang, S.-M., Kuo, C.-H., Chen, C.-A., Liu, Y.-C., & Shieh, C.-J. (2017). RSM and ANN modeling-based optimization approach for the development of ultrasound-assisted liposome encapsulation of piceid. *Ultrasonics Sonochemistry, 36*, 112–122. https://doi.org/10.1016/j.ultsonch.2016.11.016.

Ibrahim, S., Abdul Wahab, N., Ismail, F. S., & Md Sam, Y. (2020). Optimization of artificial neural network topology for membrane bioreactor filtration using response surface methodology. *IAES International Journal of Artificial Intelligence (IJ-AI), 9*(1), 117. https://doi.org/10.11591/ijai.v9.i1.pp117-125.

Karimi, F., Rafiee, S., Taheri-Garavand, A., & Karimi, M. (2012). Optimization of an air drying process for *Artemisia absinthium* leaves using response surface and artificial neural network models. *Journal of the Taiwan Institute of Chemical Engineers, 43*, 29–39.

Kashyap, P., Singh Riar, C., & Jindal, N. (2020). Optimization of ultrasound assisted extraction of polyphenols from Meghalayan cherry fruit (*Prunus nepalensis*) using response surface methodology (RSM) and artificial neural network (ANN) approach. *Journal of Food Measurement and Characterization, 15*, 119–133. https://doi.org/10.1007/s11694-020-00611-0.

Katoch, S., Singh, S., & Kumar, V. (2021). A review on genetic algorithm: past, present and future. *Multimedia Tools and Applications, 80*, 8091–8126. https://doi.org/10.1007/s11042-020-10139-6.

Keklik, N. M., Işikli, N. D., & Sur, E. B. (2017). Estimation of the shelf life of pezik pickles using Weibull hazard analysis. *Food Science and Technology, 37*(Suppl. 1), 125–130. https://doi.org/10.1590/1678-457x.33216.

Khajeh, M., Moghaddam, M. G., & Shakeri, M. (2012). Application of artificial neural network in predicting the extraction yield of essential oils of *Diplotaenia cachrydifolia* by supercritical fluid extraction. *The Journal of Supercritical Fluids, 69*, 91–96.

Khazaei, N. B., Tavakoli, T., Ghassemian, H., Khoshtaghaza, M. H., & Banakar, A. (2013). Applied machine vision and artificial neural network for modeling and controlling of the grape drying process. *Computers and Electronics in Agriculture, 98*(Supplement C), 205–213.

Klindt-Toldam, S., Larsen, S. K., Saaby, L., Olsen, L. R., Svenstrup, G., Müllertz, A., et al. (2016). Survival of *Lactobacillus acidophilus* NCFM® and *Bifidobacterium lactis* HN019 encapsulated in chocolate during in vitro simulated passage of the upper gastrointestinal tract. *LWT—Food Science and Technology, 74*, 404–410. https://doi.org/10.1016/j.lwt.2016.07.053.

Kong, F., & Singh, R. P. (2010). A human gastric simulator (HGS) to study food digestion in human stomach. *Journal of Food Science, 75*(9), E627–E635. https://doi.org/10.1111/j.1750-3841.2010.01856.x.

Krishnan, S., Hendriks, H. F. J., Hartvigsen, M. L., & De Graaf, A. A. (2016). Feed-forward neural network model for hunger and satiety related VAS score prediction. *Theoretical Biology and Medical Modelling, 13*(1), 1–12. https://doi.org/10.1186/s12976-016-0043-4.

Kundu, P., Paul, V., Kumar, V., & Mishra, I. M. (2015). Formulation development, modeling and optimization of emulsification process using evolving RSM coupled hybrid ANN-GA framework. *Chemical Engineering Research and Design, 104*, 773–790. https://doi.org/10.1016/j.cherd.2015.10.025.

Lai, K. C., Lim, S. K., Teh, P. C., & Yeap, K. H. (2016). Modeling electrostatic separation process using Artificial Neural Network (ANN). *Procedia Computer Science*, *91*(Supplement C), 372–381.

Lashkarbolooki, M., Shafipour, Z. S., & Hezave, A. Z. (2013). Trainable cascade-forward back-propagation network modeling of spearmint oil extraction in a packed bed using SC-CO2. *The Journal of Supercritical Fluids*, *73*, 108–115.

León-Roque, N., Abderrahim, M., Nuñez-Alejos, L., Arribas, S. M., & Condezo-Hoyos, L. (2016). Prediction of fermentation index of cocoa beans (*Theobroma cacao* L.) based on color measurement and artificial neural networks. *Talanta*, *161*, 31–39. https://doi.org/10.1016/j.talanta.2016.08.022.

Li, C., Yu, W., Wu, P., & Chen, X. D. (2020). Current *in vitro* digestion systems for understanding food digestion in human upper gastrointestinal tract. *Trends in Food Science & Technology*, *96*, 114–126. https://doi.org/10.1016/j.tifs.2019.12.015.

Liu, X., Jiang, Y., Shen, S., Luo, Y., & Gao, L. (2015). Comparison of Arrhenius model and artificial neuronal network for the quality prediction of rainbow trout (*Oncorhynchus mykiss*) fillets during storage at different temperatures. *LWT—Food Science and Technology*, *60*(1), 142–147. https://doi.org/10.1016/j.lwt.2014.09.030.

Liu, S., Kou, Y., & Chen, L. (2020). *Novel few-shots learning neural network for predicting carbohydrate-active enzyme (CAZyme) affinity towards fructo-oligosacacharides* (pp. 2–47). Research Square, Preprint.

Lucas-González, R., Viuda-Martos, M., Pérez-Alvarez, J. A., & Fernández-López, J. (2018). In vitro digestion models suitable for foods: Opportunities for new fields of application and challenges. *Food Research International*, *107*, 423–436. https://doi.org/10.1016/j.foodres.2018.02.055.

Mackie, A., Mulet-Cabero, A., & Torcello-Gómez, A. (2020). Simulating human digestion: Developing our knowledge to create healthier and more sustainable foods. *Food & Function*, (11), 9397–9431. https://doi.org/10.1039/D0FO01981J.

Magaletta, R. L., DiCataldo, S. N., Liu, D., Li, H. L., Borwankar, R. P., & Martini, M. C. (2010). In vitro method for predicting glycemic index of foods using simulated digestion and an artificial neural network. *Cereal Chemistry*, *87*(4), 363–369. https://doi.org/10.1094/CCHEM-87-4-0363.

Mai, N., & Huynh, V. (2017). Kinetics of quality changes of *Pangasius* fillets at stable and dynamic temperatures, simulating downstream cold chain conditions. *Journal of Food Quality*, *2017*, 1–9. https://doi.org/10.1155/2017/2865185.

Mandalari, G., Vardakou, M., Faulks, R., Bisignano, C., Martorana, M., Smeriglio, A., & Trombetta, D. (2016). Food matrix effects of polyphenol bioaccessibility from almond skin during simulated human digestion. *Nutrients*, *8*(9), 1–17. https://doi.org/10.3390/nu8090568.

Markussen, J.Ø., Madsen, F., Young, J. F., & Corredig, M. (2021). A semi dynamic in vitro digestion study of milk protein concentrate dispersions structured with different polysaccharides. *Current Research in Food Science*, (4), 250–261. https://doi.org/10.1016/j.crfs.2021.03.012.

Mennah-Govela, Y. A., & Bornhorst, G. M. (2021). Breakdown mechanisms of whey protein gels during dynamic in vitro gastric digestion. *Food & Function*, *12*(5), 2112–2125.

Minekus, M., Alminger, M., Alvito, P., et al. (2014). A standardised static in vitro digestion method suitable for food – An international consensus. *Food & Function*, (5), 1113–1124. https://doi.org/10.1039/C3FO60702J.

Mohamed, K., & Mohamed, M. (2015). Overview on the response surface methodology (RSM) in extraction processes. *Journal of Applied Science and Process Engineering*, *2*(1). https://doi.org/10.33736/jaspe.161.2015.

Mokhtarian, M., & Garmakhany, A. D. (2017). Prediction of ultrasonic osmotic dehydration properties of courgette by ANN. *Quality Assurance and Safety of Crops & Foods*, *9*(2), 161–169. https://doi.org/10.3920/QAS2015.0662.

Motavali, A., Najafi, G. H., Abbasi, S., Minaei, S., & Ghaderi, A. (2013). Microwave–vacuum drying of sour cherry: Comparison of mathematical models and artificial neural networks. *Journal of Food Science and Technology*, *50*(4), 714–722. https://doi.org/10.1007/s13197-011-0393-1.

Mulet-Cabero, A. I., Mackie, A. R., Wilde, P. J., Fenelon, M. A., & Brodkorb, A. (2019). Structural mechanism and kinetics of in vitro gastric digestion are affected by process-induced changes in bovine milk. *Food Hydrocolloids, 86*, 172–183. https://doi.org/10.1016/j.foodhyd.2018.03.035.

Neckermann, K., Claus, G., De Baere, S., Antonissen, G., Lebrun, S., Gemmi, C., & Delcenserie, V. (2021). The efficacy and effect on gut microbiota of an aflatoxin binder and a fumonisin esterase using an in vitro simulator of the human intestinal microbial ecosystem (SHIME®). *Food Research International, 145*, 1–14. https://doi.org/10.1016/j.foodres.2021.110395.

Nguyen, H. H., Shpigelman, A., van Buggenhout, S., Moelants, K., Haest, H., Buysschaert, O., Hendrickx, M., & van Loey, A. (2016). The evolution of quality characteristics of mango piece after pasteurization and during shelf life in a mango juice drink. *European Food Research and Technology, 242*(5), 703–712. https://doi.org/10.1007/s00217-015-2578-8.

Nourbakhsh, H., Emam-Djomeh, Z., Omid, M., Mirsaeedghazi, H., & Moini, S. (2014). Prediction of red plum juice permeate flux during membrane processing with ANN optimized using RSM. *Computers and Electronics in Agriculture, 102*, 1–9. https://doi.org/10.1016/j.compag.2013.12.017.

Olawoye, B., Gbadamosi, S. O., Otemuyiwa, I. O., & Akanbi, C. T. (2020). Gluten-free cookies with low glycemic index and glycemic load: optimization of the process variables via response surface methodology and artificial neural network. *Heliyon, 6*(10), e05117. https://doi.org/10.1016/j.heliyon.2020.e05117.

Pavlić, B., Pezo, L., Marić, B., Tukuljac, L. P., Zeković, Z., Solarov, M. B., & Teslić, N. (2020). Supercritical fluid extraction of raspberry seed oil: Experiments and modelling. *The Journal of Supercritical Fluids, 157*(104687), 1–11.

Peanparkdee, M., Yamauchi, R., & Iwamoto, S. (2018). Stability of bioactive compounds from Thai Riceberry bran extract encapsulated within gelatin matrix during in vitro gastrointestinal digestion. *Colloids and Surfaces A: Physicochemical and Engineering Aspects, 546*, 136–142. https://doi.org/10.1016/j.colsurfa.2018.03.021.

Pilkington, J. L., Preston, C., & Gomes, R. L. (2014). Comparison of response surface methodology (RSM) and artificial neural networks (ANN) towards efficient extraction of artemisinin from *Artemisia annua*. *Industrial Crops and Products, 58*, 15–24.

Plessis, B., Regnier, T., Combrinck, S., Steenkamp, P., & Meyer, H. (2020). Investigation of fumonisin interaction with maize macrocomponents and its bioaccessibility from porridge using the dynamic tiny-TIM gastrointestinal model. *Food Control, 113*, 1–9. https://doi.org/10.1016/j.foodcont.2020.107165.

Prakash Maran, J., Sivakumar, V., Thirugnanasambandham, K., & Sridhar, R. (2013). Artificial neural network and response surface methodology modeling in mass transfer parameters predictions during osmotic dehydration of *Carica papaya* L. *Alexandria Engineering Journal, 52*(3), 507–516. https://doi.org/10.1016/j.aej.2013.06.007.

Ramzi, M., Kashaninejad, M., Salehi, F., Sadeghi Mahoonak, A. R., & Ali Razavi, S. M. (2015). Modeling of rheological behavior of honey using genetic algorithm–artificial neural network and adaptive neuro-fuzzy inference system. *Food Bioscience, 9*, 60–67. https://doi.org/10.1016/j.fbio.2014.12.001.

Rasooli Sharabiani, V., Kaveh, M., Abdi, R., Szymanek, M., & Tanaś, W. (2021). Estimation of moisture ratio for apple drying by convective and microwave methods using artificial neural network modeling. *Scientific Reports, 11*(1), 9155. https://doi.org/10.1038/s41598-021-88270-z.

Salgaço, M. K., Perina, N. P., MorenoTomé, T., Mosquera, E. M. B., Lazarini, T., Sartoratto, A., & Sivieri, K. (2021). Probiotic infant cereal improves children's gut microbiota: Insights using the Simulator of Human Intestinal Microbial Ecosystem (SHIME®). *Food Research International, 143*, 1–9. https://doi.org/10.1016/j.foodres.2021.110292.

Sánchez-González, J. A., & Oblitas-Cruz, J. F. (2017). Application of Weibull analysis and artificial neural networks to predict the useful life of the vacuum packed soft cheese. *Revista Facultad de Ingeniería Universidad de Antioquia, 82*, 53–59. https://doi.org/10.17533/udea.redin.n82a07.

Sarkar, T., Salauddin, M., Hazra, S. K., & Chakraborty, R. (2020). Artificial neural network modelling approach of drying kinetics evolution for hot air oven, microwave, microwave convective and freeze dried pineapple. *SN Applied Sciences, 2*(9), 1621. https://doi.org/10.1007/s42452-020-03455-x.

Sebastin, J. (2018, April 23). *Artificial Intelligence: a real opportunity in food industry*. Food Quality and Safety.

Simić, V. M., Rajković, K. M., Stojičević, S. S., Veličković, D. T., Nikolić, N.Č., Lazić, M. L., & Karabegović, I. T. (2016). Optimization of microwave-assisted extraction of total polyphenolic compounds from chokeberries by response surface methodology and artificial neural network. *Separation and Purification Technology, 160*, 89–97. https://doi.org/10.1016/j.seppur.2016.01.019.

Singh, H. (2019). How AI is reshaping the food processing business. In *Customer think*. https://www.customerthink.com.

Sinha, K., Chowdhury, S., das Saha, P., & Datta, S. (2013). Modeling of microwave-assisted extraction of natural dye from seeds of *Bixa orellana* (Annatto) using response surface methodology (RSM) and artificial neural network (ANN). *Industrial Crops and Products, 41*, 165–171. https://doi.org/10.1016/j.indcrop.2012.04.004.

Somaratne, G., Ye, A., Nau, F., Ferrua, M. J., Dupont, D., Singh, R. P., & Singh, J. (2020). Role of biochemical and mechanical disintegration on β-carotene release from steamed and fried sweet potatoes during in vitro gastric digestion. *Food Research International, 136*, 1–9. https://doi.org/10.1016/j.foodres.2020.109481.

Song, Q., Zheng, Y. J., Xue, Y., Sheng, W. G., & Zhao, M. R. (2017). An evolutionary deep neural network for predicting morbidity of gastrointestinal infections by food contamination. *Neurocomputing, 226*(Supplement C), 16–22.

Stangierski, J., Weiss, D., & Kaczmarek, A. (2019). Multiple regression models and Artificial Neural Network (ANN) as prediction tools of changes in overall quality during the storage of spreadable processed Gouda cheese. *European Food Research and Technology, 245*(11), 2539–2547. https://doi.org/10.1007/s00217-019-03369-y.

Suryawanshi, B., & Mohanty, B. (2018). Application of an artificial neural network model for the supercritical fluid extraction of seed oil from *Argemone mexicana* (L.) seeds. *Industrial Crops and Products, 123*, 64–74. https://doi.org/10.1016/j.indcrop.2018.06.057.

Thuenemann, E. C., Mandalari, G., Rich, G. T., & Faulks, R. M. (2015). Dynamic gastric model (DGM). In *The impact of food bioactives on health (in vitro and ex vivo models)* (pp. 47–59). Springer International Publishing.

Thumann, T. A., Pferschy-Wenzig, E.-M., Aziz-Kalbhenn, H., Ammar, R. M., Rabini, S., Moissl-Eichinger, C., & Bauer, R. (2020). Application of an in vitro digestion model to study the metabolic profile changes of an herbal extract combination by UHPLC–HRMS. *Phytomedicine, 71*, 153221. https://doi.org/10.1016/j.phymed.2020.153221.

Torkashvand, A. M., Ahmadi, A., & Nikravesh, N. L. (2017). Prediction of kiwifruit firmness using fruit mineral nutrient concentration by artificial neural network (ANN) and multiple linear regressions (MLR). *Journal of Integrative Agriculture, 6*(7), 1634–1644.

Van de Wiele, T., Van de Abbeele, P., Ossieur, W., Possemiers, S., & Marzorati, M. (2015). The simulator of the Human Instestinal Microbial Ecosystem (SHIME). In *The impact of food bio-actives on gut health in vitro and ex vivo models* (pp. 305–318). Springer.

Van der Lugt, T., Venema, K., van Leeuwen, S., Vrolijk, M. F., Opperhuizen, A., & Bast, A. (2020). Gastrointestinal digestion of dietary advanced glycation endproducts using an in vitro model of the gastrointestinal tract (TIM-1). *Food & Function, 11*(7), 6297–6307. https://doi.org/10.1039/d0fo00450b.

Weiss, D., Stangierski, J., Baranowska, H. M., & Rezler, R. (2018). Kinetic models of quality parameters of spreadable processed Gouda cheese during storage. *Food Science and Biotechnology, 27*(5), 1387–1394. https://doi.org/10.1007/s10068-018-0377-2.

Wickham, M. J. S., Faulks, R. M., Mann, J., & Mandalari, G. (2012). The design, operation, and application of a dynamic gastric model. *Dissolution Technologies*, (19), 15–22. https://doi.org/10.14227/DT190312P15.

Wijaya, W., Zheng, H., Zheng, T., Su, S., Patel, A. R., Van der Meeren, P., & Huang, Q. (2020). Improved bioaccessibility of polymethoxyflavones loaded into high internal phase emulsions stabilized by biopolymeric complexes: A dynamic digestion study via TNO's gastrointestinal model. *Current Research in Food Science*, 2, 11–19. https://doi.org/10.1016/j.crfs.2019.11.007.

Yousefi, A. R., & Razavi, S. M. A. (2017). Modeling of glucose release from native and modified wheat starch gels during in vitro gastrointestinal digestion using artificial intelligence methods. *International Journal of Biological Macromolecules*, 97, 752–760. https://doi.org/10.1016/j.ijbiomac.2017.01.082.

Zhang, H., & Zimba, P. V. (2017). Analyzing the effects of estuarine freshwater fluxes on fish abundance using artificial neural network ensembles. *Ecological Modelling*, 359(Supplement C), 103–116.

Zheng, Z., Guo, X., Zhu, K., Peng, W., & Zhou, H. (2016). The optimization of the fermentation process of wheat germ for flavonoids and two benzoquinones using EKF-ANN and NSGA-II. *RSC Advances*, 6(59), 53821–53829. https://doi.org/10.1039/C5RA27004A.

Zhi, N.-N., Zong, K., Thakur, K., Qu, J., Shi, J.-J., Yang, J.-L., Yao, J., & Wei, Z.-J. (2018). Development of a dynamic prediction model for shelf-life evaluation of yogurt by using physicochemical, microbiological and sensory parameters. *CyTA—Journal of Food*, 16(1), 42–49. https://doi.org/10.1080/19476337.2017.1336572.

# Consumer's perception and acceptability

PART V

# CHAPTER 14

# How to assess consumer perception and food attributes of novel food structures using analytical methodologies

**Takahiro Funami and Makoto Nakauma**
*San-Ei Gen F.F.I. Inc., Texture Design Research Laboratory, Toyonaka, Osaka, Japan*

## 14.1 Introduction

Foods should provide people with the feeling of happiness and satisfaction, and enhancement of palatability or eating enjoyment must be a main motivation for all people serving in food manufacturing and product development. Food palatability is determined by human perception such as texture, flavor, appearance, sound, and temperature through five senses, and texture and flavor are said to be two major essential elements (Kohyama, 2005). Of these elements, the importance of texture is emphasized in terms of food palatability and safe eating in the recent super-aged society with which the world is facing. Safe eating is often related to swallowing difficulty, which is equal to dysphagia problem.

The texture is a whole body of mechanical, geometrical, and thermal properties of foods perceived by humans during oral processing, chewing, swallowing, etc., and corresponds to sensory characteristics via oral and oropharyngeal organs and tissues, such as teeth, tongue, palate and throat. This means that texture perception can be measured only by humans. However, sensory evaluation is sometimes difficult to carry out as it requires qualified panel to obtain creditable results, particularly in analysis-type test. It also takes time and money in performing tests and keeping the number and the level of panels, and to solve these difficulties, instrumental measurements can be in place of sensory evaluation, particularly for researchers in the food industry. A common approach is to evaluate relationships between texture perception and structural, rheological, or mechanical properties of foods, and instrumental measurements of texture-related properties with certain quantitativeness and reproducibility have been required.

With this background, this chapter outlines texture estimation and prediction through instrumental measurements. These relate to dominant physicochemical properties of foods, which govern texture and physiological conditions of food oral processing in humans. To compensate the existing gap and fortify the results from instrumental measurements, in vivo physiological measurements in human eating are also described. From the application aspect, enhanced flavor perception through texture modification and product design for dysphagia foods are presented. Readers may also know through this chapter the important role of hydrocolloids in texture study, serving as model foods/beverages or as essential ingredients themselves. This can be due to their controllable rheological properties without significantly affecting other organoleptic properties, particularly flavor and appearance (Peleg, 2006).

## 14.2 Texture perception and oral rheology and tribology

### 14.2.1 Extension rheology for texture evaluation with regard to swallowing physiology in humans

Viscoelastic fluids may exhibit different flow properties between shear and extension deformations. For food ingredients, a macromolecule with relatively rigid conformation such as xanthan gum shows some differences in flow behavior by deformation mode. Extensional flow occurs not only in food manufacturing processes such as extrusion but also in food oral processing such as feeding and swallowing. It is presumed that extensional viscosity should have effects on food texture. In fact, some studies have evidenced that extensional flow property should be important equal to shear flow property for texture evaluation of liquid and semiliquid foods during oral processing (Debruijne et al., 1993; Koliandris et al., 2011; Van Vliet, 2002). However, due to the limitation of instruments in measuring easily and precisely the extensional flow property of foods, less studies have been carried out compared to shear flow property.

Chen et al. (2008) analyzed the tensile separation behavior of commercially available fluid food products once compressed on a uniaxial compression machine. They confirmed that the maximum tensile force and work till the maximum force both are highly correlated to stickiness perceived using fingers (not orally) (Chen et al., 2008). Although their method is similar to TPA adhesiveness measurement, the importance of the study lies in the optimization of the experimental set-up, including instrumental operation conditions and selection of mechanical parameters. They discussed the possible mechanism for stickiness perception and postulated that pressure drop and cavitation due to rapid extensional flow can be critical. It was also found, using the same experimental set-up, that perceived swallowing difficulty for fluid foods is highly correlated to the maximum stretching force and the work of stretching (Chen & Lolivret, 2011). Thus, food bolus which is more difficult to deform upon extension (i.e., with low stretch-ability) and is larger in extensional viscosity tend to show longer oral residual time until swallowing, leading to the perception of a more difficult swallow.

As a progress, a novel device using real-time optical detection was proposed (Bazilevsky et al., 1990). On this device, a small amount of liquid sample sandwiched between two plates is stretched rapidly by pulling the upper plate away from the lower plate at a set speed, applying extensional deformation to the sample for the formation of liquid filament. The filament is shrunk by capillary force, in which the surface tension works as shear stress and reduces the surface area of the filament. Temporal change in diameter at the midpoint of the filament (i.e., $D_{mid}$) is monitored with a laser-type micrometer, and extensional viscosity can be calculated from this diameter change (Fig. 14.1). An instrument for extensional viscosity measurement based on this principle is commercially available as a capillary breakup extension rheometer; CaBER (Thermo Haake Gmbh, Karlsruhe, Germany), and researches using this instrument have contributed to the progress of the study.

As a representative study using polysaccharide solutions as a model liquid food, Lv et al. (2017) investigated human discrimination thresholds in perceiving extensional viscosity compared to shear

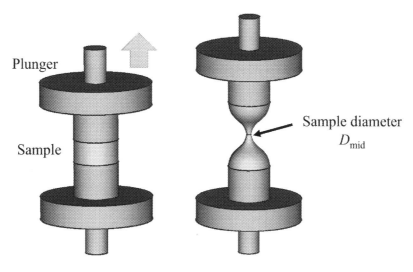

**FIG. 14.1**

Instrumental set-up for extensional viscosity measurements.

viscosity using guar gum and sodium carboxymethyl cellulose (CMC-Na) solutions (Lv et al., 2017). Results indicate that the ability to discriminate extensional viscosity is higher than that to discriminate shear viscosity, confirming that extensional viscosity is critical for textural characterization of liquid foods. He et al. (2016) studied the correlation between the texture of solutions thickened by xanthan gum and several rheological parameters (He et al., 2016). They confirmed that perceived thickness, stickiness, and mouth coating have higher correlation with extensional viscosity than shear viscosity, and that the accuracy of prediction can be increased by incorporating extensional viscosity as one of independent explanatory variables.

Extensional viscosity of food bolus may change upon mixing with saliva, in which salivary glycoprotein mucin plays an important role (Choi et al., 2014). It has been reported that extensional viscosity for xanthan gum solutions is more susceptible to saliva than that for CMC-Na solutions (Choi et al., 2014). This is attributable to a difference in the molecular conformation between xanthan gum (i.e., rigid rod molecule) and CMC-Na (i.e., flexible random coil) and also a difference in charge repulsion between two polysaccharides, both of which can affect molecular interaction between each polysaccharide and salivary mucin through structural and electrostatic effects.

### 14.2.2 Tribology for texture evaluation with regard to moisture and fat related properties of foods

Although moisture and fat related texture is a complicated sensation and difficult to express uniquely, various attempts have been made. For liquid and pasted foods, it is considered that human perceptions, including thickness, smoothness, and slipperiness, and the judgment of swallowing initiation should be largely due to surface properties of food bolus. Therefore, the tribological approach for measuring friction and lubrication between food and human tissue or organ during food oral processing should be valid.

Geometry using soft materials has been developed to recreate the friction behavior between food and soft oral surface during oral processing. Hardness and surface roughness of the tongue can be critical in oral tribology, and these attributes should be considered in establishing experimental equipment and its operation conditions. Efforts have been made to better recreate human oral physiology by using tongue tissue from animals or using polymer surface to mimic wetness and deformability of the human tongue (Carpenter et al., 2019; Dresselhuis et al., 2008), and in line with these efforts, polydimethylsiloxane (PDMS) and frosted glass were proposed as human tongue and palate alternatives, respectively (Bongaerts et al., 2007). Based on this knowledge, Anton Paar (Gratz, Austria) has upgraded a modular rheometer by introducing a geometry constituting of glass balls as a plunger and PDMS pins and plates as flooring (Fig. 14.2) for tribology measurement, and this has contributed greatly to the progress of food lubrication area (Baier et al., 2009; Biegler et al., 2016; Carvalho-Da-Silva et al., 2013; Kieserling et al., 2018; Kim et al., 2015; Krzeminski et al., 2012; Pondicherry et al., 2018; Sonne et al., 2014).

It was found using this instrument that perceived creaminess of emulsified foods is highly correlated to the friction coefficient through effects of frictional force and lubricity on food texture (Chen & Stokes, 2012). Subsequent studies contributed to evaluate the perceived richness and fattiness of foods containing fats and oils, microgels, or both (Godoi et al., 2017; Laiho et al., 2017; Liu et al., 2015; Stokes et al., 2013). For polysaccharide microgels, a correlation was found between the friction coefficient and pasty or slippery perception, while not with smoothness perception due to inhomogeneity of test samples (Krop et al., 2019), where the relationship between texture and the friction coefficient was investigated at an entrainment speed of 50 mm/s. For O/W emulsions, the correlation between the friction coefficient and perceived fattiness has been investigated using samples of equivalent shear viscosity at $50 s^{-1}$ (Malone et al., 2003a). It has been shown that lubrication behavior can be changed by the emulsifier used, particle size and distribution of oil droplets in the emulsions.

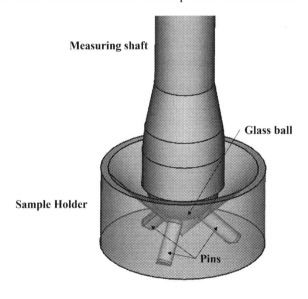

**FIG. 14.2**

Instrumental set-up for tribology measurements.

Using aqueous guar gum solutions, the friction coefficient between the substances (i.e., steel ball and elastomer with a rough surface) was measured instrumentally to investigate the relationship with human perceptions (Malone et al., 2003a). The friction coefficient at an entrainment speed of ca. 100 mm/s shows the highest correlation with perceived thickness, and this entrainment speed lies in the mixed lubrication regime in the Stribeck curve. From this result, it is considered that perceived thickness should be governed by effects from both friction and viscosity. In discussing the relationship between tribological behavior and texture, attention should be paid in the selection of measurement conditions such as entrainment speed (corresponding to tongue movement speed) and normal force (corresponding to tongue compressive force). Interaction with saliva should also be considered.

The relationship between mechanical friction behavior of simulated bolus (after homogenizing the sample in the presence of artificial saliva) and oral processing behavior in healthy subjects was investigated using hydrogel samples containing κ-carrageenan as the main ingredient but different oral tribological characteristics (Krop et al., 2020). Results indicate that a high correlation ($p<0.01$) is observed between the frictional force and the oral residence time in the mixed lubrication regime at entrainment speed of 10–100 mm/s. On the other hand, the relationship between mechanical friction behavior of simulated bolus and the sensory discrimination by untrained panelists was investigated using samples containing calcium alginate gel beads of different particle sizes sandwiched between two layers of κ-carrageenan bulk gel (Stribiţcaia et al., 2020). Results indicate that the panelists are not capable to distinguish between the samples, although friction coefficients at the boundary and mixed regimes are different. In a comprehensive review relating to tribology-sensory relationship by Shewan et al. (2020), challenge and difficulty to discuss tribology results alone broadly and universally across different types of food, and combination with other in vitro techniques such as rheology, particle sizing, and characterization of surface interactions, etc. may be a reasonable and realistic approach (Shewan et al., 2020). They also indicate that incorporation of saliva in tribological studies, the removal of confounding factors from sensory tests, and a global approach that considers all regimes of lubrication should be necessary on the premise of using well characterized tribopairs and equipment.

Astringency, one of the main quality factors of red wine, black tea, and some fruit products, can also be a friction-associated perception composed of dryness and puckering feeling. Using a mixture of human whole saliva and typical astringent compounds such as tannins, the mechanism of astringency perception was investigated and related to lubrication behavior (Brossard et al., 2016). A correlation was found between the friction coefficient at a relatively low entrainment speed; 0.075 mm/s and human astringency perception. Astringency can be perceived through mechanical stimulation and thus be determined by tribology. Recently, several subsequent tribological studies have provided various insights into astringency-related textural attributes, including dryness, puckering feeling, and roughness (Pires et al., 2020; Shewan et al., 2019).

### 14.2.3 Soft machine mechanics for texture evaluation with regard to palatal size reduction

For soft foods, bolus is formed by squeezing between the tongue and the palate without chewing. Human changes the strategy of bolus formation depending on food texture via perceived hardness as the principal attribute. Instrumental measurements usually use hard geometry, and size reduction or decomposition of foods by squeezing (i.e., tongue-palate compression) cannot be recreated fully in this way. This difficulty can be ascribed to tongue deformation in food oral processing not being considered in a conventional hard machine. In vitro texture evaluation systems using a simulated or an artificial tongue made of soft material installed into a conventional uniaxial compression machine have been established to reproduce oral conditions in squeezing (Ishihara et al., 2014, 2013; Kohyama et al., 2019). Using agar gels of various fracture stress but with equivalent fracture strain as test sample, the load was applied to the sample through compression between a soft material, simulating the tongue, and a metal probe, simulating the palate, at a crosshead speed of 10 mm/s (Fig. 14.3) to recreate squeezing (Ishihara et al., 2013). As a result, the gel was fractured when the deformation of the gel was larger than that of simulated tongue, whereas the gel was not fractured when the deformation of the gel was equal to or smaller than that of simulated tongue. In relation to sensory evaluation, when the apparent elastic modulus of the simulated tongue was ca. $5.5 \times 10^4$ Pa, whether or not the gel was fractured corresponded to the oral strategy for size reduction; squeezing or chewing. Therefore, the oral strategy for size reduction can be deduced by the presence or absence of fracture in this system. It is also suggested that humans detect a difference in deformation between food and the tongue in a relatively small strain region (i.e., ca. 10% strain) and determine the oral strategy for size reduction based on this information. Therefore, such an in vitro texture evaluation system is believed to be effective, particularly in developing foods for people with chewing and/or swallowing difficulties.

This in vitro texture evaluation system was validated using gellan gum gels (Ishihara et al., 2014). It was indicated that modification of operation conditions should be necessary in some cases to obtain high correlation with the decision of oral strategy. This may relate to physiological modulation of tongue-palate compression by food texture, occurring during consumption of highly deformable gels, including tongue-palate compression speed, tongue consistency by excitation, shear loading by the tongue, and the maximum deformation (i.e., clearance). The first factor can be realized by decreasing the crosshead speed from 10 to 5 mm/s, which is associated with the stress relaxation of the gels, a phenomenon of energy dissipation enhanced at lower deformation speed (Luyten & Van Vliet, 1995). The second factor can be realized by increasing Young's modulus of simulated tongue from ca. 55 to ca. 110 kPa. It has been elucidated from tongue pressure measurement that the slope of time course curve

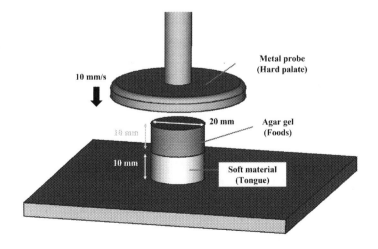

**FIG. 14.3**

Instrumental simulation of squeezing between the tongue and the hard palate, mimicking human oral processing.

of tongue pressure during the first size reduction is almost independent of physical properties (texturally brittle or deformable) or consistency (soft or hard) of polysaccharide gels (Hori et al., 2015). Based on this finding, it is reasonable to think that the decrease in tongue-palate compression speed should give rise to increased tongue pressure. Actually, lowering crosshead speed and/or increasing Young's modulus of the simulated tongue is certainly effective for texture evaluation of highly deformable samples. The incorporation of shear force into the evaluation system and increased maximum deformation should be mechanical challenges. The use of simulated tongues of larger size should be another option since the surface contact area between food and the tongue may be larger in consuming highly deformable gels. As a definition of "highly deformable," our study suggests that its boundary for the initial fracture strain should be at ca. 60–70% (Ishihara et al., 2014). Humans can decide the oral strategy for size reduction by sensing the difference in the strain of food relative to the tongue perceived dynamically during oral processing (Kohyama, 2015). Fracture strain has been emphasized as a dominant mechanical parameter for the decision of oral strategy for size reduction (Arai & Yamada, 1993), and this is the same as the determination of the biting speed (Mioche & Peyron, 1995).

Several artificial tongues with different rigidities, shapes, and sizes using transparent urethane were applied to a soft machine system for observation of fracture behavior of soft polysaccharide gels which are processed by tongue squeezing for size reduction (Kohyama, 2020; Kohyama et al., 2021). Transparency of artificial tongue enables visual inspection of the deformation behavior of either polysaccharide gel or artificial tongue. Results indicate that use of an artificial tongue with apparent elastic modulus of 20–50 kPa (i.e., lower than that of Ishihara et al., 2013) can represent the oral processing behavior of elderly people of low tongue pressure with higher accuracy. Recent studies by other research groups on the development of artificial tongue and its application to texture study are introduced. An artificial tongue which Wang et al. (2021) developed is made from PDMS and reproduces surface roughness of human tongue based on the image analysis data (Wang et al., 2021). This artificial tongue is confirmed to have almost the same lubrication properties as human tongue and to be applicable to mechanical tests in a wide range of sliding speed and load condition. In addition, Srivastava, Bosc, et al. (2021) prepared an artificial tongue using polyvinyl alcohol as an analogue to human tongue in terms of rigidity and surface roughness (Srivastava, Bosc, et al., 2021). The Young's modulus of this artificial tongue is ca. 70 kPa, which is close to that of Ishihara et al. (2013). Surface roughness is characterized by average peak height and correlation length (i.e., ca. 100 mm and ca. 400 mm, respectively), which are both close to those of human tongue. In combination with a texture analyzer, applicability of this artificial tongue to texture study is elucidated by frictional force measurement of cutting cheese, showing higher friction force for structurally inhomogeneous samples than that for structurally homogeneous samples variated by the addition level of microcrystalline cellulose. Compression behavior of hydrocolloid gels has also been analyzed using the same artificial tongue (Srivastava, Stieger, et al., 2021).

### 14.2.4 Bolus rheology

Food is transformed from its initial shape and size to form bolus before swallowing. For solid and semisolid foods, chewing or mastication is the major oral operation for size reduction, and as a result, a coherent bolus that can be swallowed safely and comfortably is formed (Alexander, 1998; Prinz & Lucas, 1995). Swallowing occurs when the criteria of particulate size, lubrication, and cohesion of food bolus are met (Malone et al., 2003a). Investigations on both food fragmentation behavior and bolus rheology should be key approaches for texture characterization during oral processing. However, such kind of investigations have targeted, in most cases, solid foods with low fluidity and deformability like vegetables and nuts (Jalabert-Malbos et al., 2007; Lucas et al., 2002; Mishellany et al., 2006; Peyron et al., 2004) but rarely viscoelastic foods like gels. It has been reported that mechanical properties of food bolus gathered from different subjects but similar in dental status are small in the inter-individual variability (Mioche et al., 2003; Peyron et al., 2004). This allows us to hypothesize that how easily and how fast a food can transfer to ready-to-swallow bolus in the mouth is certainly an important attribute for texture design of foods for specified consumer groups, like the ones with dysphagia.

Ishihara, Nakauma, Funami, Odake, and Nishinari (2011a, 2011b) studied bolus rheology using polysaccharide gels from either low-acylated gellan gum or the mixture of low-acylated gellan gum and psyllium seed gum as model foods. The texture of gellan single gel is brittle (Morris, 2006) with detectable syneresis, particularly at low concentrations. For the mixture gel, on the other hand, syneresis is not detected due to the presence of psyllium seed gum, which also increases textural deformability (Ishihara, Nakauma, Funami, Tanaka, et al., 2011). Model bolus was prepared through instrumental chewing of the gels of 2 levels of hardness in the presence or absence of simulated saliva using a mechanical simulator (Fig. 14.4), where the gels were compressed with shear in a reciprocating manner using a flat plunger (50 mm in diameter) to mimic the action of the human jaw. Model bolus from the mixture gel showed a rheologically weak gel (or structured fluid) behavior from a rheological point of view as presented by frequency-dependence of dynamic viscoelasticities. Also, dynamic viscoelasticity parameters, particularly dynamic mechanical loss tangent ($\tan\delta$), of the mixture gel were less dependent on the addition level of simulated saliva (Fig. 14.5). For the mixture gel, tan $d$ values were roughly within 0.19–0.24 with ca. 14% coefficient of variation (CV) at the lower gel hardness (i.e.,

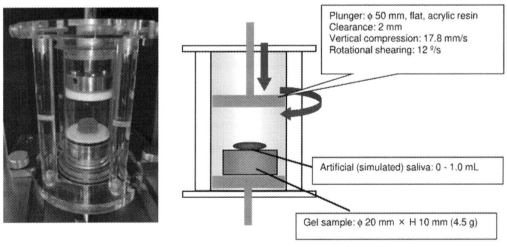

**FIG. 14.4**

Instrumental chewing simulator for preparation of model bolus.

*Reproduced from Ishihara, S., Nakauma, M., Funami, T., Odake, S. & Nishinari, K. (2011). Viscoelastic and fragmentation characters of model bolus from polysaccharide gels after instrumental mastication. Food Hydrocolloids, 25(5), 1210–1218. https://doi.org/10.1016/j.foodhyd.2010.11.008.*

## 14.2 Texture perception and oral rheology and tribology

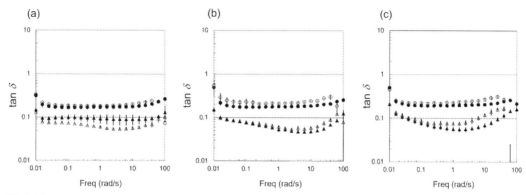

**FIG. 14.5**

Frequency-dependence of dynamic mechanical loss tangent $\tan \delta$ for model bolus. (A) Model bolus in the absence of saliva; (B) model bolus in the presence of 0.5 mL saliva; (C) model bolus in the presence of 1.0 mL saliva. Open circles: 1.0% of the mixture of low-acylated gellan gum and psyllium seed gum (#1); closed circles: 1.5% of the mixture (#2); open triangles: 0.075% of low-acylated gellan gum (#3); closed triangles: 0.15% of low-acylated gellan gum (#4). Measurements were carried out at 20°C at a fixed strain of 1% within the frequency range of 0.01–100 rad/s. Gel hardness levels for samples #1 and #3 were ca. 1000 Pa defined by the compression stress at 67% strain, whereas those for samples #2 and #4 were ca. 4000 Pa defined by the compression stress at 67% strain. Data are presented as means ± SD of triplicate.

*Reproduced from Ishihara, S., Nakauma, M., Funami, T., Odake, S. & Nishinari, K. (2011). Swallowing profiles of food polysaccharide gels in relation to bolus rheology. Food Hydrocolloids, 25(5), 1016–1024. https://doi.org/10.1016/j.foodhyd.2010.09.022*

1000 Pa) and within 0.17–0.2 with ca. 9% CV at the higher gel hardness (i.e., 4000 Pa), respectively. For gellan single gel, CV values of $\tan d$ were ca. 32% at the lower gel hardness and ca. 24% at the higher gel hardness, respectively. These differences between the mixture gel and gellan single gel are ascribed to saliva miscibility, determined by the degree of hydrophilicity of model bolus. The surface of model bolus from the mixture gel is more hydrophilic than that from gellan single gel due to the water holding function of psyllium seed gum. A boundary friction coefficient of human saliva is two orders of magnitude lower than that of water (Bongaerts et al., 2007), and saliva miscibility determines the lubricity of food bolus. Saliva can work better in the mixture gel as a binder of gel particulates and also as a lubricant during bolus formation by instrumental chewing. It can be thus concluded that rheologically weak gel and high saliva miscibility should be critical for the formation of swallowable bolus.

Bolus mechanical properties can be measured with a miniature Kramer shear cell (Álvarez et al., 2020). The Kramer shear cell consists of two compartments, the upper compartment is a 5-balded head, whereas the bottom compartment is a container of about 10 mL volume. Due to its unique structural features, a sample undergoes multiple deformations such as compression, shearing, and extrusion through a stroke of the blade. Kramer mechanical properties, including the maximum force, average force, and work required to shear the bolus, can be obtained from the force-distance curve. It is confirmed that the Kramer mechanical properties of the bolus (collected by subjects' expectoration immediately before swallowing) are highly correlated with oral processing parameters such as chewing time and number of chew. A unique self-designed mastication simulator has been developed, where the combination of a Teflon cylinder (72 mL volume, equivalent to human oral cavity) and the upper and lower Teflon lids mechanically reproduces human mouth (Pu et al., 2021). Furthermore, the bottom lid has 26 simulated teeth made of resin, including the incisors, lateral incisors, canines, and molar teeth, and the upper lid has 15 simulated teeth made of resin, which also include the 4 types of tooth. The bottom is fixed, whereas the upper is rotatable. Results from experiments using this in vitro mastication simulator confirm that mucin rapidly increases the adhesive force of bolus at the initial stage of oral processing, whereas α-amylase gradually increases the adhesive force. As indicated in these examples of mechanical mastication simulation, bolus rheology contributes to better understand the dynamic texture perception during oral processing.

## 14.3 Texture evaluation through human physiological responses

Although rheological measurements reveal mechanical and related microstructural properties of foods, both reality and complexity of oral experience may be oversimplified (Peleg, 2006). The deviation is due to lack of salivation (Stokes, 2012) and the use of stiff materials (Peleg, 2006), for instance. At least two reasons can explain the limitations of rheological measurements in food texture evaluation: the temporal difference in judgement and the complexity/simplicity of sensing between instruments and humans (Kohyama, 2015). Inconsistency between both results may be explained in a different way: sensory response to a mechanical stimulus needs to be expressed not as a characteristic value like a mean relative or an absolute score but as a distribution of the terms used by those who sense them (Peleg, 2006). Therefore, human physiological measurements should be introduced to reconcile with instrumental measurements (Foster et al., 2011; Koç et al., 2013; Wilkinson et al., 2000), where noninvasive sensors attached to human subjects are used to monitor the organ and muscle signals related to food oral processing for tracing the dynamic changes of food texture. Various kinds of in vivo measurements are often used for texture evaluation, involving palatal pressure measurements (Nakazawa & Togashi, 2000; Takahashi & Nakazawa, 1991; Takahashi & Nakazawa, 1992), electromyography (Kohyama et al., 1998, 2007; Kohyama et al., 2010; Kohyama, Hatakeyama, et al., 2005; Kohyama, Yamaguchi, et al., 2005), biting force measurements (Kohyama, Ishihara, et al., 2021; Kohyama et al., 1997, 2001, 2003; Kohyama & Nishi, 1997), and ultrasonic pulse Doppler methods (Kumagai et al., 2009; Nakazawa & Togashi, 2000). In terms of dynamics, in vivo measurements lie between instrumental measurements and sensory evaluations. Understanding the dynamic changes of foods during oral processing should be the key for texture study (Chen, 2009), and from this perspective, more in vivo measurements should be used in this research area.

### 14.3.1 Tongue pressure measurement

As the tongue plays a crucial role in a series of food oral processing, it is physiologically reasonable to link its movement to food texture. Two published works, relating to tongue pressure measurement during food oral processing and its usefulness for food texture study, are highlighted. These two studies studied the effect of food texture on tongue movement using a novel sensing system and polysaccharide gels as a model food (Hori et al., 2015; Yokoyama et al., 2014). They indicated the potential usage of tongue pressure measurement for elucidation of the dynamics of food oral processing and for visualization of food-tongue interaction. T-shaped sensor sheet with five measuring channels is the main part of the sensing system (Fig. 14.6), which makes it possible to monitor tongue pressure in real time.

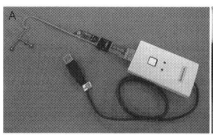 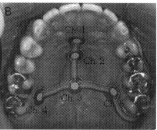

**FIG. 14.6**

Tongue pressure sensing system (A) and location of sensor sheet and channels (B). The lengths between channels 1 and 3 were 23, 25.5, 28 mm for small (S), medium (M), and large (L) size of sensor sheet, whereas those between channels 4 and 5 were 35, 38, 41 mm for S, M, and L size.

*Reproduced from Hori, K., Hayashi, H., Yokoyama, S., Ono, T., Ishihara, S., Magara, J., Taniguchi, H., Funami, T., Maeda, Y. & Inoue, M. (2015). Comparison of mechanical analyses and tongue pressure analyses during squeezing and swallowing of gels.* Food Hydrocolloids, 44, *145–155. https://doi.org/10.1016/j.foodhyd.2014.09.029.*

This device has been developed by Osaka University research team, which was originally for palatal plate design particularly for tongue cancer patients (Hori et al., 2009), and their idea to apply this to food science is novel. The benefit of the sensor sheet for texture study should be easy adaptation to oral shape without changing oral physiology nor interfering with occlusion. This situation can be realized by its thin thickness: only 0.1 mm, variety of the size, and mechanical flexibility. In addition, the sensor sheet does not cover the whole area of the palate and thus does not interfere with taste, aroma, and texture perceptions. Thus, the sensor sheet does not prevent natural eating behavior, different form a conventional system called manometer which uses a small balloon type probe (Shaker et al., 1988). Using the sensor sheet, temporal and spatial distribution patterns of tongue pressure during food oral processing was studied (Hori et al., 2015; Yokoyama et al., 2014). The same types of polysaccharide gel were prepared as in the case of bolus rheology: low-acylated gellan gum and the mixture of low-acylated gellan gum and psyllium seed gum at 3 hardness levels for each type. When subjects were asked to eat the gel samples by compressing them between the tongue and the hard palate (i.e., squeezing), temporal pattern of tongue pressure during squeezing did not change substantially upon gel texture, presenting the first onset at the mid-median part (Ch. 2), followed by the anterior-median (Ch. 1), the posterior-median (Ch. 3) parts, and the circumferential parts (Chs. 4 and 5) in this order for each type of gel sample (Fig. 14.7). Differences were noted in the offset between the two types of gel sample. It was concluded that duration of squeezing for texturally deformable gels (presented by the mixture gels) tends to be shorter than that for texturally brittle gels (presented by gellan single gels) when gel hardness is relatively low, whereas vice versa when gel hardness is relatively high. Spatial distribution pattern of tongue pressure during squeezing changed upon gel texture: the maximum amplitudes at channels 1 and 2 were larger than those at channels 4 and 5 for gellan single gels, whereas no marked difference was found between channels for the mixture gels (Fig. 14.8). Consequently, human eating behavior can be visualized by tongue movement which is changed by food texture. Foods in the form of weak gels, emulsions, and fluids are eaten by squeeze between the tongue and the palate, and in these cases, texture relates to thin film rheological behaviors of foods as well as the bulk properties (Malone et al., 2003a), including mayonnaise (Giasson et al., 1997) and chocolate (Luengo et al., 1997).

**376 Chapter 14** How to assess consumer perception and food attributes

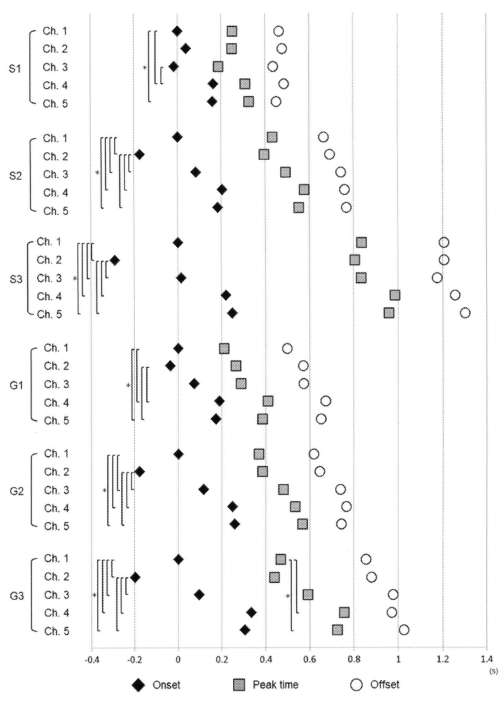

**FIG. 14.7**

See figure legend on opposite page.

The combination of shear and squeeze flows is a more realistic representation for tongue movement during food oral processing (Stokes, 2012). From these perspectives, it is expected that tongue pressure measurement should be utilized for texture evaluation, including creaminess, smoothness, sliminess, and thickness, along with tribological measurement for identifying lubrication property (Garrec & Norton, 2012, 2013; Malone et al., 2003a).

Visualization of biomechanics during food oral processing by tongue pressure measurement and its application to texture study are being advanced. For this purpose, the tongue movement was monitored with tongue pressure sensor, videofluorography, and surface electromyography in squeezing gels with different mechanical properties (varied by fracture force and fracture strain) prepared using gellan gums with different acyl contents. Using healthy adult subjects, results confirm that the larger the fracture force, the greater the tongue pressure at the initial squeeze and the longer the tongue pressure

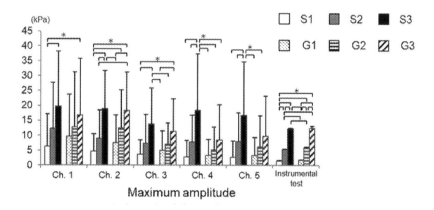

**FIG. 14.8**

The maximum amplitude of tongue pressure at each channel during the initial squeeze. *: $p < 0.05$ (one-way ANOVA test and Tukey's post hoc test). See the footnote in Fig. 14.7 for the experimental details and sample denomination.

*Reproduced from Hori, K., Hayashi, H., Yokoyama, S., Ono, T., Ishihara, S., Magara, J., Taniguchi, H., Funami, T., Maeda, Y. & Inoue, M. (2015). Comparison of mechanical analyses and tongue pressure analyses during squeezing and swallowing of gels. Food Hydrocolloids, 44, 145–155. https://doi.org/10.1016/j.foodhyd.2014.09.029.*

**FIG. 14.7, CONT'D** Time sequences of tongue pressure at each channel during the initial squeeze. The time "0" was set as the onset of channel 1. *: $p < 0.05$ (Kruskale-Wallis test and post hoc test with Bonferroni correction). S1–3 presents texturally deformable gel samples from the mixture of low-acylated gellan gum and psyllium seed gum with increased hardness in order (i.e., ca. 1600, 5500, and 12,000 Pa defined by the compression stress at 67% strain), whereas G1–3 presents texturally brittle gel samples from low-acylated gellan gum with increased hardness in order. When the number is the same, the gel hardness is equivalent between both types of gel sample. The serving volume of the gel sample was 5 mL. Data are presented as means ± SD of 15 subjects of 27.6 years old on average.

*Reproduced from Hori, K., Hayashi, H., Yokoyama, S., Ono, T., Ishihara, S., Magara, J., Taniguchi, H., Funami, T., Maeda, Y. & Inoue, M. (2015). Comparison of mechanical analyses and tongue pressure analyses during squeezing and swallowing of gels. Food Hydrocolloids, 44, 145–155. https://doi.org/10.1016/j.foodhyd.2014.09.029.*

duration are. Also, the smaller the fracture strain, the more the tongue pressure concentrates in the median line of the tongue. Oral processing behavior in squeezing the same gels was analyzed by dividing it into 4 phases: initial squeeze, middle squeeze, late squeeze, and swallowing (Murakami et al., 2021). Results confirm that the maximum tongue pressures in the initial and middle phases are larger at the median line of the tongue for gels with higher fracture force, whereas at the circumferential parts of the tongue for gels with higher fracture strain. However, no influence by fracture force or fracture strain is found in the magnitude of pressure during last squeeze and swallowing.

## 14.3.2 Electromyography

Electromyography (EMG) can noninvasively monitor the muscle activity during oral processing using electrodes attached onto the skin surface of the human muscle and has been a method for texture study (Espinosa & Chen, 2012). The EMG target should not only be the masseter and temporalis muscles, which are active in jaw-closing phase, but also the suprahyoid musculature. Suprahyoid musculature corresponds to a group of muscles which is active in jaw opening and tongue movement (Kohyama et al., 2010; Palmer et al., 1992; Shiozawa et al., 1999a, 1999b; Taniguchi et al., 2008). Although solid foods which require mastication for size reduction have been of major interest for EMG study, the number of reports on soft foods which can be processed without chewing is increasing. Relationships between some EMG variables and mechanical properties obtained by instrumental measurements or textural characteristics obtained by sensory evaluation have been clarified.

As an example for the EMG studies dealing with mastication, eight kinds of solid foods in a wide range of physicochemical properties were selected as a test food (i.e., dry hard bread, elastic konjac gel, dry sausage, soft candy, raw radish, pickled radish, boiled carrot, and raw carrot), and nine independent parameters were chosen from 28 physicochemical characteristics (i.e., stress at 10%, 50%, 70%, or 90% compression strain, breaking stress, cohesiveness, adhesiveness, density, and moisture content) (Kohyama et al., 2008). It was concluded that the mechanical properties under larger deformation should correlate well to EMG variables. Results from other studies using buckwheat noodles (Kohyama et al., 2010) and gummy candies (Hayakawa et al., 2009) support this conclusion. When correlation with EMG variables is discussed for solid foods which needs chewing for size reduction, it would be recommended in general to see food mechanical properties under large deformation conditions, sometimes beyond the fracture point. A high correlation can be found between EMG variables and mechanical properties under extremely large deformation conditions regardless the rheological nature of foods, elastic or plastic, since the upper and lower teeth almost reach contact during chewing for each type of sample. Although EMG can also detect food oral process at the early stage, it would be necessary to further investigate why there hardly seen a good correlation between EMG variables and mechanical properties under small deformation conditions or associated textural characteristics. It is not clear whether the sensitivity of EMG is not high enough or human does not care about small deformation.

All these studies used relatively hard foods consumed by chewing, but a similar principle can be applied to relatively soft foods which do not necessarily require chewing but squeezing. It was concluded from the study by Ishihara, Nakauma, Funami, Tanaka, et al. (2011) using the same types of polysaccharide gel as in the case of bolus rheology at 2 hardness levels (ca. 1000 and 4000 Pa) that the whole sequence of food oral processing is prolonged with the increase of EMG activity of the suprahyoid musculature by the hardness of each type of gel (Fig. 14.9). EMG activity of the suprahyoid

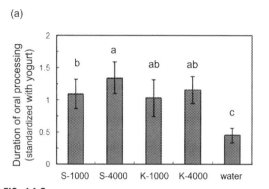
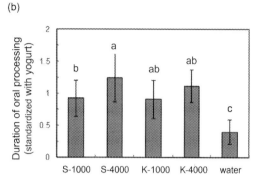

**FIG. 14.9**

Electromyographic variables for soft gels processed orally by squeezing for side reduction. (A) Duration of oral processing; (B) Activity of the suprahyoid musculature. Both EMG parameters were standardized with the corresponding data from commercial yogurt to eliminate differences in individual oral physiology and in electrode placement among subjects. Data are means ± SD of 9 subjects of 34.3 years old on average. S presents texturally deformable gel samples from the mixture of low-acylated gellan gum and psyllium seed gum with 2 gel hardness levels (i.e., ca. 1000 and 4000 Pa defined by the compression stress at 67% strain), whereas K presents texturally brittle gel samples from low-acylated gellan gum with 2 gel hardness levels. When the number is the same, the gel hardness is equivalent between both types of gel sample. The serving amount of the gel sample was 14.0 ± 1.0 g. Data with different letters are significantly different ($p < 0.05$).

*Reproduced from Ishihara, S., Nakauma, M., Funami, T., Tanaka, T., Nishinari, K. & Kohyama, K. (2011). Electromyography during oral processing in relation to mechanical and sensory properties of soft gels. Journal of Texture Studies, 42(4), 254–267. https://doi.org/10.1111/j.1745-4603.2010.00272.x.*

musculature well correlates to the compression load at extremely large strain conditions (e.g., 90% strain) and perceived hardness. Results obtained are analogous to other study using gelatin gels processed by chewing: duration of the whole sequence and EMG activity of the masseter and the temporalis increase with Instron hardness represented as the compression stress at 50% strain (Foster et al., 2006). In summary, mechanical properties at large deformation or strain relate to physiological response during food oral processing regardless the oral strategy for size reduction, chewing or tongue-palate compression.

EMG is potentially applicable not only to analyze food oral processing behavior but also to food/beverage preference (like or dislike). Using gellan gum gels flavored with 6 different aroma components (including 2 aroma components recognized as negative, neutral, and positive hedonic for each) and 3 different consistency levels, the correlation between the muscle activity of the facial muscles (corrugator supercilia and zygomatic major muscles) of healthy adult subjects during oral processing and each sensory score, including liking, wanting, valence, and arousal, was examined (Sato et al., 2020). Results confirm negative correlation between the muscle activity of the corrugator supercilii muscle and sensory rating of liking, wanting, or valence, suggesting that subjective hedonic experience, specifically the liking component, during food consumption can be objectively assessed using facial EMG.

### 14.3.3 Acoustic analysis of swallowing sound

Swallowing physiology has been inspected directly using videofluorography (VF), videoendoscopy (VE), and ultrasonic (ultrasound) method for texture study (Kumagai et al., 2009; Masashi, 2008; Saitoh et al., 2007). This is because flow velocity of bolus through the pharyngeal phase relates to swallowing ease or comfort. However, there are actually limitations in these physiological measurements. VE is usually low in quantitative performance, and VF has a risk of exposure to X-ray radiation and uses barium sulfate as a contrast medium. Contrast medium can alter the mechanical and textural properties of food sample, particularly when polyelectrolyte gels like carrageenan and gellan gum are used (Nishinari, 2006). Ultrasonic method is applicable preferably to females due to the lack of the thyroid cartilage, which may interfere with the transit of the ultrasonic pulse. Although indirect inspection, acoustic analysis can be another approach for the purpose through quantification of the swallowing sound profile.

Acoustic analysis has been used for diagnostic purposes as similar to VE, VF, and ultrasonic method in biomedical field (Lazareck & Moussavi, 2004). The principle is simple, and the sound generated by swallowing is recorded with a throat microphone placed at fixed position in proximity to the vocal cords. In the food area, on the other hand, they have been used for in vitro evaluation of crispiness and crunchiness of cereal flakes, biscuits, nuts, etc, in combination with a rupture generating mechanical tests (Varela & Fiszman, 2012) but seldom for in vivo measurements. In some studies (Ishihara, Nakauma, Funami, Odake, & Nishinari, 2011a; Nakauma et al., 2011), acoustic analysis of swallowing sound has been carried out using polysaccharide gels or solutions as model foods. Representative profile of the swallowing sound is illustrated in the case of water (Fig. 14.10), where the profile is divided into three parts, each of which is assigned to the sound associated with the closure of the epiglottis ($t_1$), flow of food bolus ($t_2$), and opening of the epiglottis ($t_3$) in order of occurrence (Hamlet et al., 1992).

For gelled foods (Ishihara, Nakauma, Funami, Odake, & Nishinari, 2011a), the relationship between acoustic parameters and sensory score was investigated during bolus swallowing using young healthy adults as subjects. As a feeding sample, the same types of polysaccharide gel were prepared at 2

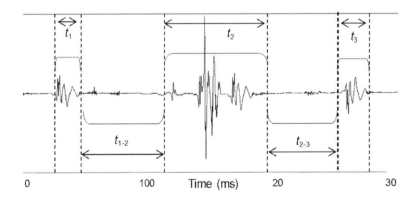

**FIG. 14.10**

Representative acoustic profile of the swallowing sound (in the case of 15 mL water fed).

*Reproduced from Nakauma, M., Ishihara, S., Funami, T. & Nishinari, K. (2011). Swallowing profiles of food polysaccharide solutions with different flow behaviors. Food Hydrocolloids, 25(5), 1165–1173. https://doi.org/10.1016/j.foodhyd.2010.11.003.*

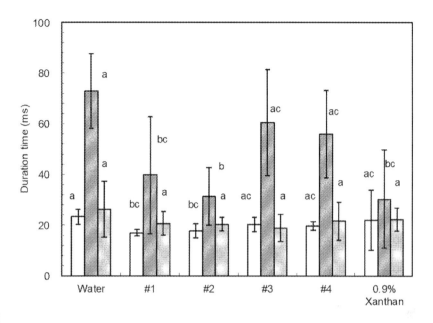

**FIG. 14.11**

Duration of each oral event during swallowing. #1: 1.0% of the mixture of low-acylated gellan gum and psyllium seed gum; #2: 1.5% of the mixture; #3: 0.075% of low-acylated gellan gum; #4: 0.15% of low-acylated gellan gum. Open bars: closure of the epiglottis $t_1$; diagonal bars: the bolus flow $t_2$; dotted bars: opening of the epiglottis $t_3$. See Fig. 14.10 for the assignment of each oral event. Gel hardness levels for samples #1 and #3 were ca. 1000 Pa defined by the compression stress at 67% strain, whereas those for samples #2 and #4 were ca. 4000 Pa defined by the compression stress at 67% strain. Each gel sample was processed by squeezing for side reduction. The serving amount of the gel sample was 7.5 g. Data are presented as means ± SD of 7 subjects of 35.1 years old on average. Data with different superscripts are significantly different ($p < 0.05$) when compared within the same oral event..

*Reproduced from Ishihara, S., Nakauma, M., Funami, T., Odake, S. & Nishinari, K. (2011). Swallowing profiles of food polysaccharide gels in relation to bolus rheology. Food Hydrocolloids, 25(5), 1016–1024. https://doi.org/10.1016/j.foodhyd.2010.09.022.*

hardness levels (ca. 1000 and 4000 Pa). As a result, duration of $t_2$ for the mixture gel (av. 39.7 and 31.3 ms in increasing hardness) was smaller than the one observed for gellan single gel (av. 60.5 and 56.0 ms in increasing hardness) and water (av. 73.0 ms), and was equivalent to that for 0.9% xanthan gum solution (av. 30.3 ms) (Fig. 14.11). Durations of $t_1$ and $t_3$ were both ca. 20 ms at each hardness level irrespective of gel type. From sensory evaluation on a 5-degree scale, the mixture gel was scored higher in perceived cohesiveness than gellan single gel (av. score 4.1 and 4.3 for the mixture gel in increasing hardness, whereas av. score 1.4 and 1.7 for gellan single gel in increasing hardness). A higher score represents easier bolus formation in the pharyngeal phase during swallowing. There are two possibilities to explain shorter $t_2$ for the mixture gel. The first one is that food bolus from the mixture gel can flow at higher velocity through the pharyngeal phase than that from gellan single gel. However, this is unlikely when the $t_2$ similarity to xanthan gum solution rather than water is taken into account. The other one is that food bolus from the mixture gel can flow as one coherent bolus rather than scattered particulates through the pharyngeal phase with a smaller variation of the flow velocity than gellan single gel. This explains well the longest $t_2$ for water, and smaller variation in flow velocity leads to higher cohesiveness sensation. For fluid foods, results will be presented in Section 14.4.

### 14.3.4 Laryngeal movement measurement

A bendable pressure sensor can intermittently monitor the laryngeal movement in swallowing by tracing the thyroid cartilage (Li et al., 2013) when it is attached with glue on skin surface. Using this system, the relationship between thyroid cartilage behavior and sensory score in swallowing xanthan gum (XG) and locust bean gum (LBG) solutions with different shear viscosities (Funami et al., 2017). Results confirm that two variables; activity of the thyroid cartilage and the maximum displacement of the thyroid cartilage highly ($|r|>0.9$) correlate negatively ($p<0.01$) to perceived cohesiveness. Similar approach has been conducted on gel samples (Matsuyama et al., 2021), where gels of different mechanical properties (varied by fracture force and fracture strain) were prepared by using XG-LBG composite preparation and low-acylated gellan gum in combination or alone. A 10 g model bolus, prepared by cutting the gel into 3 mm square cubes, was fed to healthy adult subjects who were ordered to hold the bolus for 3 s in the mouth and then to swallow at once. Results confirm that duration of laryngeal movement during swallowing highly ($|r|>0.8$) correlates negatively ($p<0.01$) to perceived cohesiveness (i.e., bolus uniformity) and swallowing ease. Also, linear multiple regression using 3 in vivo measurement variables, including duration of swallowing (from the laryngeal movement), suprahyoid musculature activity (from electromyography), and tongue activity (from tongue pressure measurement) during swallowing can predict perceived cohesiveness and swallowing ease with extremely high accuracy (presented by $r=0.987$ and $0.991$, respectively). Based on these results, suggest that bendable pressure sensor is effective in evaluating swallowability for either liquid or solid foods.

## 14.4 Texture and flavor interaction during food consumption
### 14.4.1 Flavor release control through texture modification

It has been confirmed empirically that perceived taste intensity in the mouth decreases with increasing viscosity or consistency of foods. Perceived sweetness intensity increases with decreasing fracture strain of polysaccharide gels (Morris, 1993). Polysaccharide gels which are smaller in fracture strain and more brittle in texture can be decomposed at an early stage of oral processing, and thus exposed surface area of food bolus which contacts with saliva increases in the mouth, stimulating perceived intensity of sweetness. This postulation is validated by finding that the mean particulate size of food bolus depends on the mechanical nature of foods, rather small for hard brittle foods (e.g., peanuts) but becoming larger for soft foods (e.g., gherkins) (Jalabert-Malbos et al., 2007). That is, particulate size of food bolus relates to perceived taste intensity in terms of saliva interaction. Furthermore, the effect of syneresis should be considered. The network structure for brittle gels is generally less dense than that for deformable gels, generating a larger amount of syneresis. The larger the syneresis, the more perceptive the intensity of water soluble tastant should be. In line with this, it has been reported that the flavor release of gel-like foods may be influenced by water-holding capacity (Nishinari, 2006). It is also indicated that the tastant (i.e., sucrose) amount released by free diffusion is negligible compared to that released through compression and fracture (Wang et al., 2014).

From another experiment, perceived overall flavor intensity ("overall" means sweetness and sourness in this case) decreases linearly with increasing hardness represented by fracture stress of polysaccharide gels (Clark, 2002). However, gels from gelatin and from the composite of low-acylated gellan gum, xanthan gum, and locust bean gum are positioned upward and downward the regression line; [Sensory score for overall flavor] $= 41.910 - 3.0075 \times$ [Hardness (lb-f)] ($r^2 = 0.875$), respectively. High flavor intensity for gelatin gel is due to its lower melting temperature, indicating that thermal properties are another essential attribute in determining flavor release (Nishinari, 2006). Low flavor intensity for the composite gel is due to its cohesive and deformable texture, entrapping flavor component(s) into the gel matrix and reducing contact surface area of food bolus with saliva. As a rheological parameter, fracture strain can present perceived flavor intensity better than fracture stress, and foods having smaller fracture strain and more brittle texture collapse at an early stage of oral processing and disintegrate into smaller particles. This rupture behavior of foods can increase the exposed surface area of food bolus in the mouth, stimulating perceived flavor intensity. Although these are authors' hypotheses, they may be reasonable when food breakdown behavior during oral processing is taken into account.

It has been reported that perceived aroma intensity is associated more with the release rate of aroma component(s) under a nonequilibrium condition than thermodynamic property of aroma component(s) (e.g., air-liquid partition coefficient) under an equilibrium condition (Bylaite et al., 2004). From this aspect, dynamic tests should be more suitable than static tests in discussing correlation with human perceived intensity. The study of Baek et al. (1999) using gelatin gels sweetened with sucrose and aromatized with furfuryl acetate provides supportive findings that perceived aroma intensity is more influenced by the release rate of the aroma compound than by its release amount (Baek et al., 1999). No significant difference is found in the maximum in-nose volatile concentration between different concentrations of gelatin gels, concluding that not the amount of volatile present but the rate of volatile release shows a good correlation with human perceived intensity. On the other hand, it has been shown that perceived aroma intensity is affected by texture rather than the released amount or release rate of aroma component(s) in the nasal cavity (Weel et al., 2002). Although biomechanics for explaining these findings are not yet clarified to the authors' knowledge, the presence of nerve cells in human brain for flavor perception in close proximity to those for texture perception (Rolls, 2011) may be one of the reasons.

For liquid and pasted foods, perceived flavor intensity (strawberry flavor in this case) decreases with increased viscosity thickened with polysaccharides (Morris, 1993). The diffusion coefficient of flavor component(s) decreases as viscosity increases, and accordingly perceived flavor intensity can decrease. However, the decrease in perceived flavor intensity is not expressed by a single linear correlation with viscosity. Perceived flavor intensity remains constant at low concentrations of polysaccharide added, beginning to decrease rapidly around the coil-overlap concentration $C*$ as the critical concentration for each polysaccharide. In other words, the effects of polysaccharides on perceived flavor intensity are standardized by $C*$ as long as their molecular conformation is random coil. In addition, perceived flavor intensity tends to increase at extremely low concentrations of polysaccharide added, for which increased residual time of food bolus in the oral cavity should be responsible although detailed evaluation is required. On the contrary, it has been reported that perceived sourness intensity does not present such a $C*$ dependency as described above when guar gum is used as a polysaccharide thickener (Malone et al., 2003b). It should be necessary to scrutinize the release profile of each flavor component in combination with different types of polysaccharide in terms of molecular conformation, polyelectrolyte etc.

## 14.4.2 Enhanced aroma perception through inhomogeneous spatial distribution

It has been indicated that perceived taste intensity is increased by inhomogeneous spatial distribution of tastants like sweetness (Holm et al., 2009; Mosca et al., 2013, 2010) and saltiness (Mosca et al., 2013; Noort et al., 2012, 2010) in solid foods, and similar effect can also be expected for liquid foods. As adaptation occurs when exposed continuously to taste stimulus for an extended period of time, leading to elevation of threshold, the effect of inhomogeneous spatial distribution can be attributed to prevention of adaptation (Meiselman, 1972). That is, discontinuous taste stimulus helps recovery from adaptation and enhances taste perception during oral processing. In contrast, there have been few reports on the effect of aroma inhomogeneous distribution. Using polysaccharide gels as a feeding sample, the effect of inhomogeneous aroma spatial distribution on its perceived intensity and eating behavior has been investigated (Nakao et al., 2013). Sample architecture used in the study is illustrated (Fig. 14.12), where degree of aroma inhomogeneity is variable by changing aroma concentrations in both the matrix and the dispersed gels with keeping the overall aroma concentration constant within one whole gel sample. There was no difference in bulk mechanical properties between the gel samples tested, and all the gel samples were soft enough to be processed by squeezing for size reduction without chewing. Under these experimental conditions, perceived aroma intensity increases for higher degree of aroma inhomogeneity. Also, the higher degree of aroma inhomogeneity increases both duration of oral processing and activity of the suprahyoid musculature measured by surface EMG, decreases particle size of expectorated bolus before swallowing, and increases saliva content in bolus (Fig. 14.13). These results show the importance of food structure design for the enhancement of perceived aroma intensity and human eating behavior.

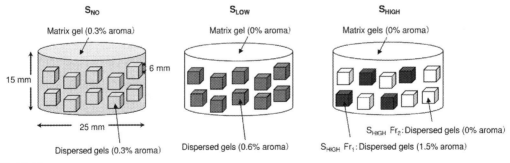

**FIG. 14.12**

Schematic drawings of the gel samples used. $S_{NO}$: homogeneous spatial distribution of aroma, where both the dispersed and the matrix gels contained 0.3% aroma compounds; $S_{LOW}$: the lower degree of inhomogeneous spatial distribution of aroma, where all dispersed gels contained 0.6% aroma compounds (0.3% in total); $S_{HIGH}$: the higher degree of inhomogeneous spatial distribution of aroma, where 40% of the dispersed gels contained 1.5% aroma compounds ($S_{HIGH}$ $Fr_1$) and remaining 60% of the dispersed gels did not contain aroma compounds ($S_{HIGH}$ $Fr_2$) (0.3% in total).

*Reproduced from Ishihara, S., Nakao, S., Nakauma, M., Funami, T., Hori, K., Ono, T., Kohyama, K. & Nishinari, K. (2013). Compression test of food gels on artificial tongue and its comparison with human test. Journal of Texture Studies, 44(2), 104–114. https://doi.org/10.1111/jtxs.12002.*

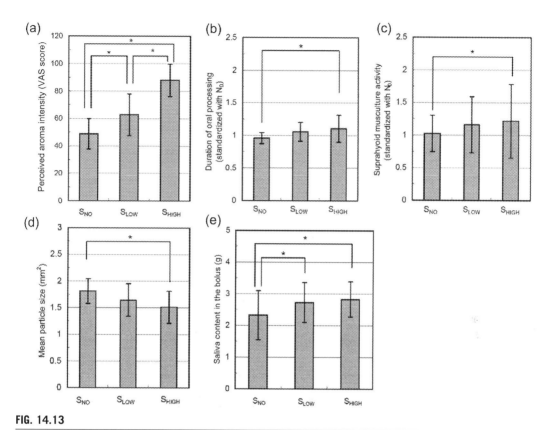

**FIG. 14.13**

The effect of the degree of aroma inhomogeneity on variables associated with food oral processing (A) Perceived aroma intensity; (B) electromyography duration of oral processing; (C) electromyography suprahyoid musculature activity; (D) mean particle size of bolus; (E) saliva content in bolus. Six subjects of 32.0 years old on average participated in these tests. Data were presented as means ± SD of the six subjects. Data with asterisk are significantly different ($p < 0.05$).

*Reproduced from Nakao, S., Ishihara, S., Nakauma, M. & Funami, T. (2013). Inhomogeneous spatial distribution of aroma compounds in food gels for enhancement of perceived aroma intensity and muscle activity during oral processing. Journal of Texture Studies, 44(4), 289–300. https://doi.org/10.1111/jtxs.12023.*

### 14.4.3 Modification of human eating behavior by aroma perception

In relation to the interaction of texture-aroma perceptions during food oral processing, one hypothesis is that texture perception (e.g., consistency or hardness) and eating behavior can be changed by the degree of aroma inhomogeneity (Funami et al., 2016). That is, the higher the degree of aroma inhomogeneity, the less the perceived hardness of foods can be. Also, eating behavior can be less consistency-dependent as the degree of aroma inhomogeneity is higher. In the study, the same sample architecture as Nakao et al. (2013) was used but with 3 different consistency levels of the gel sample. By using surface EMG, variables, including duration and activity of the suprahyoid musculature, generally increase with gel consistency at each degree of aroma inhomogeneity. At the same time, the

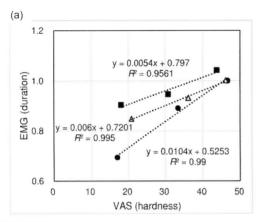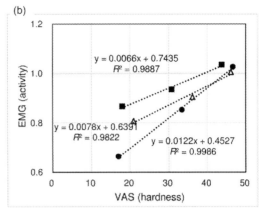

**FIG. 14.14**

Relationship between sensory and EMG (electromyography) variables. Linear regression analysis was done for the same degree of aroma inhomogeneity with various consistencies. (A) EMG duration of oral processing as a function of visual analogue scale (VAS) score of perceived hardness; (B) EMG activity of the suprahyoid musculature as a function of VAS score of perceived hardness. Closed circles: homogeneous spatial distribution of aroma; open triangles: the lower degree of inhomogeneous spatial distribution of aroma; closed squares: the higher degree of inhomogeneous spatial distribution of aroma.

*Reproduced from Funami, T., Nakao, S., Isono, M., Ishihara, S. & Nakauma, M. (2016). Effects of food consistency on perceived intensity and eating behavior using soft gels with varying aroma inhomogeneity. Food Hydrocolloids, 52, 896–905. https://doi.org/10.1016/j.foodhyd.2015.08.029.*

intensity of aroma and taste perception is decreased with gel consistency. Another important finding is that for higher aroma inhomogeneity the increases of EMG variables are less consistency-dependent, and perceived hardness intensity is lower when compared at equivalent eating effort (Fig. 14.14). No interaction is found in each EMG variable between consistency and aroma inhomogeneity. From these results, change of aroma inhomogeneity can be a strategy for promotion of oral processing without loading too much for eating on human, and this should be useful for food product development for elderly and dysphagia patients in particular.

## 14.5 Structure and formulation design of food products using hydrocolloids

### 14.5.1 Use of polysaccharides as a texture modifier in elderly foods

Polysaccharides have been used as a texture modifier in food products for a long time, and this should also be a key strategy for elderly foods development. Polysaccharides used in the food industry are all natural as their sources, found in seeds, tree sap, seaweeds, microorganism, etc. (Funami, 2011). Polysaccharides serve a variety of purposes in processed food products due to their capabilities of thickening, gelling, water holding, dispersing, stabilizing, film forming, and foaming (Funami, 2011). Through these functions, mechanical, geometrical, and surface properties of foods can change, and thus food texture is modified.

In Japan, various types of food products are being marketed for elderly. Representatives are semi-solidified (either thickened or gelled) enteral nutrition, water-supply jelly, and nutrition-supply jelly as a ready-to-eat product. Also, instant gelling agent for pasted/pureed foods and dysphagia thickener, both which are for supplement purpose and in dry powder form in most cases, work at a small addition level to existing foods and beverages. The terms "jelly", "gelling" and "thickener" all relate directly to polysaccharide functions.

Semisolidified enteral nutrition can reduce complication risks encountered in feeding thin liquid enteral nutrition via an enteral tube or percutaneous endoscopic gastrostomy. These risks include the secondary aspiration caused by gastroesophageal reflux, local skin infection as a result of leakage from gastrostomy, bedsore by prolonged feeding time, and diarrhea by rapid absorption and dumping (Goda, 2008). By selecting and combining multiple polysaccharides, intensity and quality of viscosity can be modified to achieve smooth flow through the tube, and feeding speed can be controlled to prevent diarrhea and reflux (Tanishima et al., 2010). This type of enteral nutrition is gaining market share in Japan (Nakauma et al., 2014, 2012). Regarding water-supply jelly, the creation of new texture has been tried by polysaccharide technology for mimicking not only texture but also the appearance of real fruits (Funami, 2013). Appearance can enhance human appetite as one of the dominant attributes for food palatability. Visual food recognition can influence human swallowing function (Kamiya et al., 2015), demonstrating that a larger amount of saliva is secreted when subjects see pictures of normal diets compared to those of corresponding blended foods (i.e., foods processed by blender). Another advantage of using polysaccharides instead of real fruits lies in nutrition control. For example, control of potassium intake is essential for patients with kidney dialysis treatment, and this type of food products is flexible in nutrition level by changing formulation. Instant gelling agent is used for shape retention and prevention of syneresis for pasted and pureed foods (Funami, 2013). Foods, including cooked rice, fish, meat, and vegetable, are pasted together with the gelling agent and water or soup if necessary. By remolding to the original shape, people can recognize what they are eating. Polysaccharides which hydrate or dissolve in water without heating are preferably used in this type of gelling agent.

### 14.5.2 Usefulness of xanthan gum as dysphagia thickener

Low viscosity liquids can be aspirated by dysphagia patients (Nishinari et al., 2011). To prevent this, dysphagia thickener, called toromi in Japan, has been used at hospitals and nursing-care facilities. Dysphagia thickener contains polysaccharides as a main ingredient, where xanthan gum has been replacing modified starch (Funami et al., 2009, 2006). Dysphagia thickener is usually stirred manually to disperse it in liquid foods and beverages like water, tea, juice, soup etc. Dysphagia thickener is required to exhibit functions of no lumping and rapid viscosity increase even under weak stirring conditions. Stable viscosity upon storage and usability in wide range of liquid foods are also required, and preferable organoleptic properties, including cohesive texture and no off-flavor, should be provided. Xanthan gum meets these requirements and thus is practically the best choice (Funami et al., 2006).

Most polysaccharides, not limited to xanthan, hydrate quickly in water and thus form lumps easily. Lumping is the most unfavorable attribute for dysphagia thickener since textural homogeneity cannot be ensured when this happens, and this defect is eliminated by powder processing, mostly granulation. The relationship between dispersibility and viscosity enhancement can be described as a seesaw, in which when one is improved, the other is deteriorated (Funami et al., 2006). For good balance of these

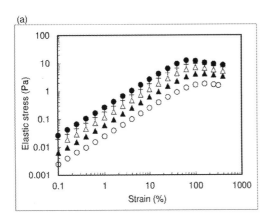 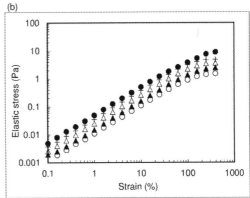

**FIG. 14.15**

Elastic stress (storage modulus × strain) of xanthan gum and locust bean gum solutions plotted as a function of strain. Measurements were carried out at 20°C. Concentration of each polysaccharide: 0.3%, 0.45%, 0.6%, 0.75% and 0.9% for xanthan gum (A) or 0.5%, 0.6%, 0.7%, 0.75%, and 0.8% for locust bean gum (B), represented by open circles, closed triangles, open triangles, crosses, and closed circles in increasing order. The yield stress and strain are estimated from the peak in the curve.

*Reproduced from Nakauma, M., Ishihara, S., Funami, T. & Nishinari, K. (2011). Swallowing profiles of food polysaccharide solutions with different flow behaviors. Food Hydrocolloids, 25(5), 1165–1173. https://doi.org/10.1016/j.foodhyd.2010.11.003.*

two attributes, technical know-how is required in the granulation process, including optimization of certain operation conditions (e.g., inlet air temperature, liquid spray rate) or usage of certain co-agents in a liquid spray, either of which aims to control hydration speed of polysaccharides.

Xanthan gum solutions are fluid, but from rheological point of view, they behave like a solid or gel (Nakauma et al., 2011). Xanthan gum solutions have elastic character represented by yield stress in contrast to locust bean gum solutions. Yield stress is detected as the peak in Fig. 14.15, in which data from strain-dependence of dynamic viscoelasticity are replotted. Yield stress is a stress at which the flow begins (Bourne, 2002), relating to the degree of internal binding and thus presenting cohesiveness of food bolus. "Structured fluid" or "weak gel" should be a preferable food structure (Nakauma et al., 2011) for swallowing ease, and this food structure can be quantified also by yield stress: ca. 7.0–9.0 Pa in steady shear viscosity range of ca. 0.9–1.2 Pa s at $10 s^{-1}$ (Nakauma et al., 2011).

It has been believed that flow speed of liquid food bolus through the pharyngeal phase is moderated by thickening and that shear viscosity can only determine flow behavior; the higher the viscosity, the slower the flow speed can be through the pharyngeal phase. Whether it is always true or not and which shear rate should be used for viscosity measurement to obtain the best correlation with physiological flow behavior should be of research interest. For the first question, the answer must be "no" when results from our studies (Funami et al., 2017; Nakauma et al., 2011) are considered. Elasticity should not be negligible or rather important in explaining physiological flow behavior of food bolus. This is supported by Chen and Lolivret (2011), concluding that extensional stretch-ability should be one of the most important mechanical parameter for swallowing ease (Chen & Lolivret, 2011). In addition, non-Newtonian fluids are reportedly safer to swallow and lower in aspiration risk than Newtonian fluids (Meng et al., 2005). The preference on non-Newtonian fluids for safe swallowing refers to a need of elasticity to ensure mechanical cohesiveness of food bolus rather than the issue of shear-rate dependence (Funami et al., 2012). For the second question, the maximum shear rate is expected to range between $400\,s^{-1}$ (Meng et al., 2005) and $3000\,s^{-1}$ or even larger $180,000\,s^{-1}$ (Nicosia & Robbins, 2001) depending on the type of simulation. However, it is still quite difficult to estimate such an in vivo characteristic by computer calculation due to irregularity of oral geometry.

Acoustic analysis of swallowing sound was carried out to investigate the correlation with sensory evaluation using young healthy adults as subjects and polysaccharide solutions from either xanthan gum or locust bean gum as a feeding sample (Funami, 2011; Nakauma et al., 2011). Technical details were mentioned before. Duration $t_2$ for xanthan gum decreases for higher concentrations despite viscosity increase (ca. 60% decrease when viscosity is increased from 0.001 to 1.61 Pa s), whereas that for locust bean gum is much less concentration-dependent (ca. 15% decrease when viscosity is increased from 0.001 to 1.53 Pa s). Also, duration $t_2$ for locust bean gum is consistently larger than that for xanthan gum when compared at equivalent shear viscosity (Fig. 14.16). This is consistent with results from ultrasonic pulse Doppler measurement, indicating that the velocity spectrum is narrower in the distribution range with increasing concentration of food polysaccharide thickeners (Kumagai et al., 2009).

From sensory evaluation on a visual analogue scale, perceived cohesiveness is scored higher with increasing concentration in general for either xanthan gum (av. score ranging from 50.4 to 86.0) or locust bean gum (av. score ranging from 32.3 to 45.3). A higher score represents easier bolus formation in the pharyngeal phase during swallowing. When compared at equivalent shear viscosities, xanthan gum is scored consistently higher than locust bean gum solutions. On the other hand, perceived swallowing ease for xanthan gum is peaked at 0.6% (av. score 80.2), whereas that for locust bean gum

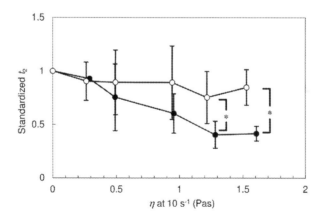

**FIG. 14.16**

Duration $t_2$ in swallowing polysaccharide solutions as a function of steady shear viscosity $\eta$ at $10\,s^{-1}$. This shear rate was selected as a condition in perceiving liquid viscosity in mouth (Shama & Sherman, 1973). Both mechanical and in vivo measurements were carried out at 20 °C. Closed circles: xanthan gum; open circles: locust bean gum. The serving volume of polysaccharide solutions was 15 mL. Each datum was standardized with control (water). Data with an asterisk are significantly different between xanthan and locust at $p < 0.05$.

*Reproduced from Nakauma, M., Ishihara, S., Funami, T. & Nishinari, K. (2011). Swallowing profiles of food polysaccharide solutions with different flow behaviors. Food Hydrocolloids, 25(5), 1165–1173. https://doi.org/10.1016/j.foodhyd.2010.11.003.*

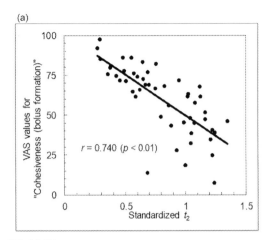 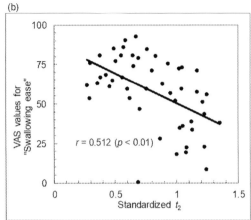

**FIG. 14.17**

Correlation between $t_2$ and score of each sensory attribute in swallowing polysaccharide solutions. The serving volume of polysaccharide solutions was 15 ml. (A) perceived cohesiveness; (B) perceived swallowing ease. Duration $t_2$ was standardized with the corresponding data for water.

*Reproduced from Nakauma, M., Ishihara, S., Funami, T. & Nishinari, K. (2011). Swallowing profiles of food polysaccharide solutions with different flow behaviors. Food Hydrocolloids, 25(5), 1165–1173. https://doi.org/10.1016/j.foodhyd.2010.11.003.*

generally decreases with increasing concentration (av. score ranging from 21.3 to 51.0). When compared at equivalent shear viscosity, xanthan gum is scored consistently higher than locust bean gum. Duration $t_2$ correlates well to either perceived cohesiveness or perceived swallowing ease (Fig. 14.17). Xanthan gum bolus flows through the pharyngeal phase as one coherent bolus with a small variation of flow velocity, and this flow behavior leads to a greater sensation of swallowing ease.

## 14.6 Conclusion

The texture is the key for food product development. The best knowledge on texture can lead to innovative product development. In the food industry, instrumental quantification of textural attributes should be important, enabling to share a guide for texture design of products in an objective manner. The authors wish that this chapter would provide food manufactures with insight into novel product development, contributing to improved Quality of Life and Activities particularly in elderlies and dysphagia patients. Digestion and adsorption controls of nutrients and healthy ingredients, similar to the concept of drug delivery system, can also be modified by texture or structural design of foods, leading to not only healthy life for humans but also reduced food loss through beneficial use of food or agricultural materials. The beneficial use of plant protein should contribute to secure alternative protein sources and to reduce greenhouse gas. Based on colloid science, the creation of new products and technologies as a solution to food associated problems should be the task for the next years.

# References

Alexander, R. M. (1998). News of chews: the optimization of mastication. *Nature, 329*, 329–329. https://doi.org/10.1038/34775.

Álvarez, M. D., Paniagua, J., & Herranz, B. (2020). Assessment of the miniature kramer shear cell to measure both solid food and bolus mechanical properties and their interplay with oral processing behavior. *Foods, 9*(5). https://doi.org/10.3390/foods9050613.

Arai, E., & Yamada, Y. (1993). Effect of the texture of food on the masticatory process. *Japanese Journal of Oral Biology, 35*(4), 312–322. https://doi.org/10.2330/joralbiosci1965.35.312.

Baek, I., Linforth, R. S. T., Blake, A., & Taylor, A. J. (1999). Sensory perception is related to the rate of change of volatile concentration in-nose during eating of model gels. *Chemical Senses, 24*(2), 155–160. https://doi.org/10.1093/chemse/24.2.155.

Baier, S., Elmore, D., Guthrie, B., Lindgren, T., Smith, S., Steinbach, A., & Läuger, J. (2009). A new tribology device for assessing mouthfeel attributes of foods. In *5th International symposium on food structure and rheology* (pp. 432–435).

Bazilevsky, A. V., Entov, V. M., & Rozhkov, A. N. (1990). *Liquid Filament Microrheometer and Some of Its Applications* (pp. 41–43). https://doi.org/10.1007/978-94-009-0781-2_21.

Biegler, M., Delius, J., Käsdorf, B. T., Hofmann, T., & Lieleg, O. (2016). Cationic astringents alter the tribological and rheological properties of human saliva and salivary mucin solutions. *Biotribology, 6*, 12–20. https://doi.org/10.1016/j.biotri.2016.03.002.

Bongaerts, J. H. H., Rossetti, D., & Stokes, J. R. (2007). The lubricating properties of human whole saliva. *Tribology Letters, 27*(3), 277–287. https://doi.org/10.1007/s11249-007-9232-y.

Bourne, M. C. (2002). In M. C. Bourne (Ed.), *Physics and texture* (2nd ed., pp. 59–106). Academic Press.

Brossard, N., Cai, H., Osorio, F., Bordeu, E., & Chen, J. (2016). "Oral" Tribological Study on the Astringency Sensation of Red Wines. *Journal of Texture Studies, 47*(5), 392–402. https://doi.org/10.1111/jtxs.12184.

Bylaite, E., Ilgūnaitė, Ž., Meyer, A. S., & Adler-Nissen, J. (2004). Influence of λ-carrageenan on the release of systematic series of volatile flavor compounds from viscous food model systems. *Journal of Agricultural and Food Chemistry, 52*(11), 3542–3549. https://doi.org/10.1021/jf0354996.

Carpenter, G., Bozorgi, S., Vladescu, S., Forte, A. E., Myant, C., Potineni, R. V., Reddyhoff, T., & Baier, S. K. (2019). A study of saliva lubrication using a compliant oral mimic. *Food Hydrocolloids, 92*, 10–18. https://doi.org/10.1016/j.foodhyd.2019.01.049.

Carvalho-Da-Silva, A. M., Van Damme, I., Taylor, W., Hort, J., & Wolf, B. (2013). Oral processing of two milk chocolate samples. *Food & Function, 4*(3), 461–469. https://doi.org/10.1039/c2fo30173c.

Chen, J. (2009). Food oral processing—A review. *Food Hydrocolloids, 23*(1), 1–25. https://doi.org/10.1016/j.foodhyd.2007.11.013.

Chen, J., Feng, M., Gonzalez, Y., & Pugnaloni, L. A. (2008). Application of probe tensile method for quantitative characterisation of the stickiness of fluid foods. *Journal of Food Engineering, 87*(2), 281–290. https://doi.org/10.1016/j.jfoodeng.2007.12.004.

Chen, J., & Lolivret, L. (2011). The determining role of bolus rheology in triggering a swallowing. *Food Hydrocolloids, 25*(3), 325–332. https://doi.org/10.1016/j.foodhyd.2010.06.010.

Chen, J., & Stokes, J. R. (2012). Rheology and tribology: Two distinctive regimes of food texture sensation. *Trends in Food Science and Technology, 25*(1), 4–12. https://doi.org/10.1016/j.tifs.2011.11.006.

Choi, H., Mitchell, J. R., Gaddipati, S. R., Hill, S. E., & Wolf, B. (2014). Shear rheology and filament stretching behaviour of xanthan gum and carboxymethyl cellulose solution in presence of saliva. *Food Hydrocolloids, 40*, 71–75. https://doi.org/10.1016/j.foodhyd.2014.01.029.

Clark, R. (2002). Influence of hydrocolloids on flavour release and sensory-instrumental correlations. In *Vol. 1. Gums and stabilisers for the food industry* (pp. 217–224). The Royal Society of Chemistry.

Debruijne, D. W., Hendrickx, H. A. C. M., Alderliesten, L., & Delooff, J. (1993). Mouthfeel of foods. *Food Colloids and Polymers: Stability and Mechanical Properties*, *113*, 204–213.

Dresselhuis, D. M., de Hoog, E. H. A., Cohen Stuart, M. A., & van Aken, G. A. (2008). Application of oral tissue in tribological measurements in an emulsion perception context. *Food Hydrocolloids*, *22*(2), 323–335. https://doi.org/10.1016/j.foodhyd.2006.12.008.

Espinosa, Y. G., & Chen, J. (2012). Applications of electromyography (EMG) technique for eating studies. In *Food Oral Processing: Fundamentals of Eating and Sensory Perception* (pp. 289–317). Wiley. https://doi.org/10.1002/9781444360943.ch13.

Foster, K. D., Grigor, J. M. V., Cheong, J. N., Yoo, M. J. Y., Bronlund, J. E., & Morgenstern, M. P. (2011). The role of oral processing in dynamic sensory perception. *Journal of Food Science*, *76*(2), R49–R61. https://doi.org/10.1111/j.1750-3841.2010.02029.x.

Foster, K. D., Woda, A., & Peyron, M. A. (2006). Effect of texture of plastic and elastic model foods on the parameters of mastication. *Journal of Neurophysiology*, *95*(6), 3469–3479. https://doi.org/10.1152/jn.01003.2005.

Funami, T. (2011). Next target for food hydrocolloid studies: Texture design of foods using hydrocolloid technology. *Food Hydrocolloids*, *25*(8), 1904–1914. https://doi.org/10.1016/j.foodhyd.2011.03.010.

Funami, T. (2013). Use of gelling agents for elderly foods. In *Stabilization and functionality of gels, Application development for the next generation* (pp. 395–401). Technical Information Institute Co, Ltd.

Funami, T., Ishihara, S., Nakauma, M., Kohyama, K., & Nishinari, K. (2012). Texture design for products using food hydrocolloids. *Food Hydrocolloids*, *26*(2), 412–420. https://doi.org/10.1016/j.foodhyd.2011.02.014.

Funami, T., Matsuyama, S., Ikegami, A., Nakauma, M., Hori, K., & Ono, T. (2017). In vivo measurement of swallowing by monitoring thyroid cartilage movement in healthy subjects using thickened liquid samples and its comparison with sensory evaluation. *Journal of Texture Studies*, *48*(6), 494–506. https://doi.org/10.1111/jtxs.12261.

Funami, T., Nakao, S., Isono, M., Ishihara, S., & Nakauma, M. (2016). Effects of food consistency on perceived intensity and eating behavior using soft gels with varying aroma inhomogeneity. *Food Hydrocolloids*, *52*, 896–905. https://doi.org/10.1016/j.foodhyd.2015.08.029.

Funami, T., Tobita, M., Hoshi, M., Toyama, Y., Sato, N., Konno, S., Hikita, H., Ito, S., Yoshihira, K., & Fujisaki, T. (2009). Methods for measuring the mechanical properties of dysphagia thickening agents: The usefulness of texture profile analysis. *The Japanese Journal of Dysphagia Rehabilitation*, *13*, 10–19.

Funami, T., Tsutsumino, T., & Kishimoto, K. (2006). Thickening and gelling agents used for dysphagia thickeners In Stabilization and functionality of gels, Application and diets. *Journal of Cookery Science Japan*, *39*, 233–239.

Garrec, D. A., & Norton, I. T. (2012). The influence of hydrocolloid hydrodynamics on lubrication. *Food Hydrocolloids*, *26*(2), 389–397. https://doi.org/10.1016/j.foodhyd.2011.02.017.

Garrec, D. A., & Norton, I. T. (2013). Kappa carrageenan fluid gel material properties. Part 2: Tribology. *Food Hydrocolloids*, *33*(1), 160–167. https://doi.org/10.1016/j.foodhyd.2013.01.019.

Giasson, S., Israelachvili, J., & Yoshizawa, H. (1997). Thin film morphology and tribology study of mayonnaise. *Journal of Food Science*, *62*(4), 640–652. https://doi.org/10.1111/j.1365-2621.1997.tb15427.x.

Goda, F. (2008). Problems and their solutions for gastrostomy tube feeding of semi-solid nutrients. *Japanese Society for Parenteral and Enteral Nutrition*, *23*, 235–241.

Godoi, F. C., Bhandari, B. R., & Prakash, S. (2017). Tribo-rheology and sensory analysis of a dairy semi-solid. *Food Hydrocolloids*, *70*, 240–250. https://doi.org/10.1016/j.foodhyd.2017.04.011.

Hamlet, S. L., Patterson, R. L., Fleming, S. M., & Jones, L. A. (1992). Sounds of swallowing following total laryngectomy. *Dysphagia*, *7*(3), 160–165. https://doi.org/10.1007/BF02493450.

Hayakawa, F., Kazami, Y., Fujimoto, S., Kikuchi, H., & Kohyama, K. (2009). Time-intensity analysis of sourness of commercially produced Gummy Jellies available in Japan. *Food Science and Technology Research*, *15*(1), 75–82. https://doi.org/10.3136/fstr.15.75.

He, Q., Hort, J., & Wolf, B. (2016). Predicting sensory perceptions of thickened solutions based on rheological analysis. *Food Hydrocolloids*, *61*, 221–232. https://doi.org/10.1016/j.foodhyd.2016.05.010.

Holm, K., Wendin, K., & Hermansson, A. M. (2009). Sweetness and texture perceptions in structured gelatin gels with embedded sugar rich domains. *Food Hydrocolloids*, *23*(8), 2388–2393. https://doi.org/10.1016/j.foodhyd.2009.06.016.

Hori, K., Hayashi, H., Yokoyama, S., Ono, T., Ishihara, S., Magara, J., Taniguchi, H., Funami, T., Maeda, Y., & Inoue, M. (2015). Comparison of mechanical analyses and tongue pressure analyses during squeezing and swallowing of gels. *Food Hydrocolloids*, *44*, 145–155. https://doi.org/10.1016/j.foodhyd.2014.09.029.

Hori, K., Ono, T., Tamine, K. I., Kondo, J., Hamanaka, S., Maeda, Y., Dong, J., & Hatsuda, M. (2009). Newly developed sensor sheet for measuring tongue pressure during swallowing. *Journal of Prosthodontic Research*, *53*(1), 28–32. https://doi.org/10.1016/j.jpor.2008.08.008.

Ishihara, S., Isono, M., Nakao, S., Nakauma, M., Funami, T., Hori, K., Ono, T., Kohyama, K., & Nishinari, K. (2014). Instrumental uniaxial compression test of gellan gels of various mechanical properties using artificial tongue and its comparison with human oral strategy for the first size reduction. *Journal of Texture Studies*, *45*(5), 354–366. https://doi.org/10.1111/jtxs.12080.

Ishihara, S., Nakao, S., Nakauma, M., Funami, T., Hori, K., Ono, T., Kohyama, K., & Nishinari, K. (2013). Compression test of food gels on artificial tongue and its comparison with human test. *Journal of Texture Studies*, *44*(2), 104–114. https://doi.org/10.1111/jtxs.12002.

Ishihara, S., Nakauma, M., Funami, T., Odake, S., & Nishinari, K. (2011a). Swallowing profiles of food polysaccharide gels in relation to bolus rheology. *Food Hydrocolloids*, *25*(5), 1016–1024. https://doi.org/10.1016/j.foodhyd.2010.09.022.

Ishihara, S., Nakauma, M., Funami, T., Odake, S., & Nishinari, K. (2011b). Viscoelastic and fragmentation characters of model bolus from polysaccharide gels after instrumental mastication. *Food Hydrocolloids*, *25*(5), 1210–1218. https://doi.org/10.1016/j.foodhyd.2010.11.008.

Ishihara, S., Nakauma, M., Funami, T., Tanaka, T., Nishinari, K., & Kohyama, K. (2011). Electromyography during oral processing in relation to mechanical and sensory properties of soft gels. *Journal of Texture Studies*, *42*(4), 254–267. https://doi.org/10.1111/j.1745-4603.2010.00272.x.

Jalabert-Malbos, M. L., Mishellany-Dutour, A., Woda, A., & Peyron, M. A. (2007). Particle size distribution in the food bolus after mastication of natural foods. *Food Quality and Preference*, *18*(5), 803–812. https://doi.org/10.1016/j.foodqual.2007.01.010.

Kamiya, M., Ota, K., Morishima, K., Sawa, S., & Kondo, I. (2015). Examination of the effect of visual food perception on swallowing. *The Japanese Journal of Dysphagia Rehabilitation*, *19*, 24–32.

Kieserling, K., Schalow, S., & Drusch, S. (2018). Method development and validation of tribological measurements for differentiation of food in a rheometer. *Biotribology*, *16*, 25–34. https://doi.org/10.1016/j.biotri.2018.09.002.

Kim, J. M., Wolf, F., & Baier, S. K. (2015). Effect of varying mixing ratio of PDMS on the consistency of the soft-contact Stribeck curve for glycerol solutions. *Tribology International*, *89*, 46–53. https://doi.org/10.1016/j.triboint.2014.12.010.

Koç, H., Vinyard, C. J., Essick, G. K., & Foegeding, E. A. (2013). Food oral processing: conversion of food structure to textural perception. *Annual Review of Food Science and Technology*, *4*(1), 237–266. https://doi.org/10.1146/annurev-food-030212-182637.

Kohyama, K. (2005). Texture characteristics. In K. Nishinari, H. Ohkoshi, & K. Kohyama (Eds.), *Handbook for Texture Creation* (pp. 281–304). Science Forum Inc.

Kohyama, K. (2015). Oral sensing of food properties. *Journal of Texture Studies*, *46*(3), 138–151. https://doi.org/10.1111/jtxs.12099.

Kohyama, K. (2020). Compression test of soft gellan gels using a soft machine equipped with a transparent artificial tongue. *Journal of Texture Studies*, *51*(4), 612–621. https://doi.org/10.1111/jtxs.12515.

Kohyama, K., Hanyu, T., Hayakawa, F., & Sasaki, T. (2010). Electromyographic measurement of eating behaviors for buckwheat noodles. *Bioscience, Biotechnology, and Biochemistry*, *74*(1), 56–62. https://doi.org/10.1271/bbb.90539.

Kohyama, K., Hatakeyama, E., Dan, H., & Sasaki, T. (2005). Effects of sample thickness on bite force for raw carrots and fish gels. *Journal of Texture Studies, 36*(2), 157–173. https://doi.org/10.1111/j.1745-4603.2005.00009.x.

Kohyama, K., Ishihara, S., Nakauma, M., & Funami, T. (2021). Fracture phenomena of soft gellan gum gels during compression with artificial tongues. *Food Hydrocolloids, 112*. https://doi.org/10.1016/j.foodhyd.2020.106283.

Kohyama, K., Ishihara, S., Nakauma, M., & Funami, T. (2019). Compression test of soft food gels using a soft machine with an artificial tongue. *Foods, 8*(6). https://doi.org/10.3390/foods8060182.

Kohyama, K., & Nishi, M. (1997). Direct measurement of biting pressures for crackers using a multiple-point sheet sensor. *Journal of Texture Studies, 28*(6), 605–617. https://doi.org/10.1111/j.1745-4603.1997.tb00141.x.

Kohyama, K., Nishi, M., & Suzuki, T. (1997). Measuring texture of crackers with a multiple-point sheet sensor. *Journal of Food Science, 62*(5), 922–925. https://doi.org/10.1111/j.1365-2621.1997.tb15007.x.

Kohyama, K., Ohtsubo, K., Toyoshima, H., & Shiozawa, K. (1998). Electromyographic study on cooked rice with different amylose contents. *Journal of Texture Studies, 29*(1), 101–113. https://doi.org/10.1111/j.1745-4603.1998.tb00156.x.

Kohyama, K., Sakai, T., Azuma, T., Mizuguchi, T., & Kimura, I. (2001). Pressure distribution measurement in biting surimi gels with molars using a multiple-point sheet sensor. *Bioscience, Biotechnology, and Biochemistry, 65*(12), 2597–2603. https://doi.org/10.1271/bbb.65.2597.

Kohyama, K., Sasaki, T., & Dan, H. (2003). Active stress during compression testing of various foods measured using a multiple-point sheet sensor. *Bioscience, Biotechnology, and Biochemistry, 67*(7), 1492–1498. https://doi.org/10.1271/bbb.67.1492.

Kohyama, K., Sasaki, T., & Hayakawa, F. (2008). Characterization of food physical properties by the mastication parameters measured by electromyography of the jaw-closing muscles and mandibular kinematics in young adults. *Bioscience, Biotechnology, and Biochemistry, 72*(7), 1690–1695. https://doi.org/10.1271/bbb.70769.

Kohyama, K., Sawada, H., Nonaka, M., Kobori, C., Hayakawa, F., & Sasaki, T. (2007). Textural evaluation of rice cake by chewing and swallowing measurements on human subjects. *Bioscience, Biotechnology, and Biochemistry, 71*(2), 358–365. https://doi.org/10.1271/bbb.60276.

Kohyama, K., Yamaguchi, M., Kobori, C., Nakayama, Y., Hayakawa, F., & Sasaki, T. (2005). Mastication effort estimated by electromyography for cooked rice of differing water content. *Bioscience, Biotechnology, and Biochemistry, 69*(9), 1669–1676. https://doi.org/10.1271/bbb.69.1669.

Koliandris, A. L., Rondeau, E., Hewson, L., Hort, J., Taylor, A. J., Cooper-White, J., & Wolf, B. (2011). Food grade Boger fluids for sensory studies. *Applied Rheology, 21*(1). https://doi.org/10.3933/ApplRheol-21-13777.

Krop, E. M., Hetherington, M. M., Holmes, M., Miquel, S., & Sarkar, A. (2019). On relating rheology and oral tribology to sensory properties in hydrogels. *Food Hydrocolloids, 88*, 101–113. https://doi.org/10.1016/j.foodhyd.2018.09.040.

Krop, E. M., Hetherington, M. M., Miquel, S., & Sarkar, A. (2020). Oral processing of hydrogels: Influence of food material properties versus individuals' eating capability. *Journal of Texture Studies, 51*(1), 144–153. https://doi.org/10.1111/jtxs.12478.

Krzeminski, A., Wohlhüter, S., Heyer, P., Utz, J., & Hinrichs, J. (2012). Measurement of lubricating properties in a tribosystem with different surface roughness. *International Dairy Journal, 26*(1), 23–30. https://doi.org/10.1016/j.idairyj.2012.02.005.

Kumagai, H., Tashiro, A., Hasegawa, A., Kohyama, K., & Kumagai, H. (2009). Relationship between flow properties of thickener solutions and their velocity through the pharynx measured by the ultrasonic pulse doppler method. *Food Science and Technology Research, 15*(3), 203–210. https://doi.org/10.3136/fstr.15.203.

Laiho, S., Williams, R. P. W., Poelman, A., Appelqvist, I., & Logan, A. (2017). Effect of whey protein phase volume on the tribology, rheology and sensory properties of fat-free stirred yoghurts. *Food Hydrocolloids, 67*, 166–177. https://doi.org/10.1016/j.foodhyd.2017.01.017.

Lazareck, L. J., & Moussavi, Z. (2004). Swallowing sound characteristics in healthy and dysphagic individuals. In *Vol. 26. Annual International Conference of the IEEE Engineering in Medicine and Biology- Proceedings* (pp. 3820–3823).

Li, Q., Hori, K., Minagi, Y., Ono, T., Chen, Y., Kondo, J., Fujiwara, S., Tamine, K., Hayashi, H., Inoue, M., Maeda, Y., & Fridman, E. A. (2013). Development of a system to monitor laryngeal movement during swallowing using a bend sensor. *PLoS One*, *8*(8), e70850. https://doi.org/10.1371/journal.pone.0070850.

Liu, K., Stieger, M., van der Linden, E., & van de Velde, F. (2015). Fat droplet characteristics affect rheological, tribological and sensory properties of food gels. *Food Hydrocolloids*, *44*, 244–259. https://doi.org/10.1016/j.foodhyd.2014.09.034.

Lucas, P. W., Prinz, J. F., Agrawal, K. R., & Bruce, I. C. (2002). Food physics and oral physiology. *Food Quality and Preference*, *13*(4), 203–213. https://doi.org/10.1016/S0950-3293(00)00036-7.

Luengo, G., Tsuchiya, M., Heuberger, M., & Israelachvili, J. (1997). Thin film rheology and tribology of chocolate. *Journal of Food Science*, *62*(4), 767–812. https://doi.org/10.1111/j.1365-2621.1997.tb15453.x.

Luyten, H., & Van Vliet, T. (1995). Fracture properties of starch gels and their rate dependency. *Journal of Texture Studies*, *26*(3), 281–298. https://doi.org/10.1111/j.1745-4603.1995.tb00966.x.

Lv, Z., Chen, J., & Holmes, M. (2017). Human capability in the perception of extensional and shear viscosity. *Journal of Texture Studies*, *48*(5), 463–469. https://doi.org/10.1111/jtxs.12255.

Malone, M. E., Appelqvist, I. A. M., & Norton, I. T. (2003a). Oral behaviour of food hydrocolloids and emulsions. Part 1. Lubrication and deposition considerations. *Food Hydrocolloids*, *17*(6), 763–773. Elsevier https://doi.org/10.1016/S0268-005X(03)00097-3.

Malone, M. E., Appelqvist, I. A. M., & Norton, I. T. (2003b). Oral behaviour of food hydrocolloids and emulsions. Part 2. Taste and aroma release. *Food Hydrocolloids*, *17*(6), 775–784. Elsevier https://doi.org/10.1016/S0268-005X(03)00098-5.

Masashi, O. (2008). Properties of rice porridge (gruel) as a meal for patients with swallowing disorder. *The Journal of Gifu Dental Society*, *35*, 7–12.

Matsuyama, S., Nakauma, M., Funami, T., Hori, K., & Ono, T. (2021). Human physiological responses during swallowing of gel-type foods and its correlation with textural perception. *Food Hydrocolloids*, *111*, 106353. https://doi.org/10.1016/j.foodhyd.2020.106353.

Meiselman, H. L. (1972). Human taste perception. *C R C Critical Reviews in Food Technology*, *3*(1), 89–119. https://doi.org/10.1080/10408397209527135.

Meng, Y., Rao, M. A., & Datta, A. K. (2005). Computer simulation of the pharyngeal bolus transport of Newtonian and non-Newtonian fluids. *Food and Bioproducts Processing*, *83*(4C), 297–305. https://doi.org/10.1205/fbp.04209.

Mioche, L., Bourdiol, P., & Monier, S. (2003). Chewing behaviour and bolus formation during mastication of meat with different textures. *Archives of Oral Biology*, *48*(3), 193–200. https://doi.org/10.1016/S0003-9969(03)00002-5.

Mioche, L., & Peyron, M. A. (1995). Bite force displayed during assessment of hardness in various texture contexts. *Archives of Oral Biology*, *40*(5), 415–423. https://doi.org/10.1016/0003-9969(94)00190-M.

Mishellany, A., Woda, A., Labas, R., & Peyron, M. A. (2006). The challenge of mastication: Preparing a bolus suitable for deglutition. *Dysphagia*, *21*(2), 87–94. https://doi.org/10.1007/s00455-006-9014-y.

Morris, E. R. (1993). Rheological and organoleptic properties of food hydrocolloids. In K. Nishinari, & E. Doi (Eds.), *Food hydrocolloids, structures, properties, and functions* (pp. 201–210). Plenum Press.

Morris, V. J. (2006). Bacterial polysaccharides. In A. M. Stephen, G. O. Phillips, & P. A. Williams (Eds.), *Food polysaccharides and their applications* (2nd ed., pp. 413–454). CRC Press.

Mosca, A. C., Bult, J. H. F., & Stieger, M. (2013). Effect of spatial distribution of tastants on taste intensity, fluctuation of taste intensity and consumer preference of (semi-)solid food products. *Food Quality and Preference*, *28*(1), 182–187. https://doi.org/10.1016/j.foodqual.2012.07.003.

Mosca, A. C., Velde, F. V. D., Bult, J. H. F., van Boekel, M. A. J. S., & Stieger, M. (2010). Enhancement of sweetness intensity in gels by inhomogeneous distribution of sucrose. *Food Quality and Preference*, *21*(7), 837–842. https://doi.org/10.1016/j.foodqual.2010.04.010.

Murakami, K., Tokuda, Y., Hori, K., Minagi, Y., Uehara, F., Okawa, J., Ishihara, S., Nakauma, M., Funami, T., Maeda, Y., Ikebe, K., & Ono, T. (2021). Effect of fracture properties of gels on tongue pressure during different phases of squeezing and swallowing. *Journal of Texture Studies, 52*(3), 303–313. https://doi.org/10.1111/jtxs.12593.

Nakao, S., Ishihara, S., Nakauma, M., & Funami, T. (2013). Inhomogeneous spatial distribution of aroma compounds in food gels for enhancement of perceived aroma intensity and muscle activity during oral processing. *Journal of Texture Studies, 44*(4), 289–300. https://doi.org/10.1111/jtxs.12023.

Nakauma, M., Ishihara, S., Funami, T., & Nishinari, K. (2011). Swallowing profiles of food polysaccharide solutions with different flow behaviors. *Food Hydrocolloids, 25*(5), 1165–1173. https://doi.org/10.1016/j.foodhyd.2010.11.003.

Nakauma, M., Nakao, S., Ishihara, S., & Funami, T. (2014). Elution profile of sodium caseinate in simulated gastric fluids using an invitro stomach model from semi-solidified enteral nutrition. *Food Hydrocolloids, 36*, 294–300. https://doi.org/10.1016/j.foodhyd.2013.09.020.

Nakauma, M., Tanaka, R., Ishihara, S., Funami, T., & Nishinari, K. (2012). Elution of sodium caseinate from agar-based gel matrixes in simulated gastric fluids. *Food Hydrocolloids, 27*(2), 427–437. https://doi.org/10.1016/j.foodhyd.2011.10.017.

Nakazawa, F., & Togashi, M. (2000). Evaluation of food texture by mastication and palatal pressure, jaw movement and electromyography. In K. Nishinari (Ed.), *Hydrocolloids part 2 fundamentals and applications in food, biology, and medicine* (pp. 473–483). Elsevier.

Nicosia, M. A., & Robbins, J. A. (2001). The fluid mechanics of bolus ejection from the oral cavity. *Journal of Biomechanics, 34*(12), 1537–1544. https://doi.org/10.1016/S0021-9290(01)00147-6.

Nishinari, K. (2006). Polysaccharide rheology and in-mouth perception. In A. M. Stephen, G. O. Phillips, & P. A. Williams (Eds.), *Food polysaccharides and their applications* (2nd ed., pp. 541–588). CRC Press.

Nishinari, K., Takemasa, M., Su, L., Michiwaki, Y., Mizunuma, H., & Ogoshi, H. (2011). Effect of shear thinning on aspiration—Toward making solutions for judging the risk of aspiration. *Food Hydrocolloids, 25*(7), 1737–1743. https://doi.org/10.1016/j.foodhyd.2011.03.016.

Noort, M. W. J., Bult, J. H. F., & Stieger, M. (2012). Saltiness enhancement by taste contrast in bread prepared with encapsulated salt. *Journal of Cereal Science, 55*(2), 218–225. https://doi.org/10.1016/j.jcs.2011.11.012.

Noort, M. W. J., Bult, J. H. F., Stieger, M., & Hamer, R. J. (2010). Saltiness enhancement in bread by inhomogeneous spatial distribution of sodium chloride. *Journal of Cereal Science, 52*(3), 378–386. https://doi.org/10.1016/j.jcs.2010.06.018.

Palmer, J. B., Rudin, N. J., Lara, G., & Crompton, A. W. (1992). Coordination of mastication and swallowing. *Dysphagia, 7*(4), 187–200. https://doi.org/10.1007/BF02493469.

Peleg, M. (2006). On fundamental issues in texture evaluation and texturization—A view. *Food Hydrocolloids, 20*(4), 405–414. https://doi.org/10.1016/j.foodhyd.2005.10.008.

Peyron, M. A., Mishellany, A., & Woda, A. (2004). Particle size distribution of food boluses after mastication of six natural foods. *Journal of Dental Research, 83*(7), 578–582. https://doi.org/10.1177/154405910408300713.

Pires, M. A., Pastrana, L. M., Fucinõs, P., Abreu, C. S., & Oliveira, S. M. (2020). Sensorial perception of astringency: oral mechanisms and current analysis methods. *Foods, 9*(8). https://doi.org/10.3390/foods9081124.

Pondicherry, K. S., Rummel, F., & Laeuger, J. (2018). Extended stribeck curves for food samples. *Biosurface and Biotribology, 4*(1), 34–37. https://doi.org/10.1049/bsbt.2018.0003.

Prinz, J. F., & Lucas, P. W. (1995). Swallow thresholds in human mastication. *Archives of Oral Biology, 40*(5), 401–403. https://doi.org/10.1016/0003-9969(94)00185-E.

Pu, D., Duan, W., Huang, Y., Zhang, L., Zhang, Y., Sun, B., Ren, F., Zhang, H., & Tang, Y. (2021). Characterization of the dynamic texture perception and the impact factors on the bolus texture changes during oral processing. *Food Chemistry, 339*, 128078. https://doi.org/10.1016/j.foodchem.2020.128078.

Rolls, E. T. (2011). The neural representation of oral texture including fat texture. *Journal of Texture Studies, 42*(2), 137–156. https://doi.org/10.1111/j.1745-4603.2011.00296.x.

Saitoh, E., Shibata, S., Matsuo, K., Baba, M., Fujii, W., & Palmer, J. B. (2007). Chewing and food consistency: Effects on bolus transport and swallow initiation. *Dysphagia, 22*(2), 100–107. https://doi.org/10.1007/s00455-006-9060-5.

Sato, W., Minemoto, K., Ikegami, A., Nakauma, M., Funami, T., & Fushiki, T. (2020). Facial EMG correlates of subjective hedonic responses during food consumption. *Nutrients*, *12*(4), 1174. https://doi.org/10.3390/nu12041174.

Shaker, R., Cook, I. J. S., Dodds, W. J., & Hogan, W. J. (1988). Pressure-flow dynamics of the oral phase of swallowing. *Dysphagia*, *3*(2), 79–84. https://doi.org/10.1007/bf02412424.

Shama, F., & Sherman, P. (1973). Identification of stimuli controlling the sensory evaluation of viscosity. II. Oral methods. *Journal of Texture Studies*, *4*(1), 111–118. https://doi.org/10.1111/j.1745-4603.1973.tb00657.x.

Shewan, H. M., Pradal, C., & Stokes, J. R. (2019). Tribology and its growing use toward the study of food oral processing and sensory perception. *Journal of Texture Studies*. https://doi.org/10.1111/jtxs.12452.

Shewan, H. M., Pradal, C., & Stokes, J. R. (2020). Tribology and its growing use toward the study of food oral processing and sensory perception. *Journal of Texture Studies*, *51*(1), 7–22. https://doi.org/10.1111/jtxs.12452.

Shiozawa, K., Kohyama, K., & Yanagisawa, K. (1999a). Food bolus texture and tongue activity just before swallowing in human mastication. *Japanese Journal of Oral Biology*, *41*(4), 297–302. https://doi.org/10.2330/joralbiosci1965.41.297.

Shiozawa, K., Kohyama, K., & Yanagisawa, K. (1999b). Influence of ingested food texture on jaw muscle and tongue activity during mastication in humans. *Japanese Journal of Oral Biology*, *41*(1), 27–34. https://doi.org/10.2330/joralbiosci1965.41.27.

Sonne, A., Busch-Stockfisch, M., Weiss, J., & Hinrichs, J. (2014). Improved mapping of in-mouth creaminess of semi-solid dairy products by combining rheology, particle size, and tribology data. *LWT - Food Science and Technology*, *59*(1), 342–347. https://doi.org/10.1016/j.lwt.2014.05.047.

Srivastava, R., Bosc, V., Restagno, F., Tournier, C., Menut, P., Souchon, I., & Mathieu, V. (2021). A new biomimetic set-up to understand the role of the kinematic, mechanical, and surface characteristics of the tongue in food oral tribological studies. *Food Hydrocolloids*, *115*, 106602. https://doi.org/10.1016/j.foodhyd.2021.106602.

Srivastava, R., Stieger, M., Scholten, E., Souchon, I., & Mathieu, V. (2021). Texture contrast: Ultrasonic characterization of stacked gels' deformation during compression on a biomimicking tongue. *Current Research in Food Science*, *4*, 449–459. https://doi.org/10.1016/j.crfs.2021.06.004.

Stokes, J. R. (2012). 'Oral' rheology. In J. Chen, & L. Engelen (Eds.), *Food oral processing fundamentals of eating and sensory perception* (pp. 227–263). Blackwell Publishing Ltd.

Stokes, J. R., Boehm, M. W., & Baier, S. K. (2013). Oral processing, texture and mouthfeel: From rheology to tribology and beyond. *Current Opinion in Colloid and Interface Science*, *18*(4), 349–359. https://doi.org/10.1016/j.cocis.2013.04.010.

Stribiţcaia, E., Krop, E. M., Lewin, R., Holmes, M., & Sarkar, A. (2020). Tribology and rheology of bead-layered hydrogels: Influence of bead size on sensory perception. *Food Hydrocolloids*, *104*. https://doi.org/10.1016/j.foodhyd.2020.105692.

Takahashi, J., & Nakazawa, F. (1991). Palatal pressure patterns of gelatin gels in the mouth. *Journal of Texture Studies*, *22*(1), 1–11. https://doi.org/10.1111/j.1745-4603.1991.tb00001.x.

Takahashi, J., & Nakazawa, F. (1992). Effects of dimensions of agar and gelatin gels on palatal pressure patterns. *Journal of Texture Studies*, *23*(2), 139–152. https://doi.org/10.1111/j.1745-4603.1992.tb00516.x.

Taniguchi, H., Tsukada, T., Ootaki, S., Yamada, Y., & Inoue, M. (2008). Correspondence between food consistency and suprahyoid muscle activity, tongue pressure, and bolus transit times during the oropharyngeal phase of swallowing. *Journal of Applied Physiology*, *105*(3), 791–799. https://doi.org/10.1152/japplphysiol.90485.2008.

Tanishima, Y., Fujita, T., Suzuki, Y., Kawasaki, N., Nakayoshi, T., Tsuiboi, K., Omura, N., Kashiwagi, H., & Yanaga, K. (2010). Effects of half-solid nutrients on gastroesophageal reflux in beagle dogs with or without cardioplasty and intrathoracic cardiopexy. *Journal of Surgical Research*, *161*(2), 272–277. https://doi.org/10.1016/j.jss.2009.03.025.

Van Vliet, T. (2002). On the relation between texture perception and fundamental mechanical parameters for liquids and time dependent solids. *Food Quality and Preference*, *13*(4), 227–236. https://doi.org/10.1016/S0950-3293(01)00044-1.

Varela, P., & Fiszman, S. (2012). Appreciation of food crispness and new product development. In J. Chen, & L. Engelen (Eds.), *Food oral processing fundamentals of eating and sensory perception* (pp. 339–356). Blackwell Publishing Ltd.

Wang, Z., Yang, K., Brenner, T., Kikuzaki, H., & Nishinari, K. (2014). The influence of agar gel texture on sucrose release. *Food Hydrocolloids, 36*, 196–203. https://doi.org/10.1016/j.foodhyd.2013.09.016.

Wang, Q., Zhu, Y., & Chen, J. (2021). Development of a simulated tongue substrate for in vitro soft "oral" tribology study. *Food Hydrocolloids, 120*, 106991. https://doi.org/10.1016/j.foodhyd.2021.106991.

Weel, K. G. C., Boelrijk, A. E. M., Alting, A. C., Van Mil, P. J. J. M., Burger, J. J., Gruppen, H., Voragen, A. G. J., & Smit, G. (2002). Flavor release and perception of flavored whey protein gels: Perception is determined by texture rather than by release. *Journal of Agricultural and Food Chemistry, 50*(18), 5149–5155. https://doi.org/10.1021/jf0202786.

Wilkinson, C., Dijksterhuis, G. B., & Minekus, M. (2000). From food structure to texture. *Trends in Food Science and Technology, 11*(12), 442–450. https://doi.org/10.1016/S0924-2244(01)00033-4.

Yokoyama, S., Hori, K., Tamine, K. I., Fujiwara, S., Inoue, M., Maeda, Y., Funami, T., Ishihara, S., & Ono, T. (2014). Tongue pressure modulation for initial gel consistency in a different oral strategy. *PLoS One, 9*(3). https://doi.org/10.1371/journal.pone.0091920.

# CHAPTER 15

# Designing and development of food structure with high acceptance based on the consumer perception

Ricardo Isaías[a], Ana Frias[a], Célia Rocha[a,b], Ana Pinto Moura[c], and Luís Miguel Cunha[a]

[a]*GreenUPorto/INOV4AGRO & DGAOT, Faculty of Sciences, University of Porto, Vila do Conde, Portugal*, [b]*Sense Test, Lda., Vila Nova de Gaia, Portugal*, [c]*GreenUPorto/INOV4AGRO & DCeT, Universidade Aberta, Rua do Amial, Porto, Portugal*

## 15.1 Introduction

The food industry is facing extremely competitive markets and it is crucial to identify critical drivers of consumer choice, purchasing patterns and consumption behaviors. In this context, the development of new food products that satisfy consumers' demands may contribute positively as a competitive tool for food companies (Bigliardi, 2013; Linnemann, 2006; Menrad, 2004).

Accumulated evidence over the last years has given major importance to the structure of foods and its relationship with more desirable sensorial attributes (such as texture, palatability, melting, etc.), nutritional properties (such as macro and micronutrient delivery and availability in our digestive tract) and resulting health benefits (Norton et al., 2014). These are particularly relevant as they are some of the main drivers of food choice (Cunha et al., 2018), expressed by the increasing consumer expectations for healthier, sustainable, and tasty food products (i.e., reduced in fat, sugar and salt, from vegetable sources, enriched in bioactive molecules), that challenges the food industry to come up with new creations (Lundin et al., 2008).

In the case of food structure, one of the most recent successful innovative approaches has emerged from the application of nanotechnology (Giordano et al., 2018). Nanotechnology (NT) is an emerging area that may involve the production, processing and application of structures, devices and systems, by controlling their shape and nanoscale size (Kuan et al., 2012). This field of studies (engineering very small particles, usually defined at a scale between 1 and 100 nm) has evidenced different properties and functionalities of materials at this scale when compared with 'conventionally sized' equivalents (Weiss et al., 2006). Due to their size, the use of nanostructures in the food industry can increase the solubility and bioavailability of compounds, change sensory properties of food, or promote a controlled release of certain compounds, especially those that have low water solubility (Kuan et al., 2012; Weiss et al., 2006). Currently, a major focus of NT applications in food is the development of nanostructured or nanotextured food ingredients and delivery systems for nutrients and supplements. Therefore, food products resulting from the application of nanotechnologies can have intentional implications on consumers' sensorial experiences, through an improved sensorial profile, providing the same sensorial profile with enhanced nutrition profile, or masking undesirable sensations (Sun et al., 2021). Additionally, it may provide healthier food choices (e.g., functional foods), promoting improved consumer acceptance of the resulting food products. However, consumer acceptance of new foods resulting from nanotechnologies, as well as others resulting from modifications in the food structure has not been unanimous: Carbohydrates, proteins, and lipids (fats) are major structural components of natural and processed foods and deliver several functional properties, contributing namely to the foods' appearance, aroma, taste, and texture. Thus, reducing these components affects not only taste, but also aroma, texture, and appearance (Conroy et al., 2018; Milner et al., 2020; Szpicer et al., 2020). This is particularly relevant in occidental societies, as sensory appeal is reported as one of the most important factors that influence individual food choices (Cunha et al., 2018). Thus, the success of efforts to design products, especially those with improved nutrition/health profile, while maintaining consumer acceptance, requires a detailed understanding of food microstructure and the interaction of food structures with physiological and behavioral processes occurring upon ingestion (Norton et al., 2014).

The main purpose of this chapter is to provide a better understanding of the determinants of acceptance of innovative food products resulting from the application of food structure design concepts.

## 15.2 Determinants of acceptance of innovative food products from food structure design

### 15.2.1 Sensory properties

The acceptance or rejection of a certain food product is influenced by a number of factors, being sensory appeal one of the most important. The food matrix influences structure and, consequently, the appearance, texture, mouth breakdown and flavor release of the food product (Aguilera, 2019). Thus, when designing new food structures it is essential to understand the interplay between the process and the food structure, their physical properties, the oral processing, and the sensory properties (Norton et al., 2014), as well as macro and micronutrient delivery and bioavailability (Mcclements, 2020).

To evaluate the sensory profile of foods, instrumental measurements often are insufficient as no instrument or combination of instruments can fully replace the human senses (Carvalho et al., 2013). Typically, machines only measure one isolated property or a set of related properties at a time; while humans can be trained and instructed to evaluate a single property, to relate properties of different nature (aroma, flavor and texture, for instance), whilst evaluating an overall hedonic sensation, such as overall liking (Rios-Mera et al., 2020). Sensory analysis tests, where humans are used as a tool, may be divided into objective (analytical) and subjective (hedonic) tests.

A central principle regarding sensory testing is that the testing method must match the objective of the test, i.e. if descriptive information is needed, a descriptive test should be used, and if the question regards consumer liking or disliking of a specific product, an affective test should be performed. In classical sensory analysis, consumers or trained panelists may be recruited to participate in a variety of different tests, which can be organized into discriminatory, descriptive, preference and hedonic trials. These varied approaches seek, one way or another, to understand the reasons behind individual liking of a product (Lawless & Heymann, 2010). Typically, a 9-point scale is used, from $1=$ dislike extremely to $9=$ like extremely, giving an overall liking characterization of the studied product or set of products (Peryam & Pilgrim, 1957). When referring to preference or hedonic trials, one can include tests in the affective sensory analysis techniques, since they access personal preferences and/or willingness to pay; or products' level of acceptance, respectively (Stone & Sidel, 2004).

Humans have five senses: touch, sight, hearing, smell and taste. Taste and smell are closely related and often interact and influence each other, either positively or negatively (Noble, 1996). In addition to odor, also temperature, pH, sight, sound, texture and sometimes pain/irritation influence flavors (Delwiche, 2004), but pH, temperature and ions have been shown to have only a small effect on sweet taste (Schiffman et al., 2000). Taste is an analytical sense, which means that the individual parts are distinguished in a mix, in contrast to a synthetic sense, like sight, where colors mix to form a new color (Shallenberger, 1998). In fact, the different taste modalities: sweet, salty, sour, bitter and umami, are influenced by interactions between water-soluble tastant molecules and taste receptors in the taste buds.

Many food properties influence the experience of it, and two important factors are taste and structure, and their relationship. The structure influences our tactile senses and determines the diffusion rate and release of taste molecules, which will affect our taste sensation. The tactile perceptions of foods are related to the food structure, something that will affect the physical feeling of the food. The consistency, may be viewed as a thickness-feeling of the food in the mouth and a perceived thickness will follow from addition of thickeners. The type of thickener and its concentration can be varied to create all possible thicknesses from slow floating, viscous solutions to very hard gels, depending on which type of food it is aimed for. Additionally, different food structures as gels, emulsions or foams will also have an impact on how the food is perceived in the mouth (Holm, 2006). It has been shown that the relationship between texture and taste is harder to elucidate for gels than for liquids. For liquid systems, it is generally considered that higher viscosities induce lower taste intensities (Izutsu et al., 1981). While for hydrocolloid solutions, when comparing two solutions, the one that is more shear thinning is often perceived as sweeter, probably because the masking effect from the hydrocolloids is reduced when viscosity drops (Holm et al., 2009).

Sweet taste can be enhanced by increasing sugar release by changing the food structure. For example, the sweetness intensity is enhanced by increased release from mixed whey protein isolate/gel and gum gels with larger pores (Sala et al., 2010). The sweetness intensity of the most brittle emulsion-filled gelatine/agar gels is almost twice as high as that of the least brittle gel (Sala & Stieger, 2013). The sweetness intensity increases with a reduced size of the fragments, which is attributed to an increase in the surface area of the fragments that promotes sugar release and increased frequency of stimulation of the receptors; taste perception can be modulated by the breakdown behavior (Mosca, Van De Velde, et al., 2015; Wang et al., 2014).

The oral processing of food structures includes biting, mastication, mixing and lubrication, bolus formation and swallowing. The food (solid or semi-solid) is broken down into small particles that are mixed and lubricated by saliva and fluids released from foods, forming the bolus that will be swallowed. During breakdown and mixing, the release of tastants from the food structure occurs, facilitating the contact with taste receptors. The release pattern of these chemicals is expected to be influenced by the type of structure. In viscous solutions, the release of tastants has shown to be affected by the mixing behavior of the solution (Ferry, Hort, et al., 2006; Ferry, Mitchell, et al., 2006; Koliandris et al., 2008). In semi-solids, the release has shown to be affected by the fracture properties of the matrix (Koliandris et al., 2008; Morris, 1994).

On another note, emulsions comprise a large part of liquid and semi-solid food products. Modifying some emulsion properties such as flavor, fat/oil content, droplet size, and type and concentration of thickening agents may in general affect the sensory perception of emulsions (Vingerhoeds et al., 2008). Fats/oils, such as medium and long-chain fatty acids, appear to have no taste; however, they contribute to texture/mouth-feel, and they interact with other components changing the chemosensory attributes of foods (Coupland & Hayes, 2014). Additionally, the formation of an oily coating of the mouth would lead to reduced mass transfer of taste molecules to the taste receptors, thus reducing taste intensity.

Salty taste was investigated on isoviscous oil/water emulsions with fat content ranging from 0% to 60%. When the product salt concentration was kept constant, the saltiness increased as oil content increased. This was explained by an increasing salt concentration in the aqueous phase upon increased oil content. However, when the salt concentration in the aqueous phase was held constant and the oil content was increased, the perceived saltiness decreased (Malone et al., 2003). This shows that apart from the salt concentration in the aqueous phase, also the aqueous phase volume, and contact surface area between sample and mouth, are important for taste intensity.

In model cheeses, the distribution of sodium chloride molecules is markedly not modified by food composition or structure (Lauverjat et al., 2009). Although this characteristic can change due to different protein matrices (e.g., gelatine, milk protein, and soy protein), the amount of sodium ions bound to the food matrix is negligible regarding its sensory perception (Lauverjat et al., 2009; Mosca, Andriot, et al., 2015; Pflaum et al., 2013a).

Salt release is also related to the food structure itself. A study investigating the influence of bread crumb texture on the intensity of perceived saltiness and the release of sodium ions during chewing, found that a significantly faster release of sodium from coarse-pored bread than from fine-pored bread was observed both in vivo and in an oral processing simulator (Pflaum et al., 2013b). Increased in vivo salt release contributes to enhanced salt perception (De Loubens et al., 2011; Lawrence et al., 2012; Pflaum et al., 2013b).

Tactile–taste interactions are a sensory integration pathway that influences taste perception. For example, the roughness/smoothness of the food significantly influenced taste perception: studies show that rough food was rated as significantly more sour than food with smooth surfaces (Slocombe et al., 2016). A gelatine gel had higher firmness and greater sodium ion mobility than soybean protein and milk protein gels, but the salty perception was rated the least (Mosca, Andriot, et al., 2015), demonstrating the importance of tactile–taste interactions on taste perception. Further research on hydrocolloid solutions, suspensions, and gels revealed that salty perception is directly affected by viscosity, gel firmness, and the mechanosensory perception of solid particles (Scherf et al., 2015). In particular, solid particles that only minimally increased the viscosity drastically reduced the perceived taste intensity, not only for saltiness but also for sour, umami, and sweet tastes, with the exception of bitterness.

### 15.2.2 Health concerns

Different studies in Western countries have shown that health and nutrition concerns are important individual food choice criteria (Cunha et al., 2018), as consumers are interested in feeling well (Carrillo et al., 2013). According to the latest data from Food and Drink Europe (FDE, 2021), health was the second driver of food innovation in Europe, standing at 31.8%. Nevertheless, there is currently still a gap between food/nutritional recommendations established internationally (WHO, 2003) and food consumption patterns, namely in Portugal (Lopes et al., 2017). To satisfy this consumer concern, in recent decades, a significant proportion of new food technologies and innovations have been oriented towards health promotion, such as the development of functional foods which, in addition to satisfying the nutritional needs of individuals, can provide health benefits (Kieling et al., 2019; Martins et al., 2019; Mouta et al., 2016; Putnik et al., 2019; Santos et al., 2017; Siró et al., 2008; Torrico et al., 2018; Zhao et al., 2019). Although there is no universally accepted definition, foods can be considered functional if they promote beneficial physiological health effects, specifically with regard to reducing the risk of developing a disease or improving a health situation, in addition to its basic nourishing function (Hasler, 2000). Considering this definition, a functional food can fit into one of the following groups (Henry, 2010):

- An unaltered natural food (e.g., oily fish);
- A food in which a component has been increased through the production process or other technologies;
- A food to which a component has been added, by technological or biotechnological means, providing benefits (e.g., spreadable cream with the addition of phytosterols);
- A food from which a component has been removed by technological or biotechnological means, so that the food provides a benefit that would otherwise not exist (e.g., yoghurt with reduced fat content);
- A food in which one component has been replaced by another with more favorable properties (e.g., soft drinks in which sugar has been replaced by sweeteners);
- A food whose bioavailability of a component has been modified (e.g., genetically modified rice in order to increase iron bioavailability).

In fact, more and more consumers believe that foods contribute directly to their health, and eating healthy products may prevent nutrition related diseases and improve physical and mental well-being (Ares et al., 2015). According to Urala and Lähteenmäki (2003), consumers connect functional foods with control over life and health, prevention of disease, being a better person and feelings of well-being. These products provide consumers a modern way to follow a healthy lifestyle, because they combine health and time saving benefits (e.g., they are easy to use and do not need previous preparation). This is particularly relevant in Western countries, where consumers perceived reward from using functional foods for their own health and well-being (Urala & Lähteenmäki, 2004).

However, one of the requirements for the consumer acceptance of functional foods is related to taste, which may be a concern in the innovation context, given that the addition of bioactive compounds or phytonutrients, with the purpose of improving functionality, can often lead to the emergence of off-flavors (Drewnowski & Gomez-Carneros, 2000; Tuorila & Cardello, 2002; Verbeke, 2006). In this context, the combination of emerging food technologies and design of food structure is in a prominent position, enabling food quality and safety maintenance, while enhancing the retention of bioactive compounds and improved sensory traits. There is an increasing trend in the food industry towards the development of fortified products with nanosized nutritional elements (Santillán-Urquiza et al., 2017). For example, phenolic compounds with high antioxidant capacity, are limited in their application in the food industry, since they are unstable in the presence of heat and light, present low bioavailability, and can produce an undesirable bitter taste (Soto-Vaca et al., 2012). However, nanoencapsulation, the use of nanotechnology that packages solid, liquid or gas nanoparticles, also known as the core or active ingredient, within a shell to form nanocapsules (Augustin & Hemar, 2009), has the potential to improve their stability and suppress or mask the unpleasant taste, having promising results regarding antioxidant activity and other health-promoting properties (Rezaei et al., 2019).

Considering that food health benefits cannot be experienced directly and immediately by consumers as they are credence attributes, health as a choice of food criteria is a question of communication and of the consumers' perception of various signals (Grunert, 2010). In this context, health claims in the labelling and advertising of foods are a very convenient tool when it comes to marketing functional foods (Pravst, 2012) that might draw attention to the health benefits of the food product. To ensure the effective information transfer, the message of health effect or of a specific compound should be communicated in an easy way (Strijbos et al., 2016; Žeželj et al., 2012) taking into account that consumers have a limited awareness and knowledge regarding the development of functional compounds and their health benefits, such as omega-3, fatty acids, selenium, carotenoids, flavonoids, xylitol, bacterial cultures, (Siró et al., 2008), and thus they doubt about their potential benefits (Annunziata & Vecchio, 2011).

### 15.2.3 Nutrition concerns

Due to consumers' awareness of their food choices and possible health benefits, subjects such as personalized nutrition, or nutrigenomics have gained greater interest (Bimbo et al., 2017; Poínhos et al., 2014; Verbeke, 2006). Additionally, it has been proven that the majority of food-related severe health conditions could be prevented if the correct lifestyle and dietary habits would have been pursued. In fact, 80% of the cardiovascular conditions and chronic diseases (i.e., coronary disease or type 2 *diabetes mellitus*) and 40% of oncological diseases can be avoided by applying the information above to daily lives (Bimbo et al., 2017; Poínhos et al., 2014).

For example, in developed countries, meat is nowadays considered a staple food, since it is consumed almost on a daily basis by the vast majority of the population. However, its nutrition facts are often considered hazardous, depending on the type of meat, since its salt and fat content could be above the nutritional recommendations. Yet, content reduction of those nutrients has proven challenging, as these compounds have a strict relationship with major food sensory attributes, such as flavor or texture. Salt, for instance, plays a pivotal role in the maintenance of meat structure, since it is involved in the interactions that bind water, protein and fat molecules together in the food matrix (Rodrigues et al., 2020). Rodrigues et al. (2020) in their experiences with reduced salt and fat-sausages, and upon sensory analysis, noted that consumers gave worse acceptance rates to products with significantly less salt content; however they concluded that a total of 27% of salt reduction could be applied, maintaining a positive sensory approval.

Taking the information above into consideration, it is clear that options should be developed in order to provide a healthier food alternative with a similar sensory experience. This can be obtained, up to a certain extent, by replacing NaCl with other salts, such as KCl or $CaCl_2$. In probiotic-dependent products, such as cheeses or yoghurts, the consequent effect of salt reduction on the starter cultures used must also be accessed, since they contribute largely to the products' texture and consistency. The starter culture's species can be replaced by a more halo-tolerant one, if that change does not affect sensory attributes in a large scale (Rysová & Šmídová, 2021).

As for lipid content, consumers tend to valorise unsaturated fats the most, while not being in favor of an extensive saturated fat percentage. This conception comes utterly from the media, since the benefits of polyunsaturated fats are usually reported, when it comes to products such as fish oil (Markey et al., 2017).

As mentioned earlier, fat reduction also has reported negative effects on textural characteristics of food products. To minimize those consequences, Campagnol et al. (2017) incorporated several functional ingredients onto Bologna-type sausages, such as fructooligosaccharides, disodium inosinate or transglutaminase. This proved to be a successful approach, since the compounds added had positive impacts on sensory features. For example, various dietary fibers have been demonstrated to stabilize emulsions in a similar way to fats, presenting themselves as promising alternatives. A variety of other studies have been carried out on this field, partly substituting fat by resistant starch or other alternatives, reaching interesting sensory approval rates (Ginani et al., 2010; Roselino et al., 2018; Sellas et al., 2021).

Finally, sugar content is another relevant factor to be pointed out. Along with salt and fat, sugar has also been associated with negative health conditions, such as *diabetes mellitus* or overweight, when consumed in excessive amounts. Thereby, reducing added sugars has become one of the food industries recent top priorities. However, this action can, once more, have a negative impact on sensory parameters, such as sweetness or texture (Mayhew et al., 2017; Oliveira et al., 2015).

Oliveira et al. (2015) did an extensive sensory characterization of possible sugar reductions in probiotic chocolate-flavored milk, having concluded that the product's overall liking was not significantly affected by a maximum of 40% reduction, which may allude to the huge amounts of added sugars practiced in some countries. Added synthetic sweeteners, even non-caloric ones, did not provoke sensory nor nutritional acceptance improvement, inferring the constant pursue for a "clean-label" product. Similarly, a study with three fruit juices has shown that consumers' hedonic reaction towards sugar reduction was product dependent, added sugar can be reduced 5–8% without affecting consumers' sensory perception and reductions in the range of 10–20% did not lead to changes in hedonic reaction (Oliveira et al., 2018).

Nevertheless, the use of natural non-caloric sweeteners seems to have a different reception from the average consumer. This can be explained by the constant thrive for naturalness in food choice, since added sweeteners like stevia seem to lead to increased acceptance rates. Mahato et al. (2021) optimized sugar/natural sweetener levels in chocolate flavored milk and concluded that the correct combination of reduced sugar with stevia and monk fruit extract led to enhanced sensory properties.

Other studies have molded food structure to minimize sensory effects caused by the lack of sugar content, such as modifying the sucrose distribution, its serum release or the fracture mechanics of foods. This set of techniques currently present the hindrance of industrial scale-up, which must be optimized in the future (Hutchings et al., 2019).

In the European Union, the nutrition and health claims made on foods are affected by the Regulation (EC) No. 1169/2011 of the European Parliament and of the Council (2011), which requires authorization of all health claims before entering the market and included in the list of authorized claims. The rules of the Regulation apply to nutrition claims (such as "low fat", "high fiber") and to health claims (such as "Vitamin D is needed for the normal growth and development of bone in children"). The objective of those rules is to ensure that any claim made on a food's labelling, presentation or advertising in the European Union is clear, accurate and based on scientific evidence. Food bearing claims that could mislead consumers are prohibited on the EU market. This not only protects consumers, but also promotes innovation and ensures fair competition. The rules ensure the free circulation of foods bearing claims, as any food company may use the same claims on its products anywhere in the European Union.

### 15.2.4 Risk perception

Novel food producing and processing technologies have grown in recent decades. However, several challenges are faced by industry and scientific individuals who lead innovative projects. One of them is the fact that consumers are mostly neophobic towards novel food technologies and industrialization (Rollin et al., 2011; Roosen et al., 2015).

Food technology neophobia is a term commonly applied when referring to consumers' fear of innovation, whereas it concerns novel foods or simply the technologies used in their production and/or processing. As it is considered a personal and emotional response to a certain subject, neophobia is a personality trait, being dependent of several factors, from socio-demographic to psychological barriers (Chen et al., 2013; Choe & Cho, 2011; Cox & Evans, 2008; Giordano et al., 2018; Pliner & Hobden, 1992). In fact, consumers are mostly neophobic towards novel food technologies and industrialization (Rollin et al., 2011, Roosen et al., 2015).

The consumer acceptance of new products and technologies is a function of the perceived risks and the perceived benefits that are associated with them (Frewer et al., 1997). Indeed, new food technologies may evoke contradicting feelings in some consumers: advancements in technologies may be perceived by consumers as a means of improving well-being and life standards, but, on the other hand, new foods and technologies may cause risks which are unobservable, unknown, delayed, out of their own control, and potentially fatal (Cardello, 2003; Cardello et al., 2007). In fact, concerns relating to new products and food processing technology are also relevant in consumers purchase decisions (Cardello, 2003; Cardello et al., 2007).

In Mediterranean countries, the interest of consumers in functional food is conditioned by the fact that they generally appreciate natural, fresh foods and consider them better for health (Menrad, 2004; Van Trijp & Van Der Lans, 2007). Annunziata and Vecchio (2011), also concluded that the perceived risks of functional foods could be a strong barrier to the consumption of such products, because according to Italian consumers, functional foods contain unnatural substances and they are suspicious towards possible harmful effects of functional foods if they are extensively used. On the other hand, consumers tend to prefer foods that are perceived as natural because the properties often attributed to natural foods are healthiness, tastiness, freshness, and eco-friendliness (Román et al., 2017). As a result, new ingredients may decrease the naturalness perceived by consumers, particularly for those that consumers are unfamiliar with (Lähteenmäki et al., 2010). Bruschi et al. (2015) studied the influence of a more natural functional food (purple wheat breads and biscuits, naturally rich in anthocyanins) in the mostly conservative population of Russia. The results showed that, even though consumers had little knowledge of the health benefits of anthocyanins, the natural origin of these foods quickly became an advantage for the consumer willingness to pay for such typology of products, therefore underlying an important food-choice driver. Despite the information above, other new strategies, such as nanotechnology is still surrounded by scientific uncertainty (Roosen et al., 2015). This gives rise to increased consumers' risk perception, a crucial parameter affecting consumer acceptance (Giordano et al., 2018; Jang & Kim, 2015; Verneau et al., 2014). This is vastly due to consumer unfamiliarity with this new technology (Giordano et al., 2018; Roosen et al., 2015).

### 15.2.5 Convenience

Convenience is a major concern during food purchases, particularly by members of urbanized societies (Cunha et al., 2018). Changes in lifestyle, occupational patterns, and urbanization have also contributed to changes in culinary practices and dietary habits, and have increased the need for safe convenience foods. Indeed, the increasing number of women in the labor market and the changes in family structure and lifestyle (for example a decreasing number of extended families and increasing number of households with single parents or double income couples) have reduced the chance to find an adult in the household with the time or the energy to prepare meals (Pinto De Moura & Cunha, 2005). Convenience is a multidimensional phenomenon, usually suggesting that some kind of effort (time, physical energy or mental energy) is saved or reduced throughout the meal production chain of a private household, this is, during shopping storage, preparation, consumption and cleaning up (Darian & Cohen, 1995; Scholderer & Grunert, 2005).

However, convenience solutions are perceived by consumers that buy these products as unhealthy and as highly caloric (Darian & Cohen, 1995; De Boer et al., 2004). Over the last century, advancements in food technology have made major contributions to the creation of convenience food solutions, through better food preservation: prolonging the shelf life of foods and enhancing their safety. For example, the UHT milk treatment has a typical shelf life of 6–9 months, until it is opened, as an alternative to pasteurized milk treatment. However, today's consumers no longer require a shelf life of several months at ambient temperature for the majority of their staple foods. As referred previously, changes in family lifestyle and increased ownership of freezers and microwave ovens, are reflected in the increasing demand for foods that are convenient to prepare, are suitable for frozen storage, or have a moderate shelf life at ambient temperature (Mishra & Sinija, 2008). Besides, there is also an increased demand for foods that have experienced fewer changes during processing and thus retain their natural flavor, color, and texture, and contain fewer additives.

Another innovative strategy to be considered is three-dimensional (3D) food printing. This approach consists in the manufacturing of foods by means of a robotically controlled process, constructing a typical product layer-by-layer. Although there is still a long road ahead for this technology, its applications are almost endless. For example, considering the dysphagia condition (the difficulty in chewing and mastication, very common in elderly people), 3D food printing could present itself as a promising solution, allowing patients to prepare and consume meals with similar appearance to normal ones, but with softened texture, thereby easing their oral processing. Consequently, individuals with these kinds of health problems would not be confronted with the pureed form of foods, which can be unpleasant and cause food-related disgust and phobias (Dankar et al., 2018; Lee et al., 2021; Pereira et al., 2021).

People with perceived time pressure may think that they do not have time to prepare meals, namely healthy meals and may seek out convenience foods (frozen main course, pre-cut food, ready-made meals, take away meals or eating out) rather than cooking from basic ingredients (Candel, 2001; Darian & Cohen, 1995; De Boer et al., 2004). In other words, consumers have become more health conscious in their food choices, but have less time to prepare healthy meals.

### 15.2.6 Price

According to neoclassical microeconomic theory, the demand for goods is interrelated due to the generally limited budget and may therefore not be considered separated from each other. In fact, whether the price of food is affordable or not depends fundamentally on household's income and socioeconomic status. Although food prices affect everyone, the issue of food cost as a barrier to dietary change is particularly relevant to low income families (Dibsdall et al., 2003; Kearney & Mcelhone, 1999; Lloyd et al., 1995). According to the 2019 EFSA's (European Food Safety Authority) report on the interest of European citizens in food safety, the data shows that, Europeans in general and Portuguese in particular, have a personal interest for the price of food (51% of Europeans and 75% Portuguese) being the most important factors when purchasing food (EFSA, 2019).

In the context of food technology, according to the study by Kuang et al. (2020) consumers are willing to buy products resulting from the application of nanotechnology, based on sensory appeal and nanotechnology benefits. When participants were asked if they would buy the foods produced with nanotechnology, the vast majority (74.5–85.7%) said they would be willing do so, with reasons split between the food's purported benefit and its sensory appeal. Interestingly, neither the food type nor the presence of nanoparticles in the food packaging or within the food itself altered this outcome. Findings from the survey of Giles et al. (2015) suggested that consumers have somewhat softened their attitudes, still expressing greater approval for nano-packaging, comparatively to other applications, but welcoming all types of applications of nanotechnology in food production if the benefits outweighed the risks and it led to healthier, tastier, and more cost-effective products. However, consumers are sensitive to product price increases resulting from the application of this technology, particularly if the benefits are not clearly evident. The factors that motivate the acceptance of new food technology focus on the competitive price and its lesser impact on the environment (Lampila & Lähteenmäki, 2007). Annunziata and Vecchio (2011) also reported that Italian consumers considered functional foods as more expensive and less easy to find than conventional products. Overall, the acceptability of new technologies lies upon the condition that the price does not increase, especially if there is no clear benefit as a trade-off.

## 15.3 Conclusions

The increasing consumers' expectations for safe, healthier, sustainable, and tasty food products (reduced in fat, sugar, salt, from vegetable sources, with bioactive molecules, etc.), challenges food scientists to meet their needs, without compromising sensory properties. The design of food structures can provide them with healthier and more convenient food options with improved sensory characteristics.

Today, emerging food technologies, such as nanotechnology or 3D printing, may offer benefits that include food safety, improved nutrition, or taste. Optimizing food structure and design, in the context of food innovation, may impact the consumers' acceptance of a new food product. Hence, it is important to consider consumer perceptions and attitudes, as these are important determinants of commercial success or failure of food products. For this reason, the consumer should be involved in the design process of new or improved food products since the early stages of development. Taking into consideration the factors influencing consumers' perception of novel food technologies and their impact on the sensory characteristics, when designing food structures, will hopefully result in higher acceptance of these products.

## Acknowledgments

This work is funded through the project AgriFood XXI (NORTE-01-0145-FEDER-000041), financed by the European Regional Development Fund (ERDF), through P2020|Norte2020—P2020|Norte2020-Projetos Integrados ICDT, *CCRN—Comissão de Coordenação da Região Norte*. Authors also acknowledge financial support from National Funds from *FCT-Fundação para a Ciência e a Tecnologia* within the scope of UIDB/05748/2020 and UIDP/05748/2020.

# References

Aguilera, J. M. (2019). The food matrix: Implications in processing, nutrition and health. *Critical Reviews in Food Science and Nutrition, 59*, 3612–3629.

Annunziata, A., & Vecchio, R. (2011). Functional foods development in the European market: A consumer perspective. *Journal of Functional Foods, 3*, 223–228.

Ares, G., Saldamando, L., Giménez, A., Claret, A., Cunha, L., Guerrero, L., Pinto De Moura, A., Oliveira, D., Symoneaux, R., & Deliza, R. (2015). Consumers' associations with wellbeing in a food-related context: A cross-cultural study. *Food Quality and Preference, 40*(Part B), 304–315.

Augustin, M. A., & Hemar, Y. (2009). Nano- and micro-structured assemblies for encapsulation of food ingredients. *Chemical Society Reviews, 38*, 902–912.

Bigliardi, B. (2013). The effect of innovation on financial performance: A research study involving SMEs. *Innovation: Management. Policy & Practice, 15*, 245–256.

Bimbo, F., Bonanno, A., Nocella, G., Viscecchia, R., Nardone, G., De Devitiis, B., & Carlucci, D. (2017). Consumers' acceptance and preferences for nutrition-modified and functional dairy products: A systematic review. *Appetite, 113*, 141–154.

Bruschi, V., Teuber, R., & Dolgopolova, I. (2015). Acceptance and willingness to pay for health-enhancing bakery products – Empirical evidence for young urban Russian consumers. *Food Quality and Preference, 46*.

Campagnol, P. C. B., Dos Santos, B. A., Lorenzo, J. M., & Cichoski, A. J. (2017). A combined approach to decrease the technological and sensory defects caused by fat and sodium reduction in Bologna-type sausages. *Food Science and Technology International, 23*, 471–479.

Candel, M. J. J. M. (2001). Consumers' convenience orientation towards meal preparation: Conceptualization and measurement. *Appetite, 36*, 15–28.

Cardello, A. V. (2003). Consumer concerns and expectations about novel food processing technologies: Effects on product liking. *Appetite, 40*, 217–233.

Cardello, A. V., Schutz, H. G., & Lesher, L. L. (2007). Consumer perceptions of foods processed by innovative and emerging technologies: A conjoint analytic study. *Innovative Food Science & Emerging Technologies, 8*, 73–83.

Carrillo, E., Prado Gascó, V., Fiszman, S., & Varela, P. (2013). Why buying functional foods? Understanding spending behaviour through structural equation modelling. *Food Research International, 50*, 361–368.

Carvalho, N. B., Minim, V. P. R., Silva, R. C. S. N., Della Lucia, S. M., & Minim, L. A. (2013). Artificial neural networks (ANN): Prediction of sensory measurements from instrumental data. *Food Science and Technology, 33*, 722–729.

Chen, Q., Anders, S., & An, H. (2013). Measuring consumer resistance to a new food technology: A choice experiment in meat packaging. *Food Quality and Preference, 28*, 419–428.

Choe, J. Y., & Cho, M. S. (2011). Food neophobia and willingness to try non-traditional foods for Koreans. *Food Quality and Preference, 22*, 671–677.

Conroy, P. M., O'Sullivan, M. G., Hamill, R. M., & Kerry, J. P. (2018). Impact on the physical and sensory properties of salt-and fat-reduced traditional Irish breakfast sausages on various age cohorts acceptance. *Meat Science, 143*, 190–198.

Council, E. P. A. (2011). *Regulation (EU) no 1169/2011*. Official Journal of the European Union.

Coupland, J. N., & Hayes, J. E. (2014). Physical approaches to masking bitter taste: Lessons from food and pharmaceuticals. *Pharmaceutical Research, 31*, 2921–2939.

Cox, D., & Evans, G. (2008). Construction and validation of a psychometric scale to measure consumers' fears of novel food technologies: The food technology neophobia scale. *Food Quality and Preference, 19*, 704–710.

Cunha, L. M., Cabral, D., Moura, A. P., & De Almeida, M. D. V. (2018). Application of the food choice questionnaire across cultures: Systematic review of cross-cultural and single country studies. *Food Quality and Preference, 64*, 21–36.

Dankar, I., Haddarah, A., Omar, F. E. L., Sepulcre, F., & Pujolà, M. (2018). 3D printing technology: The new era for food customization and elaboration. *Trends in Food Science & Technology, 75*, 231–242.

Darian, J. C., & Cohen, J. (1995). Segmenting by consumer time shortage. *Journal of Consumer Marketing, 12*, 32–44.

De Boer, M., Mccarthy, M., Cowan, C., & Ryan, I. (2004). The influence of lifestyle characteristics and beliefs about convenience food on the demand for convenience foods in the Irish market. *Food Quality and Preference, 15*, 155–165.

De Loubens, C., Saint-Eve, A., Déléris, I., Panouillé, M., Doyennette, M., Tréléa, I. C., & Souchon, I. (2011). Mechanistic model to understand in vivo salt release and perception during the consumption of dairy gels. *Journal of Agricultural and Food Chemistry, 59*, 2534–2542.

Delwiche, J. (2004). The impact of perceptual interactions on perceived flavor. *Food Quality and Preference, 15*, 137–146.

Dibsdall, L. A., Lambert, N., Bobbin, R. F., & Frewer, L. J. (2003). Low-income consumers' attitudes and behaviour towards access, availability and motivation to eat fruit and vegetables. *Public Health Nutrition, 6*, 159–168.

Drewnowski, A., & Gomez-Carneros, C. (2000). Bitter taste, phytonutrients, and the consumer: A review. *The American Journal of Clinical Nutrition, 72*, 1424–1435.

EFSA. (2019). *Special eurobarometer wave EB91.3. Food safety in the EU.*

FDE. (2021). *Data & trends EU food and drink industry [online].* Food and Drink Europe. Available: https://www.fooddrinkeurope.eu/wp-content/uploads/2021/11/FoodDrinkEurope-Data-Trends-2021-digital.pdf (Accessed).

Ferry, A.-L., Hort, J., Mitchell, J., Cook, D., Lagarrigue, S., & Pamies, B. V. (2006). Viscosity and flavour perception: Why is starch different from hydrocolloids? *Food Hydrocolloids, 20*, 855–862.

Ferry, A.-L. S., Mitchell, J. R., Hort, J., Hill, S. E., Taylor, A. J., Lagarrigue, S., & Vallès-Pàmies, B. (2006). In-mouth amylase activity can reduce perception of saltiness in starch-thickened foods. *Journal of Agricultural and Food Chemistry, 54*, 8869–8873.

Frewer, L. J., Howard, C., Hedderley, D., & Shepherd, R. (1997). Consumer attitudes towards different food-processing technologies used in cheese production—The influence of consumer benefit. *Food Quality and Preference, 8*, 271–280.

Giles, E. L., Kuznesof, S., Clark, B., Hubbard, C., & Frewer, L. J. (2015). Consumer acceptance of and willingness to pay for food nanotechnology: A systematic review. *Journal of Nanoparticle Research, 17*, 467.

Ginani, V. C., Ginani, J. S., Botelho, R. B. A., Zandonadi, R. P., Akutsu, R. D., & Araujo, W. M. C. (2010). Reducing fat content of Brazilian traditional preparations does not Alter food acceptance: Development of a model for fat reduction that conciliates health and culture. *Journal of Culinary Science & Technology, 8*, 229–241.

Giordano, S., Clodoveo, M. L., Gennaro, B. D., & Corbo, F. (2018). Factors determining neophobia and neophilia with regard to new technologies applied to the food sector: A systematic review. *International Journal of Gastronomy and Food Science, 11*, 1–19.

Grunert, K. G. (2010). European consumers' acceptance of functional foods. *Annals of the New York Academy of Sciences, 1190*, 166–173.

Hasler, C. M. (2000). The changing face of functional foods. *Journal of the American College of Nutrition, 19*, 499S–506S.

Henry, C. J. (2010). Functional foods. *European Journal of Clinical Nutrition, 64*, 657–659.

Holm, K. (2006). *The relations between food structure and sweetness: A literature review* (p. 750). SIK-report.

Holm, K., Wendin, K., & Hermansson, A.-M. (2009). Sweetness and texture perception in structured gelatin gels with embedded sugar rich domains. *Food Hydrocolloids, 23*, 2388–2393.

Hutchings, S. C., Low, J. Y. Q., & Keast, R. S. J. (2019). Sugar reduction without compromising sensory perception. An impossible dream? *Critical Reviews in Food Science and Nutrition, 59*, 2287–2307.

Izutsu, T., Taneya, S. I., Kikuchi, E., & Sone, T. (1981). Effect of viscosity on perceived sweetness intensity of sweetened sodium carboxymethylcellulose solutions. *Journal of Texture Studies, 12*, 259–273.

Jang, S., & Kim, D. (2015). Enhancing ethnic food acceptance and reducing perceived risk: The effects of personality traits, cultural familiarity, and menu framing. *International Journal of Hospitality Management, 47*, 85–95.

Kearney, J. M., & Mcelhone, S. (1999). Perceived barriers in trying to eat healthier—Results of a pan-EU consumer attitudinal survey. *British Journal of Nutrition, 81*, S133–S137.

Kieling, D. D., Barbosa-Cánovas, G. V., & Prudencio, S. H. (2019). Effects of high pressure processing on the physicochemical and microbiological parameters, bioactive compounds, and antioxidant activity of a lemongrass-lime mixed beverage. *Journal of Food Science and Technology, 56*, 409–419.

Koliandris, A., Lee, A., Ferry, A.-L., Hill, S., & Mitchell, J. (2008). Relationship between structure of hydrocolloid gels and solutions and flavour release. *Food Hydrocolloids, 22*, 623–630.

Kuan, C. Y., Yee-Fung, W., Yuen, K. H., & Liong, M. T. (2012). Nanotech: Propensity in foods and bioactives. *Critical Reviews in Food Science and Nutrition, 52*, 55–71.

Kuang, L., Burgess, B., Cuite, C. L., Tepper, B. J., & Hallman, W. K. (2020). Sensory acceptability and willingness to buy foods presented as having benefits achieved through the use of nanotechnology. *Food Quality and Preference, 83*.

Lähteenmäki, L., Lampila, P., Grunert, K., Boztug, Y., Ueland, Ø., Åström, A., & Martinsdóttir, E. (2010). Impact of health-related claims on the perception of other product attributes. *Food Policy, 35*, 230–239.

Lampila, P., & Lähteenmäki, L. (2007). Consumers' attitudes towards high pressure freezing of food. *British Food Journal, 109*, 838–851.

Lauverjat, C., Déléris, I., Tréléa, I. C., Salles, C., & Souchon, I. (2009). Salt and aroma compound release in model cheeses in relation to their mobility. *Journal of Agricultural and Food Chemistry, 57*, 9878–9887.

Lawless, H. T., & Heymann, H. (2010). *Sensory evaluation of food: Principles and practices* (2nd ed.). New York: Springer.

Lawrence, G., Buchin, S., Achilleos, C., Bérodier, F., Septier, C., Courcoux, P., & Salles, C. (2012). In vivo sodium release and saltiness perception in solid lipoprotein matrices. 1. Effect of composition and texture. *Journal of Agricultural and Food Chemistry, 60*, 5287–5298.

Lee, K. H., Hwang, K. H., Kim, M., & Cho, M. (2021). 3D printed food attributes and their roles within the value-attitude-behavior model: Moderating effects of food neophobia and food technology neophobia. *Journal of Hospitality and Tourism Management, 48*, 46–54.

Linnemann, L. (2006). The effect of government spending on private consumption: A puzzle? *Journal of Money, Credit and Banking, 38*, 1715–1735.

Lloyd, H. M., Paisley, C. M., & Mela, D. J. (1995). Barriers to the adoption of reduced-fat diets in a UK population. *Journal of the American Dietetic Association, 95*, 316–322.

Lopes, C., Torres, D., Oliveira, A., Severo, M., Alarcão, V., Guiomar, S., Mota, J., Teixeira, P., Rodrigues, S., & Lobato, L. (2017). *Inquérito Alimentar Nacional e de Atividade Física IAN-AF 2015–2016 [Online]*. Available: https://ian-af.up.pt/sites/default/files/IAN-AF%20Relat%C3%B3rio%20Resultados_3.pdf (Accessed).

Lundin, L., Golding, M., & Wooster, T. J. (2008). Understanding food structure and function in developing food for appetite control. *Nutrition & Dietetics, 65*, S79–S85.

Mahato, D. K., Keast, R., Liem, D. G., Russell, C. G., Cicerale, S., & Gamlath, S. (2021). Optimisation of natural sweeteners for sugar reduction in chocolate flavoured milk and their impact on sensory attributes. *International Dairy Journal, 115*, 104922.

Malone, M. E., Appelqvist, I., & Norton, I. (2003). Oral behaviour of food hydrocolloids and emulsions. Part 2. Taste and aroma release. *Food Hydrocolloids, 17*, 775–784.

Markey, O., Souroullas, K., Fagan, C. C., Kliem, K. E., Vasilopoulou, D., Jackson, K. G., Humphries, D. J., Grandison, A. S., Givens, D. I., Lovegrove, J. A., & Methven, L. (2017). Consumer acceptance of dairy products with a saturated fatty acid–reduced, monounsaturated fatty acid–enriched content. *Journal of Dairy Science, 100*, 7953–7966.

Martins, I. B. A., Oliveira, D., Rosenthal, A., Ares, G., & Deliza, R. (2019). Brazilian consumer's perception of food processing technologies: A case study with fruit juice. *Food Research International, 125*, 108555.

Mayhew, E. J., Schmidt, S. J., & Lee, S.-Y. (2017). Sensory and physical effects of sugar reduction in a caramel coating system. *Journal of Food Science, 82*, 1935–1946.

Mcclements, D. (2020). Future foods: A manifesto for research priorities in structural design of foods. *Food & Function, 11*(3), 1933–1945.

Menrad, K. (2004). Innovations in the food industry in Germany. *Research Policy, 33*, 845–878.

Milner, L., Kerry, J. P., O'Sullivan, M. G., & Gallagher, E. (2020). Physical, textural and sensory characteristics of reduced sucrose cakes, incorporated with clean-label sugar-replacing alternative ingredients. *Innovative Food Science & Emerging Technologies, 59*, 102235.

Mishra, H., & Sinija, V. (2008). Food technology to meet the changing needs of urban consumers. *Comprehensive Reviews in Food Science and Food Safety, 7*, 358–368.

Morris, E. R. (1994). *Rheological and organoleptic properties of food hydrocolloids. Food hydrocolloids*. Springer.

Mosca, A. C., Andriot, I., Guichard, E., & Salles, C. (2015). Binding of Na+ ions to proteins: Effect on taste perception. *Food Hydrocolloids, 51*, 33–40.

Mosca, A. C., Van De Velde, F., Bult, J. H., Van Boekel, M. A., & Stieger, M. (2015). Taste enhancement in food gels: Effect of fracture properties on oral breakdown, bolus formation and sweetness intensity. *Food Hydrocolloids, 43*, 794–802.

Mouta, J. S., De Sá, N. C., Menezes, E., & Melo, L. (2016). Effect of institutional sensory test location and consumer attitudes on acceptance of foods and beverages having different levels of processing. *Food Quality and Preference, 48*, 262–267.

Noble, A. C. (1996). Taste-aroma interactions. *Trends in Food Science & Technology, 7*, 439–444.

Norton, J. E., Wallis, G. A., Spyropoulos, F., Lillford, P. J., & Norton, I. T. (2014). Designing food structures for nutrition and health benefits. *Annual Review of Food Science and Technology, 5*, 177–195.

Oliveira, D., Antúnez, L., Giménez, A., Castura, J. C., Deliza, R., & Ares, G. (2015). Sugar reduction in probiotic chocolate-flavored milk: Impact on dynamic sensory profile and liking. *Food Research International, 75*, 148–156.

Oliveira, D., Galhardo, J., Ares, G., Cunha, L. M., & Deliza, R. (2018). Sugar reduction in fruit nectars: Impact on consumers' sensory and hedonic perception. *Food Research International, 107*, 371–377.

Pereira, T., Barroso, S., & Gil, M. M. (2021). Food texture design by 3D printing: A review. *Foods (Basel, Switzerland), 10*, 320.

Peryam, D. R., & Pilgrim, F. J. (1957). Hedonic scale method of measuring food preferences. *Food Technology, 11* (S1), 9–14.

Pflaum, T., Konitzer, K., Hofmann, T., & Koehler, P. (2013a). Analytical and sensory studies on the release of sodium from wheat bread crumb. *Journal of Agricultural and Food Chemistry, 61*, 6485–6494.

Pflaum, T., Konitzer, K., Hofmann, T., & Koehler, P. (2013b). Influence of texture on the perception of saltiness in wheat bread. *Journal of Agricultural and Food Chemistry, 61*, 10649–10658.

Pinto De Moura, A., & Cunha, L. (2005). Why consumers eat what they do: An approach to improve nutrition education and promote healthy eating. In *Consumer citizenship: Promoting new responses*. Norway: Hamar.

Pliner, P., & Hobden, K. (1992). Development of a scale to measure the trait of food neophobia in humans. *Appetite, 19*, 105–120.

Poínhos, R., Van Der Lans, I. A., Rankin, A., Fischer, A. R. H., Bunting, B., Kuznesof, S., Stewart-Knox, B., & Frewer, L. J. (2014). Psychological determinants of consumer acceptance of personalised nutrition in 9 European countries. *PLoS ONE, 9*, e110614.

Pravst, I. (2012). *Functional foods in Europe: A focus on health claims. Medicine.*

Putnik, P., Kresoja, Ž., Bosiljkov, T., Jambrak, R. E.Ž. E. K., & A., Barba, F. J., Lorenzo, J. M., Roohinejad, S., Granato, D., Žuntar, I. & Bursać Kovačević, D. (2019). Comparing the effects of thermal and non-thermal technologies on pomegranate juice quality: A review. *Food Chemistry, 279,* 150–161.

Rezaei, A., Fathi, M., & Jafari, S. M. (2019). Nanoencapsulation of hydrophobic and low-soluble food bioactive compounds within different nanocarriers. *Food Hydrocolloids, 88,* 146–162.

Rios-Mera, J. D., Saldaña, E., Cruzado-Bravo, M. L. M., Martins, M. M., Patinho, I., Selani, M. M., Valentin, D., & Contreras-Castillo, C. J. (2020). Impact of the content and size of NaCl on dynamic sensory profile and instrumental texture of beef burgers. *Meat Science, 161,* 107992.

Rodrigues, I., Gonçalves, L. A., Carvalho, F. A., Pires, M., Jp Rocha, Y., Barros, J. C., Carvalho, L. T., & Trindade, M. A. (2020). Understanding salt reduction in fat-reduced hot dog sausages: Network structure, emulsion stability and consumer acceptance. *Food Science and Technology International, 26,* 123–131.

Rollin, F., Kennedy, J., & Wills, J. (2011). Consumers and new food technologies. *Trends in Food Science & Technology, 22,* 99–111.

Román, S., Sánchez-Siles, L. M., & Siegrist, M. (2017). The importance of food naturalness for consumers: Results of a systematic review. *Trends in Food Science & Technology, 67,* 44–57.

Roosen, J., Bieberstein, A., Blanchemanche, S., Goddard, E., Marette, S., & Vandermoere, F. (2015). Trust and willingness to pay for nanotechnology food. *Food Policy, 52,* 75–83.

Roselino, M. N., De Almeida, J. F., Cozentino, I. C., Canaan, J. M. M., Pinto, R. A., De Valdez, G. F., Rossi, E. A., & Cavallini, D. C. U. (2018). Probiotic salami with fat and curing salts reduction: Physicochemical, textural and sensory characteristics. *Food Science and Technology, 38,* 193–202.

Rysová, J., & Šmídová, Z. (2021). Effect of salt content reduction on food processing technology. *Food, 10.*

Sala, G., & Stieger, M. (2013). Time to first fracture affects sweetness of gels. *Food Hydrocolloids, 30,* 73–81.

Sala, G., Stieger, M., & Van De Velde, F. (2010). Serum release boosts sweetness intensity in gels. *Food Hydrocolloids, 24,* 494–501.

Santillán-Urquiza, E., Ruiz-Espinosa, H., Angulo-Molina, A., Vélez Ruiz, J. F., & Méndez-Rojas, M. A. (2017). 8—Applications of nanomaterials in functional fortified dairy products: Benefits and implications for human health. In A. M. Grumezescu (Ed.), *Nutrient delivery* Academic Press.

Santos, L. M., Oliveira, F. A., Ferreira, E. H., & Rosenthal, A. (2017). Application and possible benefits of high hydrostatic pressure or high-pressure homogenization on beer processing: A review. *Food science and technology international = Ciencia y tecnologia de los alimentos internacional, 23,* 561–581.

Scherf, K. A., Pflaum, T., Koehler, P., & Hofmann, T. (2015). Salt taste perception in hydrocolloid systems is affected by sodium ion release and mechanosensory–gustatory cross-modal interactions. *Food Hydrocolloids, 51,* 486–494.

Schiffman, S. S., Sattely-Miller, E. A., Graham, B. G., Bennett, J. L., Booth, B. J., Desai, N., & Bishay, I. (2000). Effect of temperature, pH, and ions on sweet taste. *Physiology & Behavior, 68,* 469–481.

Scholderer, J., & Grunert, K. G. (2005). Consumers, food and convenience: The long way from resource constraints to actual consumption patterns. *Journal of Economic Psychology, 26,* 105–128.

Sellas, M. C., De Souza, D. L., Vila-Marti, A., & Torres-Moreno, M. (2021). Effect of pork back-fat reduction and substitution with texturized pea protein on acceptability and sensory characteristics of dry fermented sausages. *CyTA-Journal of Food, 19,* 429–439.

Shallenberger, R. S. (1998). Sweetness theory and its application in the food industry. *Food Technology, 52,* 72–76.

Siró, I., Kápolna, E., Kápolna, B., & Lugasi, A. (2008). Functional food. Product development, marketing and consumer acceptance—A review. *Appetite, 51,* 456–467.

Slocombe, B., Carmichael, D., & Simner, J. (2016). Cross-modal tactile–taste interactions in food evaluations. *Neuropsychologia, 88,* 58–64.

Soto-Vaca, A., Gutierrez, A., Losso, J. N., Xu, Z., & Finley, J. W. (2012). Evolution of phenolic compounds from color and flavor problems to health benefits. *Journal of Agricultural and Food Chemistry, 60*, 6658–6677.

Stone, H., & Sidel, J. L. (2004). *Sensory evaluation practices*. San Diego: Elsevier Academic Press.

Strijbos, C., Schluck, M., Bisschop, J., Bui, T., De Jong, I., Van Leeuwen, M., Von Tottleben, M., & Van Breda, S. G. (2016). Consumer awareness and credibility factors of health claims on innovative meat products in a cross-sectional population study in the Netherlands. *Food Quality and Preference, 54*, 13–22.

Sun, R., Lu, J., & Nolden, A. (2021). Nanostructured foods for improved sensory attributes. *Trends in Food Science & Technology, 108*, 281–286.

Szpicer, A., Onopiuk, A., Półtorak, A., & Wierzbicka, A. (2020). The influence of oat β-glucan content on the physicochemical and sensory properties of low-fat beef burgers. *CyTA—Journal of Food, 18*, 315–327.

Torrico, D. D., Hutchings, S. C., Ha, M., Bittner, E. P., Fuentes, S., Warner, R. D., & Dunshea, F. R. (2018). Novel techniques to understand consumer responses towards food products: A review with a focus on meat. *Meat Science, 144*, 30–42.

Tuorila, H., & Cardello, A. V. (2002). Consumer responses to an off-flavor in juice in the presence of specific health claims. *Food Quality and Preference, 13*, 561–569.

Urala, N., & Lähteenmäki, L. (2003). Reasons behind consumers' functional food choices. *Nutrition & Food Science, 33*, 148–158.

Urala, N., & Lähteenmäki, L. (2004). Attitudes behind consumers' willingness to use functional foods. *Food Quality and Preference, 15*, 793–803.

Van Trijp, H. C., & Van Der Lans, I. A. (2007). Consumer perceptions of nutrition and health claims. *Appetite, 48*, 305–324.

Verbeke, W. (2006). Functional foods: Consumer willingness to compromise on taste for health? *Food Quality and Preference, 17*, 126–131.

Verneau, F., Caracciolo, F., Coppola, A., & Lombardi, P. (2014). Consumer fears and familiarity of processed food. The value of information provided by the FTNS. *Appetite, 73*, 140–146.

Vingerhoeds, M. H., De Wijk, R. A., Zoet, F. D., Nixdorf, R. R., & Van Aken, G. A. (2008). How emulsion composition and structure affect sensory perception of low-viscosity model emulsions. *Food Hydrocolloids, 22*, 631–646.

Wang, Z., Yang, K., Brenner, T., Kikuzaki, H., & Nishinari, K. (2014). The influence of agar gel texture on sucrose release. *Food Hydrocolloids, 36*, 196–203.

Weiss, J., Takhistov, P., & Mcclements, D. J. (2006). Functional materials in food nanotechnology. *Journal of Food Science, 71*, R107–R116.

WHO. (2003). Diet, nutrition and the prevention of chronic diseases. In *Report of a Joint WHO/FAO Expert Consultation [Online]*. Available: http://whqlibdoc.who.int/trs/WHO_TRS_916.pdf (Accessed January 2008).

Žeželj, I., Milošević, J., Stojanović, Ž., & Ognjanov, G. (2012). The motivational and informational basis of attitudes toward foods with health claims. *Appetite, 59*, 960–967.

Zhao, Y.-M., De Alba, M., Sun, D.-W., & Tiwari, B. (2019). Principles and recent applications of novel nonthermal processing technologies for the fish industry—A review. *Critical Reviews in Food Science and Nutrition, 59*, 728–742.

# Index

Note: Page numbers followed by *f* indicate figures and *t* indicate tables.

## A

Absolute deviation values (ADD), 335
Acceptable daily intake (ADI), 225
α-linolenic acid (ALA), 251–252
ARCOL (Artificial colon) system, 323
Artificial intelligence (AI), 334
Artificial neural networks (ANN), 334–335, 352–354
   digestion modelling, 344–354
   food processing, 334–343, 334*f*, 340–343*t*
Artificial tongues, 371
Astringency, 369

## B

Bioavailability, 327

## C

Capillary breakup extension rheometer (CaBER), 366
Cardiovascular diseases (CVDs), 243
Cholesterol, 245
Chyme, 281
Clean label, 9–10
Cocoa butter (CB), 255–256
Cocoa butter equivalents (CBE), 255–256
Coefficient of variation (CV), 372–373
Colloidal delivery systems (CDSs)
   food functionalization, 68–98, 87*f*, 89–90*t*
   loading performance, 81–86
Compound annual growth rate (CAGR), 234–235
Consumer acceptance, 399–400
Consumer behavior, 13
Controlled release, 64–65
Cotyledon tissue, 284
Cross-validation, 344
Cumulative extraction yield (CEY), 336

## D

DELTA model, 8–9
Diacylglycerols (DAG), 255
DiDGi system, 321–322, 322*f*
Dietary types, 8
Diet shift, food consumption trends, 8–10, 9–10*f*
Digestion, 280–282, 280*f*, 316
Docosahexaenoic acid (DHA), 251–252
Dynamic gastric model (DGM), 321, 345–346

## E

EAT-lancet commission, 8
Electric fields (EF), 43–45, 50–51
EFSA. *See* European food safety authority (EFSA)
Eicosapentaenoic acid (EPA), 251–252
Electrical conductivity (EC), 339
Empirical models (EM's), 336
Emulsions, 401–402
Encapsulated embryos, 277–279
Encapsulation, 26
Enzymatic, sugar reduction methods, 231–234
Enzymatic interesterification, 255
Epigallocatechin gallate (EGCG), 339
EU market, 405
European food safety authority (EFSA), 225, 245, 407
Extensional viscosity, 367

## F

Fat mimetics, 244–245
Fats, 19
Fat substitutes, 244–245
Fatty acids (FAs), 243
Fibrous, 277–279
Fish matrices, 297–298
Flavors with modifying properties (FMPs), 225
Flexitarian (FLX), 6
Food, 315
   fat digestibility, 261–263
   sodium reduction, 206–208
Food and drug administration (FDA), 225
Food applications, 172–174
   ionic gelation, 173–174
   spray chilling, 172–173
   spray drying, 172
Food colloids, gastronomy, 142–146, 143*f*
   aerosols, 144
   emulsion, 143
   foam, 143–144, 144*f*
   gel, 143
   sol, 143
Food microstructures, 146–149
   enzymes, 146–147, 147*t*
   fermentation, 146, 148–149, 148*f*
   fermented food, 149
Food processing, 12
   technologies, 21, 113–114, 114–115*t*

Food reformulation, free sugar intake, 222–225, 223–224t
Food structure, 115–120, 116f, 277–279, 278f, 323
   food macromolecules, 118–120
   nutritional quality, 126–130
   physical properties, 120–130
   sensorial, 122–125, 123f
   solid foods, 117
   structural modifications, particulate foods, 117–118
Food structuring, 11–12, 17–18, 150–151
   aerated food, 151, 152f
   fat reduction/replacement
      emulsions, hydrogels, oleogels, 150–151
      mayonnaise, 151
4D printing, 22
Free fatty acids (FFAs), 261–262
Fructo-oligosaccharides (FOS), 233–234
Functional food, 4, 32–33, 402–403

## G

Galacto-oligosaccharides (GOS), 232
Gastrointestinal tract (GIT), 279
   absorption models, 323–325
      cellular models, 323–324
      organoids, 325
      Ussing chambers (UC), 324–325
   animal-based food structures modification, 293–303
      meat, 293–298, 295f
      milk, 298–303, 300–301f
   in vivo digestion models, 325–327
      animal models, 325–326
      human subjects, 326–327
   plant based food structures modification, 284–293
      cereals, 287–288
      starchy legumes, 284–286, 286f
      tree nuts, 290–293
      vegetables and fruits, 288–290, 290f
Gastronomy, 140–142, 142t, 144t, 145–146
Gelled foods, 380–382
Gelled/structured emulsions, 256–257, 257–258t
Generally recognized as safe (GRAS), 225
Genetic algorithm (GA), 337
Gluco-oligosaccharides (GuOS), 234
Glucose–fructose oxidoreductase (GFOR), 234
Glycemic index (GI), 223–225, 352
Glycemic load (GL), 353
Gold standard strategy, 315
GRAS. See Generally recognized as safe (GRAS)

## H

Health, 19, 402–403
Healthy diets, 6
Heat transfer phenomena-based models (HTPBM's), 336

High-density lipoproteins (HDL), 245
Homogenized milk, 299–300
Homogenized UHT milk, 300–301
Human gastric simulator, 321, 346
Human taste modalities, 401
Hydrothermally processed, 284

## I

INFOGEST network, 321
Innovative food products acceptance, 400–408
   convenience, 406–407
   health concerns, 402–404
   nutrition concerns, 404–405
   price, 407–408
   risk perception, 405–406
   sensory properties, 400–402
In vitro oro-gastro-intestinal digestion models, 316–323
   dynamic in vitro models, 319–323
      gastric acidification modelling, 320, 320f
      gastric & intestinal emptying modelling, 319–320
   dynamic models, 321–323
      colonic models, 323
      dynamic gastrointestinal models, 321–323, 322f
      dynamic stomach models, 321
   static in vitro models, 316–319, 317f, 318t
      gastric phase, 316–317
      intestinal phase, 317–319
      oral phase, 316
Isomalto-oligosaccharides (IMO), 233

## L

Levenberg-Marquardt back propagation (LMBP), 353
Linoleic acid (LA), 251–252
Lipid fraction modification, 244–260
   fat replacers, 246–249, 248–250t
   healthy fatty acid profile, 250–260, 251–252t, 257–258t, 261f
Locust bean gum (LBG), 382
Long-chain n-3 PUFAs (LC n-3 PUFAs), 251–252
Low-density lipoproteins (LDL), 245

## M

Macronutrients digestion, 282–284
   lipids, 282–283
   proteins, 282
   starch, 283–284
Malto-oligosaccharides (MOS), 233
Mass transfer phenomenon-based models (MTPBM's), 336
Mastication, 288–289
Mean absolute deviation (MAD), 335–336

Mean square error (MSE), 339
Mediterranean countries, 406
Medium and long chain triglycerides (MLCT), 255–256
Membrane bioreactor (MBR), 338
Microbiota, 281–282
Microencapsulation
　materials, 168–172
　　core, 168–172
　　wall materials, 166–168, 168t
　strategies, 160–166, 160t, 161f
　　ionic gelation, 164–166, 165–166f
　　spray chilling, 162–164, 163f
　　spray drying, 161–162, 162f
Microwave assisted extraction (MAE), 335–336
Milk coagulation, 299
Milk protein concentrate (MPC), 352
Moderate electric fields (MEF), 45–51, 45f, 46t, 49f, 55f
Monoacylglycerols (MAG), 255, 283
Multiple linear regression (MLR), 339

## N

Nanotechnology (NT), 399–400
Neophobia, 406
New food structures, 20–29, 20f
　customized, 22–23
　encapsulated, enhanced bioavailability, 26–27
　energy density food, 20–22
　specially designed, digestion modulation, 23–26, 24f
　tailored, 27–29, 28f, 29–30t
New food structures, influencing factors, 30–32
　manufacturing process, 31
　market analysis, 32
　post-production, 32
　product research, development, 30–31
New food structuring techniques, 19–20
Non communicable diseases (NCDs), 220
Non-nutritive sweeteners (NNS), 223–225
NOVA classification system, 11
Nutraceutical encapsulation, 65–68, 66f, 74t
Nutrition, 5, 19, 277

## O

Oil bodies (oleosomes), 290
Oil bulking agents, 258–259
Oil-in-glycerol, 256
Oil structuring, 254
Oleic acid (OA), 251–252
Optical microscopy, 285
Oral processing, 284–285, 401
Organogels/oleogels, 259–260

## P

Palm oil mill effluent permeate flux (POME), 338
Parenchyma tissue, 288
Particle size distribution (PSD), 284–285
Particle technology, 165–166
Perceived aroma intensity, 383
Pescatarian (PSC), 6
Plant proteins, 246
Postmortem tenderization, 294
Processed foods (PF), 11
Proteins, 18–19
Pulsed electric fields (PEF), 296–297

## R

Raw meat, 294
Resistant starch (RS), 352–353
Response surface methodology (RSM), 335
Root mean square error (RMSE), 335–336

## S

Salt, sodium, 188–189
Saturated fatty acids (SFAs), 243
Scintigraphy, 327
Sensory analysis, 400–401
Sensory evaluation (SA), 339
Shelf life time (SLT), 339
Short chain fatty acids (SCFA), 323
Simulated human intestinal microbial ecosystem model (SHIME), 323, 346–350
Sodium carboxymethyl cellulose (CMC-Na), 366–367
Sodium chloride reduction, 187–188, 188f
Sodium reduction strategies, 195–206, 196–199t
Sodium role, 189–195, 190t
　flavor enhancer, 193
　microbial stability, 194–195
　protein functional properties, 190–192
Solid fat content (SFC), 255–256
Square sum error (SSE), 336–337
Squeezing, 369–370
Starch, 24–25
Sugar, 21, 221–222, 226–231, 228–230t
Supercritical fluid extraction (SCFE), 336
Sustainability, 4
Sustainable development goals (SDG), 3
Swallowing, 366
Sweetness intensity, 401

## T

Tactile-taste interactions, 402
Tenderization, 293–294
Texture, 365

Texture evaluation, human physiological responses, 374–382
    electromyography (EMG), 378–379, 379f
    laryngeal movement measurement, 382
    swallowing sound acoustic analysis, 380–382, 380–381f
    tongue pressure measurement, 374–378, 375–377f
Texture, flavor interactions, 382–386
    aroma perception, inhomogeneous spatial distribution, 384, 384–385f
    flavor release control, texture modification, 382–383
    human eating behavior modification, 385–386, 386f
Texture perception, oral rheology, tribology, 366–374
    bolus rheology, 371–374, 372–373f
    extension rheology, swallowing physiology, 366–367, 367f
    soft machine mechanics, palatal size reduction, 369–371, 370f
    tribology, moisture & fat, 367–369, 368f
3D printing, 22–23
TNO's gastrointestinal model (TIM), 346
Total aerobic counts (TAC), 339
Total phenolic content (TPC), 335–336
Total phenolic compound, 335
Trans-fatty acids (TFAs), 243
Transmembrane pressure (TMP), 338
Triacylglycerols (TAGs), 282–283
    enzymatic modification, 254–256

## U

Ultra-processed foods (UPF), 11
Ultrasound-assisted extraction method (UAE), 337

## V

Vegan (VGN), 6
Vegetarian (VEG), 6
Videoendoscopy (VE), 380
Videofluorography (VF), 380
Viscoelastic fluids, 366
Visual analog scale (VAS), 353

## W

Western diet, 8
World Health Organization (WHO), 220

## X

Xanthan gum (XG), 382

Printed in the United States
by Baker & Taylor Publisher Services